BEAM INSTRUMENTATION WORKSHOP

AIP CONFERENCE PROCEEDINGS 333

BEAM INSTRUMENTATION WORKSHOP

VANCOUVER, B. C., CANADA OCTOBER 1994

EDITORS: GEORGE H. MACKENZIE
BILL RAWNSLEY
JANA THOMSON
TRIUMF

American Institute of Physics **New York**

L.C. Catalog Card No. 95-79635
ISBN 1-56396-352-3
DOE CONF-9410219

Printed in the United States of America.

CONTENTS

APPENDICES

PREFACE

The sixth workshop on Beam Instrumentation was held in Vancouver, Canada, 3–6 October 1994. The workshop was hosted by the TRIUMF laboratory and attended by 138 registrants, 39 from overseas. Eleven countries were represented—the most to date in this series of workshops.

The programme was modeled on the successful 1993 meeting in Santa Fe. Early morning tutorials were followed by invited talks; later in the day contributed papers were presented orally or by poster. Seven vendors exhibited their wares during the poster session. Six discussion topics were distributed between two separate sessions.

The Coast Plaza Hotel, where the sessions were held and where most delegates were able to stay, is close to restaurants, beaches and parks, and the weather was especially good for October. Nevertheless, all sessions were very well attended, proof of the relevance of the topics chosen by the Programme Committee and the speakers' reputations. The weather and the roof garden adjacent to the session rooms provided a wonderful combination for informal discussions.

The banquet was held at Science World. Following the meal, Julien Bergoz presented the 1994 Faraday Cup Award, sponsored by Bergoz Inc., to Edouard Rossa of CERN for his work in imaging the electron and positron bunches in LEP in real time and in three dimensions. A paper describing this work was presented the following day and is included in these proceedings.

The workshop ended with a tour of the TRIUMF laboratory. Since the cyclotron was operating, its instrumentation could not be inspected; however, we were able to interrupt beam delivery for brief periods to demonstrate some beam measurements, data analysis and display. Several TRIUMF groups arranged displays of their devices and equipment and the proton therapy treatment area and the TR13 mini-cyclotron were visited.

Many people at TRIUMF contributed to the success of the workshop. Members of the Local Arrangements Committee are listed below. We would also like to thank members of the probes and engineering physics technical groups (Doug Clayton, Dan Gray, Ken Lukas, John Stewart) who displayed equipment, colleagues who manned stations during the tour or served as guides, and members of the secretarial and design office staff who helped with publicity and mailouts. Special mention should go to Ron Balden (data base), Richard Lee (graphics), Eddie Knight (poster), and Dave Ross (logos). We thank the TRIUMF Director, Alan Astbury, for opening the workshop and Gerardo Dutto, Head of the Cyclotron Division, for his support over the year.

We acknowledge gratefully, grants from the High Energy Physics Division of the Department of Energy (U.S.) to defray part of the publishing costs and from Varian Canada Inc. for financing some refreshment breaks.

Finally, we thank those who contributed papers for their cheerful cooperation.

The next workshop in this series will be hosted by the APS Division of Argonne National Laboratory and is expected to be held in the Spring of 1996.

George Mackenzie
Bill Rawnsley

PROGRAMME COMMITTEE

George Mackenzie, TRIUMF, Chair
Bill Rawnsley, TRIUMF, Co-chair
Claude Bovet, CERN
Jean-Claude Denard, CEBAF
Robert Hettel, SSRL-SLAC
James Hinkson, LBL
Georges Jamieson, SSCL
Heribert Koziol, CERN
Alex Lumpkin, ANL
Ralph Pasquinelli, FNAL
John Perry, CEBAF
Mike Plum, LANL
Robert E. Shafer, LANL
Gary A. Smith, BNL
Greg Stover, LBL
Robert C. Webber, SSCL and FNAL
Richard Witkover, BNL
Jim Zagel, FNAL

LOCAL ARRANGEMENTS COMMITTEE

Elly Driessen
Maria Freeman
Andy Hurst
Shane Koscielniak
George Mackenzie
Bill Rawnsley
Jana Thomson
Yan Yin
Milos Zach

BEAM INSTRUMENTATION WORKSHOP 2 – 6 OCTOBER 1994

	SUNDAY	MONDAY	TUESDAY	WEDNESDAY	THURSDAY
8:00		OPENING	LINSCOTT TUTORIAL	HAHN TUTORIAL	McGINNIS TUTORIAL
8:15		WEBBER TUTORIAL			
9:30		BREAK	BREAK	BREAK	BREAK
10:00		WILKE INVITED	RAWNSLEY INVITED	BORDEN INVITED	CHU INVITED
11:00		SMITH TUTORIAL	DISCUSSION GROUPS	ROSSA INVITED	WRAP - UP
12:00				LUNCH	LUNCH
12:30		LUNCH	LUNCH		
13:30				CONTRIBUTED ORAL PRESENTATIONS	TOUR OF TRIUMF
14:00		CONTRIBUTED ORAL PRESENTATIONS	POSTER PRESENTATIONS and VENDOR EXHIBITION		
15:45		BREAK		BREAK	
16:00		DISCUSSION GROUPS		FREE	
17:30					
18:00	REGISTRATION STARTS				
19:00	RECEPTION				
21:00			BANQUET		

SPEAKER	TOPIC
• Michael Borden (LANL)	- Radiation Effects in Accelerator Components
• William Chu (LBL)	- Instrumentation for Medical Beams
• Alan Hahn (FNAL)	- Statistical Data Analysis
• Ivan Linscott (STANFORD)	- Digital Signal Processing
• David McGinnis (FNAL)	- The Design of Beam Pickup and Kickers
• Bill Rawnsley (TRIUMF)	- TRIUMF Beam Instrumentation
• Edouard Rossa (CERN)	- Real Time Single Shot Three-Dimensional Measurement of Picosecond Photon Bunches
• Gary Smith (BNL)	- The Prosaic Laplace and Fourier Transforms
• Bob Webber (SSCL)	- Beam Current Monitoring
• Mark Wilke (LANL)	- Optical and X-ray Imaging of Electron Beams Using Synchrotron Emission

Participants at the 6th Beam Instrumentation Workshop gather in the Garden Terrace of the Coast Plaza Hotel, Vancouver. (photo Marcello Pavan)

Elly Driessen and Maria Freeman welcome registrants.
(photo Marcello Pavan)

Front row: Edouard Rossa, recipient of the 1994 Faraday Cup award (right), and Julien Bergoz, sponsor of the award (left).
Back row: Some members of the BIW Programme Committee which adjudicates the competition; left to right - Bob Webber, Jim Zagel, George Mackenzie (Chairman 1994 Workshop), Jim Hinkson, Alex Lumpkin (Chairman 1996 Workshop) Jean-Claude Denard, Claude Bovet, Gary Smith and Greg Stover.
(photo Bill Rawnsley)

INVITED PAPERS

Charged Particle Beam Current Monitoring Tutorial

Robert C. Webber
Fermi National Accelerator Laboratory, P.O. Box 500, Batavia, Illinois 60510 USA*

Abstract

A tutorial presentation is made on topics related to the measurement of charged particle beam currents. The fundamental physics of electricity and magnetism pertinent to the problem is reviewed. The physics is presented with a stress on its interpretation from an electrical circuit theory point of view. The operation of devices including video pulse current transformers, direct current transformers, and gigahertz bandwidth wall current style transformers is described. Design examples are given for each of these types of devices. Sensitivity, frequency response, and physical environment are typical parameters which influence the design of these instruments in any particular application. Practical engineering considerations, potential pitfalls, and performance limitations are discussed.

INTRODUCTION

Particle accelerators are constructed and operated for a growing variety of applications from high energy physics research to cancer treatment, isotope production, biological studies, lithography, and even food preservation. One fundamental measure of an accelerator's or beam line's performance and a necessary parameter in the application of it's particle beam is the quantity or intensity of particles accelerated and transported to the final point of utilization.

Intensity measurements may be made with intercepting or non-intercepting monitors. Intercepting monitors operate by introducing sufficiently dense material directly into the particle beam's path to produce measurable signals from the resulting ionization or nuclear processes. Non-intercepting monitors rely on macroscopic scale interaction with the electromagnetic fields of the beam for signal energy.

Intercepting devices are not necessarily destructive of the beam being measured, but most implementations result in absorption of a large amount of energy from the beam particles or, in the case of ion beams, a change of the charge state. A large variety of intercepting devices exist and are used in applications where these side-effects are acceptable, most often in single pass situations where the beam particles each pass through the device only once. Intercepting monitors have the advantage that they may be designed to measure neutrally charged particle beams.

---

* Work supported by the U.S. Department of Energy under contract No. DE-AC02-76CH03000.

Intercepting monitors are rarely acceptable in applications, such as synchrotrons, where each particle of the circulating beam passes through a monitoring device many times. The utilization of non-intercepting instruments, sensitive to the electromagnetic fields due to the particles' charge and its motion, is quite natural, since the accelerators themselves rely on interactions with those same fields for acceleration and steering. The flow of charged particles through an accelerator or beam line constitutes, in the strictest sense, an electric current, proportional to the number of particles and to the charge carried by each. As a result non-intercepting beam intensity instruments are often referred to as beam current monitors. They exhibit same properties and are designed on the same principles as current monitors in the electrical circuit sense.

This paper shall focus on monitors of the non-intercepting type. Even among these, the diversity of designs and performance parameters is large.(1, 2, 3) Different applications demand devices covering a broad range of current sensitivities and time or frequency response. Current magnitudes to be measured vary from kiloamperes in induction linacs to microamperes or less in heavy ion accelerators. The wide range of time responses necessary is indicated by application examples including measurement of low duty cycle picosecond current pulses in electron machines like LEP, high duty cycle nanosecond pulses in hadron accelerators like the Tevatron in fixed target mode, video frequency pulses like a macropulse of beam through a proton linac, and the direct current component of circulating beam in storage rings and colliders. Frequently a single machine needs instruments to measure several of these extremes. For example, the Fermilab Tevatron, in collider mode, requires high accuracy and high stability dc current measurement for quick determination of beam storage lifetime, as well as fast and accurate measurement of individual nanosecond bunch intensities to differentiate between proton and antiproton bunches.

FUNDAMENTALS

Associated with the particle beam are electric and magnetic fields, due respectively to the charge carried by the particles and the fact that the charges are in motion. Coupling to either one or a combination of both these fields will provide a signal related to the beam intensity. Devices which couple primarily to the electric field are occasionally used to measure high frequency beam signal components for intensity monitoring purposes. It is difficult, however, to obtain satisfactory low frequency response with such capacitive sensors without resorting to high impedance circuitry which can lead to noise susceptibility. As will be shown later, a severe penalty, in terms of available signal power, may be incurred with the use of these electric field devices, especially in the video frequency band. The discussion in this article will mainly concentrate on utilization of the beam's magnetic field as a signal source.

A Particle Beam as an Electric Current

In the purest sense, a beam of charged particles in motion is an electric current. The magnitude of that current is simply

$$I_b = q\lambda_N v \tag{1}$$

where q is the beam particle's charge, λ_N the number of particles per unit length, and v the velocity at which the particles are traveling. In a circuit sense, the beam appears as an ideal current source manifesting a nearly infinite source impedance. That this is the case can be readily seen by considering that source impedance is simply that change in voltage which must be applied to the source to force a given change in current at its terminals. Since the beam current is determined by the velocity of the charged beam particles, a large enough voltage must be applied to significantly alter this velocity in order to cause any change in current. For all but the lowest energy beam, this must typically be a voltage comparable to the beam's energy (in eV units). Clearly the impedance of the beam is normally extremely large.

This should not be construed to imply however that all beam sensors exhibit a large source impedance. Any device constructed has inherent shunting impedances which contribute source impedance terms as observed by an external circuit. The beam itself simply contributes nothing to this impedance. The simplest electric field sensing devices (capacitive probes or pick-ups) will exhibit a capacitive source impedance equal to the capacitance between the electrode and its surroundings. Magnetic type pick-ups in their simplest design present an inductive source impedance.

A particle beam usually has a complex spatial charge distribution along the axis of motion, resulting in a current with a broad frequency spectrum. The typical example is a pulsed or periodic large scale longitudinal distribution comprised on a finer scale of multiple individual bunches. In a synchrotron or storage ring this pulse is periodic with the revolution frequency resulting in signal components from dc to frequencies corresponding to the inverse of the bunch length. A useful reference on frequency domain analysis of bunched beam signals has been written by Siemann.(4) A rare example of a beam with very few ac current components is the debunched anti-proton beam stored in the Fermilab accumulator ring where particles uniformly fill the circumference. Nevertheless, the discrete particle nature of the beam results in wideband shot noise current components which are the source of the signals essential for beam emittance reduction by stochastic cooling in that machine.

A dc beam current component, that is, a net charge transfer in one direction via the beam, will nearly always be present. This must be the case in a beam transport system with any magnetic bending elements. Such a system will correctly steer and transport only beams with one sign of current, i.e. particles of opposite charge can both travel through that system only if they travel in opposite directions. A positive charge to the right and a negative charge to the left both appear as current of the same sign. In the Fermilab Tevatron, for example, the clockwise circulating protons and the counter clockwise circulating antiprotons each produce signals which add with the same sign in a dc current monitor. To cancel each other's dc component both would need to travel the same direction which is not possible.

The rich spectrum of the typical beam currents provides both advantages and potential problems. The availability of signal nearly anywhere in the spectrum permits much design flexibility and a wide range of options for current monitoring instruments. However, all spectral components outside the design bandwidth of an instrument represent potential noise sources. These signals can feed through to the output via unsuspected resonances, by overloading active circuitry, by aliasing within sampled signal type circuits, or by any other non-linear process. It is quite easy, for example, to construct a dc current monitor that will indicate the presence of a dc current component in response to only an ac stimulation. Carelessness in understanding and controlling an instrument's response to all frequencies to which it may be exposed can lead to unexpected and difficult to interpret output signals. Even for in-band signals, it is often necessary to understand the relationship between the average beam current and the charge distribution responsible for that current. This is especially true in the case of narrowband monitors. The signal observed at the fundamental frequency for a given amount of charge distributed with 100% sinusoidal modulation will be one amplitude; that for the same quantity of charge distributed as a train of delta functions with the same periodicity will be two times that amplitude. This follows simply from the Fourier decomposition of the distributions.

In the consideration of time and frequency aspects of beam current monitoring, an interesting observation presents itself in proton accelerators (e.g. Fermilab Linac and Booster) where the range of energies is non-relativistic and the particle velocity increases as the beam is accelerated. Down the Linac, even though particle velocity is a strong function of location, each beam current monitor (assuming no lost particles) is observed to read the same current as all others in the Linac. Yet in the Booster as the particle velocity increases a current monitor indicates a proportionally increasing current.

How is this paradox explained? The Linac is arranged to function as a fixed frequency system. The spacing between accelerating cells is adjusted to be proportional to the velocity corresponding to the particle energy at each point in the Linac. This results in a constant transit time between cells to match the fixed rf accelerating frequency. Physically the beam is a group of bunches each containing some constant charge and always separated in time by the cell to cell transit time (or some fixed multiple.) The beam current averaged over the time interval between bunches is simply that bunch charge divided by the interval. This is a conserved quantity everywhere in the Linac and all monitors read the same current. Consistency is maintained with Eq.1 because as the velocity increases λ_N, the number of particles per unit length, actually decreases! The spatial charge distribution is stretched as the beam progresses through the Linac. You can convince yourself of this by considering the spacing between two automobiles (with the same initial velocity) before and after the acceleration of coasting down a hill on the roadway.

In the Booster synchrotron, the beam path length around the machine remains closely fixed while the particle velocity increases by about 50%. If the total charge in the ring is constant but passes any current monitor location once per decreasing revolution period, then the current observed at that location indeed increases with the particle velocity. To first order, the spatial charge distribution in a synchrotron is preserved as the revolution frequency increases. This increasing beam current effect becomes negligible as the particles become highly relativistic and the limiting velocity is reached.

The Magnetic Field and Magnetic Induction

The magnetic field produced by the beam current is the source of signal energy for the current monitor. The magnitude of this field in the vicinity of the particle beam in a vacuum is calculated using *Ampere's Law*

$$\oint (\boldsymbol{B} \cdot d\boldsymbol{l}) = \mu i \tag{2}$$

in the same manner as elementary text books derive the field around a wire.(5) The result is that the magnitude of B at a distance r from the beam is

$$B(r) = \mu \frac{i}{2\pi r} \tag{3}$$

with the lines of induction forming concentric circles around the beam. For our purposes the temporal variations of B can be taken as identical to those of the current itself. For every spectral component of the current, there is a corresponding spectral component of B, including dc.

Magnetic induction provides the necessary coupling between the beam and the signal pick-up in the classical current monitor. *Faraday's Law of Induction*

$$\oint (\boldsymbol{E} \cdot d\boldsymbol{l}) = -\frac{d\Phi_B}{dt} \tag{4}$$

where $\Phi_B = \int \boldsymbol{B} \cdot d\boldsymbol{S}$ is the magnetic flux due to the beam current, provides the basis for design of magnetic sensors and allows calculation of available signal. *Faraday's Law* simply states that the induced emf around any path, e.g. a loop of wire, is proportional to the time rate of change of magnetic flux through the area enclosed by that path. Note that a simple induction loop can provide no information as to the dc component of the magnetic field and therefore of the dc beam current since that flux contribution has zero time derivative.

In circular accelerators or storage rings, the beam travels around the circumference forming an approximately circular current loop. Such a loop is a classical magnetic dipole with lines of induction cutting normal to the plane of the loop.(Fig. 1a) Accelerator beams, of course, travel through evacuated regions of space, typically enclosed by metallic tubes. In the simplest geometry the beam current loop is concentrically enclosed by a conducting beam tube. This seems quite natural and benign from a vacuum and mechanical viewpoint. However, when subjected to inspection from an electrical point of view, it is observed that the beam tube forms a 'pick-up loop' intercepting nearly all the beam induced flux penetrating the ring's center area. The beam and its vacuum tube form a transformer of the simplest design with the beam acting as the primary 'winding'

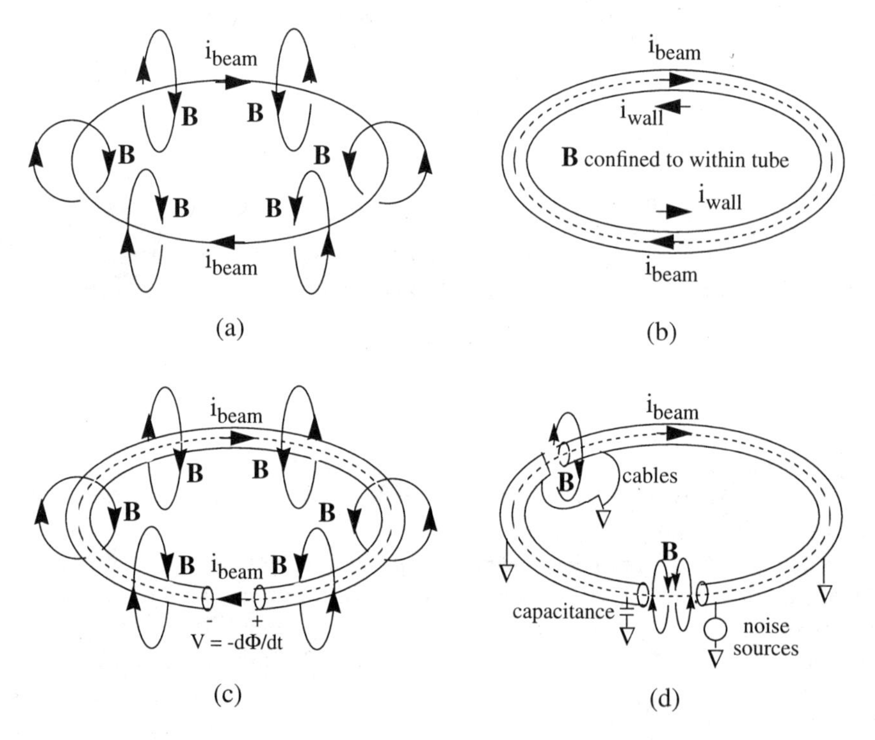

Figure 1. (a) Lines of magnetic induction around circulating beam. (b) Wall currents induced in beam tube attenuate external **B**. (c) Break in tube impedes wall currents permitting external **B** and appearance of induced voltage. (d) Layout of typical accelerator shows complex and distributed paths available to induced currents.

and the tube, the secondary. In this manner, so called 'wall currents' are induced in the beam tube when the secondary circuit is completed.(Fig. 1b)

Lenz's Law, in one form, states that the *induced currents will flow in the opposite sense of the current responsible for their induction.* Applied in this case it means that the wall currents flow opposed to the beam current. These currents produce flux of their own with a sign so as to reduce the primary (beam) current's flux within the enclosed loop area. This effectively 'shields' the space outside the beam tube from the beam's magnetic fields. To the extent that the wall current and the beam current may be equal in magnitude this shielding is perfect and the beam's field is confined within the tube.

Attempts to measure the beam current using an inductive pickup loop outside the beam tube become futile due to this flux cancellation effect. At audio frequencies this can be overcome by inserting a short non-conducting section in the beam tube to interrupt the induced wall current and inhibit the flux cancellation outside the beam tube.(Fig. 1c) The voltage induced across such a gap can and indeed has been be used as a beam monitor. The typical accelerator however is not so simple. The vacuum chamber contains kicker magnets, accelerating cavities, and numerous dc breaks for various reasons (like someone else's beam monitor). In addition, the beam tube is generally grounded, intentionally or by chance, at numerous locations forming shunt current paths around the gap. At even moderate frequencies, the effects of capacitance to ground and across the gap become significant. These effects tend to 'localize' the induction current paths with the result that what is happening on the other side of the ring is of little matter and it is not useful to think of the beam tube as a simple global induction loop.(Fig. 1d)

At high frequencies the skin effect constrains all beam induced currents to the inner wall of the beam tube. At frequencies where the skin depth becomes small compared to the tube wall thickness the magnetic shielding effect of the tube becomes nearly perfect and only severely attenuated magnetic information about the beam current exits outside the beam tube. However, any break in the pipe will interrupt even these high frequencies and act as a 'window' permitting the high frequency currents and their associated fields to 'leak' outside. This situation is frequently used to advantage, and introduced by design, in current monitors, rf cavities, etc. Unfortunately the same 'leaks' occur at breaks introduced by chance or carelessness and become a source of noise to un-shielded circuits.

No dc or zero frequency component is permitted by Faraday's Law to exist in the wall current. That component of the beam's field penetrates the beam tube unaffected (provided the tube is not made of a magnetic material.) Complete information of the beam's direct current component exists external to the beam tube, though no simple inductive loop can detect it. Slightly more complicated devices, such as the DCT described later, can serve effectively in this situation.

All beam current monitors operate on the principal of intercepting lines of magnetic induction produced by the beam current. Any successful design must have access to some component of that magnetic flux and must take complete account of the environment determining the distribution of that flux in the sensitive volume of the monitor. All resistive and reactive shunting current paths, all shunting magnetic paths, and all external field sources (magnetic and electric) must be controlled or the consequences of unpredictable performance may be suffered.

BEAM CURRENT MONITOR DESIGNS

The commonly used beam current monitor configurations fall into four general categories: classical ac transformers, wall current monitors, dc monitors, and narrowband rf monitors.

Toroidal transformers and variations thereof are generally used for monitoring beam currents in the audio to tens of megahertz frequency range. To intercept the beam's magnetic flux, these transformers are built within the metal vacuum housing or around a dc break in the beam pipe. Electrostatic shielding is generally included to prevent unwanted signals due to capacitive coupling to the beam charge.

A special variation of the toroidal transformer called a wall current monitor is used for wide bandwidth current measurements up to a few gigahertz. It can be thought of as a device which produces a signal by directly intercepting the induced wall currents. Since it is designed to function at high frequencies, it requires access to the inner wall of the beam tube either directly or through an appropriate 'window.'

DC beam current monitors, though having access to the beam's dc magnetic field even outside a conducting beam tube, are generally built around a break in the tube to permit response to audio and higher frequency signal components. Practical dc monitors are hybrid instruments combining a dc sensitive magnetic modulator section and a classical ac transformer to offer moderately high bandwidth with dc response.

Narrow band current monitors are designed to be sensitive to selected spectral components of the beam current signal. Devices of this type can offer the high transfer impedance necessary for measurement of small beam currents. These monitors for begin to look like accelerating cavities, though they are not generally required to handle high power. They can operate within and as an inherent part the vacuum vessel or external to it with an appropriate 'window.'

Classical AC Transformers

The classical ac transformer consists of a primary and one or more secondary windings typically on a toroidal form as depicted in Fig. 2a. *Faraday's Law* gives the voltage on any winding as

$$V_k = -N_k \frac{d\Phi_T}{dt} \tag{5}$$

where Φ_T is the total flux linking that winding.

The contribution to the total flux within the toroid due to a wire current i flowing in any N turn winding k is

$$\Phi_k = \int \boldsymbol{B}_k \cdot d\boldsymbol{S} = \int_a^b \mu \frac{N_k i_k}{2\pi r} h dr = \mu \frac{N_k i_k h}{2\pi} \ln \frac{b}{a}. \tag{6}$$

This follows from the definition of flux and Eq. 3. It is customary and often more convenient from a circuit viewpoint, rather than applying the flux concept directly, to use the quantity known as self inductance, L. The relationship between the self inductance of winding k and its associated flux is defined by

$$\left(V_k = -N_k \frac{d\Phi_k}{dt} = L_k \frac{di_k}{dt} \right) \Rightarrow \left(L_k = \frac{-N_k \Phi_k}{i_k} \right). \tag{7}$$

Combining this with Eqs. 6 yields

$$L_k = \mu \frac{N_k^2 h}{2\pi} \ln \frac{b}{a} = N_k^2 L_0 \approx \mu \frac{N_k^2 A}{l} \tag{8}$$

where $L_0 = (\mu h / 2\pi) \ln (b/a)$ is simply the single-turn inductance. This reveals that inductance is purely a property of the geometry of the winding, the number of turns, and the relative permeability of the toroid core. The final approximate form of Eq. 8 is valid for a thin-walled toroid geometry, where $\boldsymbol{B}$ can

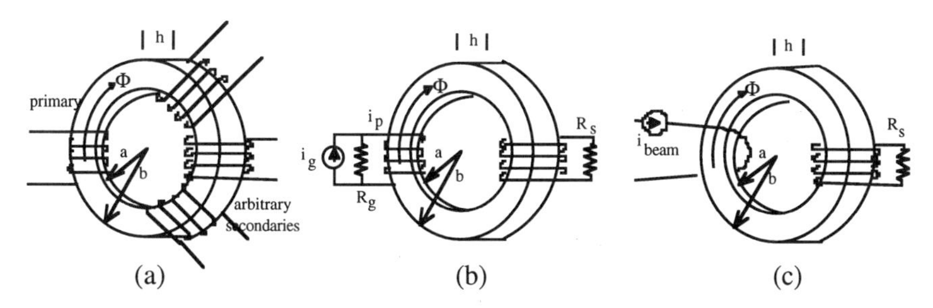

Figure 2. (a) Generic toroidal transformer. (b) Typical two winding current transformer. (c) Beam current transformer.

be taken as constant between radii a and b. A is the cross-sectional area of the toroid, $A = (b-a)h$, and l is the magnetic path length around the toroid, $l = \pi(b+a)$. This form is handy in transformer design since magnetic material vendors frequently specify their toroid products in terms of effective cross section and effective magnetic length.

Given these fundamentals, the performance of a transformer in any circuit configuration may be calculated. Assuming all windings link the total flux in the toroidal core, that flux is simply the sum of the contributions from all windings:

$$\Phi_T = \sum_{m=1}^{k} \Phi_m = \sum_{m=1}^{k} \frac{L_m i_m}{N_m}. \tag{9}$$

Solving simultaneously with Eq. 5 for all windings, the current and voltage solutions for all coupled circuits may be obtained.

Applying this to the configuration in Fig. 2b, find

$$\Phi_T = \frac{L_p i_p}{N_p} + \frac{L_s i_s}{N_s} \tag{10}$$

and

$$\frac{d\Phi_T}{dt} = -\frac{V_p}{N_p} = -\frac{V_s}{N_s}. \tag{11}$$

At this point it is convenient to equate the N turn primary to the single turn beam, $i_p = i_b$ and $N_p = 1$, (Fig. 2c) and to be satisfied with the steady state frequency domain solution permitting replacement of the derivative operator with the LaPlace variable $s = j\omega$. Given $i_s = V_s/R_s$, Eqs. 10 and 11 respectively become

$$\Phi_T = L_0 i_b + \frac{L_s i_s}{N_s} \tag{12}$$

$$s\Phi_T = -\frac{i_s R_s}{N_s}. \tag{13}$$

Simultaneous solution yields

$$i_s = \frac{(s\tau)}{(1+s\tau)}\left(\frac{i_b}{N_s}\right) \tag{14}$$

where $\tau = L_s/R_s$. This is the response of the simple beam current transformer.

Inspection of this result reveals that for frequencies $\omega \gg 1/\tau$ the secondary winding current is simply $i_s = i_b/N_s$, equal in magnitude to the beam current divided by the number of turns and in phase with it. This is the characteristic of an ideal current transformer. For frequencies $\omega \ll 1/\tau$, the secondary current is $i_s = (s\tau)\, i_b/N_s$, proportional in magnitude to the derivative (that is, the frequency) of the beam current and 90^0 out of phase. Operation in this regime is that of the so-called B-dot coil, a small pick-up loop often used to measure the field in pulsed magnets. Note that only the ratio of L and R determine which regime and 'way of thinking' apply for any given signal frequency.

Available signal power is another important aspect of current monitors. Signal power determines the ability of the signal to compete with noise sources. To achieve operation in the ideal current transformer mode for all frequencies of interest set $\tau > 1/\omega_{\mathrm{min}}$, where ω_{min} is the lowest signal frequency. The available signal power is then

$$P_{\mathrm{signal}} = i_s^2 R_s = \frac{i_b^2}{N_s^2} R_s . \tag{15}$$

Normalizing to i_s^2 and substituting $N_s^2 L_0/\tau$ for R_s find:

$$\frac{P_{\mathrm{signal}}}{i_s^2} = \frac{L_0}{\tau} = \frac{\left(\mu h \ln\frac{b}{a}\right)}{2\pi\tau} \approx \mu\frac{A}{l} . \tag{16}$$

Available signal power is proportional to the permeability of the media and the toroid's cross-sectional area, and inversely proportional to the toroid's effective radius. This is the reason low loss, high permeability cores are commonplace in current transformer designs. Subject to the constraint that $L_s/R_s \gg 1/\omega_{\mathrm{min}}$, available signal power is independent of the load resistance and the number turns.

Energy loss in the core material and the effects of capacitance between the winding turns and to ground limit the high frequency performance of classical ac transformers. Proper selection of core material and careful construction techniques are required to achieve optimal performance. Resonant responses may be damped by the use of distributed shunt resistances across the winding. These damping resistors absorb some of the available signal power, but can provide the benefit of serving as a back termination when a long cable is to be driven. Metallic shielding around the signal winding is appropriate to avoid signal contamination by capacitive coupling to the beam or external noise sources. The shielding or housing must be constructed so as not to act as a shorted secondary winding.

A monitor to quantify the charge content of very short, isolated beam bunches has been described by Unser.(6, 7) Named the integrating current transformer, this device relies on the impulse response of a modified toroidal transformer to produce an output proportional to the bunch charge. All information on the temporal characteristics of the bunch is lost.

A Signal Power Comparison: Magnetic vs. Capacitive Pick-ups

Now that the ground work is laid for a quantitative understanding of magnetic sensors, a digression is in order to appreciate the prevalence of such monitors over electric field devices.

Imagine a beam tube allowing a limited cylindrical volume of space for a beam monitor, either capacitive or magnetic. The toroidal current monitor just discussed serves as the magnetic monitor when the available volume is described by ID and OD dimensions $2a$ and $2b$ respectively and by axial height h. The corresponding capacitive sensor takes the form of a thin-walled conducting cylinder of length h and radius a inside a beam tube of radius b. (Fig. 3)

For relativistic beam particles, where all electric field components are perpendicular to the axis of motion, the capacitive electrode will intercept only and all the electric field lines of those beam particles within its length. A charge, equal and opposite to that of the contained beam, is induced on the inner surface of the

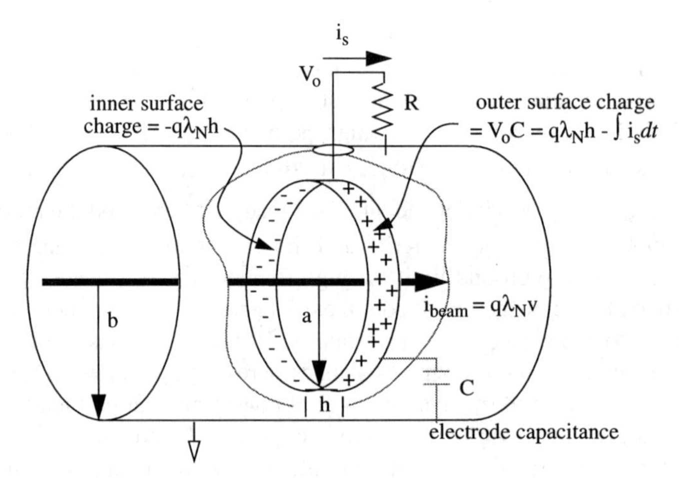

Figure 3. Capacitive beam current monitor.

electrode. Using the variable definitions given for Eq. 1, this quantity of charge (for constant λ_N within the electrode length) is

$$q_{eb} = -\int_{-\frac{h}{2}}^{\frac{h}{2}} q\lambda_N dl = -q\lambda_N h \tag{17}$$

An equal and opposite charge is induced on the outside wall of the electrode. This is the charge available to be drained off as i_s through an external load resistance R_s to provide signal power. The charge available on the electrode is given by

$$q_e = -q_{eb} - \int_{-\infty}^{\infty} i_s dt. \tag{18}$$

The electrode capacitance for this coaxial geometry is

$$C = \frac{q_e}{V} = \frac{q_e}{\int_a^b E(r)\, dr} \tag{19}$$

where *E(r)* is found from *Gauss' Law*

$$\oint (\boldsymbol{E} \cdot d\boldsymbol{S}) = \frac{q_e}{\varepsilon} \tag{20}$$

to be $E(r) = q_e / (2\pi\varepsilon r h)$. The resulting capacitance is

$$C = \frac{2\pi\varepsilon h}{\ln(b/a)}. \tag{21}$$

This capacitance, in combination with the load resistance, forms an RC circuit driven by the beam induced charge currents with a signal voltage $V = q_e/C = i_s R_s$. Making substitutions into Eq. 18 and switching to LaPlace notation for the integral yields

$$i_s C R_s = q\lambda_N h - \frac{i_s}{s}. \tag{22}$$

Solving for i_s with $i_b = q\lambda_N v$ leads to:

$$i_s = \frac{s i_b h}{v(1 + s\tau)} \tag{23}$$

where $\tau = R_s C$.

As with the magnetic current monitor, this response becomes frequency independent for frequencies $\omega \gg 1/\tau$. Under this condition $i_s = i_b h / v\tau$ and the available signal power is

$$P_{\text{signal}} = i_s^2 R_s = \left(\frac{i_b h}{v}\right)^2 \frac{R_s}{\tau^2} = \left(\frac{i_b h}{v}\right)^2 \frac{1}{\tau C}. \tag{24}$$

Normalizing to i_b^2 and substituting Eq. 21 for C results in:

$$\frac{P_{\text{signal}}}{i_b^2} = \frac{h \ln \frac{b}{a}}{2\pi\varepsilon\tau v^2}. \tag{25}$$

This can now be compared to the signal power available from the corresponding magnetic sensing current monitor with the same time constant, Eq. 16. The ratio of available power between the monitors is

$$\frac{P_M}{P_C} = \varepsilon\mu v^2 = \varepsilon\mu\beta^2 c^2 \tag{26}$$

where $\beta = v/c$.

Noting that $\mu = \mu_0 \mu_r$, $\varepsilon = \varepsilon_0 \varepsilon_r$ and $c^2 = 1/\mu_0 \varepsilon_0$, this becomes simply

$$\frac{P_M}{P_C} = \varepsilon_r \mu_r \beta^2. \tag{27}$$

For relativistic beams, $\beta \approx 1$, two rather remarkable observations follow this result: 1) the available signal power is identical for either type of device where vacuum is the only media, 2) for any medium other than vacuum the magnetic monitor offers superior performance. The advantage is often orders of magnitude at moderate frequencies where relative permeabilities > 10,000 can readily be obtained. This justifies the prevalence of magnetic type beam current monitors. Capacitive monitors, in this regime, have their place where signal power is not an important concern and at high frequencies where most magnetic materials no longer exhibit high permeability and low loss.

Eq. 27 indicates that the magnetic monitor loses its advantage for low energy beams. For example, it predicts, for a beam of $\beta = 0.01$, that the signal power from a capacitive monitor will be equal to that from a magnetic monitor with a core of permeability 10,000. Indeed, if the particles are at rest, the beam current is zero and a magnetic current transformer measures exactly that. However, caution is necessary in the quantitative interpretation of Eq. 27 for non-relativistic beams. This is because its derivation is based on a prior assumption, valid only for highly relativistic beams, that the monitor intercepts only and all the electric field lines of those beam particles within its length (see paragraph leading to Eq. 17.)

AC Transformers with Hereward Feedback

Electronic feedback may be used to extend the low frequency response of the ac beam current transformer. The need for such artificial enhancement may result from spatial constraints limiting the size of a magnetic toroid or from the expediency of utilizing a standard commercial device which may fall short of the required low frequency response. The former case often prevails in the low energy section of a Linac.(8) Space there is usually at a premium, yet high fidelity reproduction of a beam pulse several hundred microseconds in duration is required with minimal signal droop. The latter case exists in the Fermilab Booster where feedback means the difference between utilizing a standard commercial ac current transformer or resorting to a higher cost dc responding device. In the Booster satisfactory performance with less than 1% droop over a 35 millisecond beam current pulse is achieved.

A combined active/passive network which provides the benefits of Hereward feedback without sacrificing high frequency response due to amplifier bandwidth limitations is described by Unser.(1) A simplified circuit arrangement employing Hereward feedback is shown in Figure 4. Response of this simple circuit can be calculated in much the same manner as for the passive circuit in Fig. 2.

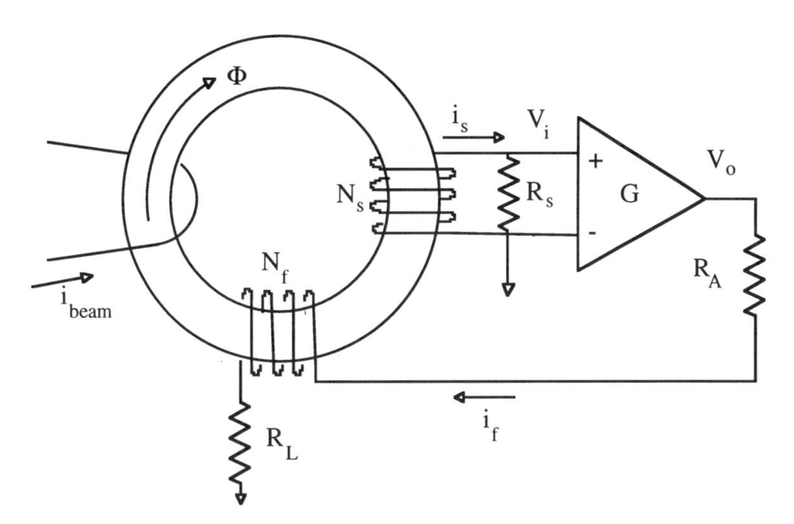

Figure 4. Beam current transformer circuit with Hereward feedback.

Here $\Phi_T = L_0 i_b + \frac{L_s i_s}{N_s} + \frac{L_f i_f}{N_f}$, $V_i = -sN_s\Phi_T$, $i_s = V_i/R_s$, and $V_o = GV_i = i_f(R_A + R_L) + sN_f\Phi_T$. Solving these equations simultaneously yields

$$\frac{V_o}{i_b} = -\frac{sN_s G L_0}{1 + s\tau}. \tag{28}$$

This is similar to the solution obtained for the circuit without feedback, except that τ now takes the more complicated form

$$\tau_H = L_0\left(\frac{N_s^2}{R_s} + \frac{N_f^2}{R_A + R_L} + \frac{GN_f N_s}{R_A + R_L}\right) = \tau_s + \tau_f + \tau_c \tag{29}$$

where $\tau_c = \frac{GL_0 N_f N_s}{R_A + R_L} = \frac{G\sqrt{L_s L_f}}{R_A + R_L}$. In the usual case where the amplifier gain is large, the τ_c term dominates. Thus, the circuit is able to extend the low frequency response (increase the effective τ) by a large factor, the order of magnitude of the amplifier gain. For frequencies $\omega \gg 1/\tau_c$ the response simplifies to

$$\frac{V_o}{i_b} = \frac{R_A + R_L}{N_f}. \tag{30}$$

Such performance enhancement appears attractive, but is not obtained without penalty. A look at the noise performance of the circuit reveals the trade-off. Imagine a noise voltage source in series with the amplifier input. V_i becomes $V_i = -sN_s\Phi_T + V_{\text{noise}}$ and the circuit's solution is obtained in the same manner. The result appears in the form

$$\frac{V_o}{V_{\text{noise}}} = \frac{G(1 + s\tau_f)}{1 + s\tau_H} \tag{31}$$

where τ_H and τ_f are as defined above and, for convenience, the beam current has been set to zero. At frequencies well above $1/\tau_f$ this becomes $V_o/V_{\text{noise}} = N_f/N_s$. Feedback via the coupled windings suppresses noise amplification and the noise sensitivity is frequency independent. At frequencies well below $1/\tau_H$ the noise sensitivity becomes $V_o/V_{\text{noise}} = G$. The noise appears at the output amplified by the full gain of the amplifier. This occurs because at such low frequencies there is effectively no coupling between the feedback and sense windings; the magnetic transformer no longer plays any role in the circuit operation. This is typically evidenced in such circuits by high sensitivity to amplifier drift and signal baseline wander that is difficult to control. These problems usually limit the maximum acceptable gain and the low frequency cut-off enhancement that is possible.

At the amplifier output, the signal to noise ratio is then

$$\mathrm{SNR} = \frac{-\left(\dfrac{sN_s G L_0}{1+s\tau_H}\right) i_b}{\left(\dfrac{G(1+s\tau_f)}{1+s\tau_H}\right) V_{\mathrm{noise}}} = -\frac{sL_0 N_s i_b}{V_{\mathrm{noise}}(1+s\tau_f)}. \quad (32)$$

Above the frequency corresponding to τ_f (i.e., the natural time constant of the feedback winding circuit) the SNR is constant at $\mathrm{SNR} = \left(\dfrac{N_s}{N_f^2}\right)\left(\dfrac{i_b(R_A+R_L)}{V_{\mathrm{noise}}}\right)$. Below that frequency the SNR degrades at 20db per decade, even though the signal sensitivity remains flat down to $1/\tau_H$! This is depicted in Fig. 5. At $\omega = 1/\tau_H$, where signal roll-off begins, the signal to noise ratio has deteriorated by a factor equal to the feedback circuit gain *G*. The enhanced low frequency response is obtained at the expense of increased noise.

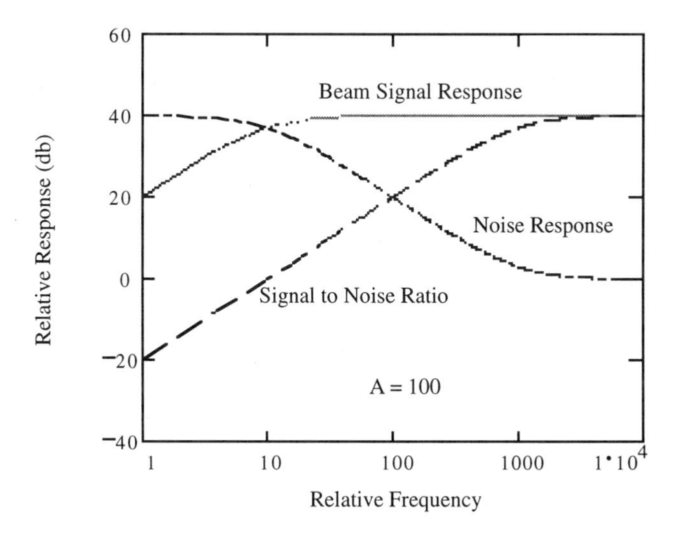

Figure 5. Response of transformer circuit with Hereward feedback, A=100.

WALL CURRENT MONITORS

Devices known as wall current monitors are usually employed for measuring beam current signals with bandwidths greater than about 100 Mhz. They are able to resolve the temporal shape of beam bunches to better than 100 picoseconds, typical of those found in proton and other hadron accelerators.(9) A wall current monitors function as single turn transformer, but may be simply thought of as a monitoring resistance placed directly in the path of the wall currents. Typical transfer impedances for wall current monitors (i.e., output volts per unit beam current) are from several hundred milliohms to several ohms.

The art of wall current monitor design and construction involves smoothly channeling the wall currents on the inner surface of the beam tube through the monitoring resistance, tapping onto that resistance in a manner so as to yield a signal voltage insensitive to beam position, and properly controlling all shunt impedances; this, all with flat response over gigahertz bandwidths. A detailed discussion of wall current monitors has been written by Webber.(10)

DIRECT CURRENT TRANSFORMERS

Direct Current Transformers (DCTs) are able to provide accurate and high precision measurement of circulating beam currents over a dynamic range of 10^5 or greater. Sensitivities of better than 1 μA are achievable and DCTs built at Fermilab have operated over periods of years with baseline drifts of no more than 5 μA. A great deal of literature is available describing the operation and design details of DCTs.(11, 12, 13, 14, 15) High quality devices are commercially available.

Since the dc beam current provides no time varying flux component to generate a signal by magnetic induction, an ac flux component is 'brought to the beam' via the action of a magnetic modulator circuit. The operation of a magnetic modulator, and hence a DCT, is based on the non-linear characteristics of high quality tapewound magnetic cores. Figure 6 shows a simplified DCT magnetic modulator layout. Two toroidal cores are switched between flux saturation levels, first one polarity then the other, by counter-phased windings powered by an external source. In the absence of any dc beam current and to the extent that the two cores exhibit matched and symmetric B-H characteristics, sense windings of a common polarity around each core produce equal and opposite signals. The output of either winding is non-zero only during that time in which the flux in the cores is changing (i.e., not saturated.) The sum of the two sense winding signals is zero. This is depicted in Fig. 7a.

A dc beam current through the two cores biases each with flux of the same polarity, while the flux in one core due to the modulator drive remains out of phase with that in the second.(Fig. 7b) Each core reaches its saturation flux level at a

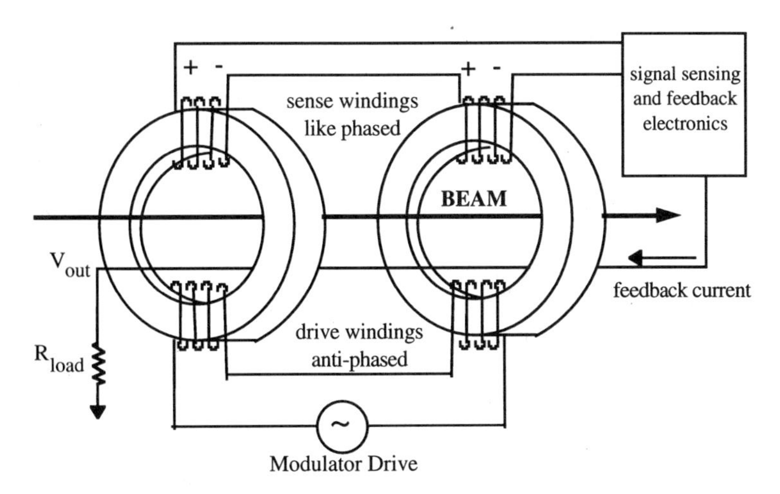

Figure 6. Magnetic modulator section of DC Transformer with flux nulling feedback.

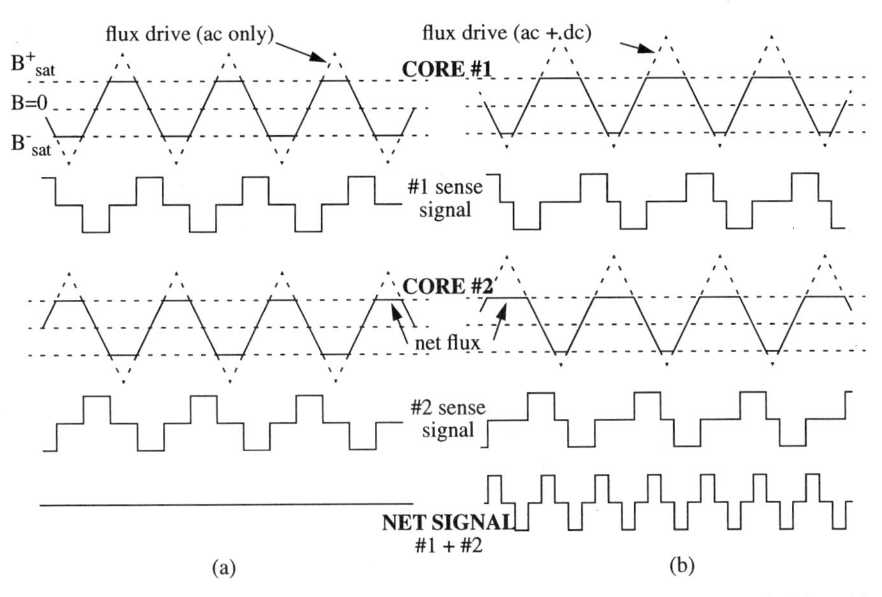

Figure 7. Magnetic modulator signals, (a) with no dc present and (b) with exaggerated dc presence.

different point in the excitation cycle for one alternation than for the other alternation. The net result is a flux imbalance between the two cores, producing an output signal at even harmonics of the excitation frequency.

The magnetic modulator section of the DCT is essentially a magnetic 'mixer', translating dc signals to a different location in the frequency spectrum. Functioning as a sampling device, it will produce aliasing of signal frequencies greater than half the excitation frequency. For example, a modulator operating at 1 Khz drive frequency will produce an identical output for a 1 Khz beam current component as for an equal magnitude dc current. This problem is usually avoided in a beam current monitoring DCT by coupling the dc modulator with an ac transformer in a hybrid network with a crossover frequency well below one half the excitation frequency. DCTs are normally designed to operate in a feedback configuration as shown in Fig. 6. Current is forced through a feedback winding on the cores to oppose the beam current and maintain a dc flux null.

REFERENCES

1. Unser, K. B., "Toroidal AC and DC Current Transformers for Beam Intensity Measurements," Atomkernenergie - Kerntechnik, Vol. 47, 1985, No.1.
2. Borer, J. and Jung, R., "Diagnostics," presented at CERN Accelerator School on Antiprotons for Colliding Beam Facilities, October 11-21, 1984, Geneva, also CERN/LEP-BI/84-14.
3. Lambertson, G. R., "Electromagnetic Detectors," presented at Joint US-CERN School on Particle Accelerators, October 20-26, 1988, Isola di Capri, also LBL-26075.
4. Siemann, R. H., "Bunched Beam Diagnostics," in *Physics of Particle Accelerators*, (1987 and 1988 U. S. Particle Accelerator School proceedings), AIP Conference Proceedings No. 184, ISBN 0-88318-384-6, 1989.
5. Halliday, D. and Resnick, R., Fundamentals of Physics, 3rd Ed.(John Wiley & Sons, New York,1988), Ch. 31, p. 714 ff.
6. Unser, K.B.,"Measuring Bunch Intensity, Beam Loss and Bunch Lifetime in LEP," presented at EPAC 90, Nice, France, June 12-16, 1990, also CERN/SL/90-27 (BI).
7. Unser, K. B., "Design and Preliminary Tests of a Beam Intensity Monitor for LEP," 1989 IEEE Particle Accelerator Conference Proceedings, IEEE Catalog No. 89CH2669-0, pp.
8. Power, J. F., "Compact Cryogenic Toroid for Beam Current Measurements," submitted to Nuclear Particle Beam Technology Symposium, Monterey, CA, July 17-21, 1989, also LA-CP-89-383.

9. C. D. Moore, et al., "Single Bunch Intensity Monitoring System Using an Improved Wall Current Monitor," 1989 IEEE Particle Accelerator Conference Proceedings, IEEE Catalog No. 89CH2669-0, pp. 1513-1515.
10. Webber, R. C., "Longitudinal Emittance: An Introduction to the Concept and Survey of Measurement Techniques, Including Design of a Wall Current Monitor," in *Accelerator Instrumentation*, (Proceedings of the First Annual Accelerator Instrumentation Workshop, 1989, Upton, New York), AIP Conference Proceedings No. 212, ISBN 0-88318-645-4, 1990, pp. 85-126.
11. Unser, K., "Beam Current Transformer with D.C. to 200 Mhz Range," IEEE Trans. Nuc. Sci., NS-16, June, 1969, pp. 934-938.
12. Unser, K. B., "The Parametric Current Transformer, A Beam Current Monitor Developed for LEP," in *Accelerator Instrumentation*, (Proceedings of the Third Accelerator Instrumentation Workshop, 1991, CEBAF, Newport News, VA), AIP Conference Proceedings No. 252, ISBN 0-88318-934-8, 1992, also CERN SL/91-42 (BI).
13. Talman, R., "Beam Current Monitors," in *Accelerator Instrumentation*, (Proceedings of the First Annual Accelerator Instrumentation Workshop, 1989, Upton, New York), AIP Conference Proceedings No. 212, ISBN 0-88318-645-4, 1990, pp. 1-25.
14. Schutte, W., "Mechanical Design of the Beam Current Transformers for the HERA Proton Ring," 1991 Particle Accelerator Conference Proceedings, IEEE Catalog No. 91CH3038-7, pp.
15. Burtin, G., et al., "Mechanical Design, Signal Processing and Operator Interface of the LEP Beam Current Transformers," presented at EPAC 90, Nice, France, June 12-16, 1990, also CERN/SL/90-30 (BI).

THE PROSAIC LAPLACE AND FOURIER TRANSFORM*

G. A. Smith
Brookhaven National Laboratory, Upton NY

ABSTRACT

Integral Transform methods play an extremely important role in many branches of science and engineering. The ease with which many problems may be solved using these techniques is well known. In Electrical Engineering especially, Laplace and Fourier Transforms have been used for a long time as a way to change the solution of differential equations into trivial algebraic manipulations or to provide alternate representations of signals and data. These techniques, while seemingly overshadowed by today's emphasis on digital analysis, still form an invaluable basis in the understanding of systems and circuits. A firm grasp of the practical aspects of these subjects provides valuable conceptual tools.

This tutorial paper is a review of Laplace and Fourier Transforms from an applied perspective with an emphasis on engineering applications. The interrelationship of the time and frequency domains will be stressed, in an attempt to comfort those who, after living so much of their lives in the time domain, find thinking in the frequency domain disquieting.

INTRODUCTION

Laplace and Fourier Transforms are, as Nahin [1] suggests, the mathematical signature of the modern electrical engineer. Even before the development of a rigorous mathematical structure, and in the face of heavy criticism from pure mathematicians, Oliver Heaviside developed his Operational Analysis, a true predecessor of the Laplace Transform applied to electric circuits. In the preface to his book, Network Analysis [2], Van Valkenberg states that;

> "The modern EE must be bilingual when speaking of network response. He must speak the language of the time domain and must also be trained in the language of the frequency domain. He may think in either language, but must be able to translate from one to the other at a moment's notice."

In a recent conversation with a vendor of wide bandwidth RF power amplifiers I asked about the pulse response of one of their units. After a long

*Work performed under the auspices of the U.S. Department of Energy, Contract No. DE-AC02-76CH00016

pause they said that they simply did not know. No such data had ever been taken. Since this amplifier was being considered for our wide band damper system in the AGS, I felt that it was important to know that pulses could be amplified by the unit with reasonable fidelity so I asked them if they might supply with some Bode plots of the unit thinking that I might infer a pulse response from that. They said they had no such thing. They could talk about gain flatness and maximum phase deviation in the pass band and 3dB points and such but why anyone would even be interested in pulse response seemed to be a mystery to them. In the end we borrowed one of their units and looked at the pulse response ourselves.

In another conversation with an engineer who was designing a high frequency amplifier, I asked how the circuit he was building responded to very fast pulses. "I really don't know", he said, "I've done all my testing using the network analyzer. Never looked at that!" On the other hand, one RF engineer I met recently said something like, "Yeah, every time I say db you can see their eyes get glassy".

In no way do I mean these comments as criticisms of the people involved. But perhaps it is clearer what the motivation for this tutorial paper was. It has begun to become apparent to me that within this world of specialists, myself included, stuck in one domain or the other, we are perhaps overdue for a review of the ways in which we are related. Its a big subject, and I know that what follows does not completely fill in every detail, in fact it only scratches the surface, but I hope that this review does stimulate the curious to further thought.

INTEGRAL TRANSFORMS

An Integral Transform $f^T(p)$, of a function , $f(q)$ is defined by the integral;

$$f^T(p) = \int_\Omega K(p,q) f(q)\, dq \tag{1}$$

Where $K(p,q)$ is a chosen function of p and q, and is called the *kernel* of the transform, and $f^T(p)$ is called the *transform* of $f(q)$. The symbol **Ω** indicates the domain over which the integral is to be evaluated. This domain can be finite or, as in the cases we will study here, one or both of the limits of integration can be infinite.

Now it is implied in mathematical treatments of transforms that there exists an inversion formula, that has the form

$$f(q) = \int_\Omega K_I(p,q) f^T(p)\, dp \tag{2}$$

This inverse "undoes" the first operation in the sense that it gives us back the original function. There is a mapping of one function space onto the transform function space that preserves "almost" everything about the original function yet puts it into a different, sometimes hardly recognizable form. In much the same way we can expand a periodic function $f(t + nT)$ into a Fourier series. The series expansion does not particularly resemble the original function, yet they are the same.

Common examples of integral transforms are [4,5],

The Laplace transform *(where p is complex, q is real)*

$$f^T(p) = \int_{-\infty}^{\infty} e^{-pq} f(q)\, dq \tag{3}$$

The Hankel transform *(of order n, where p and q are both real)*

$$f^T(p) = \int_0^{\infty} q\, J_n(pq) f(q)\, dq \tag{4}$$

The Mellin transform

$$f^T(p) = \int_{-\infty}^{\infty} q^{p-1} f(q)\, dq \tag{5}$$

The Fourier sine transform

$$f^T(p) = \int_0^{\infty} \sin(pq) f(q)\, dq \tag{6}$$

The complex Fourier transform

$$f^T(p) = \int_{-\infty}^{\infty} e^{-ipq} f(q)\, dq \tag{7}$$

The transforms shown change the initial function $f(q)$, which may be real or complex, into a new function in another real or complex variable. The result of this "transformation" is to simplify the original function with respect to some operation. In cases involving partial differentiation for example one variable can be removed. In the case of the Laplace and Fourier Transforms, which we will focus on here, a differential equation of order n reverts to a polynomial of the same order making the problem of finding a solution an algebraic one.

Of course implicit in these transform approaches is the fact that once the "solution" is found in the transform space, it must be possible to revert to the original problem space by "inverting" the result. This may not be trivial.

THE LAPLACE TRANSFORM

In most texts on this subject, the Fourier Transform is introduced first. In reality the Fourier Transform is younger than the Laplace Transform by about 30 years. Perhaps the reason for introducing it first is because to truly appreciate the Laplace Theory requires a greater understanding of complex analysis than does Fourier Analysis. In addition the development of Fourier Analysis from Fourier Series may give one a better sense of connection to the idea of a spectrum and hence a feeling of still being tethered to reality. As an EE student however, I was mercilessly thrown the equations below, with no initial knowledge of how they might relate to Fourier Series (Fourier Transforms had never been discussed) and it was not until much later that I learned that there was a connection. We will return to this.

The Laplace Transform of a function of time $f(t)$ is defined as,

$$F(s) = \int_{-\infty}^{\infty} f(t)\, e^{-st}\, dt \tag{8}$$

This is the *bilateral* Laplace Transform. In practice, unlike the Fourier Transform which we will consider later, the Laplace Transform is most often used with time dependant signal functions which are defined for time $t = 0^+$ often written as $f(t)=u(t)\,f(t)$ where $u(t)$ is the unit step function. This allows us to write equation (8) as,

$$F(s) = \int_{0+}^{\infty} f(t)\, e^{-st}\, dt \tag{9}$$

and this is called the *unilateral* transform.

The inversion formula is

$$f(t) = \frac{1}{2\pi i} \oint_{-\infty}^{\infty} F(s)\, e^{st}\, ds \tag{10}$$

where $s = \sigma + iw$, the general complex variable. As is often the case with a terse mathematical formula as shown above, much is implied.

Now it is not obvious under what the conditions the integral in (8) exists. The question is one of whether the integral converges as the limits approach infinity. This will surely depend on the choice of $f(t)$. It can be shown that, in general, convergence of the upper limit is assured so long as $f(t)$ is piecewise continuous and if for some complex number s_0 ,

$$\int_0^{\infty} f(t)\, e^{-s_0 t}\, dt \tag{11}$$

converges, the integral then converges for all Re $s >$ Re s_0 [4]. In addition convergence of the lower limit is assured if

$$\int_0^{\infty} f(-t)\, e^{s_1 t}\, dt \tag{12}$$

converges, then convergence is assured for all Re $s <$ Re s_1. Provided Re $s_1 <$ Re s_0. So for our function $f(t)$, a vertical strip is defined in the s-plane within which (8) exists. This is called the strip of convergence of $F(s)$.

Now the next step is to realize that the function, $F(s)$, found by performing the integration in (8) will be *regular* in its strip of convergence. By the principal of *analytic continuation* then, we know that this is sufficient for the function to exist everywhere outside of the strip of convergence. This may seem like black magic, but it is the result of some powerful theorems from complex function theory.[4,8,9]

Of course we still have to face the problem of finding the inverse, that is, we must solve (10) to recover $f(t)$ from our $F(s)$ if we are truly to have something worthy of being called a transform.

Now it might be more enlightening to write the inversion formula as.

$$f(t) = \frac{1}{2\pi i} \oint_{c-i\infty}^{c+i\infty} F(s)\, e^{st}\, ds \tag{13}$$

We need to remember that integration of a function of a complex variable is defined for contours in the s-plane and that the integral of an *analytic* function is independent of the contour we choose. The limits of integration shown above highlight the fact that in evaluating the integral we need to pay attention to the contour in the complex plane (sometimes referred to as a Bromwich contour) if the inverse integral is going to exist. What we must insure here is that the contour we choose to integrate over in the s-plane is inside the region of convergence of the Laplace integral of $f(t)$. This information is not contained in $F(s)$ and must be known by some independent means.

To evaluate the integral, we first do a partial fraction expansion and then invoke Cauchy's theorem of residues. We can use any contour we wish, but since the regions of convergence are all vertical strips a vertical line inside this strip makes the most sense. Using contour 1 in figure (1) and letting R become infinite, will give us $f(t)$ for $t > 0$ and using contour 2 gives us $t < 0$. This concept shows why the bilateral Laplace Transform is not too useful. Since we must have independent knowledge of the strip of convergence to uniquely accomplish the inversion, and since it is conceivable that after some algebraic manipulations in the Laplace domain this information is obscured, then the bilateral case may not give us a clearly unique result. On the other hand the unilateral case always has all the singularities to the left of some real number and so the strip of convergence always occupies the entire right half plane above some Re $s > 0$. Moving our contour over to the right sufficiently far (so that all the singularities of $F(s)$ are to the left) gives us a unique result.

Table 1 - Some Laplace Transform Pairs

$f(t)$	$F(s)$
$\delta(t)$	1
$u(t)$	$\frac{1}{s}$
$e^{-at}u(t)$	$\frac{1}{s+a}$
$\sin(bt)\,u(t)$	$\frac{b}{s^2+b^2}$
$\cos(bt)\,u(t)$	$\frac{s}{s^2+b^2}$
$e^{-at}\sin(bt)\,u(t)$	$\frac{b}{(s+a)^2+b^2}$

But we are drawn into a strange world with this definition. So far this is strictly a mathematical statement, and the fact that $F(s)$ is a complex function of a complex variable leads to the question of whether or not any real physical meaning can be attached to the operations here. As we continue we will try to understand what this all means in applies to physical reality.

Engineers or system analysts rarely calculate a Laplace Transform from (9). For all practical purposes tables of transforms exist that contain almost all the results anyone faced with a practical problem would ever need. Listed in Table 1 are a few important ones. We will need to refer to these later. In addition to the transform pairs in Table 1, it is helpful to know how certain operations in the time domain are reflected in the complex frequency domain. Some are shown in Table

3, and these are easily proved.

THE FOURIER TRANSFORM

The Fourier Transform of a function of time, $f(t)$ may be defined as

$$F(w) = \int_{-\infty}^{\infty} f(t) e^{-iwt} dt \tag{14}$$

And its inversion formula as

$$f(t) = \frac{1}{2\pi} \int_{-\infty}^{\infty} F(w) e^{iwt} dw \tag{15}$$

This is the bilateral Laplace Transform with $s = iw$, that is, $\sigma = 0$, so long as the imaginary axis lies within the region of convergence of the Laplace integral. The function $F(w)$ will, in general, be complex, that is it will have the form

$$F(w) = R(w) + i X(w) \tag{16}$$

where $R(w)$ and $X(w)$ are *real* functions of the *real* variable w. This may appear at first to make it less general than the Laplace Transform, but when it comes to the inversion integral the requirement that (13) possesses a known strip of convergence is actually more restrictive than that (15) exists for all real w.

Now as in the Laplace Transform we are faced with the problem of the existence of an integral, in this case (14). In order that the integral in (14) converge, one set of conditions on $f(t)$, known as the Dirichlet conditions: (1) the function is continuous almost everywhere; (2) at any point of discontinuity t_0, that $f(t_0 +)$ and $f(t_0 -)$ exist; (3) The function $f(t)$ is absolutely integrable. that is;

$$\int_{-\infty}^{\infty} | f(t) | \; dt \tag{17}$$

exists. These requirements are, simply stated, that physically realistic signals always possess a Fourier Transform. It must also be pointed out that they are sufficient, not necessary. The function

$$\frac{\sin (wt)}{t} \tag{18}$$

is not absolutely integrable but has a Fourier Transform. On the other hand, $sin(t)$ and the unit step, $u(t)$ do not, strictly speaking, have Fourier Transforms. Neither does DC. Dirac delta function is a question mark.

Table 2 - Some Fourier Transform Pairs

$f(t)$	$F(w)$
$\delta(t)$	1
$u(t)$	$\pi\,\delta(w) + \frac{1}{iw}$
$\cos w_0 t$	$\pi[\delta(w - w_0) + \delta(w + w_0)]$
$u(t+1) - u(t-1)$	$2\,\frac{\sin(w)}{w}$
$e^{-a\|t\|}$	$\frac{2a}{w^2 + a^2}$

I will not deal in detail with the delta function here, only say that it has always seemed a shame to me that a such an important piece of applied mathematics is still on such thin ice. I believe that by now the mathematical world has managed to embrace this entity under the theory of distributions or in linear operational calculus where it is treated nicely as an ideal point added to a space of ordinary functions. The problem is, however, as it seems always to have been, that the "function" as treated in these mathematical formalisms, seems to be bereft of a real physical meaning. Anyway, we will go on to use it as usual, and to assume that a large, short, pulse is a good approximation to the mathematical entity. Those who are interested can read further [6,7,8].

Table 2 shows some typical and interesting Fourier Transform pairs. Note that these pairs are for the bilateral case. For the case where we have an $f(t)$ which is limited to $t > 0$, that is, $f(t)u(t)$ the pairs become those in Table 1 substituting $s = iw$.

OPERATOR - TRANSFORM PAIRS

No discussion of Transforms would be complete without at least a look at how some fundamental operations, for example differentiation, integration, etc., are reflected in each domain. These relationships form the basis for understanding the way some behavior or characteristic in one domain is mirrored in the other.

They are the result of many of the theorems about the transform, form the basis for deriving new transform pairs, and provide the justification for other conclusions.

Table 3 Operator - Transform Pairs for Laplace and Fourier Transforms

Operation	f(t)	F(s)	F(w)
Linearity	$af_1(t) \pm bf_2(t)$	$aF_1(s) \pm bF_2(s)$	$aF_1(w) \pm bF_2(w)$
Differentiation	$\frac{d^n f(t)}{dt^n}$	$s^n F(s)$	$(iw)^n F(w)$
Integration	$\int f(t)\, dt$	$\frac{F(s)}{s} + \frac{f^{(-1)}(0+)}{s}$	$\frac{F(w)}{iw}$
Convolution	$f_1(t) * f_2(t)$	$F_1(s)\, F_2(s)$	$F_1(w)\, F_2(w)$

LINEAR SYSTEMS

Laplace and Fourier Transforms are particularly powerful when applied to the analysis of time invariant linear systems. *Linear systems* are those in which, given a system response described by the operator *T*, two inputs $i_1(t)$ and $i_2(t)$, and arbitrary constants *a* and *b*, then,

$$T[a\, i_1(t) + b\, i_2(t)] = a\, T[i_1(t)] + b\, T[i_2(t)] \quad \textbf{(19)}$$

Time invariance is a characteristic of a system with constant, parameters and states that the same input produces the same output no matter when, in time, it is applied. This is a fundamental feature of most analog circuits with fixed components, for example.

Now the following shows the impact of these two conditions. Let *h(t)* be the response of the system above to a unit impulse $\delta(t)$. (In fact we will reserve the notation *h(t)* and its corresponding transforms, *H(s)* and *H(w)* as call them the system response, system function or impulse response interchangeably.) Time invariance assures that,

$$h(t - t_0) = T[\delta(t - t_0)] \quad \textbf{(20)}$$

for any value of t_0. An arbitrary input $f(t)$ can always be expressed as a sum of unit impulses in the following way [8].

$$f(t) = \int_{-\infty}^{\infty} f(t_0)\, \delta(t - t_0)\, dt_0 \tag{21}$$

Invoking linearity, which should allow us to take the operator under the integral in the above, this gives us,

$$T[f(t)] = \int_{-\infty}^{\infty} f(t_0)\, T[\delta(t - t_0)]\, dt_0 = \int_{-\infty}^{\infty} f(t_0)\, h(t - t_0)\, dt_0 \tag{22}$$

and,

$$T[f(t)] = \int_{-\infty}^{\infty} f(t_0)\, h(t - t_0)\, dt_0 = \int_{-\infty}^{\infty} f(t - t_0)\, h(t_0)\, dt_0 \tag{23}$$

This result is called the *convolution theorem* and is a direct result of linearity and time invariance. It is also a statement of superposition. In any event, it can be interpreted to be a direct statement of the principal that once we know the response of a linear, time invariant system to an impulse, we can find the response to any other input.

The next step is to see how this convolution integral behaves under both transformations. This will lead us to a result that is as important and as powerful in the analysis of our systems as the transforms and their inversion formulas. It is hinted at in Table 3. For the case of the Fourier Transform we consider the transform of (23).

$$O(w) = \int_{-\infty}^{\infty} e^{-iwt} \left[\int_{-\infty}^{\infty} f(t_0)\, h(t - t_0)\, dt_0 \right] dt = \int_{-\infty}^{\infty} f(t_0) \left[\int_{-\infty}^{\infty} h(t - t_0)\, e^{-iwt}\, dt \right] dt_0$$

Now as a result of the shifting theorem (not proved here but its easy), see Table 3 as well,

$$O(w) = \int_{-\infty}^{\infty} f(t_0)\, [H(w) e^{-iwt_0}]\, dt_0 = F(w)\, H(w) \tag{24}$$

So convolution in the time domain corresponds to multiplication in the Fourier domain (and vice versa). Similarly, with the bilateral and unilateral Laplace Transform.

As an additional bit of information, it is worthy of note that most physical systems are *causal.* This simply means that an output can only occur in response to an input. Causality allows us to assume that $h(t) = 0$ for $t < 0$ and, also that if $o(t_1)$ is the output of a system for an input $f(t_1)$, and if $f(t_1) = 0$ for $t < t_1$, then $o(t_1) = 0$ for $t < t_1$.

Causality also allows us to rewrite (22) as,

$$T[f(t)] = \int_{-\infty}^{t} f(t_0)\, h(t - t_0)\, dt_0 \tag{25}$$

because $h(t - t_0) = 0$ when $t_0 > t$. This statement says that only the past counts in finding $o(t)$. In general, when we find the transform of an input we find it for all time. It might happen, although it is rare in practice, that an input does not have a Laplace Transform that is defined for all time, e^{t^2} is one. We can still find solutions to this input by cutting off the function (with a clever use of time shifted step functions) at some time greater than one we would be interested in. This create no loss of generality because response of the system up to that time is only a function of history.

RELATING FOURIER TO LAPLACE

We are now in a position to look briefly at the relationship that exists between the two transform types. We have already begun this by noting that the defining integral definitions for the Fourier Transform and the Laplace Transform differ in the substitution of the complex variable s for iw. This of course makes a more than subtle difference in the outcome chiefly because, in the Laplace case, we end up dealing with complex variables. With Fourier we still remain in the domain of the real numbers. But the fact is that, given $H(s)$ for some system, so long as the region of convergence includes the iw axis, $H(iw) = H(w)$. That is, we know the Fourier Transform given the Laplace Transform. This is what allows us to put the Fourier Transform of a signal into a system designed and characterized by Laplace techniques and come out with meaningful result. This fact is also used, in network synthesis for example, together with the principal of analytic continuation in inferring the Laplace Transform from the Fourier Transform. This is a fancy way of saying, for example, that we can design a filter around a desired frequency and phase characteristic.

One thing we have not addressed, although it has been referred to several times is how we arrive at the notion that the variable w, in either case refers to something we call frequency, or how it comes about that $H(w)$ is "frequency" response as we might understand the term as the response of a physical system to a purely sinusoidal excitation. This notion, at least in the Fourier Transform case, can be easily verified if we accept the transforms shown in Table 2. From this it can easily seem that;

$$o(t)=\frac{1}{2\pi}\int_{-\infty}^{\infty}F(w)H(w)e^{iwt}dw \quad \textbf{(26)}$$

substituting

$$o(t)=\frac{1}{2\pi}\int_{-\infty}^{\infty}H(w)e^{iwt}\pi\,[\delta(w-w_0)+\delta(w+w_0)]\,dw \quad \textbf{(27)}$$

and

$$o(t)=\frac{1}{2}\,[H(w_0)e^{-w_0t}+H(-w_0)e^{w_0t}] \quad \textbf{(28)}$$

This shows, even though we have no idea what the values of $H(\pm w_0)$ will be, that the *time* dependance will always be a sine or cosine at the frequency of the input, w_0. This is of course what is commonly meant by frequency response. In addition, the fact that the Fourier Transform of sines and cosines in time result in impulses in the frequency domain, forms the final link to the sense of the frequency domain as we commonly speak of it.

The same result can be demonstrated for the Laplace Transform although the situation is different. The unilateral Laplace transform will give us a result as if the sinusoidal input was impressed at t =0+, so the response will include the transients generated by the application of the input signal. to proceed, we will make the following observations about *H(s)*. For a typical system made up of models that are described by integral and differential equations with constant coefficients, *H(s)* will always be a ratio of two polynomials in the complex variable *s*. The coefficients, a_n and b_n, will be real constants. That is, it will have the form,

$$H(s)=\frac{a_0s^n+a_1s^{n-1}+\dots\dots+a_{n-1}s+a_n}{b_0s^m+b_1s^{m-1}+\dots\dots+b_{m-1}s+b_m}=\frac{a_0}{b_0}\frac{(s-z_1)(s-z_2)\dots\dots(s-z_n)}{(s-p_1)(s-p_2)\dots\dots(s-p_m)} \quad \textbf{(29)}$$

where the last term in the equation shows the polynomials in their generalized factored form.

Now when a sinusoidal input (see Table 1) is applied to such a system we get the following response (still in the Laplace domain),

$$O(s)=H(s)\frac{w_0}{s^2+w_0^2} \quad \textbf{(30)}$$

We would normally find the inverse of this response by a expanding the result in using a partial fractions expansion of (30). Terms associated with the roots of the denominator of H(s) will represent the transient response of the system. On some scale, however long that may be, these will eventually die out leaving only a term associated only with the sinusoidal signal. This is usually called the steady state response. Since it is this part we are looking for we can say that,

$$O_{ss}(s) = \frac{K_a}{(s - iw_0)} + \frac{K_b}{(s + iw_0)} \tag{31}$$

where

$$K_a = H(s)\frac{w_0}{s - iw_0}\Big|_{s=-iw_0} = \frac{H(-iw_0)}{2i} \quad ; \quad K_b = H(s)\frac{w_0}{s + iw_0}\Big|_{s=iw_0} = \frac{H(iw_0)}{2i}$$

now we invoke another property of real system functions. Since *H(s)* it is a *real function* of a complex variable (i.e. *H(x)* is real number if *x* is real) then

$$H(s^*) = H^*(s) \qquad \textit{and so} \qquad H(-iw_0) = H^*(iw_0)$$

writing the above in terms of the real and imaginary parts of *H(iw)*,

$$H(iw_0) = R(w_0) + iX(w_0) \qquad \textit{and} \qquad H(-iw_0) = R(w_0) - iX(w_0)$$

and with some perseverance, we find that,

$$O(s) = \frac{X(w_0)s + w_0R(w_0)}{s^2 + w_0^2} = X(w_0)\frac{s}{s^2 + w_0^2} + R(w_0)\frac{w_0}{s^2 + w_0^2} \tag{32}$$

using Table 1 in the reverse we see that,

$$o(t) = X(w_0)\cos(w_0t) + R(w_0)\sin(w_0t) \tag{33}$$

and so we have the result that the steady state response to a sinusoidal input is a sinusoid of frequency w_0 with amplitude and phase equal to

$$|H(w_0)| = [R(w_0)^2 + X(w_0)^2]^{\frac{1}{2}} \qquad \varphi = \arctan\frac{X(w_0)}{R(w_0)} \tag{34}$$

In summary we note that, in fact, the Laplace Transform gives, in the steady state, the same result as the Fourier Transform.

APPLYING THE LAPLACE TRANSFORM

When I first studied the Laplace Transform, I remember that the instructor drew an analogy with the action of logarithms and the operations of multiplication and division of real numbers. At the time I dismissed the analogy as an oversimplification of what seemed to me to be a considerably more complicated operation, and thought very little more about it. In my recent revisit to the subject I realized how closely the two do resemble each other. Under the transformation of the real numbers to logarithms, the operations of multiplication and division become addition and subtraction, which, when all you have is a pencil and a piece of paper, can make those operations much easier, especially when the numbers are large and there is a need for accuracy . Of course what you lose in this transformation is the ease of doing addition and subtraction on the numbers. In order to do those operations you are forced to return to the original number space by inverting the original process. In fact without some "feel" for the inversion process, it is very difficult to appreciate the meaning of the result one has obtained.

Well, this is analogous to the situation encountered with the Laplace transform. We simplify the problem by transforming into a "complex frequency" space. Here the operations of convolution become multiplication of polynomials in our new domain. But we have lost the explicit sense of the time dependance, trading it in for a variable space, in a sense, a mathematical artifact. In order to fully understand the result of our analysis we must be able to return to the time domain or at least to get a "feel" for how the solution we see relates to that original domain.

In the analysis of networks, system functions analogous to the generic $h(t)$, often called a transfer impedance or admittance functions are summations under the rules of Kirchoffs Laws from terms such as,

$$R\, i(t) \quad , \quad L\frac{di(t)}{dt} \quad , \quad \frac{1}{C}\int i(t)\, dt \quad , \quad G\, v(t) \quad , \quad C\frac{dv(t)}{dt} \quad , \quad \frac{1}{L}\int v(t)\, dt$$

under the Laplace Transform, and invoking linearity, these can be replaced at the component level with terms like,

$$R\, I(s) \quad , \quad L\, s\, I(s) \quad , \quad \frac{1}{C\, s} I(s) \quad , \quad G\, V(s) \quad , \quad Cs\, V(s) \quad , \quad \frac{1}{L\, s} V(s)$$

From this, under whatever technique is used, transfer functions are generated in composite systems that are the ratios of polynomials in the complex variable s just like equation (29).

Now in the last term in equation (29), the polynomials have been factored into their roots. The roots of the denominator (Hurwitz) polynomial are known as poles and are the singular points of the $H(s)$, the numerator roots are called the zeros of the function. Depending on the relative magnitude of m and n, poles or zeros are also said to exist at infinity, the order of the pole or zero being equal to the difference of m and n. In this sense it is always true that the total number of poles always equals the total number of zeros.

Some things are known about these roots when they result from ordinary passive circuit components. For a system to be stable its poles must be in the open left half of the complex plane. Clearly this avoids terms which would produce growing exponentials in the time domain (see Table 1). If they occur off the real axis they must occur in complex conjugate pairs. This is because the coefficients of the polynomials must always be real numbers. Poles must also be simple if they are located on the imaginary axis. There are restrictions on how the numerator and denominator can differ. For driving point functions (impedances and admittances) m and n can differ by no more than one. For transfer functions, numerator and denominator may differ by more than unity but the order of the denominator is always greater. Driving point function zeros must also be in the left half-plane.

Now, one very powerful tool for analysis of is a simple plot in two dimensions of the pole and zero locations. This is known as a *pole-zero* diagram. Illustrated in figure (2), it is a graphical representation of the network function. If we include the scale factor it is a complete description of the function it represents. For a transfer function, it is a representation in the Laplace domain of the impulse response of the network. Interestingly enough, these diagrams are far more often referred to than drawn, still the usefulness of the concept in relating time to frequency is unequaled. For example, as illustrated in the figure, the magnitude and phase response at an arbitrary point on the iw axis is easily seem. If one draws a vector from each finite pole and zero, the value of $|H(w)|$ is the ratio of the magnitudes of the zero vectors to the pole vectors and the phase angle is the sum of the angles defined to the same point by the vectors associated with the zeros minus the pole angles. This is much easier to visualize than it is to say.

Now a plot of $20 \log_{10} |H(w)|$ and arg $H(w)$ constitute what is called a Bode diagram of the system function. It is a graphical representation of the Fourier Transform of the function which, if it is known in analytical form gives the Laplace Transform by substituting $w = s/i$ (remember analytic continuation). If as often happens, the magnitude and phase response is known only by measurement, or only over part of the iw axis, then it must be fit to some algebraic approximation by some means. For systems with small numbers of simple poles and zeros this is not difficult since these roots cause certain specific behavior in the Bode plot that is easy to see. Because the magnitude plot is done on a log scale it makes behavior associated with both the pole and zeros additive, hence easier to spot and interpret.

A TIME FOR UNDERSTANDING

Figure (3) shows an illustration of how various simple pole-zero patterns reflect in the time domain. The thing to remember here is that it is the poles that determine the time response, the zeros effect only the relative magnitude and sign (remember the final form of the partial fraction expansion). A zero at the origin in a transfer function means no DC.

Figure (4) shows a simple RC circuit that is a good model of one of our accelerator pick-up electrodes. Capacitor C1 represents the coupling of the electrode to the beam, capacitor C2 is the capacitance of the electrode to the beam pipe. The resistor R1 represents the input resistance to any electronics that might be connected to the output of the PUE. The equation on the figure represents the voltage transfer function from the beam to the resistor R1. Notice that it has a zero at the origin and one simple pole on the real axis. The pole position is a function of R1, and it moves away from the origin as the resistance gets lower. This alters the low frequency response of the circuit, causing the characteristic "walking" of the signal seeking its average value (remember no DC) to take less time. We can see the base line shift of the upper waveform in figure (4). This would not be troublesome if the beam pulses were always the same size, but when one is looking for beam position or betatron tune information, which normally shows up as modulating pulse amplitudes over many beam pulses, a pole in the circuit response near the frequency of interest can cause havoc in attempting to properly detect those effects.

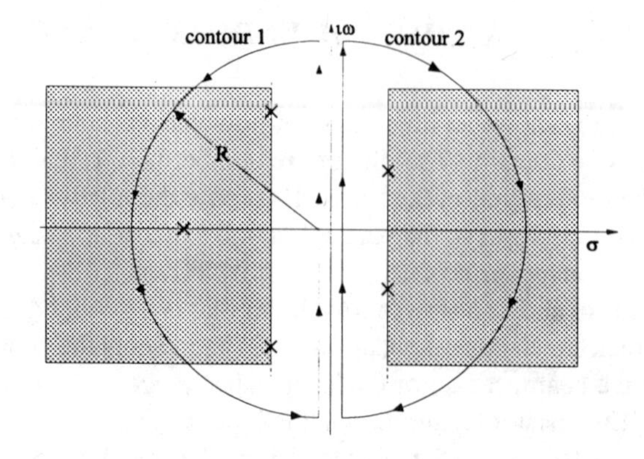

Figure 1 - Complex integration contours used to find the inverse Laplace transform. The shaded areas represent regions in which the integrals do not converge.

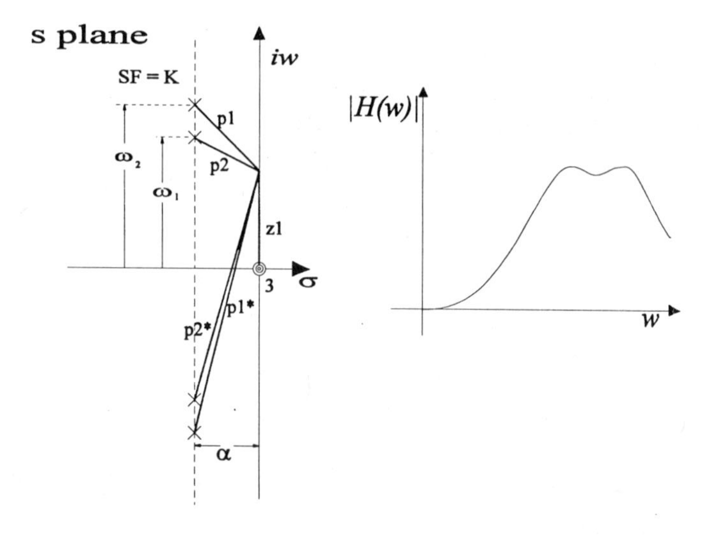

Figure 2 - Using the pole-zero diagram of $H(iw)$ to find $|H(w)|$

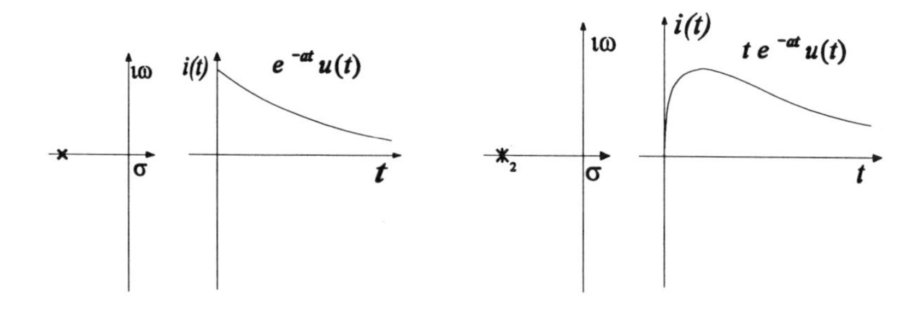

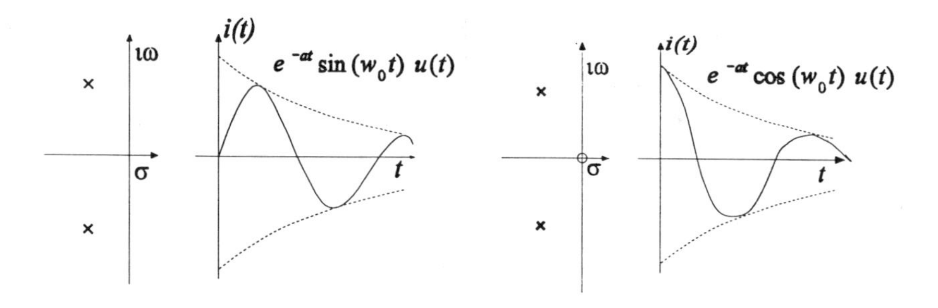

Figure 3 - This illustrates the time domain response coresponding to several simple pole - zero configurations. If this were a network function then they would be the impulse response of that system.

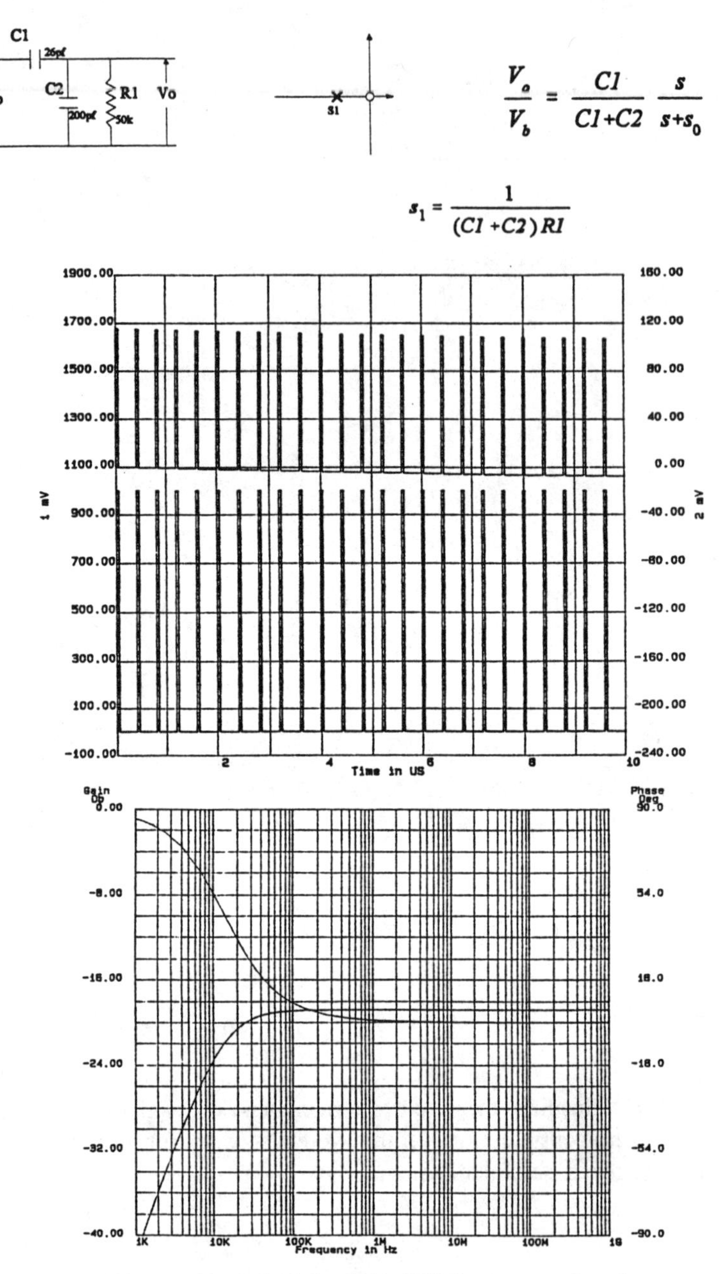

Figure 4 - Model of the PUE output circuit

REFERENCES

1. Paul J. Nahin, Behind the Laplace Transform, IEEE Spectrum, March 1991.

2. M. E. Van Valkenberg, Network Analysis, Prentice-Hall, 1955

3. Ernst Weber, The Evolution of Scientific Electrical Engineering, IEEE Antennas and Propagation Magazine, Feb. 1991

4. Kurt Bernado Wolf, Integral Transforms in Science and Engineering, Plenum Press, 1979

5. C. J. Tranter, Integral Methods in Mathematical Physics, John Wiley and Sons, Inc. 1966

6. Murray F. Gardner, John L. Barnes, Transients in Linear Systems, Volume 1, John Wiley and Sons, Inc. 1957.

7. Wilber R. LePage, Complex Variables and the Laplace Transform for Engineers, Dover Publications, Inc. 1980.

8. Athanasios Papoulis, The Fourier Integral and Its Applications, McGraw-Hill Book Company 1962

9. Ronald N. Bracewell, The Fourier Transform and Its Applications, 2d ed. rev., McGraw-Hill Book Company, New York, 1986.

10. R. E. Shafer, Beam Position Monitoring, AIP Conf. Proc. 212 (Accelerator Instrumention), pp30-33

11. Edna E. Kramer, The Nature and Growth of Modern Mathematics, Princeton University Press, 1981.

12. Paul R. Grey, Robert G. Meyer, Analysis and Design of Analog Integrated Circuits, 3d ed. John Wiley and Sons, Inc. 1993.

13. Kenneth K. Clark, Donald T. Hess, Communication Circuits, Analysis and Design, Addison-Wesley Publishing Co., 1971.

Digital Signal Processing

A Review of DSP Formalism, Algorithms and Networks for the Beam Instrumentation Workshop Vancouver Canada October 4, 1994

Ivan Linscott
Stanford University and SLAC, 235 Durand,Stanford CA, 94305

Abstract

The formalism of Digital Signal Processing (DSP), is reviewed with the objective of providing a framework for understanding the utility of DSP techniques for Beam Instrumentation and developing criteria for assessing the merits of DSP applications.

Outline of the Presentation

1. Discrete-Time Signals and Sampling
2. Discrete-Time Systems
3. Linear, Time-Invariant Systems
4. Frequency Domain Representation of Sequences
5. The Fourier Transform
6. Reconstruction from Sampled Sequences
7. Fourier Transform of Band-Limited Functions
8. The Z-Transform
9. Networks and State Representation
10. Transformation Among Networks
11. Summary

1.0 Discrete-Time Signals†

A Signal is a measurable function of a physical variable that conveys information via an established protocol. Signal processing strategies arise from the need to extract this information and obtain symbolic abstractions in a reliable and efficient manner. Of particular interest is the desire to find processing strategies that produce compact, high fidelity representations of the signals. The most attractive processing strategies are efficient and thus cost effective. The emphasis on compact representation and efficient processes, or algorithms, will be a common theme throughout this presentation.

To begin, Continuous-time Signals are signals that are continuous in time, and represented functionally as, $y = x(t)$, where t is continuous. These signals are commonly described as analog waveforms. When the time continuum is sampled at discrete intervals the corresponding samples of the signal are identified as Discrete-time Signals.

† For a more detailed discussion see, for example, "Discrete-Time Signal Processing," by Oppenheim & Schafer

The discrete-time signals will then be represented as,

$$y_n = x(t_n) = x_n,$$

where t_n is evaluated at the discrete intervals in time. When all time intervals have the same duration, τ, then $t_n = n\tau$, and the sampling is said to be uniform. A Digital Signal, y_n, is a Discrete-time signal where the value of the sample itself is quantized, or assigned to one of a finite number of levels.

$$y_n = Q_p(y_n),$$

where Q is a quantizer, with precision p that assigns y_n to one of 2^p levels. Thus,

$$y_n = x_n$$

A Sequences of samples, $\{x_n\}$, is a collective entity that can be an effective means of representing a signal as well as the information it contains. With x(t) describing a signal then a sequence of its samples, $\{x_n\}$, can be represented as,

$$x_n = \int_{-\infty}^{\infty} x(t)\delta(t - t_n)dt = \sum_{k=-\infty}^{\infty} x_k \delta(n-k) \qquad (1.0.1)$$

The sum of impulses, $\Sigma\ \delta(n - k)$, can be thought of as the sampling function and will be given greater attention later.

Examples of a simple function that is uniformly sampled are presented in Figure 1.1, where the sample interval is first much smaller than the wavelength λ, and then greater than $\lambda/2$. In the latter case, the samples can be associated with an alternative waveform of larger wavelength λ_2. This ambiguity is called aliasing and is related to the ambiguities of reconstructing a waveform using only its samples. Notice that the alternative, or aliased, waveform appears reflected in time and lower in frequency. Half the aliased wavelength, $\lambda_2/2$, is greater than the sample interval. This apparent reflection in time is further illustrated in the aliased sampling example in Figure 1.2, where the parabolic trajectories of the water droplets from a dripping faucet are undersampled, resulting in an aliased alternate representation that has the appearance of making the droplets drip back up into the faucet.

2.0 Discrete-Time Systems

A Discrete-Time System is a process that transforms a sequence into a sequence.

$$y_n = T\{ X_n \} \qquad (2.0.1)$$

Examples of this very simple but quite general transformation are the Ideal Delay and Running Average. An important sub-class of these transformations are the Linear

Systems. The transformation is linear when the transform of the sum of sequences is equivalent to the sum of the transform of the sequences. Thus a system T is linear if for two sequences, x_1 and x_2,

$$\mathbf{T\{\ x_1 + x_2\} = T\{x_1\} + T\{x_2\} = y_1 + y_2\ .} \tag{2.0.2}$$

Time Invariant Systems are likewise interesting for in these systems the elements of T do not change with time. Thus if a sequence is shifted, i.e. delayed, by n_o, with the result that $x_s = x(n - n_o)$, then T is time invariant if,

$$\mathbf{T\{x_s\} = y(n - n_o) = y_s}, \text{ for all } \mathbf{n_o}. \tag{2.0.3}$$

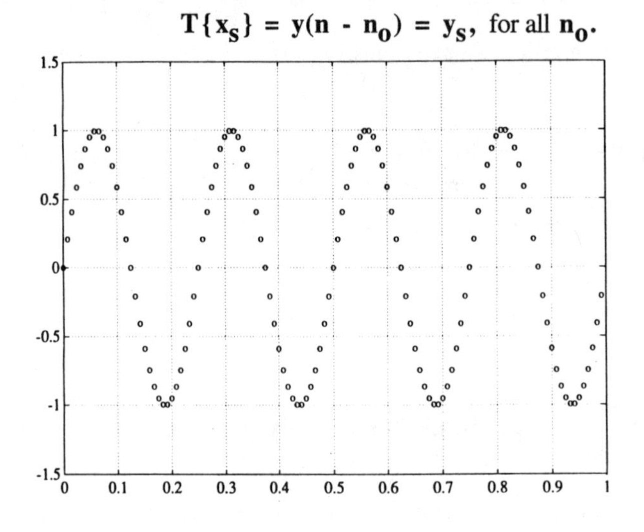

Oversampled Sequence with 30 samples/cycle, (where minimum for reconstruction is 2 samples/cycle)

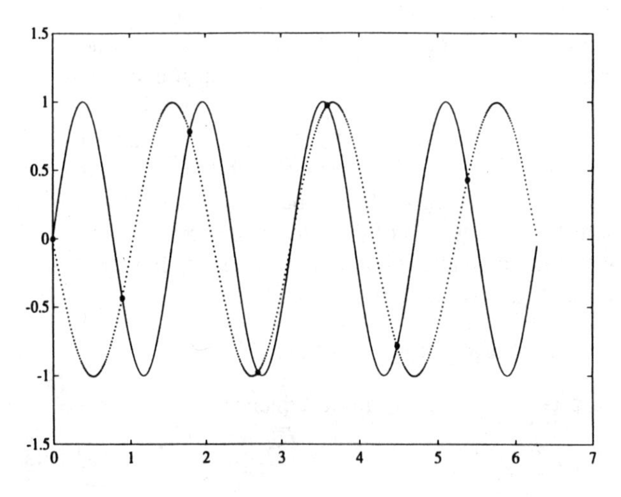

Undersampled, or Aliased with 7 samples in 4 cycles, or 1.75 samples/cycle. Dotted line is alternative waveform reconstructed from samples.

Figure 1.1. Sampled Sequences

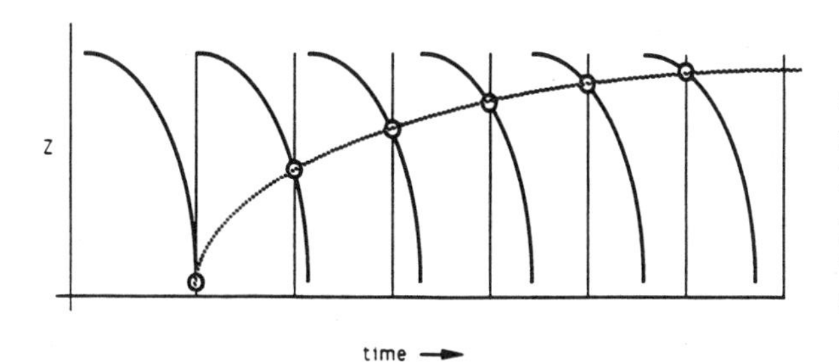

Aliased Sampling of Dripping Faucet. Dotted line is alternative trajectory reconstructed from samples. Note that time appears reversed and the water drips up.

Figure 1.2. Undersampled Aliased Parabolic Waveform.

The concepts of causality and stability feature strongly in the development of effective signal processing strategies. Here, these are defined in terms of sequences.

Causality: a system is causal if y_n depends only on x_k, for $k < n$.
Stability: a system is stable if the inputs, x_n , are finite and the outputs, y_n , are finite.

3.0 Linear, Time-Invariant Systems

Systems that are both linear and time-invariant play a central role in signal processing. The response of such systems to any input can be specified in terms of the response of the system to an impulse. Thus, the Impulse Response, $h_k(n)$ is the response of system T, to an impulse $\delta(n\text{-}k)$, at n=k.

$$\begin{aligned} h_k(n) &= T\{\ \delta(n-k)\ \} \\ &= h(n-k), \text{ if T is Time Invariant.} \end{aligned} \tag{3.0.1}$$

The response of the system T to the input sequence x_n is,

$$y_n = T\{\ x_n\ \}, \quad \textbf{remember} \quad x_n = \Sigma\, x_k\, \delta(n-k).$$

$$y_n = T\Big\{ \sum_{k=-\infty}^{\infty} x_k \delta(n-k) \Big\}$$

$$= \sum_{k=-\infty}^{\infty} x_k T\{\ \delta(n-k)\ \}\ , \textbf{ if T is Linear}$$

$$= \sum_{k=-\infty}^{\infty} x_k h(n-k) \quad = x * h \tag{3.0.2}$$

Thus, the response, or output, from a linear, time-invariant system is the convolution of the impulse response of the system with the input sequence. This simple convolution rule is the cornerstone of Digital Signal Processing techniques and technologies.

3.1 Linear, Time-Invariant Systems: Properties

A few properties of linear, time-invariant systems will be useful. The first is that convolution is associative.

x_n → h_{1n} → h_{2n} → y_n

$$(x * h_1) * h_2$$

$$= x * (h_1 * h_2)$$

$$= x * h$$

x_n → h_n → y_n $\qquad h = h_1 * h_2$

(3.1.1)

The convolution of an impulse response with an input sequence is commonly referred to as a filter, where the filter can itself be produced through the convolution of multiple components.

Next, convolution is distributive $x * (h_1 + h_2) = x{*}h_1 + x{*}h_2 = y_1 + y_2$.

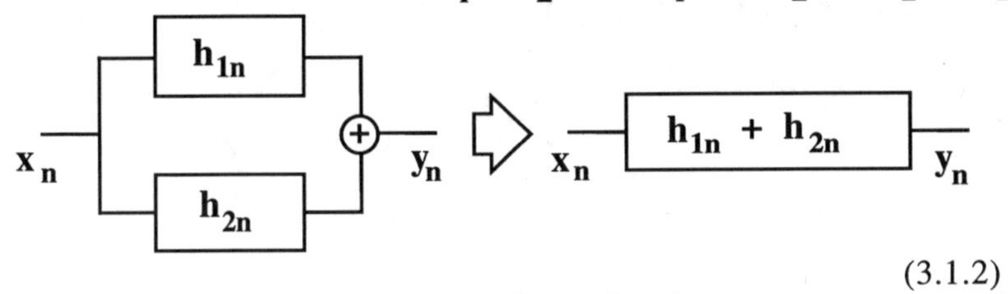

(3.1.2)

A convolution is almost always present in performing a measurement. The measuring instrument acts as a filter, in effect combining the representation of a process with the effects of the instrument. This combination has the form of a convolution whenever the measurement process is both linear and time-invariant. Then the effects of the measurement can be compensated in principle by inverting the convolution. This involves convolving the measured output with a response that can be thought of as the inverse of the measuring instrument's impulse response, h^{-1}. The inverse impulse response has the property $h^{-1} * h = 1$, and this property is the basis of a means of obtaining h^{-1}, as will be discussed in the later section on the Fourier Transform. A particular example of the deconvolution of a distorted waveform is shown in Figure 3.1, and is the profile of a radio signal transmitted through the rings of Saturn as the Voyager Spacecraft flew in occultation behind the planet. The radio signal received on earth was distorted by diffraction at Saturn's rings and this distortion was removed by applying a deconvolution filter with the inverse response of diffraction and resulted in the occultation profile, shown in Fig. 3.1, that revealed a wealth of phenomena, like the chirped oscillation from nearby moonlets, that was hidden in the received, diffracted signal.

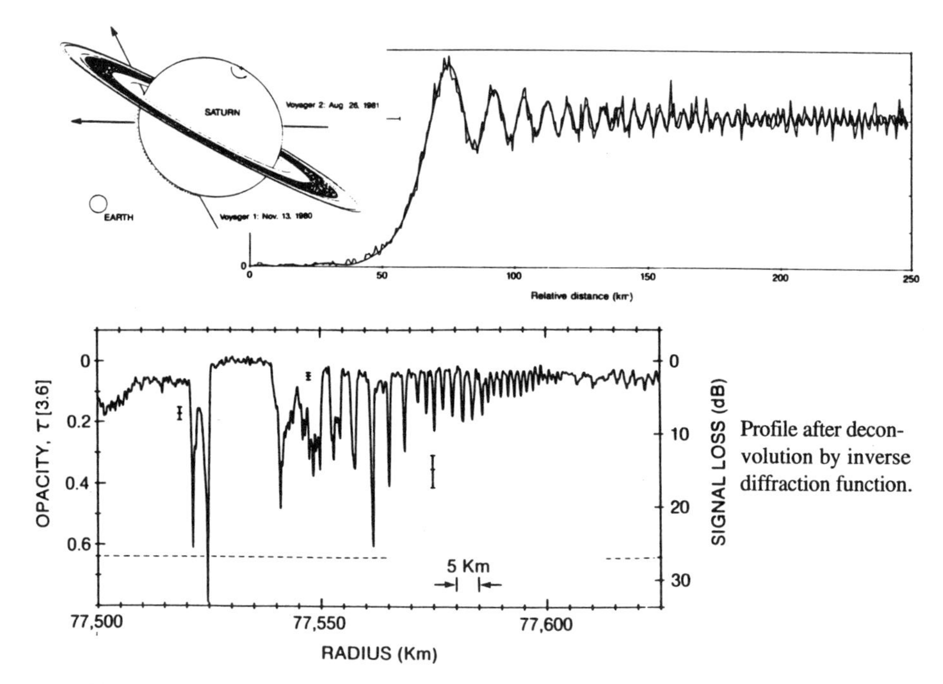

Figure 3.1. Radio signal received in occultation by Saturn's Rings.

4.0 Frequency Domain Representation of Discrete-Time Signals

Discrete-time signals can be represented in the frequency domain by using a discrete-time version of the Fourier transform. As a prelude to developing the discrete Fourier transform, the utility of complex exponentials in linear time-invariant systems will first be reviewed. Consider that x_n, is a signal in the form of a complex exponential, that is a signal whose real part is proportional to $\cos(\omega n)$, and whose imaginary part is proportional to $\sin(\omega n)$. Such signals contain a single frequency ω, and form a linearly independent set when the value of ω is restricted to the rational fractions m/N, m < N. Then x_n can be represented as,

$$x_n = e^{j\omega n} \quad \text{for } -\infty < n < \infty, \text{ and } e^{j\omega n} = \cos \omega n + j \sin \omega n$$

The response of a linear time-invariant system h, to the complex exponential x is,

$$\begin{aligned} y_n = x * h &= \sum_{k=-\infty}^{\infty} h_k e^{j\omega(n-k)} \\ &= e^{j\omega n} \sum_{k=-\infty}^{\infty} h_k e^{-j\omega k} \\ &= e^{j\omega} H(e^{j\omega}) \end{aligned} \tag{4.0.1}$$

Thus, the response y is proportional to x, which means that x is an eigenfunction of h. The proportionality constant H is,

$$H(e^{j\omega}) = \sum_{k=-\infty}^{\infty} h_k \, e^{-j\omega k} \qquad (4.0.2)$$

5.0 The Fourier Transform

A version of the Fourier transform will now be developed that is appropriate for discrete-time signals. First we will start with the definition of the Fourier transform for continuous time.

$$X(f) = \int_{-\infty}^{\infty} x(t)\, e^{-2\pi i f t}\, dt \qquad \textbf{forward transform} \qquad (5.0.1)$$

$$x(t) = \int_{-\infty}^{\infty} X(f)\, e^{+2\pi i f t}\, df \qquad \textbf{inverse transform} \qquad (5.0.2)$$

Consequences of Linearity: $\mathbb{F}(x_1 + x_2) = \mathbb{F}(x_1) + \mathbb{F}(x_2)$

$\mathbb{F}$ denotes Fourier Transform

Symmetry Properties:	x real	$\leftrightarrow \mathbb{F} \leftrightarrow$	re X even, im X odd
	x real, even	$\leftrightarrow \mathbb{F} \leftrightarrow$	X real, even
	x real,odd	$\leftrightarrow \mathbb{F} \leftrightarrow$	X imag,odd

5.1 Fourier Transform Theorems

Several Theorems for the Fourier transform relating to the behavior of functions under Fourier transformation are stated here for later use, but without proof. For their proof see e.g. Bracewell.

Scaling theorem:

$$F(x(at)) = \frac{1}{|a|} X(f/a) \quad \textbf{for any } a \neq 0. \qquad (5.1.1)$$

Shift Theorem:

$$F(x(t-a)) = e^{-2\pi i a f}\, X(f) \quad \textbf{for any a.} \qquad (5.1.2)$$

Modulation Th'm: $$\mathcal{F}\left(x(t) \cos(2\pi f_1 t)\right) = \frac{1}{2} X(f - f_1) + \frac{1}{2} X(f + f_1) \quad (5.1.3)$$

Derivative Th'm: $$\mathcal{F}(x') = 2\pi i f \; X \qquad \text{where } x' = \frac{dx}{dt} \quad (5.1.4)$$

Convolution Th'm: (5.1.5)

$$\mathcal{F}\left(x(t) * y(t)\right) = X \bullet Y \quad \text{, where } X = \mathcal{F} x, \text{ and } Y = \mathcal{F} y$$

The convolution theorem provides attractive, intuitive insight into the interpretation of convolution as a filter. According to this theorem, $y = x * h = \mathbf{F}^{-1}(X H)$, and if H is frequency selective, i.e. has the response of a filter, then only those frequencies present both in the sequence {x}, and H survive the product XH, and pass through the filter to become the output y For example, if h has the impulse response of a bandpass filter, then H has the form of Figure 5.1, and the convolution x * h retains only those frequencies present both in x and within the passband of H.

Figure 5.1, Frequency Spectra of Impulse Response

5.2 Example Fourier Transforms

Here are a few examples of Fourier transforms that will be especially useful in the remaining sections.

5.2.a Fourier Transform of a Gaussian, g(x). (notice here that x is the dependent variable)

$$g(x) = e^{-\pi x^2}$$

$$G(s) = \int_{-\infty}^{\infty} e^{-\pi x^2} e^{-2\pi i x s} \, dt$$

$$= e^{-\pi s^2} \quad (5.2.1)$$

i.e. The Fourier transform of a Gaussian is a Gaussian, however,

$$\mathcal{F}\left(g(x / \sigma)\right) = \sigma \, G(\sigma s)$$

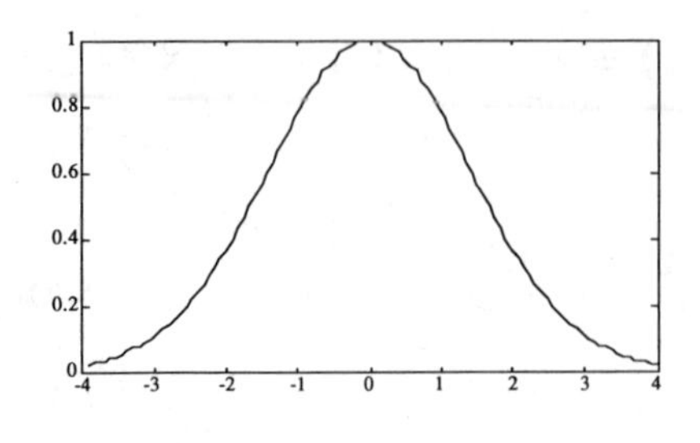

g(x/σ), Gaussian of width σ.

G(σs), Gaussian of width 1/σ.

5.2.b Fourier Transform of a Rectangular Window, $\Pi(t)$. (notice here that t is the dependent variable)

$$\mathbf{x(t)} = \Pi(t) = \begin{cases} 1 & |t| < 1/2 \\ 0 & \text{elsewhere} \end{cases}$$

$$\mathcal{F}(\Pi(t)) = \int_{-1/2}^{1/2} 1 \; e^{-2\pi i f t} \, dt$$

$$= \frac{e^{-2\pi i/2} - e^{-2\pi i/2}}{2\pi i f} = \frac{i2 \sin(2\pi f/2)}{2\pi i f}$$

$$= \text{sinc}(f) \qquad 5.2.2)$$

5.3 Fourier Transform of the the Sampling Function

An important step in developing a Fourier transform for discrete-time sequences is to understand the Fourier transform of the sampling function itself. A sampling function III(t) will be defined to help formalize the sampling process. Recall, that a sequence of samples is obtained from a continuous function x(t), by,

$$x_n = \int_{-\infty}^{\infty} x(t) \, \delta(t_n - t) \, dt \qquad (5.3.1)$$

To help exhibit the explicit, periodic sampling of x, define **III(t)**, the Shah Function:

$$\text{III}(t) = \sum_{n=-\infty}^{\infty} \delta(t - n) \qquad (5.3.2)$$

then the product, **III(t) x(t)**, samples **x(t)** at unit intervals in time.

Note that by scaling:

$$\mathbf{III(at)} = \frac{1}{|a|} \sum_{n=-\infty}^{\infty} \delta(t - n/a)$$

this scaling samples x at intervals of n/a.

The Shah is its own Fourier transform. **F(III(t)) = III(f)**. The proof of this equivalence is such an attractive example of the use of the Fourier transform theorems that the proof is sketched here to illustrate the utility of these simple rules. For a more detailed proof see e.g Bracewell, "The Fourier Transform and Its Application,"Ch. 5.

5.3.1 The Fourier Transform of the Shah Function.

define

$$\mathrm{III}(x) = \lim_{\substack{\sigma_1 \to \infty \\ \sigma_2 \to 0}} \left(g(x/\sigma_1) \sum_{k=-\infty}^{\infty} g\left(\frac{x - x_k}{\sigma_2}\right) \right)$$

$g(x/\sigma_1)$ → **1 at limit**; $g\left(\frac{x - x_k}{\sigma_2}\right)$ → $\delta(x - x_k)$ **at limit**

using Theorems scaling, convo'n, shift, scaling

$$F(\mathrm{III}(x)) = \lim_{\substack{\sigma_1 \to \infty \\ \sigma_2 \to 0}} \sigma_1 G(\sigma_1 s) * \sum_{k=-\infty}^{\infty} \sigma_2 e^{-2\pi i x_n s} G(\sigma_2 s)$$

$$= \lim_{\substack{\sigma_1 \to \infty \\ \sigma_2 \to 0}} \sigma_1 \sigma_2 \sum_{k=-\infty}^{\infty} \int_{-\infty}^{\infty} e^{-2\pi i x_n s} G(\sigma_2 s)\, G(\sigma_1(s-s'))\, ds$$

$G(\sigma_2 s) \to 1$, $\lim \sigma_2 \to 0$; $G(\sigma_1(s-s')) \to \delta(s - s')$, $\lim \sigma_1 \to \infty$

$$= \sum_{k=-\infty}^{\infty} e^{-2\pi i x_n s} = \begin{cases} \infty & x_n = n \\ 0 & \text{otherwise} \end{cases} = \sum_{k=-\infty}^{\infty} \delta(s - n_k)$$

$$= \mathrm{III}(s), \text{ the Shah function}$$

6.0 Reconstruction of a Continuous Function from its Samples

The notation, X(f) = F[x(t)], represents the Fourier transform of a continuous function x(t) into the frequency domain. When x is sampled, resulting in the sequence $\{x_n\}$, then the Fourier transform of the sequence is, formally,

$$F\{ x_n \} = F(III(at)x(t)).$$

The product of III(at) with x becomes, in the frequency domain, the convolution of $III(f_s f)$ with X(f), a process that replicates X(f) at intervals of the sample frequency. This process is illustrated in Figure 6.1 for the case where X is band limited and has a bandwidth that is less than the sample frequency. Thus the replicas of X are separated by small frequency intervals where the spectrum goes to zero.

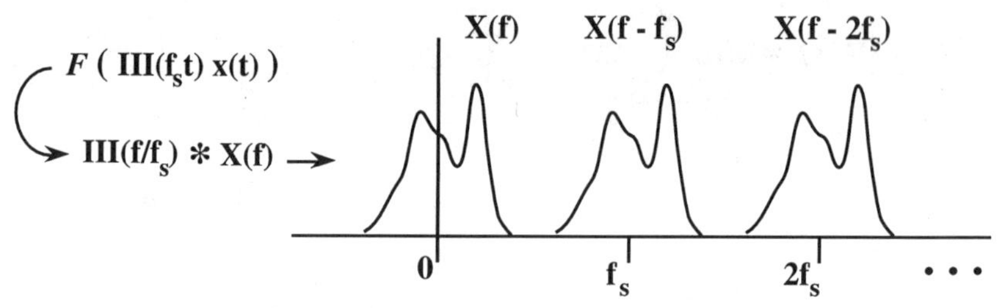

Figure 6.1. Frequency Domain Representation of Sampled Function.

If x(t) is band-limited, with cutoff frequency $f_c < f_s$, and $f_s = a$, then x(t) is just the inverse Fourier transform of the reflica X(f) at f=0, or dc. i.e.,

$$\begin{aligned}
x(t) &= F^{-1}(X(f - 0)) \\
&= F^{-1}\left(\Pi(f/f_c) \sum_{n=-\infty}^{\infty} X(f - na) \right) \\
&= F^{-1}(\Pi(f/f_c)\, X(f) * III(f/a)) \\
&= (F^{-1} \Pi(f/f_c)) * F^{-1}(X(f) * III(f/a)) \\
&= (\text{sinc}(f_c t)) * (x(t)\ III(at)) \\
&= \text{sinc}(f_c t) * \left(\sum_{n=-\infty}^{\infty} x(t)\delta(t - n/a)\right) \\
&= \sum_{n=-\infty}^{\infty} x(t - n/a)\ \text{sinc}(t - f_c n/a)
\end{aligned}$$

The Shannon Reconstruction Th'm (6.0.1)

This reconstruction is exact. The enitre continuous function is recovered from only its samples. The samples are therefore an exact representation. As an example of this reconstruction process, a bipolar pulse is approximated by a sequence of more and more samples in Figure 6.2, where as more samples are included in the reconstruction the resultant waveform converges rapidly to the expected shape.

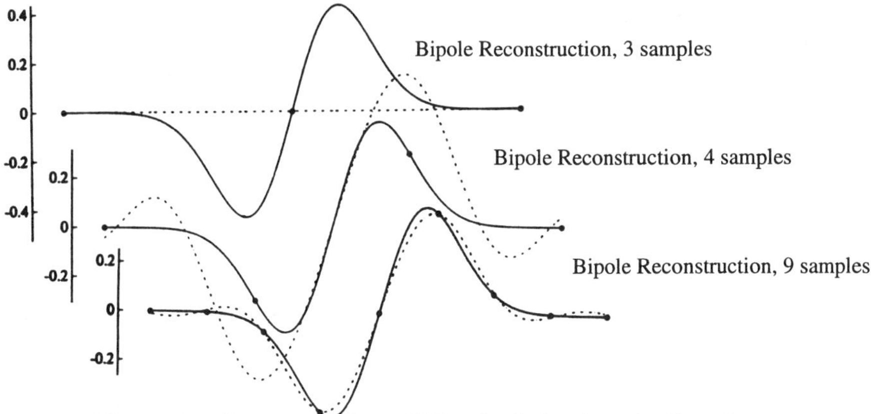

Figure 6.2. Reconstruction of Bi-polar Pulse from its Samples.

7.0 Fourier Transform of Samples from Band-Limited Functions

The next step in developing a Discrete Fourier Transform is to replace the expression of a continuous function with its Shannon reconstruction equivalent. Then the Fourier transform of a function will operate only on the samples of that function. Start with the Fourier transform of a continuous function,

$$\mathbf{X(f)} \;=\; \mathbf{F[\; x(t)\;]}, \tag{7.0.1}$$

now replace x(t) with the expression 6.0.1, which is the reconstructed continuous function x(t) generated from its samples $\{x_n\}$.

$$\begin{aligned}
\mathbf{X(f)} &= \mathcal{F}\left(\operatorname{sinc}(f_c t) * \underbrace{\left(\sum_{n=-\infty}^{\infty} x(t)\delta(t - n/a)\right)}_{\longleftarrow\ \text{samples}\ \longrightarrow} \right) \\
&= \left(\mathcal{F}\mathcal{F}^{-1}\prod(f/f_c)\right) \cdot \mathcal{F}\left(\sum_{n=-\infty}^{\infty} x(t)\delta(t - n/a)\right) \\
&= \prod(f/f_c) \cdot \left(\sum_{n=-\infty}^{\infty} x_n\, e^{-2\pi i n f/a}\right) \longleftarrow \int_{-\infty}^{\infty} e^{-2\pi i f t}\, x(t)\,\delta(t - n/a)\,dt
\end{aligned} \tag{7.0.2}$$

When x(t) is periodic, x(t - p) = x(t), then

$$\mathbf{X(f)} \;=\; \prod(f/f_c) \cdot \left(\sum_{n=0}^{N-1} x_n\, e^{-2\pi i n f/a}\right) \tag{7.0.3}$$

where p = Na, or N = m p/a and m can be any integer.

By choosing the cutoff frequency to be half the sample frequency,

$$f_c = a/2, \text{ i.e. } f = f_s/2,$$

then with f = k, f_s/N = (ka)/N, k=0,1,...N-1,
This results in the desired expression for the Discrete Fourier Transform (DFT).

$$X(k) = \sum_{n=0}^{N-1} x_n \, e^{-2\pi i n k/N} \quad \textbf{The Discrete Fourier Transform} \tag{7.0.4}$$

8.0 The Z Transform

The tools of convolution and the DFT are the means by which algorithms can be developed to meet the needs of signal processing tasks. However, to implement an algorithm in a signal processing network a new tool is needed to help transform an algorithm into a network. The z transform is introduced for this purpose. The z transform is a discrete-time process whose continuous time equivalent is the Laplace transform.

The z Transform,

$$X(z) = \sum_{n=-\infty}^{\infty} x_n \, z^{-n} \quad \text{where z is a complex variable} \tag{8.0.1}$$

Region of convergence R, is an annular ring in the z plane bounded by r_- and r_+.
There are three possible forms for R:

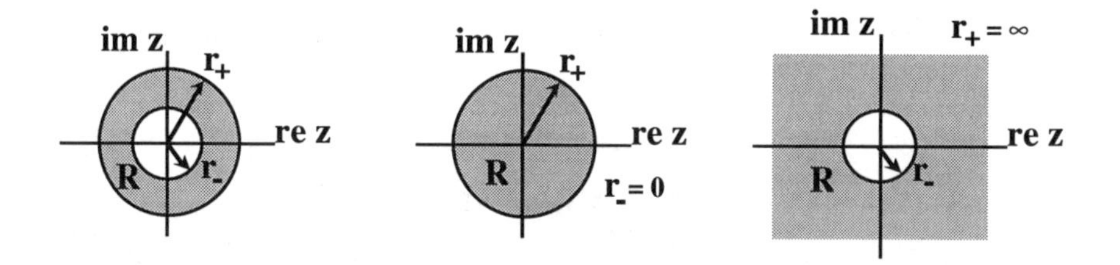

A Special Class of z Transforms is of the form,

$$X(z) = \frac{A(z)}{B(z)} \tag{8.0.2}$$

where A(z) and B(z) are polynomials in z.

The roots z_n of A(z) are the "zeros" and the roots z_m of B(z) are the "poles" of X(z).

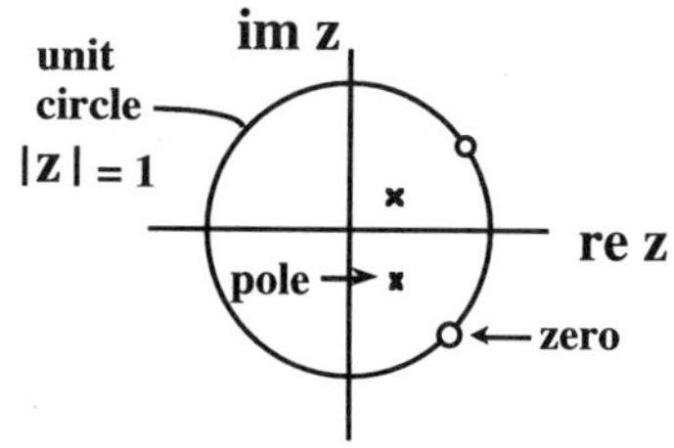

Example pole/zero diagram

Example z-Transform of step function:

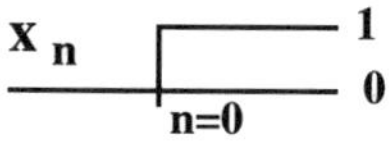

$$X(z) = \sum_{n=0}^{\infty} z^{-n} = \frac{1}{1 - z^{-1}} = \frac{z}{z-1} \qquad |z| > 1$$

Thus. the Z Transform has a zero at z = 0, and a pole at z = 1.

8.1 Properties of the z Transform

Linearity:	$Z(ax + by) = Z(ax) + Z(by)$		$= aX + bY$	(8.1.1)
Delay	$Z(x(n-k)) =$		$= z^{-k} X$	(8.1.2)

when k = 1, $X(n\text{-}l) = z^{-1} X$, thus z^{-1} is called the unit delay operator

A convolution rule exists for the convolution of the z transform of sequences:

$$Z(\ x*y\ ) = X \bullet Y \tag{8.1.3}$$

A sketch of the proof of the convolution rule starts with the definition of convolution

$$x*y = \sum_{k=-\infty}^{\infty} x(k)\, y(n-k)$$

$$Z(x*y) = \sum_{n=-\infty}^{\infty} \sum_{k=-\infty}^{\infty} x(k)\, y(n-k)\, z^{-n}$$

$$= \sum_{k=-\infty}^{\infty} x(k) \sum_{n=-\infty}^{\infty} y(n-k)\ z^{-n} = \sum_{k=-\infty}^{\infty} x(k) \left\{ \sum_{m=-\infty}^{\infty} y(m)\, z^{-m} \right\} z^{-n}$$

$$m = n - k$$

$$= X \bullet Y$$

9.0 Networks and Correspondence Rules

The input/output relation of a system is but a processing specification, or algorithm,

The algorithm is a processing strategy which specifies how a discrete-time input sequence is transformed by the system and becomes the output. Convolution is a good example of this process and the processing strategy of convolution can be inferred by inspecting the formal i/o relationship. However many processing strategies are possible and many algorithms can be found that produce the same result. The differences between these choices often reflect differences in implementation. Some algorithms may be more efficient, others may require fewer resources. The implementation of an algorithm consists of a network of data paths connecting processing nodes where numerical operations are performed. The data processing network is then the realization of an algorithm. As with the case for algorithms, many choices may be possible among network architectures that support the same algorithm. The problem of finding the best algorithm to use for a system and then the best architecture for the implementation network may be quite complex.

The z transform of the system response affords a great deal of insight into the relationship between the system and its associated algorithms, as well as the relationship between an algorithm and its networks. It is often possible to determine an optimum algorithm as well as an effective implementation network for a system. The objective now is to develop a correspondence rule for transforming first a system specification into an algorithmic description and then an algorithm into a processing network. This correspondence rule can be found using the z transform.

To start, restrict the choice of systems to just those that are linear, time-invariant. From the earlier development, the response of a Linear Time-Invariant Systems is,

$$\mathbf{y(n) = \Sigma\, h(k)\;\; x(k\text{-}n)} \qquad (9.0.1)$$

using the z Transform and Convolution Theorem this response in the z transform domain becomes,

$$\mathbf{Y(z) = H(z)\;\; X(z),} \qquad (9.0.2)$$

which implies $\mathbf{H(z) = Y(z) \;/\; X(z)}$. When both y(n) and x(n) are known, then for example, the system's frequency response is obtained by evaluating H(z) on the unit circle,

$$\mathbf{H(f)} = \mathbf{H}(e^{-2\pi i f T})$$

When Y(z) and X(z) are polynomials in z, then H is composed of poles and zeros, and this is the case when the i/o process that transforms x(n) to y(n) has the form:

$$\sum_{k=0}^{N} a_k\; y(n\text{-}k) \;=\; \sum_{m=0}^{M} b_m\; x(n\text{-}m) \qquad (9.0.2)$$

because the z Transform of this process is

$$\sum_{k=0}^{N} a_k\, z^{-k}\, Y(z) \;=\; \sum_{m=0}^{M} b_m\, z^{-m}\, X(z) \qquad (9.0.3)$$

Thus, the ratio of the input and output polynomials becomes,

$$Y(z)/X(z) = \sum_{m=0}^{M} b_m z^{-m} \Big/ \sum_{k=0}^{N} a_k z^{-k}$$

$$= H(z) \qquad (9.0.4)$$

9.1 Correspondence to Networks

To obtain a Direct Implementation of a system specified by polynomials in x and y, of the form,

$$\sum_{k=0}^{N} a_k \; y(n-k) = \sum_{m=0}^{M} b_m \; x(n-m)$$

let

$$y(n) = \sum_{m=0}^{M} b_m \; x(n-m) - \sum_{k=1}^{N} a_k \; y(n-k)$$

Referring to (3.2.3) for the z transform of the system, the following signal processing network can be constructed directly from the inspection of the z transform.

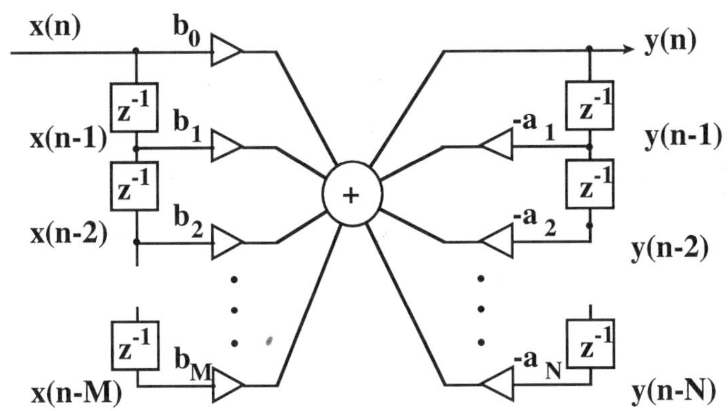

9.2 State Representation for Networks†

The Direct Implementation may not be the most suitable implementation. Alternative networks can be found using a transformation rule. This rule is formulated using a representation of the processing nodes or "state" of the network. These representations are the State Variables and are a vector sequence whose elements are the values of the network's nodes at each time step. As an example, here is a typical 2nd order filter that will be used to develop the state variable representation and network transformation rule.

† See Jackson for details of this discussion.

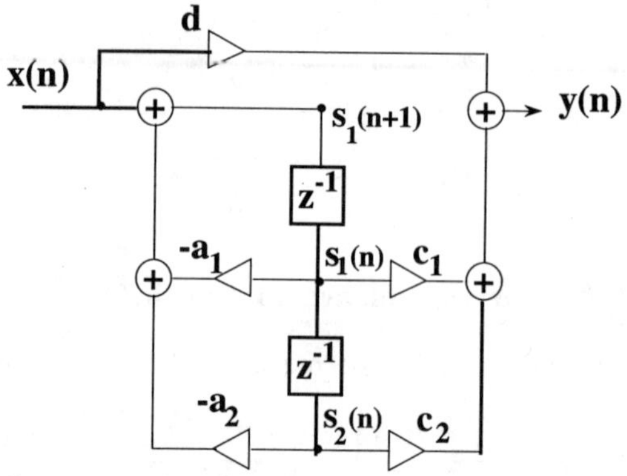

By associating the concept of "state" with each node in the network, the operation of the network can be expressed as a set of linear equations in the state variables, which are the values produced at each of the nodes at each step in time. The states in the network are then related by the operation of the nodes and a set of state equations can be specified. The state equations for the example network are,

$$s_1(n+1) = -a_1 s_1(n) - a_2 s_2(n) + x(n) \qquad (9.2.1)$$

$$s_2(n+1) = s_1(n)$$

$$y(n) = c_1 s_1(n) + c_2 s_2(n) + d\, x(n)$$

These state equations can in general be expressed in matrix form by first organizing the state variables into a vector with the order of the vector's elements determined by the depth of delay of each state. Then the state vectors are related to the input and output sequences as follows,

$$\vec{s}(n+1) = \mathbb{A}\,\vec{s}(n) + \vec{b}\,x(n)$$

$$y(n) = \vec{c}\,'\,\vec{s}(n) + d\,x(n)$$

$$\vec{s}(n) = \begin{pmatrix} s_1(n) \\ s_2(n) \end{pmatrix} \qquad (9.2.2)$$

where

$$\mathbb{A} = \begin{pmatrix} -a_1 & -a_2 \\ 1 & 0 \end{pmatrix} \qquad \vec{b} = \begin{pmatrix} 1 \\ 0 \end{pmatrix}$$

$$\vec{c}\,' = (c_1 \;\; c_2)$$

10.0 State Transformations Among Networks

State Variables are generally NOT unique. Often, many alternate state vectors are possible. These alternatives can be generated, for example, from a set of linear transformations T:

$$s(n) = T\, s(n)$$

T is any non-singular N x N matrix. For the example network,

$$\hat{\vec{s}}(n+1) = T\mathbb{A}\vec{s}(n) + T\vec{b}\,x(n)$$
$$= T\mathbb{A}T^{-1}\vec{s}(n) + T\vec{b}\,x(n)$$
$$= \hat{\mathbb{A}}\,\hat{\vec{s}}(n) + \hat{\vec{b}}\,x(n)$$

$$y(n) = \vec{c}'\,T^{-1}\hat{\vec{s}}(n) + d\,x(n)$$

$$= \hat{\vec{c}}'\,\hat{\vec{s}}(n) + \hat{d}\,x(n) \qquad (10.0.1)$$

Produces a transformed filter,

$$\hat{\mathbb{A}} = T\mathbb{A}T^{-1} \qquad \hat{b} = T b$$

$$\hat{c}' = c'\,T \qquad \hat{d} = d \qquad (10.0.2)$$

3.6 Diagonal Decomposition

When $\mathbb{A}$ has eigenvalues, so that:

$$\mathbb{A} = P\Lambda P^{-1}$$

where

$$\Lambda = \mathrm{diag}(\lambda_k), \text{ and } \mathbb{A}\vec{v}_k = \lambda_k\vec{v}_k$$

$$P = (\vec{v}_1\vec{v}_2 \ldots \vec{v}_N)$$

and **P** is a matrix whose columns are eigenvectors of the transformation **T**, Then the natural modes of the system can be decoupled. Let $\mathbf{T} = \mathbf{P}^{-1}$, and transform the system accordingly,

$$\hat{\mathbb{A}} = T\mathbb{A}T^{-1} = P^{-1}P\Lambda P^{-1}P = \Lambda$$

$$\hat{\vec{b}} = T\vec{b} = P^{-1}\vec{b}$$

$$\vec{c}' = \vec{c}'\,T^{-1} = \vec{c}'\,P \qquad (10.0.1)$$

This process of diagonal decomposition produces a new network with the following form,

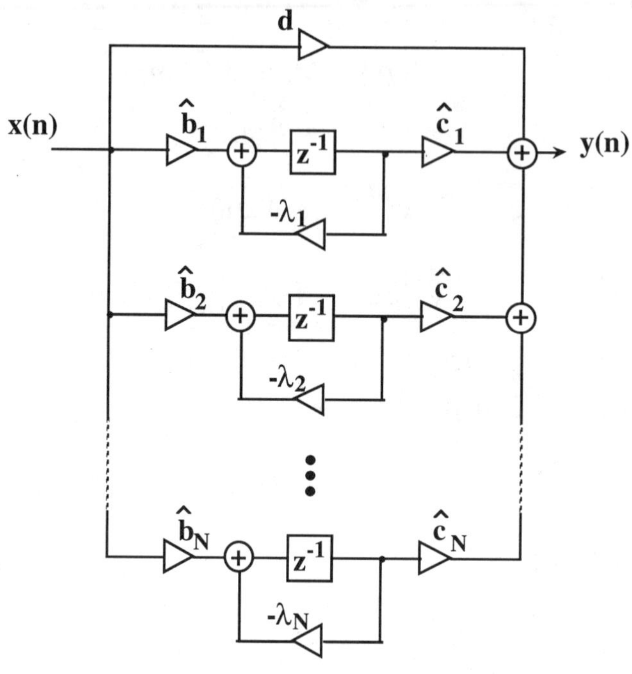

Summary

This presentation was organized around a simple idea of extraordinary value. This central concept is that a continuous function can be exactly and entirely reconstructed from its samples. This perfect reconstruction is possible if the continuous function is band limited and the function is sampled periodically with a sampling interval no larger than half the reciprocal of the cutoff frequency. Under these conditions the samples themselves are an exact representation of the sampled function or waveform. All the information contained in the waveform is present in those samples. Therefore the sequence of samples, obtained this way, can be processed to extract the information contained in the signal or waveform. The strategies that can be implemented for processing the sampled signals was the second focus of this presentation. The tools of convolution and Fourier transform were introduced to provide the basis for formulating these signal processing strategies, or algorithms. Then the Z-transform was used to establish a connection between the algorithms and their implementation as signal processing networks. Finally, alternative network architectures were found using state space representation and similarity transforms.

This then is the path from sampled functions to network architectures. There was not space available here to touch on such essential concepts as efficient algorithms, high performance networks, quantization error, signal to noise ratio assessment, signal detection and false alarm rate estimation. For these topics the texts cited in the Bibliography afford some excellent discussion. However, the hope is that the path has been illuminated and the missing steps are an invitation to travel.

Bibliography

1. Principal Texts used in the preparation of this presentation:

Bracewell, R.N., "The Fourier Transform and Its Application", 2nd ed., McGraw-Hill, 1986.

Jackson, L.B., "Digital Filters and Signal Processing", 2nd ed., Kluwer, 1989.

Oppenheim, A.V, and Schafer, R.W., "Discrete-Time Signal Processing" Prentice-Hall, 1989.

2. Supplemental Texts:

Franklin, G.F., Powell, J.D., Emami-Naeini, A., "Feedback Control of Dynamic Systems", Addison-Wesley, 1991.

Kailath, T., "Linear Systems", Prentice-Hall, 1980.

Peled, A., Liu, B., "Digital Signal Processing", John Wiley, 1976.

Strum, R.D., and Kirk, D.E., "Discrete Systems and Digital Signal Processing", Addison-Wesley, 1988.

3. Additional References:

Allen, P.,E. and Holberg, D.R., "CMOS Analog Circuit Design", Harcourt Brace Jovanovich, 1987.

Cowan, J.D., Tesauro, G., and Alspector, J., ed. " Advances in Neural Information Processing System 6", Morgan-Kaufman, 1994.

The Design of Beam Pickup and Kickers

D. P. McGinnis
Fermi National Accelerator Laboratory[*]

Abstract

The behavior of beam pickup and kickers subjected to the electromagnetic fields of relativistic charged particles is examined. The concept of image currents is explained. The frequency domain response of a simple stripline pickup is derived and the behavior of kickers is explained through Lorentz reciprocity.

INTRODUCTION

Electromagnetic beam position pickups are an important component in the instrumentation of particle accelerators. Also, pickups and kickers are important to the design of beam feedback systems such as dampers and stochastic cooling systems. Pickups are usually designed to measure a particle's transverse or longitudinal position in a beam pipe by intercepting a portion of the electric and magnetic field that results from a moving charged particle. Kickers are basically pickups used in reverse. The electromagnetic field in a kicker is used to change a particle's transverse or longitudinal momentum.

The design of pickups is somewhat forgiving. Just about anything that is stuck in a beam pipe will intercept some portion of a beam's electromagnetic field. This accounts for the wide variety of pickups and kickers found in particle accelerators. However, the careful consideration of pickup and kicker parameters such as bandwidth, sensitivity, and impedance are important to the optimization of the design of beam position or damper systems. This paper will try to illustrate some of the fundamental principles in beam pickup and kicker designs. To keep the discussion clear and simple, this paper will for the most part consider the conventional stripline pickup and kicker.

IMAGE CURRENT

For most cases, the beam energy is much greater than the energy that is siphoned away in the pickup signal. That is, the pickup will not significantly decelerate the beam. With this in mind we can treat the beam as a current source. However, the beam charge will not directly flow through a pickup (unless the beam hits the wall of the beam pipe). The pickup will intercept some portion of the electromagnetic fields that accompany the beam.

Any charged particle will exhibit an electromagnetic field. When the particle is at rest the field is electric and points radially outward in all directions as shown in Fig. 1a. As a particle moves it has both electric and magnetic fields. The way to calculate these fields is to transform the electric field of a particle in its own rest frame to the lab reference frame by using the relativistic electromagnetic tensor. Fortunately we do not need to go into this kind of detail in this paper. A qualitative view of a moving particle's field is shown in Fig. 1b. As a particle's velocity increases, the

---

[*] Operated by the Universities Research Association under contract with the United States Department of Energy.

amount of electric field pointing in the direction of motion decreases and the azimuthal magnetic increases. The resulting field pattern starts looking like a flattened pancake whose width is inversely proportional to the particle energy. As the particle velocity approaches the speed of light or as the kinetic energy of the beam becomes much greater then the particle's rest energy, the direction of the electric and magnetic fields becomes transverse to the direction of motion(TEM).(1) To keep things simple, this paper will restrict its attention to high energy particle beams.

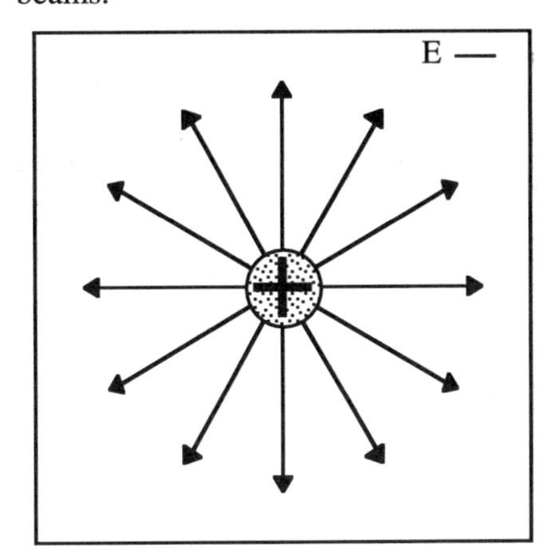

Figure 1a. Electric field of a charged particle at rest.

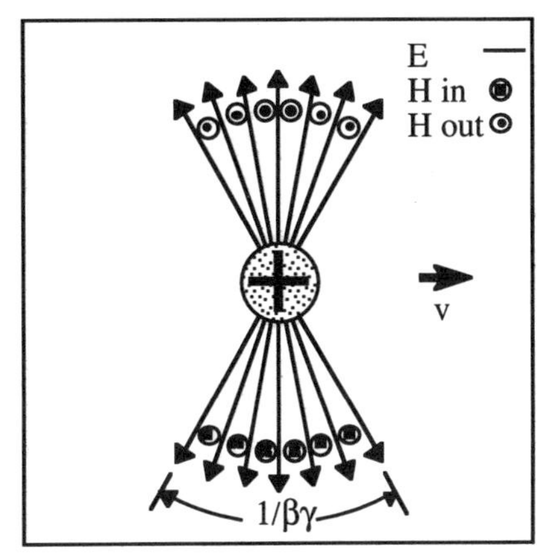

Figure 1b. Electromagnetic fields of a relativistic charged particle. (γ>>1)

When a charged particle is formed at a source, a particle of equal an opposite charge is also usually created. For example, a proton is extracted from a hydrogen atom by stripping the electron from the atom. This electron is still attracted to the proton and would like to follow the proton on its journey. As the proton is injected into the beam pipe (which we will assume is a metal for simplicity), the electron would like to "hop" onto the metal beam pipe. A metal can be considered as a sea of electrons that float around a lattice of positively charged ions. The velocity of these electrons is much smaller than the speed of light so that the electron that was injected onto the metal could not possibly keep up with the faster moving proton. (Note that due the nature of quantum mechanics one cannot really keep track of a particular electron in this "sea" of electrons.) The energy of the disturbance caused by the first electron "jumping" onto the metal is transmitted from electron to electron by means of an electromagnetic wave that follows the moving proton. This disturbance can be thought of as an image current flowing along the surface of the beam pipe that follows the proton. The magnitude of the image current is equal to the beam current but has the opposite sign.

The limiting case of a TEM pulse for a high energy particle beam electromagnetic field allows one to use Ampere's law to solve for the distribution of image current flowing on the walls of the beam pipe. Ampere's law with displacement current is:

$$\oint \vec{H} \bullet d\vec{l} = I_{beam} + \varepsilon \frac{\partial}{\partial t} \iint_{surface} \vec{E} \bullet \bar{z} \cdot dA \tag{1}$$

where z is a unit vector that runs parallel to the beam and dl is a vector that is in a plane normal to the beam direction as shown in Fig 2. Since the electric field of the TEM wave is perpendicular to the beam direction, the displacement current term on the right side of Eq. 1 is zero. The transverse magnetic field can be found from the solution of the static two dimensional version of Ampere's law

$$\vec{\nabla} \times \vec{H} = \vec{J}_{beam} \tag{2}$$

The transverse electric field is equal to the magnetic field multiplied by the wave impedance of free space:

$$\vec{E} = -\sqrt{\frac{\mu_o}{\varepsilon_o}} \bar{z} \times \vec{H} \tag{3}$$

Since there are no electromagnetic fields in the metal of the beam pipe, an image current density must flow on the surface of the beam pipe to cancel out the magnetic field tangent to the metal surface.

$$\vec{J}_{image} = \bar{n} \times \vec{H} \tag{4}$$

where J_S has the units of Amperes/meter and **n** is a unit vector that is normal to the beam pipe surface.(2)

Commercial computer programs that can solve the static two dimensional Ampere's equation are readily available. However, we will discuss two solutions in which the analytical solution is well known. For a cylindrical beam pipe with the beam current passing through the center of the pipe:

$$J_{image} = -\frac{I_{beam}}{2\pi b} \tag{5}$$

If a pencil thin beam passes off center as shown in Fig. 3 (3):

$$J_{image}(\phi) = -\frac{I_{beam}}{2\pi b}\left[\frac{b^2 - r^2}{b^2 + r^2 - 2br \cdot \cos(\phi - \theta)}\right] \tag{6}$$

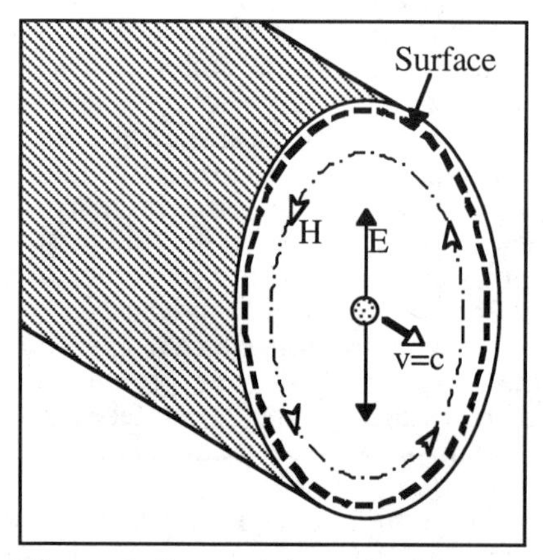

Figure 2. Beam pipe configuration for application of Faraday's Law.

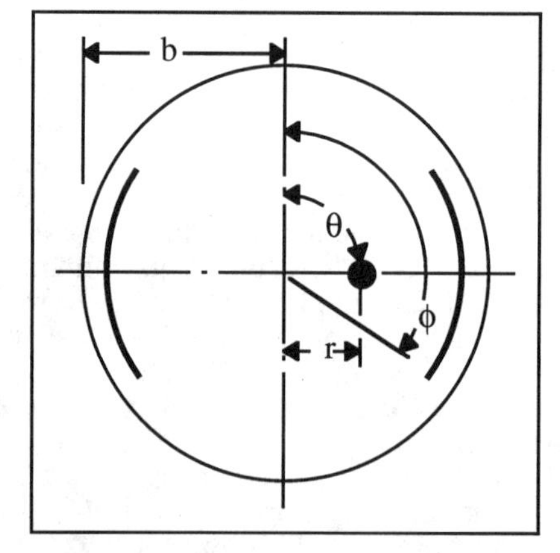

Figure 3. Beam going off center through the beam pipe. The dark lines on the sides are the pickup electrodes.

The image current density as a function of angular position for an offset pencil thin beam is shown in Fig. 4. If two electrodes are fashioned to intercept only a fraction of the image current as shown in Fig. 3, the electrode closer to the beam will intercept more image current compared to the further electrode. The difference in intercepted image currents between the two electrodes will be proportional to the beam position for small beam displacements. However, for large beam displacements, the difference signal will deviate from being linear with beam position as shown in Fig. 5. This situation can be corrected by fashioning the width as a function of electrode length.(3) We will define the sensitivity of the electrode as equal to the ratio of image current captured to the total image current.

$$s = \frac{\int_{\text{width}} J_s dw}{I_{\text{beam}}} \tag{7}$$

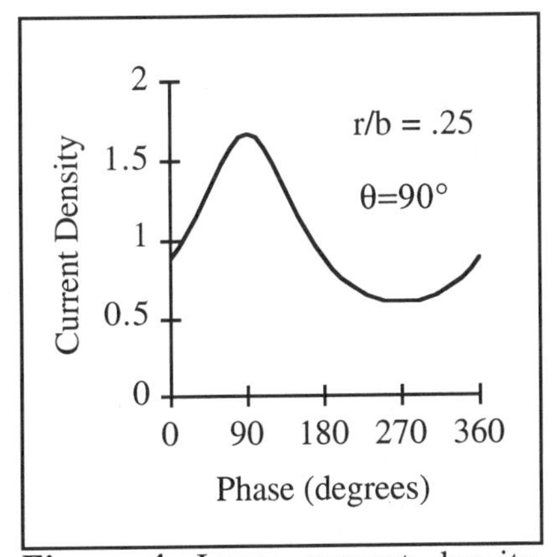

Figure 4. Image current density versus angular position for the electrode configuration shown in Fig. 3.

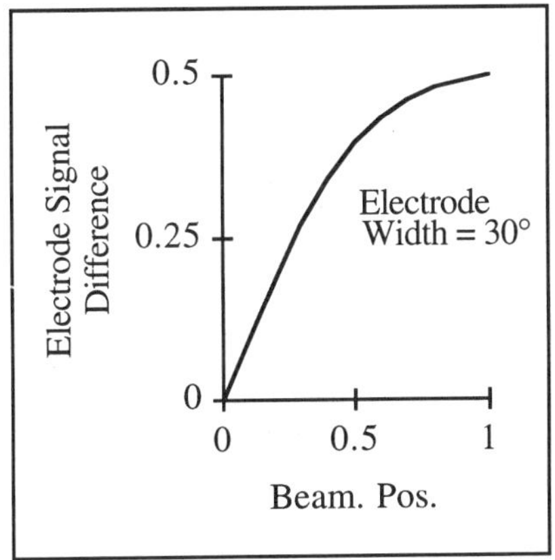

Figure 5. Difference in intercepted image current between the two electrodes shown in Fig. 3. versus actual beam displacement.

TIME DOMAIN RESPONSE

The following discussion will describe the time domain response of an extremely simple beam pickup. We will describe the electromagnetic fields in terms of integral quantities such as voltage and current.

Consider a moving charged particle in the center of a circular beam pipe traveling at a speed very close to the velocity of light as shown in Fig. 6. The observer is located at z=0. Assume that the particle will cross z=0 at time t=0. Because we are assuming that the particle's energy is much greater than its rest mass, the width of the pancake pattern of the EM fields as shown in Fig. 1b will shrink to infinitesimally narrow. The electrical current viewed by the observer can be approximated as:

$$i(t) = q \cdot \delta(t) \tag{8}$$

where δ is the Dirac delta function. The Dirac delta function δ(t) is zero everywhere outside t=0, infinite at t=0., and:

$$\int_{-\infty}^{\infty} \delta(t)dt = 1 \tag{9}$$

Because the δ function as it is written in Eq. 9, has units of 1/time, the current described by Eq. 8 has the correct units of charge/time. The image current which flows along the inner wall of the beam pipe is equal in magnitude but opposite in sign to the particle current. Also by symmetry, the current density is uniform around the entire circumference of the beam pipe.

Imagine that the beam pipe is sawed in half at z = 0. The gap between the two halves of the beam pipe looks like a capacitor to the image current(for infinitely long beam pipes this capacitance will be infinite). The only means for the image current to pass through the gap is by means of a displacement current which is analogous to an AC current flowing through a capacitor. If one could measure the voltage drop across the gap without disturbing the uniformity of the image current density, the voltage drop across the gap would be zero before t=0 and for t>0:

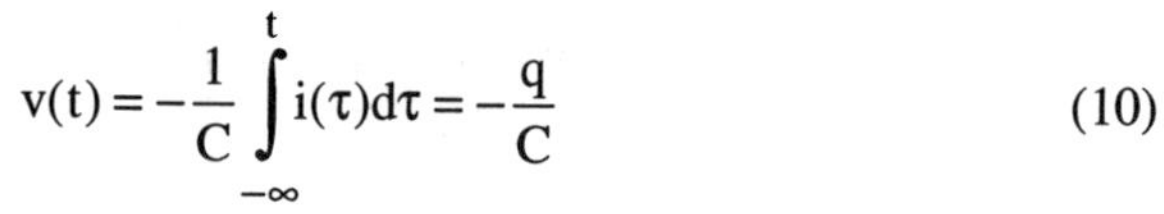

$$v(t) = -\frac{1}{C}\int_{-\infty}^{t} i(\tau)d\tau = -\frac{q}{C} \tag{10}$$

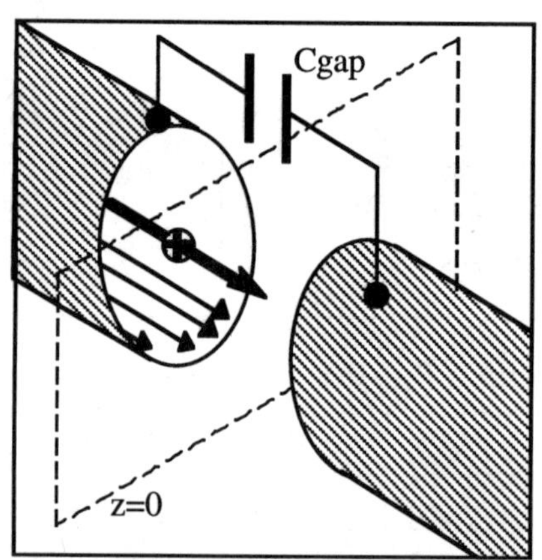

Figure 6. Beam pipe with capacitive gap cut at center.

THE FREQUENCY DOMAIN RESPONSE

The result of Eq. 10 is a simple result that could have been easily derived without going through the steps of Eqs. 8-10 but; the geometry of most pickups is not as simple as the configuration shown in Fig. 6. In general the voltage at the output of a pickup due to a single point charge is a function of time:

$$v(t) = q \cdot z(t) \tag{11}$$

where the units of z(t) is Ohms/second. For another particle that crosses the observation point at $t = t_p$, the voltage due to that particle is:

$$v_p(t) = q \cdot z(t - t_p) \tag{12}$$

If the beam contains more than one particle, the voltage is the sum of all the individual voltages generated by each particle.

$$v(t) = q \sum_p N_p z(t - t_p) \tag{13}$$

where N_p is the number of particles that cross z=0 at $t=t_p$. Up to this point we have been considering particles on a one by one basis. For most beams, there will be a large number of particles. The number of particles in a small slice of time called dt_p is:

$$N_p = \frac{i(t_p)}{q} dt_p \tag{14}$$

As the length of the slice approaches zero, the sum in Eq. 13 can be rewritten as an integral:

$$v(t) = \int_{-\infty}^{t} z(t - t_p) \cdot i(t_p) dt_p \tag{15}$$

Equation 15 is the convolution of the impulse response of the pickup with the beam current. Equation 15 is an important result. It implies that to completely describe the pickup all we need to know is the impulse response of the pickup z(t) and the charge distribution of the beam.

Coupled with the fact that convolution integrals are in general not easy to do and the availability of RF spectrum analyzers and network analyzers, it is much easier to solve Eq 15. in the frequency domain.(4) The Fourier transform of the voltage is:

$$V(\omega) = \int_{-\infty}^{\infty} v(t) e^{-j\omega t} dt \tag{16}$$

Since $V(\omega)$ is the spectral density of v(t), the units of $V(\omega)$ is Volts/Hertz. Applying the Fourier transform to Eq 15., we get Ohms law for pickups:

$$V(\omega) = Z(\omega) \cdot I(\omega) \tag{17}$$

where $I(\omega)$ is the spectral density of i(t) and has units of Amperes/Hertz. $Z(\omega)$ has units of Ohms and is called the pickup impedance. The rest of the paper will concentrate an how to calculate or measure $Z(\omega)$.

BEAM SPECTRUM

Before a pickup design can begin, the frequency spectrum of the beam current should be known. This section will discuss the spectrum of two simple beam profiles. These profiles are the square pulse and the gaussian bunch as shown in Fig. 7.

For the square pulse:

Time Domain | Frequency Domain

$$i(t) = \frac{N_b q}{2\sigma} \quad \text{for } |t| < \sigma$$

$$i(t) = 0 \quad \text{for } |t| > \sigma \qquad\qquad I(\omega) = N_b q \frac{\sin(\omega\sigma)}{\omega\sigma} \tag{18}$$

where N_b is the number of particles in the bunch. For a gaussian pulse:

Time Domain | Frequency Domain

$$i(t) = \frac{1}{\sqrt{\pi}} \frac{N_b q}{\sigma} e^{-\left(t/\sigma\right)^2} \qquad\qquad I(\omega) = N_b q e^{-\left(\omega\sigma/2\right)^2} \tag{19}$$

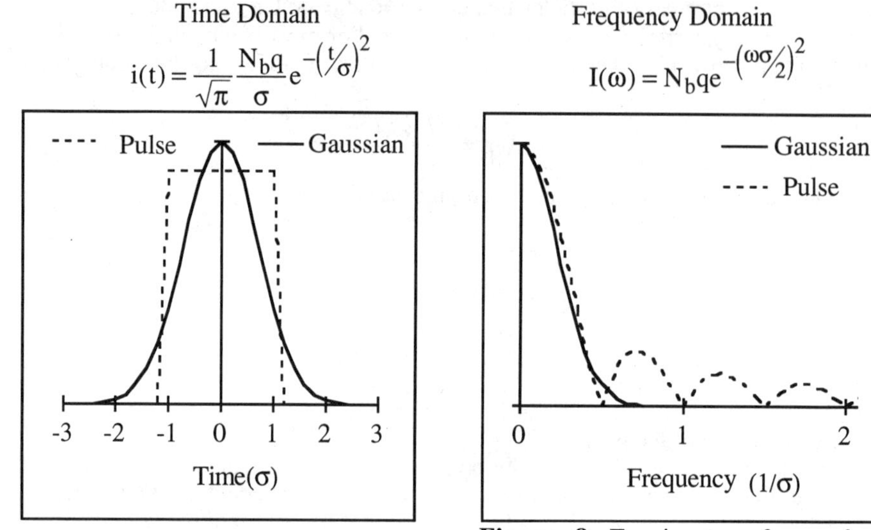

Figure 7. Time domain picture of a square pulse and gaussian waveform.

Figure 8. Fourier transform of a square pulse and a gaussian waveform.

The frequency response of these pulses is shown in Fig. 8. Note that the DC value for both cases is the total amount of charge in the bunch. The square wave pulse falls off much slower than the gaussian bunch at high frequencies. This is due to the sharp edges of the square pulse in the time domain. Although a perfect square pulse might be a simple approximation, one should also realize that the steep drop-off at high frequencies for a gaussian pulse is due to the fact that the tails of the gaussian in the time domain extend to infinity! In reality all beam pulses have a finite extent in time. Therefore the use of a gaussian bunch as a model for the bunch structure will not be accurate at high frequencies. In general any kind of edges in the distribution will cause the beam current spectral density to fall off as $1/\omega$ at high frequencies.

TRANSMISSION LINE THEORY

Because many pickup designs are based on transmission lines that couple to the beam, this section will give a brief over view of transmission line theory. A TEM transmission line consists of at least two conductors that run parallel two each but are separated by some small distance. Between the conductors there is stored magnetic energy which is proportional to the inductance of the conductors and there is electrical energy which is proportional to the capacitance between the two conductors. A small section of transmission line can be modeled as a series inductance with a shunt capacitance as shown in Fig. 9. The current going through the inductor is proportional to the voltage drop across the inductor:

$$v-(v+\Delta v)=L\cdot\Delta z\frac{\partial i}{\partial t}\quad\xrightarrow{\Delta z\to 0}\quad -\frac{\partial v}{\partial z}=L\frac{\partial i}{\partial t} \tag{20}$$

where L is the inductance per unit length. The current exiting the transmission line segment is the difference between the input current and the capacitance current (5):

$$i+\Delta i=i-C\cdot\Delta z\frac{\partial v}{\partial t}\quad\xrightarrow{\Delta z\to 0}\quad -\frac{\partial i}{\partial z}=C\frac{\partial v}{\partial t} \tag{21}$$

where C is the capacitance per unit length. Equations 22 -23 govern how the voltage and the current "bootstraps" with each other along the transmission line. This "bootstrapping" can be written down as forward and reverse propagating waves:

$$v=v^{+}\left(t-\frac{z}{vel}\right)+v^{-}\left(t+\frac{z}{vel}\right)$$

$$i=\frac{v^{+}}{Z_o}\left(t-\frac{z}{vel}\right)-\frac{v^{-}}{Z_o}\left(t+\frac{z}{vel}\right) \tag{22}$$

The (+) superscript indicates a wave in the +z direction and the (-) superscript indicates a wave traveling in the -z direction. The symbol **vel** is the phase velocity of the waves and is equal to:

$$vel=\frac{1}{\sqrt{LC}} \tag{23}$$

The symbol Z_o is called the characteristic impedance of the transmission line and is the ratio between the wave voltage and the wave current. It can be written in terms of the line inductance and capacitance as:

$$Z_o=\sqrt{\frac{L}{C}} \tag{24}$$

The first thing to note about Eq. 22 is that the form of the solution is very general. Thus, if a pulse is initiated on a transmission line its shape will not distort as it travels down the line. Also, the forward and reverse waves are independent modes. That is, if a forward wave is initiated on a transmission line it will remain a forward traveling wave as long as no discontinuity occurs on the transmission line.

However if a discontinuity occurs on the line, then both forward and reverse traveling waves must exist to support the boundary conditions governed by the discontinuity. For example, assume that a transmission line is terminated in some resistance R_t and a forward traveling wave has been initiated on the transmission line and is traveling towards the termination. At the termination, the ratio between the voltage and the current must be R_t. Therefore, the ratio between the forward and reverse waves can be found by dividing the top equation by the bottom equation in Eq. 22 and setting it equal to R_t:

$$\frac{v^-}{v^+} = \frac{R_t - Z_o}{R_t + Z_o} = \Gamma \tag{25}$$

This ratio is known as the reflection coefficient. After the forward wave hits the termination a reverse wave is set up to match the boundary condition and this wave heads back to the initial source of the waves. There are three limiting conditions. The first situation is when the termination is open or $R_t=\infty$, the reflection coefficient is +1. A reverse wave of the same magnitude and polarity is sent back to the source. Next, if the termination is shorted or $R_t=0$, the reflection coefficient is -1. A reverse wave of the same magnitude but opposite polarity is launched back towards the source. Finally, if the termination $R_t=Z_o$, the reflection coefficient is zero and no reverse wave is created. An observer looking into the load could not distinguish between a transmission line of infinite length and a line terminated with a resistor equal to its characteristic impedance.

In this paper we will be working mostly in the frequency domain so we will consider only sinusoidal waves. A forward traveling wave can be written as:

$$v^+(t,z) = V_o^+ \cdot \cos\left(\omega\left(t - \frac{z}{vel}\right)\right) = RE\left\{A \cdot e^{j\omega t}\right\} \tag{26}$$

Because of this simple time dependence, the operations of integration and differentiation become the simple operations of division and multiplication of $j\omega$:

$$\int dt \longrightarrow \frac{1}{j\omega} \qquad\qquad \frac{\partial}{\partial t} \longrightarrow j\omega \tag{27}$$

The quantity A is a complex number and can be thought of as a phasor where:

$$|A| = V_o^+ \qquad\qquad \arg(A) = -\omega\frac{z}{vel} \tag{28}$$

The negative sign for the phase is the result of choosing the positive sign for the time dependence $j\omega t$. The derivative of the phase with respect to frequency gives the time delay through the transmission line of length z:

$$-\frac{\partial}{\partial\omega}\arg(A) = \frac{z}{vel} = \tau_{delay} \tag{29}$$

Equation 28 holds for a simple transmission line. In general, the phasor "**A**" can be a complicated function of frequency. A vector network analyzer can be used to measure the variation of these wave phasors versus frequency. Most network analyzers can measure a circuit with one or two ports as shown in Fig 10. At each port, there is an ingoing wave "**A**" and an outgoing wave "**B**". The network analyzer measures the ratio of the outgoing waves to the ingoing waves:

$$\begin{aligned} B_1 &= S_{11}A_1 + S_{12}A_2 \\ B_2 &= S_{21}A_1 + S_{22}A_2 \end{aligned} \tag{30}$$

The matrix elements are called "S" parameters and are complex numbers. The diagonal matrix elements S_{11} and S_{22} are similar to the reflection coefficient described above. The off diagonal matrix elements S_{12} and S_{21} can be thought of transmission coefficients. (6)

THE STRIPLINE PICKUP

The stripline pickup shown in Fig. 11a is one of the most common types of beam pickup.(3) The metal electrode forms a transmission line along the direction in which the beam travels. The stripline is terminated at both ends with resistors equal to the characteristic impedance of the stripline. Note that both resistors do not have to reside inside the beam pipe. They can be brought outside by transmission lines with the same characteristic impedance and terminated outside the vacuum chamber.

As discussed in section on the frequency domain response, the pickup behavior can be characterized by its impulse response. As the impulse image charge follows the beam down the beam pipe, it will encounter the upstream edge of the stripline as shown in Fig. 11b. This edge looks similar to the gap discussed in section on the time domain response. However because the width of the stripline is only a fraction of the beam pipe circumference, the stripline will not intercept all of the image charge. The fraction that it does accept will be designated as **s**. This fraction of image charge that is intercepted by the loop will travel across the upstream gap as a displacement current. This displacement current will give rise to a voltage pulse on the upstream end of the stripline. The magnitude of the pulse is equal to the fraction of image current captured times the characteristic impedance of the stripline. Because the stripline is terminated at the upstream and downstream ends with the same characteristic impedance, the voltage pulse will split into two equal pulses with one of the pulses traveling to the upstream termination and the other pulse traveling to the downstream termination.

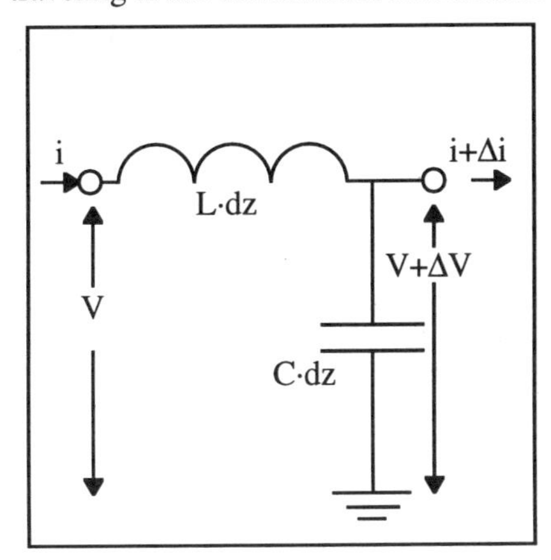

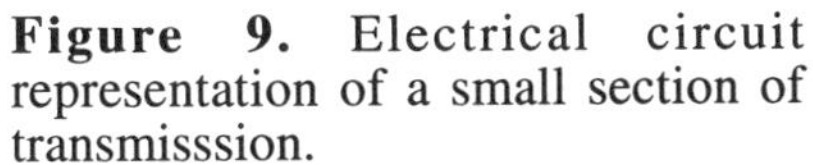
Figure 9. Electrical circuit representation of a small section of transmisssion.

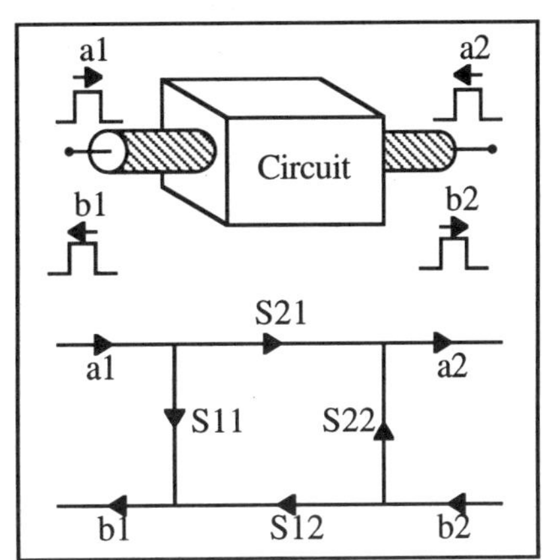

Figure 10. Definition of two port S parameters.

The fraction of image current traveling downstream will travel along the top of the stripline until it encounters the downstream gap. At the downstream gap another pulse is created. This pulse is equal to the upstream pulse but has the opposite polarity. This change in polarity is due to the fact that the electric field lines at the downstream gap start on the edge of the stripline and terminate on the ground plane. Whereas at the upstream gap, the electric field lines started at the ground plane and terminated on the upstream edge of the stripline.

The pulse at the downstream gap also splits into two pulses. However, if the velocity of the beam and the phase velocity of the stripline are equal (as in the case for high energy beams), the pulse that was created at the upstream gap and headed towards the downstream termination will arrive at the downstream gap at the same time the image current is inducing the downstream voltage pulse. Since the pulses heading towards the downstream termination have opposite polarity, these pulses will cancel each other and there will be no energy dissipated in the downstream termination! The pickup behaves very much like a microwave contra-directional coupler in that the direction of the energy flow in the pickup is in the opposite direction in which the beam is traveling. In practice some small fraction of energy will be dissipated because of small mismatches in velocity and discontinuities along the stripline.

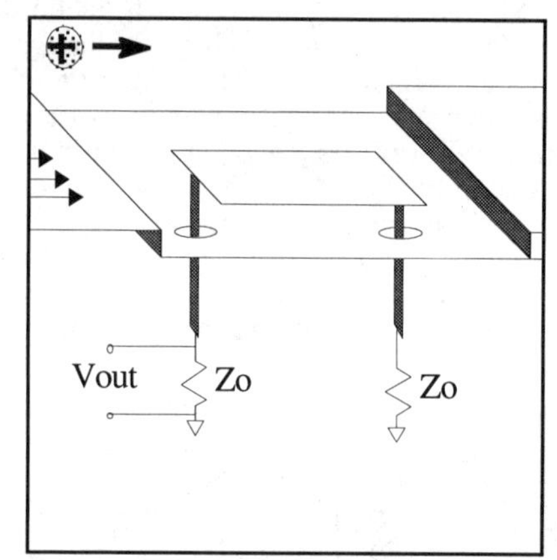

Figure 11a. A stripline pickup.

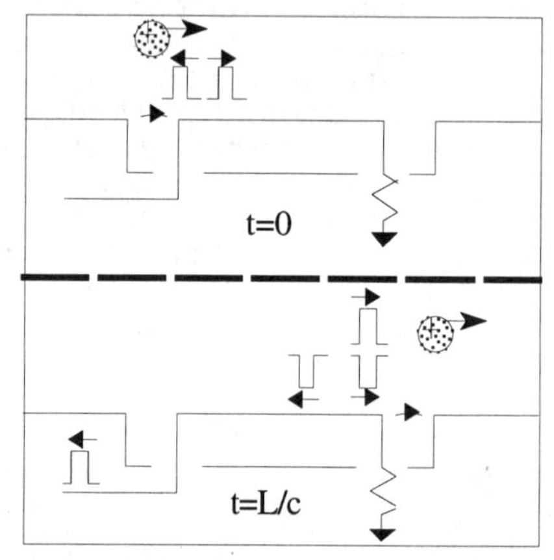

Figure 11b. Snapshot of induced voltage pulses on a stripline pickup.

The impulse response in the time domain which is sensed at the upstream termination consists of the first half pulse created at the upstream gap followed by a half pulse in the opposite polarity and delayed by twice the transit time of the stripline:

$$z(t) = \frac{s \cdot Z_o}{2}\left(\delta(t) - \delta\left(t - \frac{2 \cdot \text{len}}{c}\right)\right) \tag{31}$$

The Fourier transform of the impulse response is:

$$Z(\omega) = s \cdot Z_o \cdot e^{j\frac{\pi}{2}} \cdot e^{-j\frac{\omega \cdot \text{len}}{c}} \sin\left(\frac{\omega \cdot \text{len}}{c}\right) \tag{32}$$

which is plotted in Fig. 12. The delay of the response is equal to the length of the stripline. The phase intercept (phase when ω->0) is 90°. The magnitude of the response has a maximum at frequencies where the length is an odd multiple of quarter wavelengths:

$$f_{\text{center}} = \frac{1}{4}\frac{c}{\text{len}}(2n-1) \tag{33}$$

where n=1,2,3.... . The stripline pickup is usually designed to operate in the first lobe (n=1). The 3dB points of this lobe is:

$$f_{lower} = \frac{1}{2} f_{center} \qquad\qquad f_{upper} = 3 \cdot f_{lower} \tag{34}$$

which is greater than an octave of bandwidth! A simple frequency domain circuit model for this type of pickup is shown in Fig. 13.

Finally, at low frequency the response approaches:

$$Z(\omega) \xrightarrow{\omega \frac{len}{c} << 1} s \cdot j\omega L \cdot len \tag{35}$$

where L is the inductance per unit length of the stripline. Equation 35 tells us that the pickup actually looks like an <u>inductor</u> to the image current at low frequencies.

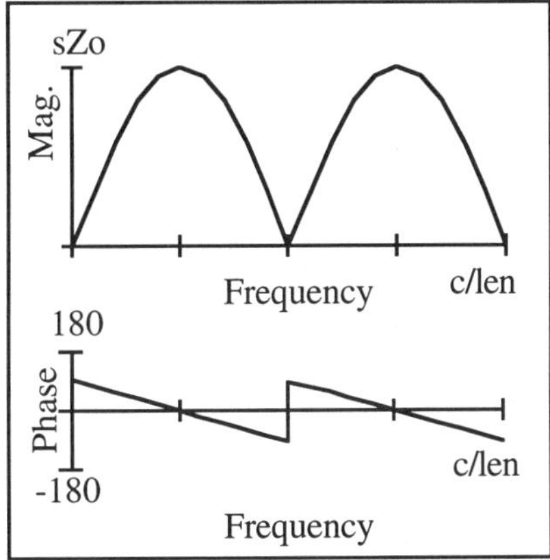

Figure 12. Frequency domain response of a stripline pickup.

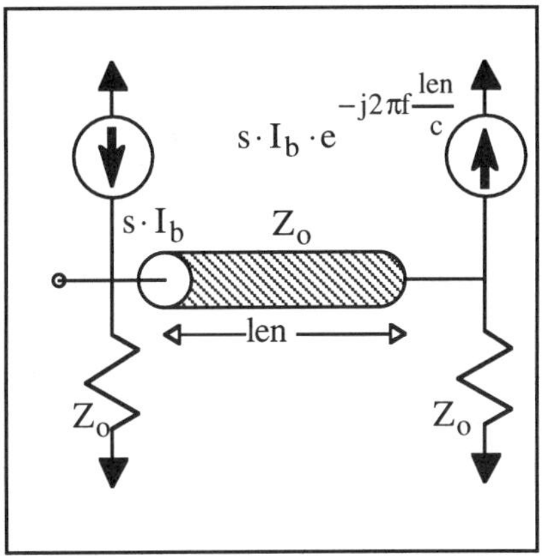

Figure 13. Equivalent electrical circuit of a stripline pickup.

HIGH FREQUENCY EFFECTS

As shown earlier, the pickup sensitivity increases as the width increases to intercept more image current. Also, the pickup sensitivity is proportional to the pickup impedance. However, the impedance of stripline decreases as the width increases. To combat this drop in impedance, the height of the stripline above the ground plane needs to be raised. The net result of increasing the pickup sensitivity and impedance is to make striplines that are wide and tall. In the analysis of the previous section, it was assumed that the image current distribution that was intercepted at the gaps formed a voltage pulse simultaneously and was represented as a delta function. That is, the current intercepted on the outside edges of the gap reached the termination at the same time as current that was intercepted in the middle of the gap. If the pickup width is an appreciable fraction of a wavelength at the frequency of interest, these assumptions are no longer valid.(7) An extremely simple model of a impulse response for a pickup that is very wide is shown in Fig. 14a. This model assumes that the image current at the outside edges of the gap take a longer time to reach the termination as compared to image current intercepted at the centers of the gap. The frequency response is given as:

$$Z(\omega) = s \cdot Z_o \cdot e^{j\frac{\pi}{2}} \cdot e^{-j\omega\left(\frac{len}{c} + \sigma\right)} \frac{\sin(\omega\sigma)}{\omega\sigma} \sin\left(\frac{\omega \cdot len}{c}\right) \tag{36}$$

where σ is equal to the delay time between the center and the outside edges of the gaps. This response function is plotted in Fig. 14b. The response is equal to the frequency response of the square pulse described in Eq. 18 times the impulse response of Eq. 32. One should note that Eq. 38 collapses down to the original impulse response described in Eq. 36 when σ approaches zero. As shown in Fig. 14b, the spreading of the image charge on the gap has the effect of lowering the bandwidth of the response. So, as common to many different types of electrical engineering, there is a upper limit on the gain-bandwidth product of the performance of the pickup.

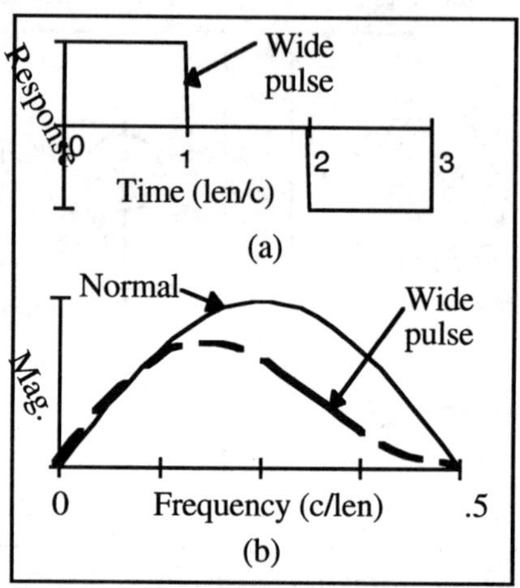

Figure 14. Simple time and frequency domain impulse responses of a pickup with transit time delay along gaps included.

For extremely high frequency pickups such as stochastic cooling arrays, the band limiting effect of the transit time delay due to electrode width is a serious limitation. Also, the appropriate transit time delay is difficult to calculate due to the three dimensional behavior of the stripline pickup. One should also note that we have completely avoided the discussion of parasitic inductances and capacitances due to end effects of the stripline electrode. However, some of these problems may be made more tractable by changing the three dimensional geometry of the stripline pickup to a planar, two dimensional, pickup as shown in Fig. 15 (8).

The gaps for this electrode can be modeled approximately by slotline transmission lines. Slotline transmission lines do not support TEM waves so that a unique characteristic impedance cannot be defined. However, a power impedance of the form:

$$Z_o \equiv \frac{\left| \int_{gap} \vec{E} \cdot d\vec{l} \right|^2}{2P} \tag{37}$$

where E is the electric field across the gap and P is the power transmitted along the slotline. A circuit model for this structure is shown in Fig. 16 where the weighted image current is separated by transmission lines. Note that if the transmission lines on the transverse gaps were omitted, this circuit model would reduce to the simple stripline model shown in Fig. 13. The overall effect of this model is to describe the electrode with an effective length.

$$\text{Length}_{\text{effective}} = \text{Length} + \text{Width} \tag{38}$$

Pickups with bandwidths up to 4 GHz have been fabricated with this geometry. At these high frequencies, the physical length shrinks almost to zero and most of the delay is made up in the width of the electrodes. As the length shrinks to zero, the electrode becomes the magnetic dual of a folded dipole antenna.

KICKERS AND RECIPROCITY

Up to this point there has been no discussion on the behavior of kickers. However, using the Lorentz reciprocity theorem, the behavior of kickers can be understood from the pickup behavior. The Lorentz reciprocity theorem is a useful and well known theorem for describing antenna systems. Lorentz reciprocity is derived by manipulating two solutions of Maxwell's equations. One solution describes the transmitter (kicker) and the other solution describes the receiver (pickup) (9).

Consider the receiver or pickup configuration in Fig. 17a. The beam which is represented as a current source density J_p induces an electric field at the pickup E_p. The current source also induces fields E_p,H_p throughout the beam pipe volume which is surrounded by a surface S. Figure 17b describes the transmitter or kicker configuration. The kicker is powered by a current source density J_k and produces a field E_k that acts on the beam. Also there are fields throughout the volume designated by E_k, H_k. By manipulating Maxwell's equations describing the two configurations, the identity:

$$\oint\!\!\!\oint_{\text{surface}} \left(\vec{E}_k \times \vec{H}_p - \vec{E}_p \times \vec{H}_k\right) \bullet d\vec{s} = \iiint_{\text{volume}} \left(\vec{E}_p \bullet \vec{J}_k - \vec{E}_k \bullet \vec{J}_p\right) d\text{vol} \tag{39}$$

is derived. The contribution to the surface integral on the left hand side of Eq. 39 at the beam pipe walls is zero because the electric fields for both cases vanish. Also, the surfaces in which the beam enters and exits are far enough away from the electrode so that the fields are normal to the beam direction and this contribution to the surface integral vanishes. Equation 39 can be rewritten as:

$$\iint \vec{J}_k \left(\int \vec{E}_p \bullet d\vec{l} \right) \bullet d\vec{S} = \iint \vec{J}_p \left(\int \vec{E}_k \bullet d\vec{l} \right) \bullet d\vec{S} \tag{40}$$

which can be re-written in integral form as:

$$I_k V_p = I_p V_k \tag{41}$$

The pickup impulse response is the ratio of the voltage induced on the pickup to the beam current. The kicker impulse response is the ratio the longitudinal energy given to the beam to the current in the kicker. Equation 41 can be re-arranged

$$Z_p = Z_k \tag{42}$$

Which states that to know the impulse response of the kicker all one has to do is measure (or derive) the impulse response of the pickup! However, it should be noted that due to the "dot" products in Eq. 40, this theorem holds only for kickers that give an accelerating kick in the direction of beam travel (i.e.. longitudinal kickers). Also as in the case of the stripline pickup above, a stripline kicker will be contra-directional. That is the voltage wave in the kicker starts at the downstream end of the electrode and travels to the upstream end.

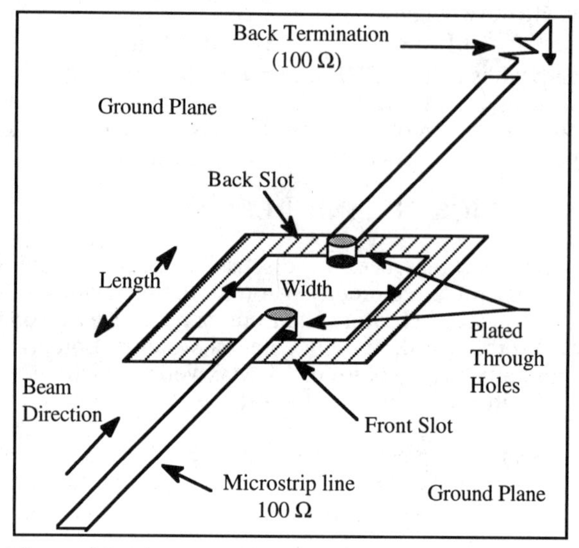

Figure 15. Simple geometric model of a planar loop.

For transverse kickers another theorem using Maxwell's equations can be derived. However, we will use some simple arguments to derive the transverse kicker response for a stripline kicker.

A longitudinal stripline kicker is shown in Figure 18a. Both the top and bottom electrodes are wired in the sum mode. The pickup response and therefore the kicker response is zero at DC as discussed in Eq. 32. This can be simply seen by noting that at low frequencies there is no phase delay through the stripline so that the electric field along the beam direction at the upstream gap cancels the electric field at the downstream gap. There is no net acceleration. However, for a transverse kicker the electrodes are wired in the difference mode as shown in Fig. 18b so there is a net transverse electric field at DC. To account for this DC transverse field, Eq. 32 must be modified:

$$Z_{\perp}(\omega) = s \cdot Z_o \cdot e^{-j\frac{\omega \cdot len}{c}} \frac{\sin\left(\frac{\omega \cdot len}{c}\right)}{\left(\frac{\omega \cdot len}{c}\right)} \tag{43}$$

This is formally derived by the Panofsky - Wenzel theorem which relates the transverse impedance to the longitudinal impedance with a factor of 1/jω (10). The response of a longitudinal and transverse kicker is shown in Fig. 19. Note that although the kicker has a DC response, the upper -3dB frequency is 40% lower than the upper -3dB frequency for the longitudinal kicker because of the 1/jω factor of the transverse field.

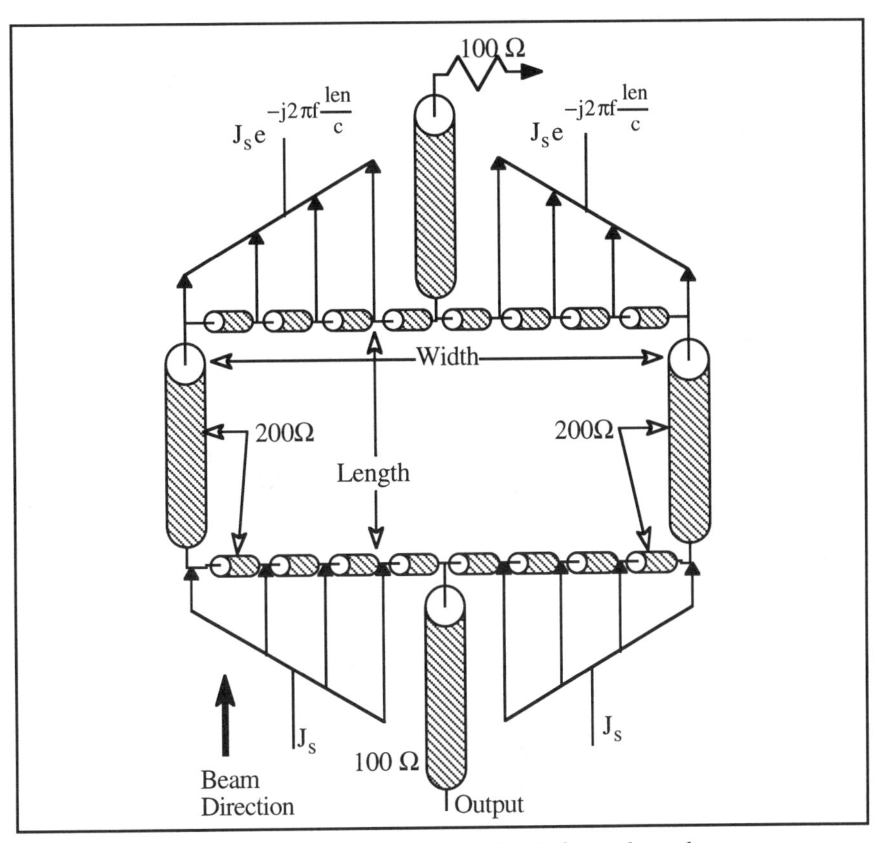

Figure 16. Equivalent electrical circuit for a planar loop.

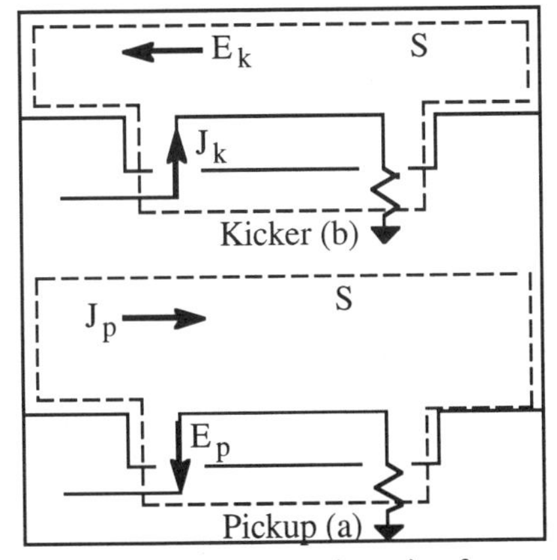

Figure 17. Lorentz reciprocity for a longitudinal pickup and kicker.

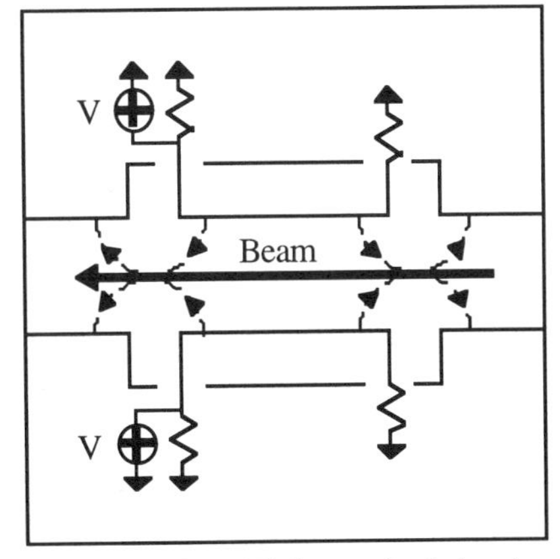

Figure 18a. Kicker wired in the sum mode (longitudinal).

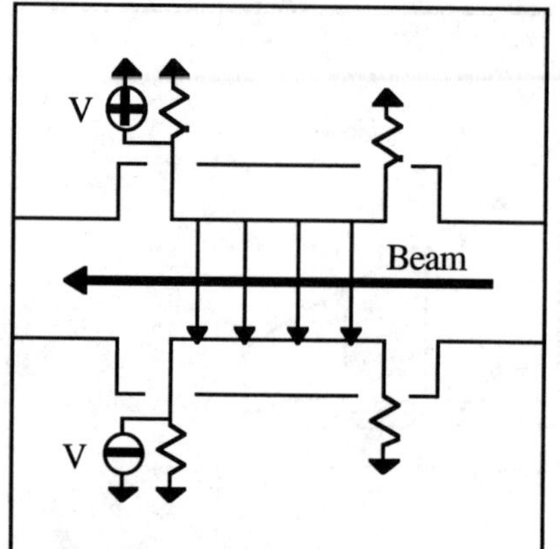

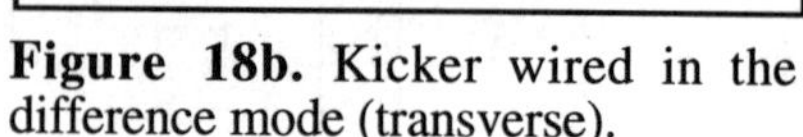

Figure 18b. Kicker wired in the difference mode (transverse).

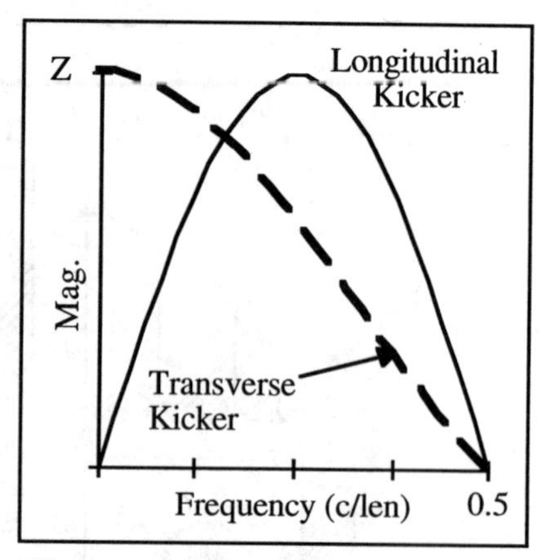

Figure 19. Response of a longitudinal and transverse kicker.

WIRE MEASUREMENTS

After an pickup or kicker electrode has been designed, it is often desirable to test the performance of the electrode before it is installed in the accelerator. The transverse electromagnetic (TEM) fields of the beam can be simulated by stretching a wire over the electrode in the direction of beam travel. The wire forms a TEM transmission line with the ground plane of the electrode with a characteristic impedance of Z_o which is a function of the wire diameter and its distance away from the ground plane. Downstream of the pickup, the wire is terminated with the ground plane with a resistor equal to Z_o. The electrical schematic of a wire measurement is shown in Fig. 20.

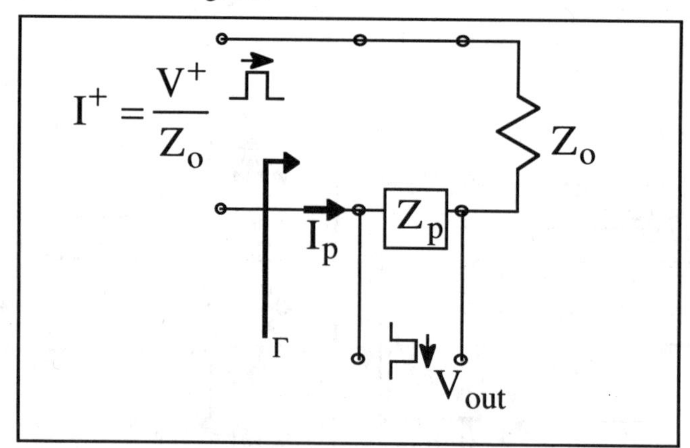

Figure 20. Electrical schematic of a wire measurement.

The impedance seen by the forward traveling wave at the pickup is the pickup impedance plus the characteristic impedance of the transmission line. Because this net impedance is not equal to the characteristic impedance of the wire transmission line, a portion of the incident wave will be reflected at the pickup backwards towards the source. In the case of the TEM wave due to an actual charged particle beam, this reflected wave could not exist because a charged particle beam cannot support a TEM wave traveling in the reverse direction! The only waves that travel in the reverse direction for a particle beam are the higher order waveguide modes of the beam pipe. The frequencies of usual interest are usually below the cutoff of these higher order modes.

The effect of this reflected wave for the wire measurement can be determined by calculating the network analyzer transfer function S21.

$$S_{21} = \frac{V_{out}}{V^+} \tag{44}$$

The current flowing into the electrode for the wire model is:

$$I_p = I^+ \cdot (1 - \Gamma) \tag{45}$$

where:

$$\Gamma = \frac{(Z_p + Z_o) - Z_o}{(Z_p + Z_o) + Z_o} \tag{46}$$

The transfer response is:

$$\frac{V_{out}}{V^+} = S_{21} = \frac{1}{Z_o + \frac{1}{2} Z_p} Z_p \tag{47}$$

which can be inverted to find the electrode impedance:

$$Z_p = \frac{S_{21}}{1 - \frac{1}{2} S_{21}} Z_o \tag{48}$$

For simplicity, we have assumed that the pickup electrode presents a lumped impedance Z_p to the wire transmission line. In reality, this is a gross oversimplification because the pickup is often operated at frequencies in which the electrode is a quarter wavelength long. Therefore, the terms in the denominators of Eqs. 47 and 48 are not accurate.

However, the denominator terms of Eqs. 47 and 48 can be neglected if the characteristic impedance of the wire is made much larger than the impedance of the electrode. This is difficult to due in practice. The characteristic impedance of the wire transmission line only increases logarithmically as the diameter of the wire is decreased so characteristic impedances greater than 400Ω are difficult to obtain. Also the measuring ports of network analyzers have a characteristic impedance of 50Ω. In order to avoid reflections caused by an abrupt change in characteristic impedance between the network analyzer cable and the wire transmission line, a matching network must be built between the network analyzer and the wire transmission line. The bandwidth of the matching networks are usually inversely proportional to the difference in characteristic impedance between the network analyzer and the wire transmission line so that broadband measurements are difficult to make.

Another method to avoid the denominator terms in Eqs. 47 and 48 is to reduce the electromagnetic coupling (S_{21}) between the wire and the electrode. This is fairly easily done for the measurement of beam position electrodes. To measure the beam position, two electrodes are placed on opposite sides of the beam pipe and the difference between the two signals is taken as shown in Fig. 21a. Because of the odd symmetry of the measurement, the measurement will not be sensitive to any longitudinal component of the electric field at the center plane between the two electrodes. Therefore one of the electrodes can be replaced with a ground plane inserted between the two electrodes as shown in Fig. 21b. The ground plane in Fig. 22 preserves the odd symmetry of the measurement shown in Fig. 21a. The beam in Fig. 22 can be modeled with a wire over the ground plane as shown in Fig. 21c. The wire size and spacing is chosen to make a 50Ω transmission line. The 50Ω transmission line configuration of Fig. 21c avoids the mismatches between the network analyzer and the wire. If the wire is placed sufficiently close to the ground plane, the coupling will be reduced so that:

$$S_{21} << 1 \tag{49}$$

The parameter of interest for a beam position pickup is the ratio of the impedance to the transverse displacement of the wire which is shown in Fig. 21c as **x**.

$$Z_p(\Omega / \text{mm}) = \frac{S_{21}}{x} Z_o \tag{50}$$

One important mistake that is often made with wire measurements is to forget to connect the ground of the network analyzer coaxial cable to the ground plane of the electrode. This will cause a serious disturbance to the image currents flowing on the electrode ground plane which will distort the measurement. This connection is emphasized by heavy lines in Fig. 21c.

TIME DOMAIN GATING

Even when there is a good ground connection between the network analyzer and the wire transmission line, there will be reflections due to the mismatch in geometry between the analyzer cables and the wire transmission line. These reflections will introduce some errors in the measurements. The response shown in Fig. 22a is typical of a measurement were there are some unwanted reflections. The reflections show up as "noise" or ringing added on top of the desired signal. The spacing in frequency between the notches of the ringing is inversely proportional to the distance between the electrode and the source of mismatch.

Most network analyzers have a time domain option built in. The impulse response in the time domain is obtained by performing the inverse Fourier transform on the frequency response of S_{21}. This response is shown in Fig. 22b. Besides the normal doublet response as described in Eq. 31, there is a small reflection of the response that comes at a much later in time. In the time domain option of certain network analyzers, a "gate" can be put around the desired signal as shown in Fig. 22b. Then, when the Fourier transform is applied to the gated signal the ringing is removed from the frequency response as shown in Fig. 22c.

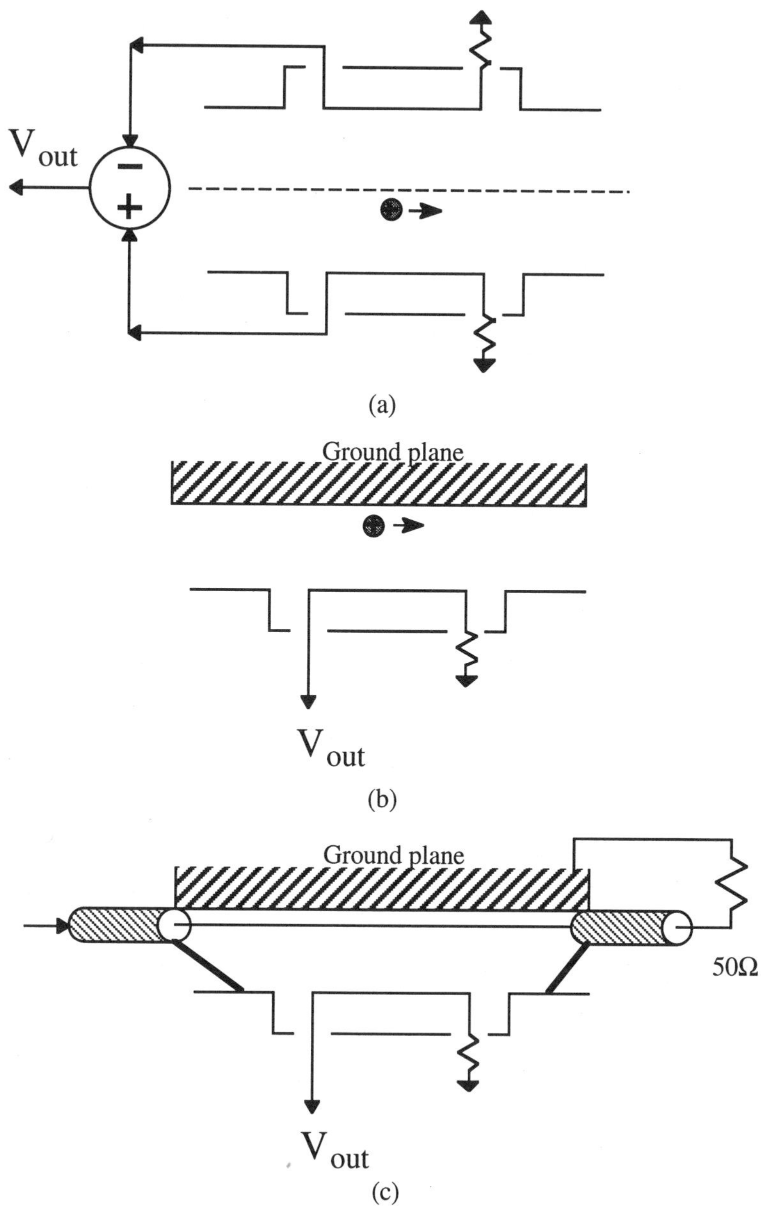

Figure 21. Low coupling difference mode wire measurement.

SUMMARY

A simple understanding of beam pickup and kickers can be obtained as considering a relativistic beam as a current source. The current source that impinges on the electrodes can be thought of as an image current flowing along the beam pipe walls or a TEM wave that flows through the beam pipe. The response of a pickup to any input signal can be calculated by using the impulse response of pickup. The difficult operation of convolution in the time domain turns out to be multiplication in the frequency domain. The high frequency behavior of pickups can often be explained in terms of transmission line concepts. The simple one dimensional model of a stripline electrode breaks down when the width of the stripline is on the order of a wavelength for the frequency of interest. Extremely high frequency electrodes can be made using two dimensional planar electrodes which account for transit time delays along the edges of the pickup. Using Lorentz reciprocity, the behavior of longitudinal kickers can be understood studying the pickup response. However, for transverse kickers, a division of jω must be applied to the pickup response in order to get to true kicker response.

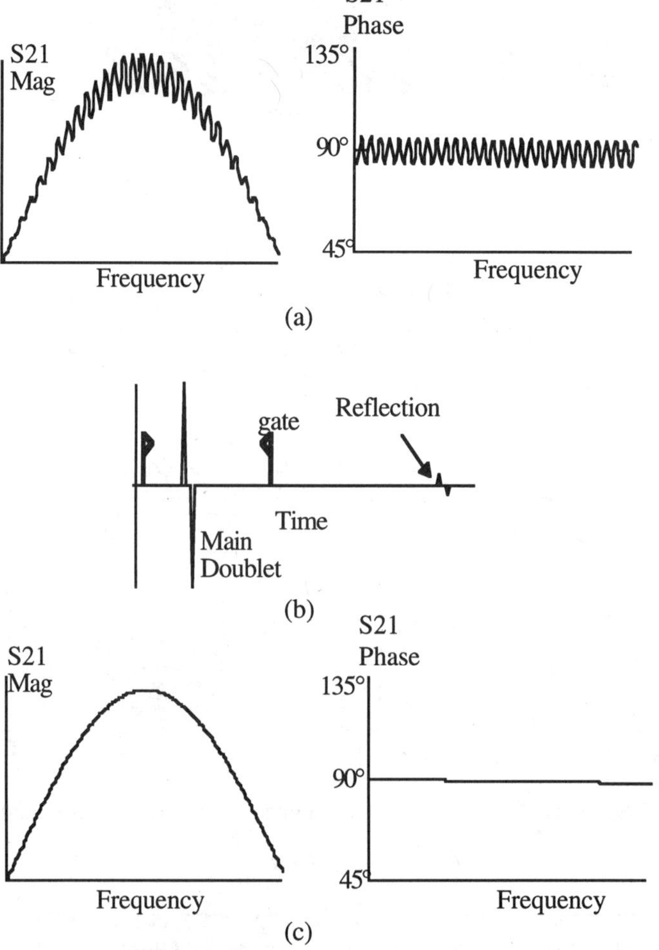

Figure 22. Time domain gating.

REFERENCES

1. J. D. Jackson, *Classical Electrodynamics* (John Wiley and Sons, Inc. 1962), Ch. 11, pp. 555.
2. R. F. Harrington, *Time Harmonic Electromagnetic Fields* (McGraw-Hill, 1961), Ch. 1, pp. 34.
3. R. E. Schafer, "Beam Position Monitoring," in *AIP Conference Proceedings on Accelerator Instrumentation* (AIP, Upton, NY. 1989) pp. 26-55.
4. W. H. Hayt, Jr. and J. E. Kemmerly, *Engineering Circuit Analysis* (McGraw-Hill, 1978), Ch. 19, pp. 664-668.
5. R. G. Brown, R. A. Sharpe, W. L. Hughes, and R. E. Post, *Lines Waves and Antennas* (John Wiley and Sons, Inc. 1973), Ch. 2, pp. 14-32.
6. R. G. Brown, R. A. Sharpe, W. L. Hughes, and R. E. Post, *Lines Waves and Antennas* (John Wiley and Sons, Inc. 1973), Ch. 5, pp. 112-122.
7. D. P. McGinnis, J. Petter, J. Marriner, S. Y. Hsueh, "Frequency Response of 4-8 GHz Stochastic Cooling Electrdodes," *Proceedings of the 1989 IEEE Particle Accelerator Conference*, Vol. 1, 639, March 1989.
8. D. P. McGinnis, "Theory and Design of Microwave Planar Electrodes for Stocahstic Cooling of Particle Beams," Microwave and Optical Technology Letters, Vol. 4, Number 11, Oct. 1991, pp 439-443
9. R. F. Harrington, *Time Harmonic Electromagnetic Fields* (McGraw-Hill, 1961), Ch. 3, pp. 116-120.
10. D. A. Edwards, M. J. Syphers, An Introduction to the Physics of High Energy Accelerators, (John Wiley and Sons, Inc. 1993), Ch. 6, pp 199.

Statistical Data Analysis

A.A. Hahn
Fermi National Accelerator Laboratory, Batavia, IL 60510 *

Abstract

The complexity of instrumentation sometimes requires data analysis to be done before the result is presented to the control room. This tutorial reviews some of the theoretical assumptions underlying the more popular forms of data analysis and presents simple examples to illuminate the advantages and hazards of different techniques.

INTRODUCTION

The function of this tutorial is to reintroduce the concepts of statistical data analysis to Instrumentation Engineers. I assume that everyone has already had an introduction sometime earlier in their course work. I hope to emphasize some practical considerations along with the theoretical underpinnings of the analysis. Another motivation is prompted by the availability of software packages which contain quite powerful statistical packages.

The primary reference is a graduate student text by Bevington (1). Before we get too deep into the subject, it is useful to remind ourselves why this is an important topic. All the examples given below are taken from work which has been done in the Accelerator Division Instrumentation Department at Fermilab.

Data Reduction and Precision Measurements

Several instruments produce from one to hundreds of kilobytes of raw data per measurement cycle. Some examples at Fermilab are the Synchrotron Light Detector which images the bunch by bunch transverse beam shape and is read out by a video camera, the Sample Bunch Display (SBD) which measures the individual bunch intensity and length in the Main Ring and Tevatron (2), the Booster Ion Profile Monitor which measures the transverse profile of the Booster beam (3), the Collision Point BPM system which measures the space-time collision point of the proton and antiproton beams at the two colliding points in the Tevatron, and the Flying Wire System which measures the bunch by bunch transverse beam profile in the Main Ring and Tevatron. What the Main Control Room wants from our systems is at most a few tens of words of summarized beam information. It is up to analysis to provide accurate and timely information.

Several of the aforementioned systems are also being asked to provide automated measurements of the beam to the several percent level. It has been necessary to use sophisticated active noise subtraction techniques.

*Operated by the Universities Research Association under contract with the United States Department of Energy.

Calibration of Instruments

Figure 1 shows the calibration of the Tevatron DCCT Toroid (The plotted y value is the number which appears in the Main Control Room. The advantage of a fit over simply drawing a straight line is that given the data, anyone can get the same answer.

Figure 1. Calibration of Tevatron DCCT (T:IBEAM)

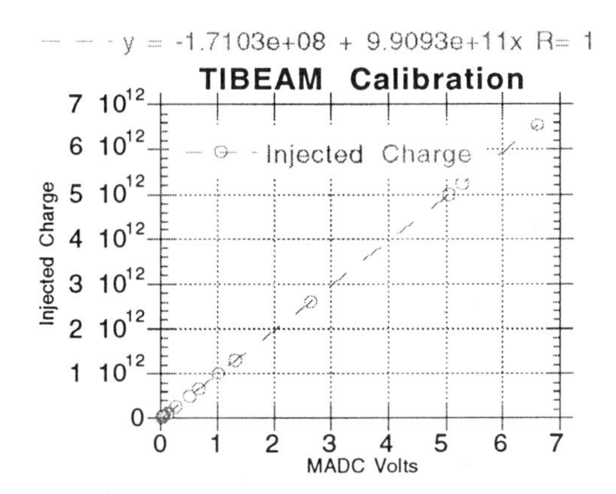

Parameterization of complex data by simple functions

Several times I have parameterized empirical curves by a relatively straightforward function. An example was the shape of the end field of a Tevatron dipole magnet. I needed a functional form which could be plugged into a formula to calculate the synchrotron photon yield. An error function shape fitted to the data was accurate enough for early prototyping.

SIMPLE STATISTICS-A REFRESHER

A fundamental requirement when a measurement is made is that a true value exists. This true value is often called the mean. If we make repeated measurements, we will find a distribution of the measurements about this mean. The width of the distribution tells us how precise our measurement is. A narrow distribution gives us confidence that we can determine the mean value well, while a wide distribution causes us worry. For an underlying probability distribution (either discrete, P_i, or continous, $P(x)$) we can define the mean μ and the width σ in the following manner:

$$\mu = \sum_{i=1}^{\infty} iP_i \xrightarrow{continuous} \int_{-\infty}^{\infty} xP(x)dx$$

$$\sigma = \sqrt{\sum_{i=1}^{\infty} (i-\mu)^2 P_i} \xrightarrow{continuous} \sqrt{\int_{-\infty}^{\infty} (x-\mu)^2 P(x)dx}$$

$$= \sqrt{(\sum_{i=1}^{\infty} i^2 P_i) - \mu^2} \xrightarrow{continuous} \sqrt{(\int_{-\infty}^{\infty} x^2 P(x)dx) - \mu^2}$$

I will sometimes refer to σ as the rms (root-mean-square) width. σ^2 is also known as the variance. The distributions are normalized, i.e.

$$1 = \sum_{i=1}^{\infty} P_i \xrightarrow{continuous} \int_{-\infty}^{\infty} P(x)dx$$

Sample mean and variance

If we make n measurements of data, we can define a sample mean and width by:

$$\overline{x} = \left(\sum_{i=1}^{n} x_i / n \right)$$

$$s = \sqrt{\overline{(x-\mu)^2}} = \sqrt{\overline{x^2} - \mu^2} \xrightarrow{\mu \to \overline{x}} \sqrt{(\overline{x^2} - \overline{x}^2)\left(\frac{n}{n-1}\right)}$$

The last step was a consequence of replacing the population mean by the sample mean. One degree of freedom in the data has been removed since the data set has already been used once in calculating the sample mean. In the limit of an infinite number of measurements, $\mu = \lim_{n \to \infty} \overline{x}$ and $\sigma = \lim_{n \to \infty} s$.

For example if we measure the resistance of a collection of 5% 1 MΩ resistors, we expect the mean to be 1 MΩ. (Is this really true?) We also expect to find a distribution of the individual resistors about the mean value. The exact width depends upon what the 5% specification really means. Figure 2 shows an actual distribution of 51 measurements of 1 MΩ resistors. This example illustrates the difference between experimental precision and measurement accuracy. The experimental precision of measuring was better than the resistor width distribution. This was confirmed by measuring the same resistor many times and noting the fluctuations were much smaller than 1%. But what of the absolute accuracy of the measurement? A summary of the data analysis is:

Mean	1021 kΩ
Std Deviation	22 kΩ
Std Deviation of Mean	3.1 kΩ.

Figure 2. Histogram of measurements of 1MΩ 5% resistors.

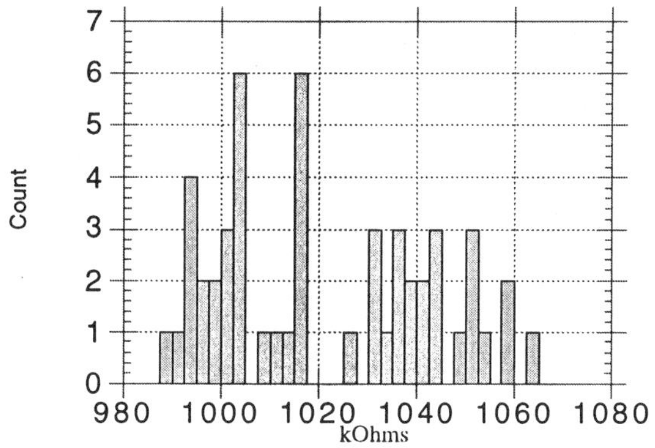

Statistical fluctuations due to the underlying processes (radioactive decay, industrial production methods, measurement techniques) are easily handled by traditional statistical methods. Systematic or calibration errors are much more difficult to deal with. For example, the resistance measurements gave reproducible readings at the 2 kΩ (0.2%) level. If we make ten measurements of the same resistor ,we can claim that our measurement is good (for that resistor) to 0.7 kΩ (we will show this in a following section). However the Ohmmeter may only be absolutely calibrated to the 1% level. What can we do to get a handle on systematic errors? Obviously we could buy a more accurate Ohmmeter or repeat the measurements with other brand Ohmmeters. (Why not use five of the same brand?). Changing measuring devices or techniques is equivalent to converting the systematic error into a statistical error.

Common Probability Distributions

Binomial

$$P(m,p,N) = \frac{N!p^m(1-p)^{N-m}}{m!(N-m)!}$$

The binomial distribution gives the probability for m successes out of N independent trials with the probability of a single success being p. It is known as a discrete distribution since the observables (m) are integers. The mean value and σ are $\mu = Np$ and $\sigma = \sqrt{Np(1-p)}$. An example of this distribution would be the number of times (m) one would expect to roll doubles on a pair of dice for N throws. The probability (p) per throw = 6/36 = 1/6.

Poisson

$$P(m,\mu) = \frac{\mu^m e^{-\mu}}{m!}$$

The Poisson distribution can be derived from the binomial distribution in the limit that p->0, and N-> infinity in a manner that their product (the mean) Np -> μ, a finite number. The observables "m" are integer and >0, although μ can be any positive real number. A unique and dearly beloved feature of this distribution is $\sigma = \sqrt{\mu}$. The Poisson distribution occurs in the counting statistics of radioactive decay.

Gaussian or Normal

$$P(x,\mu,\sigma) = \frac{1}{\sqrt{2\pi}\sigma} e^{-\frac{1}{2}\left(\frac{(x-\mu)}{\sigma}\right)^2}$$

Another limiting form of the binomial distribution when N >>1 is the Gaussian Distribution. The Gaussian distribution is a continuous distribution - x can vary continuously over the entire real axis. The Gaussian distribution is characterized by its mean μ and width σ. This distribution occurs everwhere!

Uniform

$$P(x,\mu,w) = \frac{1}{w} \text{ if } |x - \mu| \le \frac{w}{2}$$

$$= 0 \text{ if } |x - \mu| \ge \frac{w}{2}$$

The ability of calculators and computers to generate (pseudo)random numbers has made this distribution easily accessible to most people. This is a continuous distribution which has equal probability anywhere within the total window width "w" and identically zero probability outside. It can easily be shown to have a $\sigma = \sqrt{w^2/12}$. It serves nicely as a poor man's Gaussian if this value of σ is used.

The Binomial (12 trials, p=1/12), Poisson μ =1, and Gaussian (μ =1, σ = 1) distributions are plotted in fig. 3a. The Gaussian is plotted only for positive x values. Figure 3b shows the case for the binomial (N=240, p=1/12), Poisson (μ =20), and Gaussian (μ =1, σ = 1). The similarity of the distributions for these varied conditions shows why the Gaussian is used so often.

Figure 3(a,b). Plots of Binomial, Poisson , and Gaussian distributions for $\mu = 1$ and $\mu = 20$ respectively. The curve is the Gaussian Distribution. The plots illustrate the similarities of the three distributions.

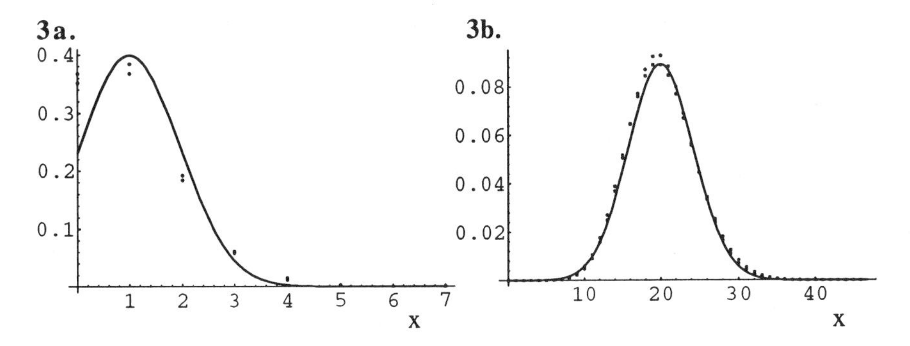

PROPAGATION OF ERRORS

Analytic approach

If a quantity "y", itself is a function of other variables which have errors, how can we determine σ_y? First one takes the derivative of the function with respect to the independent variables. Then the sum is squared and the cross terms dropped since they average to zero for independent measurements.

$$\sigma_R^2 = \left\langle \left(\sum_i \frac{\partial R}{\partial r_i} dr_i \right)^2 \right\rangle \Rightarrow \sum_i \left\langle \left(\frac{\partial R}{\partial r_i} d_{r_i} \right)^2 \right\rangle = \sum_i \left(\frac{\partial R}{\partial r_i} \sigma_{r_i} \right)^2$$

Some common functions are:

$$R = r_1 + r_2, \qquad \sigma_R = \sqrt{\sigma_{r_1}^2 + \sigma_{r_2}^2},$$

$$R = r_1 * r_2, \qquad \sigma_R / R = \sqrt{\left(\sigma_{r_1}/r_1\right)^2 + \left(\sigma_{r_2}/r_2\right)^2},$$

$$R = \frac{r_1}{r_2}, \qquad \sigma_R / R = \sqrt{\left(\sigma_{r_1}/r_1\right)^2 + \left(\sigma_{r_2}/r_2\right)^2},$$

$$R = r_1 r_2^2, \qquad \sigma_R / R = \sqrt{\left(\sigma_{r_1}/r_1\right)^2 + \left(2\sigma_{r_2}/r_2\right)^2}, \text{ and}$$

$$R = \cos(r), \quad \sigma_R / R = |\tan(r)| \sigma_r.$$

Example: Error in the mean from n measurements

We can calculate the error in the mean on n measurements by noting we can use the propagation equation for a sum of terms:

$$\bar{x} = \sum_{i=1}^{n} \frac{x_i}{n} \text{ and } d\bar{x} = \sum_{i=1}^{n} \frac{dx_i}{n} dx_i, \text{ giving}$$

$$\sigma_{\bar{x}} = \frac{\sqrt{\sum_{i=1}^{n} \sigma_{x_i}^2}}{n}, \text{ if } \sigma_{x_i} \text{ are the same} = \sigma_x, \text{ then}$$

$$\sigma_{\bar{x}} = \frac{\sigma_x}{n} \sqrt{\sum_{i=1}^{n} 1} = \frac{\sigma_x}{n} \sqrt{n} = \frac{\sigma_x}{\sqrt{n}}$$

This is a very important result. It says that the uncertainty in the mean value decreases as the square root of the number of measurements. An important corollary is that the error does not decrease as fast as the number of measurements.

Monte Carlo approach to propagating errors

The analytic approach to error propagation assumes that all higher derivatives of R are negligible compared to the first derivative. This is obviously false whenever R is at a relative maximum or minimum since here the first derivative is zero. An example is the case R=cos(r) when r is 0. Notice that in the analytic description, $\sigma_R = 0$! At these points it is necessary to employ a Monte Carlo approach to error determination. Sometimes it is easier than calculating a complicated derivative, even if the analytic approach is in principle fine. The method is quite simple. One generates n random numbers. If a Gaussian generator is available, so much the better. However a uniform distribution with window w = 3.46σ works fine. So r = r_{mean}+(Ran-0.5)*w will do the trick (Ran is distributed from 0->1). Just calculate the function R using each of the generated r values. The resulting distribution represents the uncertainty in R. Note that it may be asymmetric due to the higher derivative terms which are always left out in the analytic approach. Figure 4 illustrates data generated for R=cos(r), with r = 0.0±0.1.

Figure 4. Histogram of Cos(r) with r randomly generated in the interval r = 0.0±0.1.

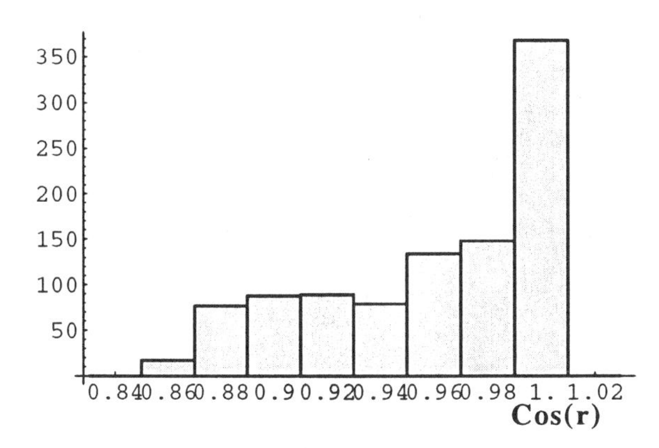

Central Limit theorem-or Why Daddy is everything a Gaussian?

Given an arbitrary distribution which has a mean and variance defined, the Central Limit Theorem tells us that the distribution of the AVERAGE of N measurements is Gaussian with a variance σ^2/N. The way this is typically interpreted is that most measurements we make are due to averages of processes which are ongoing at the microscopic level. For example, if we measure the current through a circuit, it is the sum of 10^{23} moving electrons. The individual electrons probably have a Maxwell-Boltzmann Distribution, but our measurements of the current will resemble a Gaussian distribution.

Figure 5. Illustration of Central Limit Theorem.

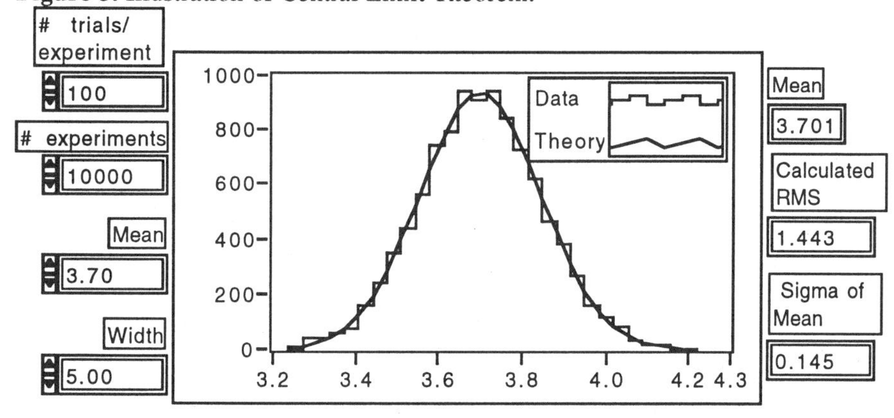

Figure 5 illustrates the Central Limit Theorem. The data on the left of the plot are inputs into the calculation and the data on the right are the outputs. An experiment consists of 100 random numbers (trials) generated uniformly in a window of *Width* =5.0 (rms=1.443), and centered at 3.70 (*Mean*). The sample mean was calculated from the data of each experiment. A total of 10000 experiments were run and the mean from each was histogrammed in the plot. The smooth curve is a Gaussian with $\mu = 3.7$, σ= rms/sqrt(#trials)=0.1443, and area equal to 10000. The curve is in excellent agreement with the histogram. The quantities on the right are calculated from the actual histogram.

ESTIMATION OF PARAMETERS FROM DATA.

Principle of Maximum Likelihood

Suppose we have n independent data measurements y_i taken at n points x_i. These points x_i could all be the same point or they could just be the order in which the measurement was made. How can this data be used to learn something about the underlying functional relationship of y versus x? Clearly we have learned how to do this already in the case of an calculating a mean.

Figure 6. The Probablity of measuring y_i at x_i

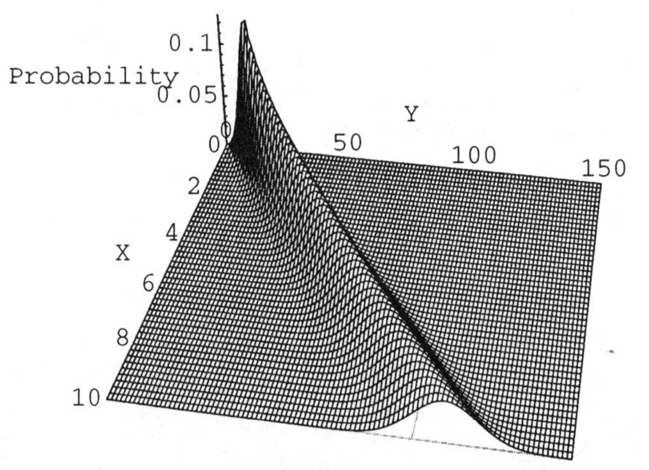

Let's assume we know the form of the functional relationship, if not all the details. An slightly more complicated example than the mean would be $y(x_i)=a_1 + a_2 * x_i$, where a_1 and a_2 are unknown parameters that we would like to determine. At each independent point x_i, the measurement y_i is distributed about the "true" value $y(x_i)$ with a width σ_i. Figure 6 illustrates this concept. At any point x the measured y_i is distributed according to the Gaussian distribution. The true value *y(x)* is represented by the ridge *y(x)*=10+10*x* on the figure. σ(x) was made to vary as the squareroot of *y(x)* to illustrate the effect of a non-constant σ. Thus the Gaussian distribution is narrower and "peakier" at low values of *x*. However the area under each Gaussian along the *y* direction is equal to one. The probability for a particular measurement y_i is P_i which depends on y_i, $y(x_i)$, and σ_i. Figure 7 shows Monte Carlo data generated using the above assumptions. The data are plottted with the window error bars (3.46σ). The other curves will be explained in the next section.

Figure 7. Data generated along the line *y(x)* = 10 + 10 *x* . The data points are shown with window errors (3.46σ). The Curves are explained in the text.

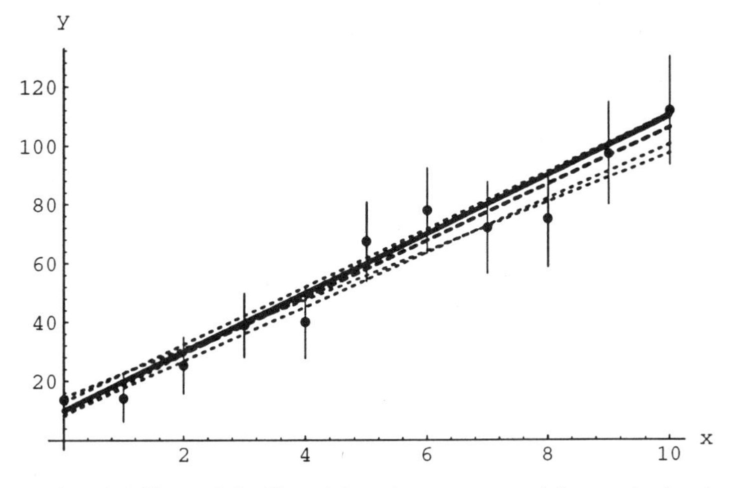

The total probability or Likelihood, L to have measured the particular data set of y_i values is $L = \prod_i^n P_i$, where P_i is the probability for each particular y_i . Given this data, we would like to estimate values for the parameters *a* of *y(x)*. Coming to our rescue is the Principle of Maximum Likelihood stating that the best estimate we can make for the parameters will be the ones which maximize the Likelihood (or total probability). I like to think of this as the Principle that Nature Plays Fair.

Uniform and Gaussian examples

If the underlying statistical distribution is known, one can calculate the probabilities for different values of the unknown parameters. Figure 8 plots the Uniform and Gaussian Likelihood distribution (using the same equivalent rms (Sqrt(*y(x)*))) for the data shown in fig.7 by varying the parameters a_1 and a_2 and recalculating L. Notice that the Uniform Likelihood Distribution is flat - all values for a_1 and a_2 in the plateau are equally likely. All values outside are excluded (L=0). Without invoking a new principle we cannot go further with this distribution. The Gaussian Likelihood is peaked about a particular value of $(a1,a2) = (10.1,9.6)$. These values represent the most likely values for $(a1,a2)$. It will turn out that the Gaussian Distribution will have some very convenient mathematical features which will save us from having to do all this work. In fig. 7, the dotted lines are the $(a1,a2)$ values from the corners of the Uniform Likelihood distribution. The heavy solid line is the actual theoretical curve $(a1,a2)$=(10,10). The heavy dashed line is determined from the peak of the Gaussian Likelihood Distribution

Figure 8. The Likelihood distributions for the data. The left plot is from a uniform probability function. The right plot from a Gaussian probability function. Both use the same data and rms width.

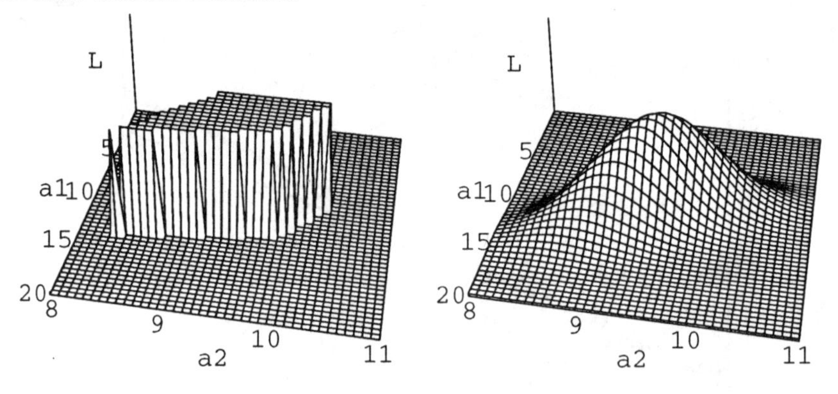

Development of Least Squares Fits

If the individual $y(x)$ distributions are Gaussian, some wonderful things happen. The Likelihood function is:

$$L = \prod_i^n P(y_i, y(x), \sigma(x_i)) = \prod_i^n \frac{1}{\sqrt{2\pi}\sigma(x_i)} e^{-\frac{1}{2}\left(\frac{(y_i - y(x_i))}{\sigma(x_i)}\right)^2}$$

$$= \left(\prod_i^n \left(\frac{1}{\sqrt{2\pi}\sigma(x_i)}\right)\right) e^{-\frac{1}{2}\sum_i^n \left(\frac{(y_i - y(x_i))}{\sigma(x_i)}\right)^2}$$

Notice that once we have our data set, and have determined the errors at each point, the only dependence in L is in the parameters a, which appear only in the function $y(x,a)$. Therefore maximizing the Likelihood is equivalent to minimizing sum of the squares in the exponent. This exponent is known as chisquare,

$$\chi^2 = \sum_i^n \left(\frac{(y_i - y(x_i))}{\sigma(x_i)} \right)^2$$

Linear fit to parameters

We can gain some insight by expanding χ^2 in a second order Taylor's series about about its minimum,

$$\chi^2(a) = \chi^2(a^0) + \sum_j \frac{\partial \chi^2(a^0)}{\partial a_j} da_j + \frac{1}{2} \sum_{j,k} \frac{\partial^2 \chi^2(a^0)}{\partial a_j \partial a_k} da_j da_k .$$

First derivative terms

The first term in the expansion is just the value of χ^2 at the minimum. The second term is

$$\frac{\partial \chi^2(a^0)}{\partial a_j} = \frac{\partial \sum_i \left(\frac{y(x_i,a) - y_i}{\sigma_i} \right)^2}{\partial a_j} = 2 \sum_i \left(\frac{y(x_i,a) - y_i}{\sigma_i^2} \right) \frac{\partial y(x_i,a)}{a_j} .$$

The nomenclature is that "*j*" refers to the jth parameter, and "*i*" to the ith data point.

Since we want to minimize χ^2 ,we set the each of the first derivative terms to zero. If $y(x,a)$ can be written as a linear function of the parameters a_j,

$$y(x_i,a) = \sum_j f_j(x_i) a_j ,$$

the mathematics simplifies even further and we have what is known as a Linear Least Squares Fit. With this assumption the equations simplify to:

$$0 = \sum_k \alpha_{jk} a_k - \beta_j \text{ , where}$$

$$\alpha_{jk} = \sum_i \frac{f_j(x_i) f_k(x_i)}{\sigma_i^2}, \quad \text{and } \beta_j = \sum_i \frac{y_i f_j(x_i)}{\sigma_i^2} .$$

Notice that α_{jk} , which is an element of the curvature matrix (see next section), depends only on the functional form of $y(x)$ and the error values but NOT the actual y_i data values. All information about the data are contained in the β_j . These equations simpify in form into a matrix equation $\{\alpha\}(a) = (\beta)$. This equation can be inverted to solve for the a parameters giving

$$(a) = \{\alpha\}^{-1}(\beta) = \{\varepsilon\}(\beta) \text{ or}$$
$$a_j = \varepsilon_{jk}\beta_k \ .$$

ε_{jk} is an element of what is known as the error matrix.

Second derivative terms

With $y(x_i,a)$ a linear function of the a, it is easy to show that the coefficient of the second derivative of χ^2 is α_{jk}. Therefore around the minimum,

$$\chi^2(a) = \chi^2\left(a^0\right) + \sum_{j,k} \alpha_{jk} da_j da_k .$$

This is exact for the linear function case. α_{jk}.represents the curvature of χ^2. We will see that a steep curvature of χ^2 implies small uncertainties in the determination of the parameters a.

Errors for the parameters a

Now that we have a mathematical prescription to find the parameters, it is reasonable to ask what are the rms errors of the parameters themselves? The uncertainty can be propagated back to the source of the uncertainty - the errors in y_i. Since $a_j = \varepsilon_{jk}\beta_k$,

$$da_j = \sum_i \frac{\partial a_j}{\partial y_i} dy_i = \sum_i \frac{\partial\left(\sum_k \varepsilon_{jk}\beta_k\right)}{\partial y_i} dy_i = \sum_k \varepsilon_{jk} \sum_i \frac{\partial \beta_k}{\partial y_i} dy_i ,$$

$$da_j = \sum_k \varepsilon_{jk} \sum_i \frac{\partial\left(\sum_l \frac{y_l f_k(x_l)}{\sigma_l^2}\right)}{\partial y_i} dy_i = \sum_k \varepsilon_{jk} \sum_i \frac{f_k(x_i)}{\sigma_i^2} dy_i . \text{ Therefore,}$$

$$\sigma_{jl}^2 = \langle da_j da_l \rangle = \left\langle \left(\sum_k \varepsilon_{jk} \sum_i \frac{f_k(x_i)}{\sigma_i^2} dy_i \right) \left(\sum_m \varepsilon_{lm} \sum_n \frac{f_m(x_n)}{\sigma_n^2} dy_n \right) \right\rangle$$

$$= \sum_{k,m} \varepsilon_{jk} \varepsilon_{lm} \sum_i \frac{f_k(x_i) f_m(x_i)}{\sigma_i^4} \sigma_i^2 = \sum_{k,m} \varepsilon_{jk} \varepsilon_{lm} \alpha_{km} = \sum_k \varepsilon_{jk} \delta_{lk} = \varepsilon_{jl}$$

This explains why ε_{jk} is called the error matrix.

Explicit example

This section calculates the fit and errors for the sample generated data which has been used throughout this paper. With the function $y_{actual}(x) = 10.0 + 10.0x$, the data are

$$x_i = [0,\ 1,\ 2,\ 3,\ 4,\ 5,\ 6,\ 7,\ 8,\ 9,\ 10]$$
$$y_i = [13.6,\ 14.1,\ 25.3,\ 38.9,\ 40.0,\ 67.3,\ 77.8,\ 72.0, 75.0,\ 97.1,\ 115.3]$$
$$\sigma_{y_i} = [3.2,\ 4.5,\ 5.5,\ 6.3,\ 7.1,\ 7.8,\ 8.9,\ 9.5,\ 10.0,\ 10.5]$$

With these values the matrices are:

$$\{\alpha\} = \begin{Bmatrix} \sum_{i=1}^{11} \frac{1}{\sigma_{y_i}{}^2} & \sum_{i=1}^{11} \frac{x_i}{\sigma_{y_i}{}^2} \\ \sum_{i=1}^{11} \frac{x_i}{\sigma_{y_i}{}^2} & \sum_{i=1}^{11} \frac{x_i^2}{\sigma_{y_i}{}^2} \end{Bmatrix} = \begin{Bmatrix} 0.302 & 0.798 \\ 0.798 & 4.702 \end{Bmatrix} \text{ and,}$$

$$(\beta) = \begin{pmatrix} \sum_{i=1}^{11} \frac{y_i}{\sigma_{y_i}{}^2} \\ \sum_{i=1}^{11} \frac{y_i x_i}{\sigma_{y_i}{}^2} \end{pmatrix} = \begin{pmatrix} 10.67 \\ 53.00 \end{pmatrix} \qquad \{\varepsilon\} = \{\alpha\}^{-1} = \begin{Bmatrix} 6.00 & -1.02 \\ -1.02 & 0.38 \end{Bmatrix}.$$

Solving for the parameters and their errors gives

$$(a) = \{\varepsilon\}(\beta) = \begin{pmatrix} 10.1 \\ 9.56 \end{pmatrix} = \begin{pmatrix} \text{constant} \\ \text{slope} \end{pmatrix} \qquad \sigma_a = \begin{pmatrix} \sqrt{\varepsilon_{11}} \\ \sqrt{\varepsilon_{22}} \end{pmatrix} = \begin{pmatrix} 2.4 \\ 0.62 \end{pmatrix}, \text{ with}$$

$\chi^2 = 10.4 / 9$ degrees of freedom at the minimum.

Calibration errors from results

Now that we have determined our parameters it would be useful to use the results. Suppose $y(x,a)$ represent a calibration of a system. How do we state the error of the calibration? It is incorrect (in principle) to claim the uncertainty of the calibration as being given by the data fluctuations from the curve. Why? Because if (a big if!) the data are really described by our chosen function, we should be able to do better since the fit is using all the points. A good example would be the error in the mean. If we make 100 measurements of a value, the error in the mean should be a factor of 10 lower than the rms width of the actual data distribution. If we know the rms width, an easy check is calculating the χ^2 value. It should be (as we will see) approximately equal to the number of data points for a good fit. Likewise in the case we have just calculated, our knowledge of the equation of the line is much better than the fluctuations of the data around it. But how well do we know it? The uncertainty can be calculated by propagating the errors. However we will propagate the error only to the parameters instead of going all the way back to the data. The error in $y(x)=a_1+a_2x$ is

$$dy(x,a) = \sum_j \frac{\partial y(x,a)}{\partial a_j} da_j = \sum_j f_j(x) da_j .$$

$$\sigma^2_{y(x,a)} = \left\langle \sum_j f_j(x) da_j \sum_k f_k(x) da_k \right\rangle = \sum_{j,k} f_j(x) f_k(x) \langle da_j da_k \rangle$$

$$= \sum_{j,k} f_j(x) f_k(x) \varepsilon_{jk}$$

In our particular example, the last equation becomes

$$\sigma_{y(x,a)} = \sqrt{f_1(x)f_1(x)\varepsilon_{11} + 2f_1(x)f_2(x)\varepsilon_{12} + f_2(x)f_2(x)\varepsilon_{22}}$$

$$= \sqrt{\varepsilon_{11} + 2x\varepsilon_{12} + x^2\varepsilon_{22}}$$

$$= \sqrt{6.00 - 2.04x + .38x^2}$$

Notice that the parameters are correlated. We cannot ignore the cross terms. The reason for this is that they have been determined from the the same set of data.

$\sigma_{y(x,a)}$ is plotted as a function of x in fig. 9. Notice how the error is smaller at all x values than our data estimations. Also note how the error grows very fast for x values outside our data set (x>10). This tells us to beware extrapolations!

Figure 9. The error in the fit $y(x)= a_1+ a_2x$ as a function of x. The parameters a are the fit values.

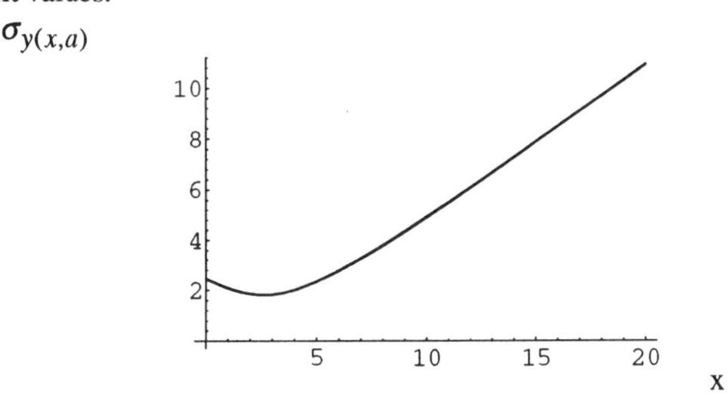

Non-Linear Least Square Fits

Log and other endruns

Some functions with non-linear parameters lend themselves to a linear fit by a remapping of the independent and dependent variables. The most common are

$$y = a_1 e^{a_2 x} \rightarrow \log(y) = \log(a_1) + a_2 x \text{ and}$$

$$y = a_1 x^{a_2} \rightarrow \log(y) = \log(a_1) + a_2 \log(x).$$

The new variables are log(y) and x, and log(y) and log(x) respectively. This was an important feature, especially when most data analysis took place on graph paper (remember semi-log, logarithmic, and probability paper ?). Even now the technique is useful due to the simplicity and availability of software on computers and calculators. However caution should be applied since the error bars are certainly not equal (generally what the software assumes), even if they were for the original y values. Typically the smaller values of y are overweighted if one assumes equal errors for log(y). This will cause the fit to overemphasize the small y values. In addition the error bars are no longer symmetric about the “true” value. For example if the error is ± 10 and y=10, y_{data} might range from 0 to 20. On a log scale, the error bar should range from -infinity to log(20).

Linearization

If *y(x)* is not a linear function of the parameters, what can be done? Due to the success of the Linear Least Squares Fit mathematics, the answer is obvious-linearize the function! Expanding *y(x,a)* about the parameters *a* gives:

$$y(x,a) = y\left(x,a^0\right) + \sum_j \frac{\partial y\left(x,a^0\right)}{\partial a_j} da_j, \text{ and}$$

$$\chi^2 = \sum_i \frac{\left(y(x_i,a) - y_i\right)^2}{\sigma_i^2} \to \sum_i \frac{\left(y\left(x_i,a^0\right) - y_i + \sum_j \frac{\partial y\left(x,a^0\right)}{\partial a_j} da_j\right)^2}{\sigma_i^2}.$$

The entire formalism for the linear least squares fit can be taken over with the substitution of $(y(x_i) - y_i)$ for $(-y_i)$ and realizing that the parameters are now *daj*. The a^0 's are considered as constants, and hopefully close to their true values. The implemented algorithms usually require the user to make an initial guess at the a^0. In addition one usually supplies the fit with the analytic partial derivatives, again computed at the a^0 starting points. Some algorithms will compute the derivatives numerically for you, but this triples the number of calculations of a complicated function (which is already being evaluated at every x value). The fit calculates the *da*'s which then are added to the a^0's and the process starts over again, this time using the new a^0's to calculate *y(x)* and its derivatives again. The process can be numerically intensive if there are a lot of data points. Generally the user uses some requirement to stop the iteration such as Chisquare not changing. The definition of errors in the parameters is the same ε_{jj} as in the linear case.

As has been mentioned, many packages exist in spreadsheet, statistical, and lab data acquisition software which will fit non-linear functions. The art of the process is starting the fit off at the right point. It is one thing to fit a curve by hand-specifying the starting parameters, and quite another to automate the sequence under all conditions the aaccelerator provides. If the signal to noise of the data is good, it will usually be easy to automate the process. Most of our work has been done with functions which are close to being Gaussian in shape and on a reasonable background:

$$y(x,a) = a_1 + a_2 x + a_3 e^{-\frac{1}{2}\left(\frac{x-a_4}{a_5}\right)^2}.$$

We typicallly estimate the background a_1 at the ends of the data array, set a_2 = 0, a_3 = (maximum value - a_1), $a_4 = x$ value of the maximum value, and finally find the full width half maximum (FWHM) peak heights on either side of the maximum and set a_5 =(FWHM/2.35). This method is reasonably insensitive the poor signal to noise ratios.

Chisquare Phenomenology

Chisquare Contours

If the contours of χ^2 are rotated ellipses with respect to the parameter axes (easiest to visualize in a two parameter fit), the error matrix is not diagonal (normal case). If the ellipse's major and minor axes are aligned with the parameter axes, then the errors in the parameters have no correlations (and the error matrix is diagonal). A simple example (case 1) can be had from our simple Linear Least Squares Fit to the line $y(x)= 10+10x$. (assume constant errors for this example). The contour of χ^2 is shown in fig.10a. It is rotated. If we use a function (case 2) y(x)=a1+a2*(x-5), the χ^2 contour is shown fig.10b. What we have done is replaced the y intercept of the first function with a constant which is now equivalent to the average y value (recall that x = 5 is the center x value of our data). Since the average y value is independent of the slope,we have "decoupled" the two parameters. If one recalls the formula for the off-diagonal element of $\{\alpha\}$, the two parameter version has the value

$$\alpha_{12} = \sum_{i=1}^{n} \left(\frac{\left(\frac{\partial y(x_i)}{\partial a_1} \frac{\partial y(x_i)}{\partial a_2} \right)}{\sigma_i^2} \right).$$

For case 1 and 2 respectively, these values are $\alpha_{12} = 0.55$ and $\alpha_{12} = 0$. The values of x = (0, 1, 2, 3,..10), are the values used in generating the data. This example suggests how to "orthogonalize" the function, even in the case where σ_i is not constant. Set the off-diagonal elements of $\{\alpha\}$=0 and solve for the constant. The actual goodness of the fit is exactly the same in both cases. The difference is that it is easier to understand the errors of the parameters in case 2. Practically speaking, this technique is almost never used. However I plan to use it to make a point in the next paragraph.

If we look at the contours in fig. 10, the first contour drawn is the one which is one (1) higher than the minimum value of χ^2. This means that the probability for our data to have been generated by the values of the pararameters which lie on this contour is

$$L \propto e^{-\frac{1}{2}\left(\chi_{\min}^2 + 1\right)} = L_{\max} e^{-\frac{1}{2}} = 0.61 L_{\max}.$$

In case 2 with the "orthogonalized" function, the error matrix diagonal elements are just the inverse of the curvature matrix diagonals (since $\{\alpha\}$ is diagonal). Thus it is true that the one sigma error in the parameter is given by the increment which increase χ^2 by one unit since $\chi^2(a) = \chi^2(a^0) + \alpha_{jj}\varepsilon_{jj} = \chi^2(a^0) + 1$. This can shown to be true even in case 1, as long as the other parameters are re-optimised. As a matter of fact, the technique of finding the 1 sigma contour is the best way to define the parameter errors since it works even in the case when χ^2 is not well described by the second order expansion - the case in some nonlinear fits. It should be noted that the second order expansion of χ^2 is exact for all functions which are linear in their parameters. Also note again that $\{\alpha\}$, the curvature, is completely set by the estimate of the data error (once the function has been set and the data measured).The actual data points y_i have nothing to do with $\{\alpha\}$. This means that the parameter errors are very sensitive to the estimation of errors. As a side note, the parameters are also sensitive to the x_i values that the y_i are measured at. One cannot expect a fit to work without sampling the data where the function is significant.

Figure 10. Contour Plots of χ^2 for the Linear Fit. The horizontal axis is a_1, and the vertical axis is a_2. (a) is using the function $y(x) = a_1 + a_2 x$ while (b) uses the function $y(x) = a_1 + a_2 (x - 5)$.

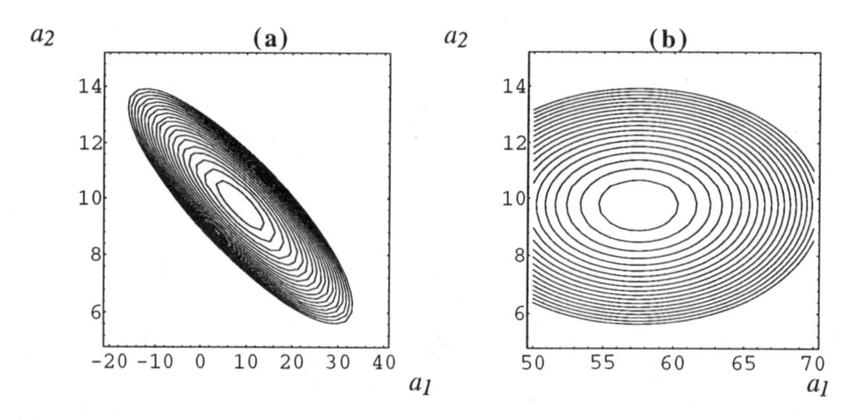

After all this work it would be nice to get error bars from our fits. Unfortunately most software packages leave out this step. This is fine if the error bars are the same magnitude for all data points, because at least the formulae for the the parameters are correct. Even still most packages don't calculate either the curvature or error matrix so they leave it to you anyway. Calculating the curvature matrix is not such a big deal if you really need it, and most packages do have matrix inversion routines which will get you the error matrix. If the errors are really varying dramatically across the data set, these packages may not even calculate the parameters very well.

Of course the major difficulty is estimating the error bars in the first place. I suspect most of us do this empirically by checking the repeatability of our measurements. If one is fitting curves automatically under widely varying conditions, this option isn't generally available. A trick which does work if one can assume equal errors is to measure the point to point fluctuations in the background region of the fit with a simple rms calculation. If it works, use it!

Chisquare and goodness of Fit:

Most of us have heard about the goodness of the fit or χ^2. If we look back at the definitions of χ^2 we see that if the theoretical function $y(x)$ is close to reality, and if we have estimated the errors at each point correctly, $(y(x)-y_i)$ should sometimes be less than σ and sometimes greater, since a Gaussian distribution has been assumed. Therefore over a data set we should expect that each term in the sum should contribute one unit to χ^2, and χ^2 will approximately equal the total number of data points N. Usually we divide χ^2 by the number of degrees of freedom $\nu = (N-n)$ to give what is known as the "reduced" chisquare or $\chi^2_{\nu} = \chi^2/(N-n)$. χ^2_{ν} by the preceding arguments should now be approximately 1 if the fit is good. Probability tables exist for χ^2_{ν} which can be used to check the goodness of the fit.

What if χ^2_{ν} is too ridiculously big or small? The first question should deal with the σ estimations. Are they too small (large χ^2_{ν}) or too big (small χ^2_{ν}). If this looks fine then one wonders about the function and/or the data themselves. Unfortunately there is no easy way to disentangle the two except by looking. Is the data asymmetric about the mean?, ...Is the background handled correctly?, ...Are there obviously bad data points? This is the black magic of the process.

Fitting versus simple statistical analysis

Suppose one is trying to extract the mean and width of a gaussian-looking peak, which is riding on a noise floor, i.e.

$$y(x) = Ae^{-\frac{1}{2}\left(\frac{x-\mu}{\sigma}\right)^2} + B.$$

When is a simple moment analysis appropriate to get the centroid and width of the peak?

I have found that when the signal to noise is 10/1 or better, it is possible to forego fancy fitting and do a simple moments analysis. By this is meant

$$\mu = \frac{\sum_{i=1}^{n}(y_i - B)x_i}{\sum_{i=1}^{n}(y_i - B)} \quad \text{and} \quad \sigma = \sqrt{\frac{\sum_{i=1}^{n}(y_i - B)(x_i - \mu)^2}{\sum_{i=1}^{n}(y_i - B)}}\,.$$

These are simply the weighted μ and σ of the x values, using the y_i 's as weights Before one can make this calculation it is necessary to strip away the noise floor. The resulting (y_i-B) (in the background region) will fluctuate both positively and negatively around zero, and on the average cancel. However if the fluctuations are large compared to A, the peak amplitude, the results (especially σ) will be very erratic. One must always be on guard under these conditions.

CONCLUSION

Statistical Analysis of data is within the reach of all Instrumentation Engineers. This tutorial has tried to stress the assumptions which are implicit in most analyses. One should realise that we have only scratched the tip of the iceberg.

1(1) P.R.Bevington, Data Reduction and Error Analysis for the Physical Sciences,McGraw-Hill Book Co.(1969).
(2) E.L.Barsotti Jr., "A Longitudinal Bunch Monitoring System Using LabVIEW and High-speed Oscilloscopes", Contributed Poster this Workshop
(3) J.R.Zagel, D.Chen, and J.Crisp, " Fermilab Booster Ion Profile Monitor System Using LabVIEW", Contributed Poster this Workshop

Beam Diagnostics at TRIUMF

W. R. Rawnsley
TRIUMF, 4004 Wesbrook Mall, Vancouver, B.C., Canada, V6T 2A3

Abstract

This paper will discuss the development of some selected beam probes for the TRIUMF 520 MeV H^- cyclotron and some beamline monitors. The cyclotron has 7 movable probes that make use of vertical, differential and crossed wire heads. Isochronism is obtained by time of flight scans rather than by the use of phase probes. Recent developments include the use of a stripline monitor to measure the quadrupole moment of a beam, construction of a low current beam phase monitor, and the use of solid state diodes by the proton therapy group for dosimetry. A series of wire and blade scanners has been used to measure beam profiles at currents from 15 nA to 150 μA.

INTRODUCTION

The cyclotron accelerates an H^- ion beam in a train of bunches 0.5 ns to 5 ns wide at 23 MHz.(1,2) Average intensities of 0.1 nA to 200 μA are available i.e. 26 ppb to 5.3×10^7 ppb. The normal beam is CW but we have a pulser in the injection system which can eliminate some of the bunches. It can be pulsed at $1/(5\times2^{12})$ of the RF (about 1 kHz) with a duty cycle of 0.1 to 99.2% for diagnostic purposes and control of the average current, see fig. 1.

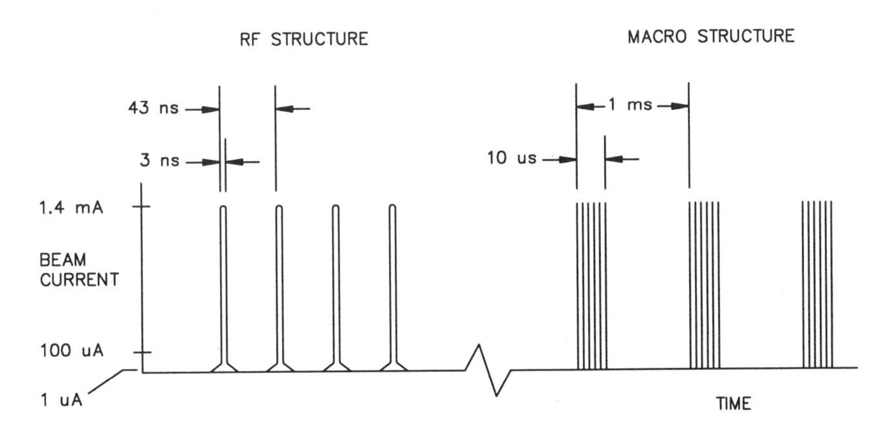

Figure 1. The time structure of a 1 μA beam at 1% duty cycle.

CYCLOTRON DIAGNOSTICS

Cyclotron Probes and Signals

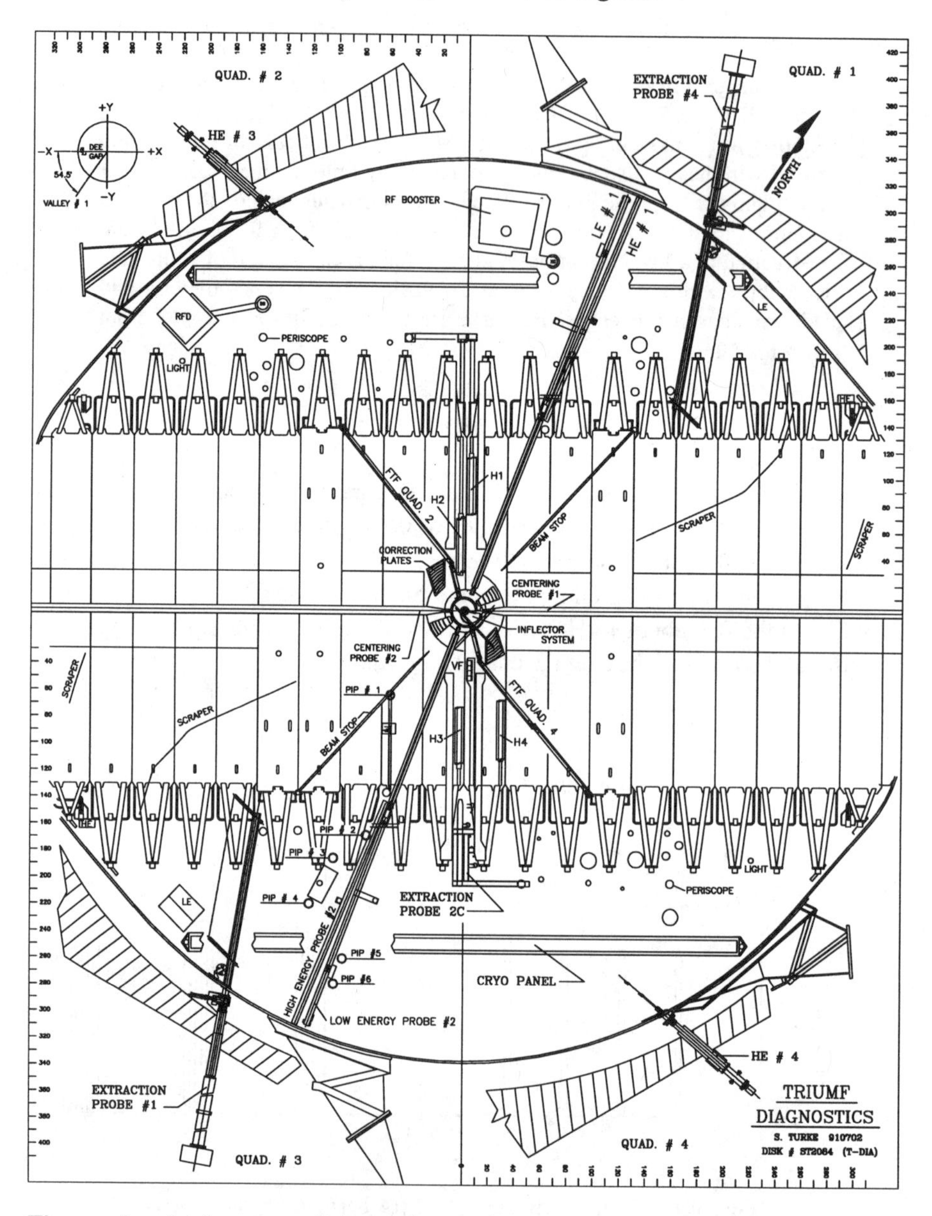

Figure 2. A plan view of the cyclotron showing the beam instrumentation. The machine has 7 probes which travel radially through the beam plane.

Two low energy probes, LE1 and 2, on opposite sides of the cyclotron run from 0.3 MeV to 86 MeV, fig. 2. Two high energy probes, HE1 and 2 run from 67 MeV to 516 MeV. HE3 and 4 run from 460 MeV to 516 MeV. The fingers of a LE probe are 8 mm thick copper and intercept the beam whereas those of a HE probe are 0.13 mm tantalum and only strip the H^- ions. In the magnetic field the stripped electrons recirculate through the fingers until they are stopped. A HE vertical head consists of 5 horizontal fingers spaced vertically while a differential head has 2 radial fingers with a separation of 1.3 mm. One of the LE probes has a 3 finger vertical head projecting 2 mm past a large block, the sum of the 3 fingers gives differential information. A pop-in-probe (PIP6) has a single foil that can be scanned 3 in. vertically and 7.25 in. radially about an energy of 430 MeV.(3)

The ac components of the signals are used to measure the energy gain per turn and the machine isochronism while the dc components are used to measure the average beam current, height and the radial density. The signals are brought out of the cyclotron vault on balanced twinax cables, about 70 m long, to differential ac and dc amplifiers connected in parallel. This system allows simultaneous measurement of dc and ac signals, optimization of their amplifiers, and avoids multiplexing of the low level currents. A simplified diagram of the signal separation scheme used in the probe electronics is shown in fig. 3. The amplifiers are true differential input current amplifiers, though this is not shown here.(4) C1 and R2 form a network which passes the high frequency components of the probe signal to the ac amplifier and the low frequency components to the dc amplifier. The crossover frequency is 5.3 Hz. The rise time of the dc section is 66 ms, allowing data samples to be recorded every 0.05 in. of radial travel at a probe speed of 0.5 in/s. Six gain ranges are provided by selecting the feedback resistors with low leakage CMOS switches. The ac section need not respond to signals of less than the 1 kHz pulser frequency. The droop is 1.7% at a 50% duty cycle.

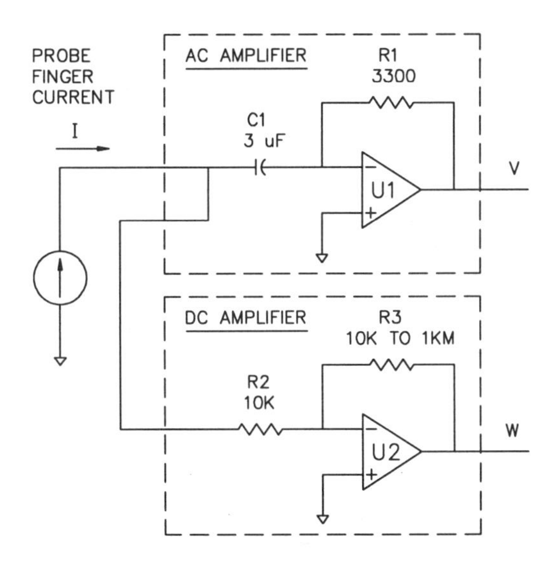

Figure 3. The current from a probe electrode is separated into ac and dc components.

Crossed Wire Head

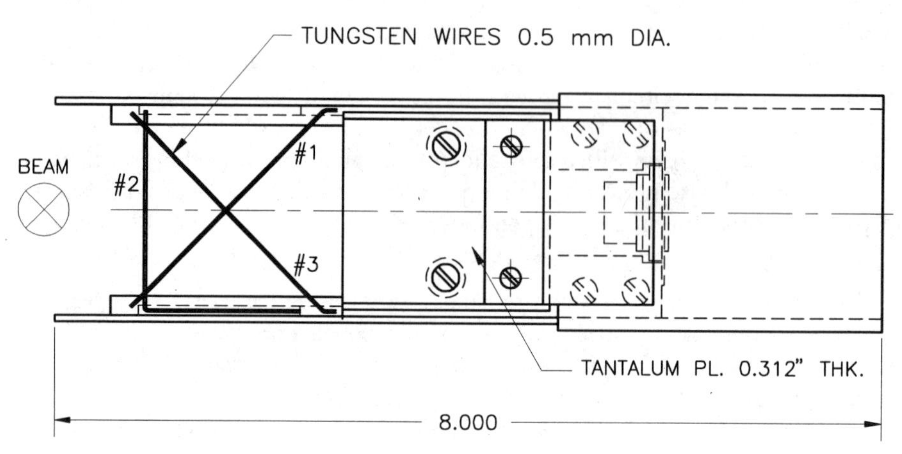

Figure 4. The crossed wire head.

A crossed wire head, modeled on a PSI design (5), consists of three 0.5 mm diameter tungsten wires, two crossing at 45° and the third vertical, fig. 4. A scan begins with the head inside the first turn and it is run to higher radius. The data processing program keeps track of individual turns from the centre out. The vertical and radial spatial extent of each turn is estimated by back projections of the thresholds where the intercepted current falls to 5, 25 and 85 % of its peak value, forming 3 hexagons. One such projection is shown in fig. 5. This algorithm provides a simplistic tomographic reconstruction of the turn cross section in a vertical–radial plane, fig. 6.

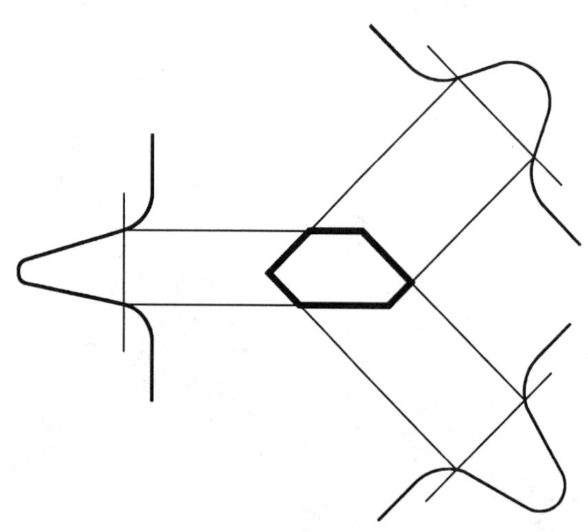

Figure 5. A simple tomographic reconstruction algorithm gives the beam envelope in terms of hexagons

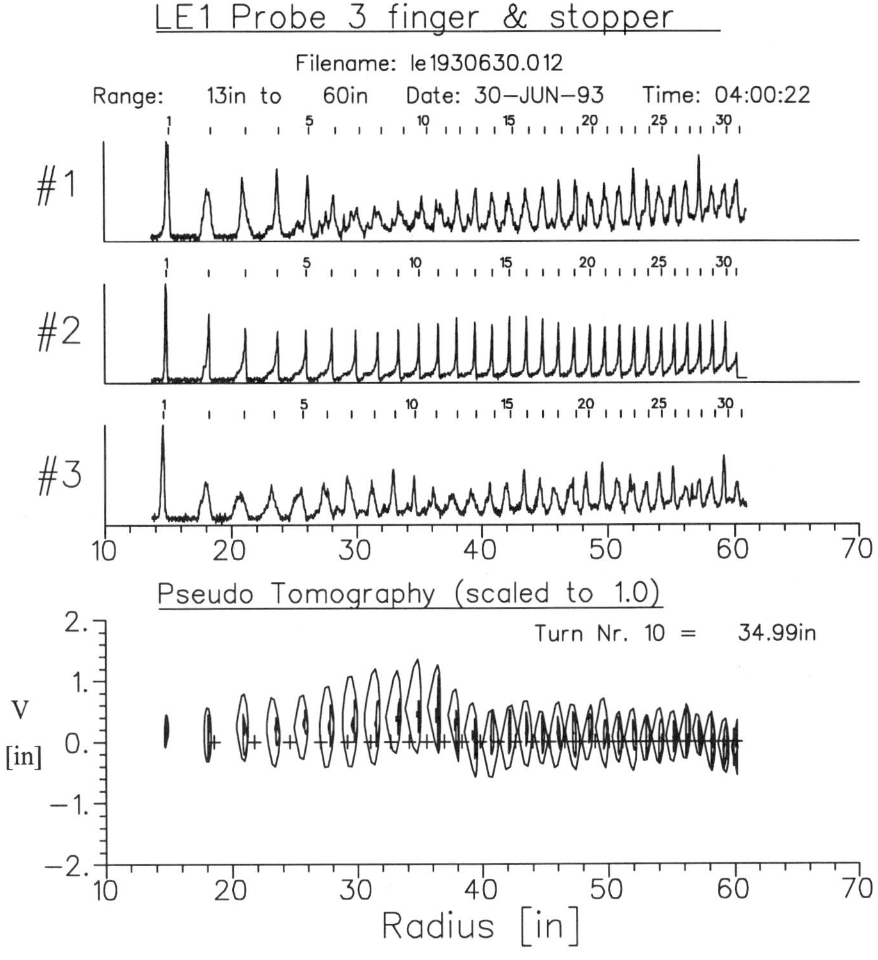

Figure 6. The signals from the three LE head wires can be used to construct a tomograph of the centre region turn pattern. One can see progressively closer radial spacing of turns and an excursion in the vertical plane at a radius of ~32 in. which would give rise to a coherent betatron oscillation but is eliminated by electrostatic correction plates at ~38 in. radius.(6)

Time of Flight Measurement

Most large cyclotrons have an array of phase detectors spaced radially to measure isochronism, e.g. RIKEN.(7) In the early days the stray electric field from the TRIUMF dees was very large - swamping the beam induced signal, consequently an alternative system was developed. The time of flight (TOF) of a beam macro-pulse through the injection system and cyclotron can be measured by starting a time to digital converter (TDC) with the pulser signal and stopping with the arrival of the first beam on a probe head.

The TOF signal provides the total number of turns to a given radius, i.e. energy, and hence the energy gain per turn as a function of radius. A set of slits and flags in the central region of the cyclotron may be used to restrict the phase acceptance of the machine; 150 ps beam bunches are possible. A phase acceptance of 8° is easy to obtain using a radial flag and one slit, H2. This narrow $\Delta\phi$ should be used for the most accurate measurements. The TOF system triggers on the first part of the arriving pulse. It is advantageous to keep the frequency response as low as possible to reject 23 MHz RF pickup and to reduce system noise. The signal paths are not ideal. The probe head signals pass through about 10 m of flat, shielded ribbon cable inside the probe and through low pass filters with a bandwidth of 1.7 MHz at the vacuum tank feedthroughs. Nevertheless, the measured system rise time is close to that of the amplifier; 0.2 μs.

Heating of the probe head and tank activation (stripped ions from the HE probes are deflected into the tank wall) restrict usual average beam currents to approximately 1 μA at the probes. As a result TOF measurements are carried out at low duty factors. Typically a 1% duty factor at 1 μA is used, equivalent to an operational current of 100 μA. Under these conditions the system noise is equivalent to about 0.5 % peak to peak of beam. The main source of noise is electrical interference, as shown by disconnecting the balancing line at the amplifier which increases the noise by a factor of 10.

The ac signals from the centre 3 fingers are summed in the ac amplifier while the dc signals are digitized separately. For a differential head, both fingers would be summed. The oscilloscope display of the TOF signal is used to optimize the overall machine isochronism. The cyclotron RF frequency is fine tuned to achieve the fastest rise time and to minimize the TOF delay.

The stripped electrons yield an "inverted", negative going signal on the display. Figure 7 shows four overlapping shots of the TOF signal taken about 5 seconds apart from the HE3 probe at 500 MeV with a 5.8% duty cycle 398 nA beam; equivalent to about 6.9 μA.

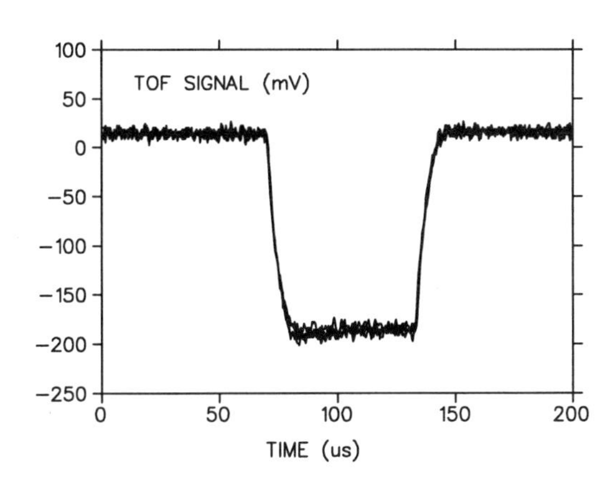

Figure 7. A TOF signal from HE3. The beam was on during the lower portion of the trace.

A general purpose data acquisition computer program, called Probe Scan Utility (PSU), is used to record the TOF and finger currents vs. radius as a probe is run through the machine. Analysis produces a V·cos ϕ scan; shown in fig. 8. V is the peak dee gap voltage and ϕ is the phase of the beam w.r.t. the RF. The radius is first converted to energy using a look up table. V·cos ϕ is proportional to the reciprocal of the derivative of the TOF with respect to energy and is 1/4 of the energy gain per turn. Since we do not measure sin ϕ, a second measurement with altered magnetic field or frequency is needed to determine the sign. In this example, the isochronism was poor near the outside of the machine but was later improved by tuning the trim coils to alter $B_z(R)$ locally and maximize V·cos ϕ averaged over the scan.

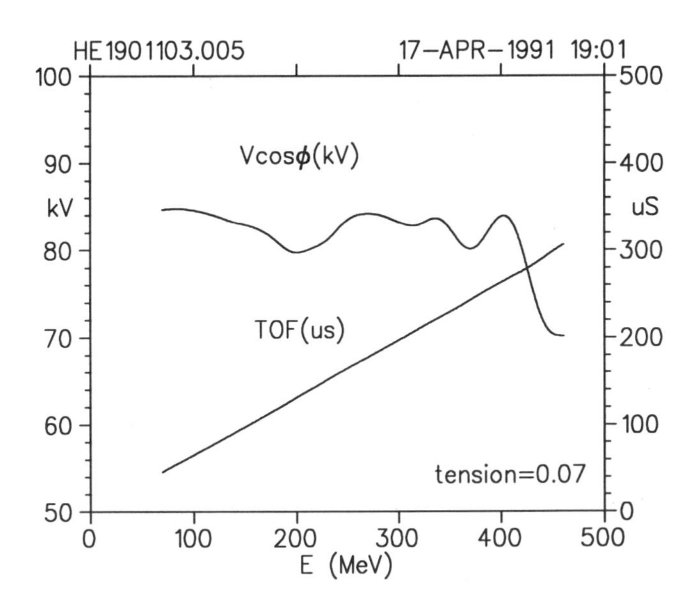

Figure 8. Time of flight and resulting V·cos ϕ are functions of energy.

NON-INTERCEPTING MONITORS

Striplines

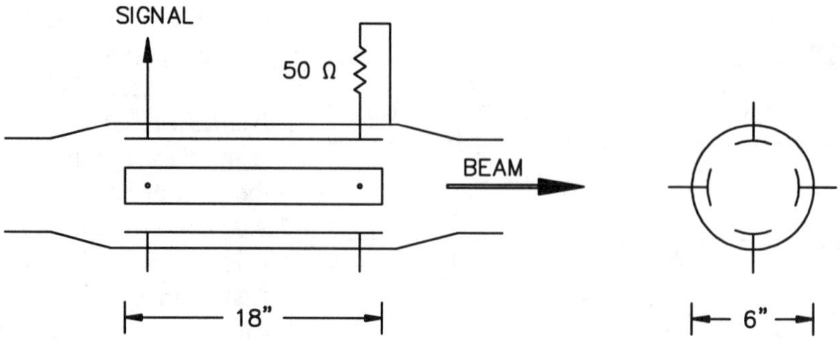

Figure 9. The stripline monitor has 4 strips in an enlarged section of the pipe.

TRIUMF has developed a number of fast nonintercepting monitors, including magnetic pick up loops, wall gaps and a multistrip monitor. A stripline monitor has been used to measure the position, intensity, and time structure of a proton beam. The device consists of 4 striplines mounted inside the wall of a cylindrical beam pipe, fig. 9. The strips are 0.062 in. thick and 18 in. long and subtend 45°. The cylinder inner radius is 2.96 in. and the strips are spaced 0.409 in. from the wall.(8) Signals are taken from the upstream ends of the strips while the downstream ends are terminated in resistive loads.(9)

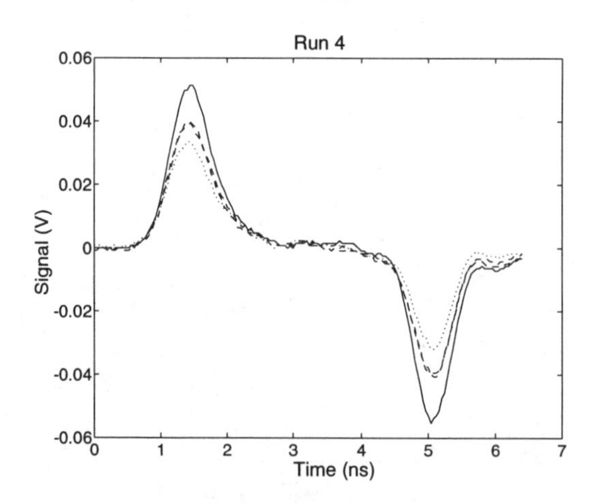

Figure 10. Stripline signals for a case where a slit is used to restrict the phase width transmitted and to select the trailing portion of the beam bunch.

The stripline monitor is normally used as a simple beam position monitor using an AM/PM system purchased from Fermilab and modified to operate at 46 Mhz.(10) The signals have also been useful for occasionally checking the length of short beam bunches. Strip signals are shown for a phase selected

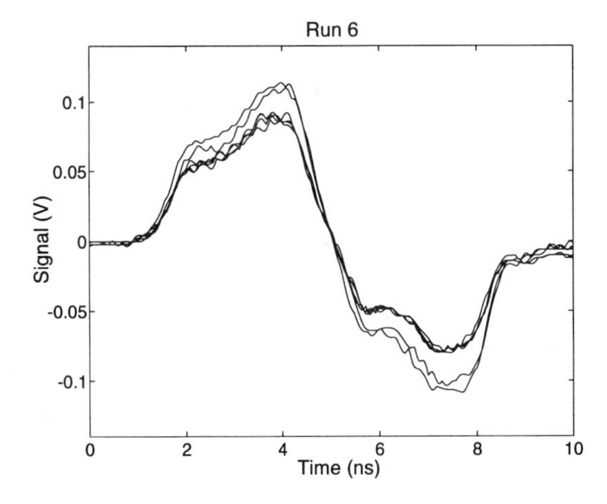

Figure 11. Full phase band.

tune, where slits were used to reduce the bunch length to less than twice the electrical length of the strips, fig. 10. Figure 11 shows strip signals for a full acceptance tune. In this case the trailing edge of the noninverted pulse partially cancels the leading edge of the inverted pulse.

A multistrip monitor has been used to measure the position and higher order moments of the beam transverse shape along the bunch.(11) Recently, attempts have been made to confirm some of these measurements using the stripline monitor. Miller has shown that the 4 signals can be used to measure the intensity, x and y centroid positions and the quadrupole moment of a Gaussian shaped beam.(12) He expressed the image currents at 4 points around the wall circumference in terms of the beam characteristics. For our case, we integrated the image currents over the 45° strips and then solved for the beam characteristics in terms of the strip currents. This results in some simple coefficients entering the equations. Note that the quadrupole moment with respect to the centre of the monitor has two contributions; that due to the offset of the beam centroid and that due to the beam shape. The formula eliminates the former component to yield the quadrupole moment due to the shape of the beam:

$$\frac{\sigma_x^2 - \sigma_y^2}{a^2} = \frac{k_1\,(l\,r - d\,u) + k_2\,(d^2 - l^2 - r^2 + u^2)}{(d + l + r + u)^2} \tag{1}$$

Where σ_x and σ_y are the rms half widths in the x and y directions, a is the effective strip radius (taken here as the average of the inner strip radius and the wall radius). l. r. d and u are the left, right, down and up strip currents. k_1 and k_2 are 3.217 and 0.4977, respectively (or 3.0 and 0.5 for very narrow strips).

The measurements were taken with 6 to 11 μA of beam. The beam shape at the stripline was varied by changing upstream magnet settings. Figure 12 shows the stripline measurements averaged over a beam bunch compared to those measured using a nearby multiwire chamber monitor. The multiwire chamber measures the shape (or moments) projected in the longitudinal direction. The conditions for the runs are shown in table 1.

Table 1. Quadrupole moment measurement conditions.

Run	Beam Shape	Beam Phases	Duty Cycle	Protons/Bunch
2	slightly tall	leading	98 %	5.2×10^6
3	slightly tall	trailing	98 %	$<5.2\times10^6$
4	slightly tall	trailing	99 %	3.0×10^6
5	round	trailing	50 %	2.8×10^6
6	wide	all	10 %	1.7×10^7
7	slightly tall	all	10 %	1.7×10^7

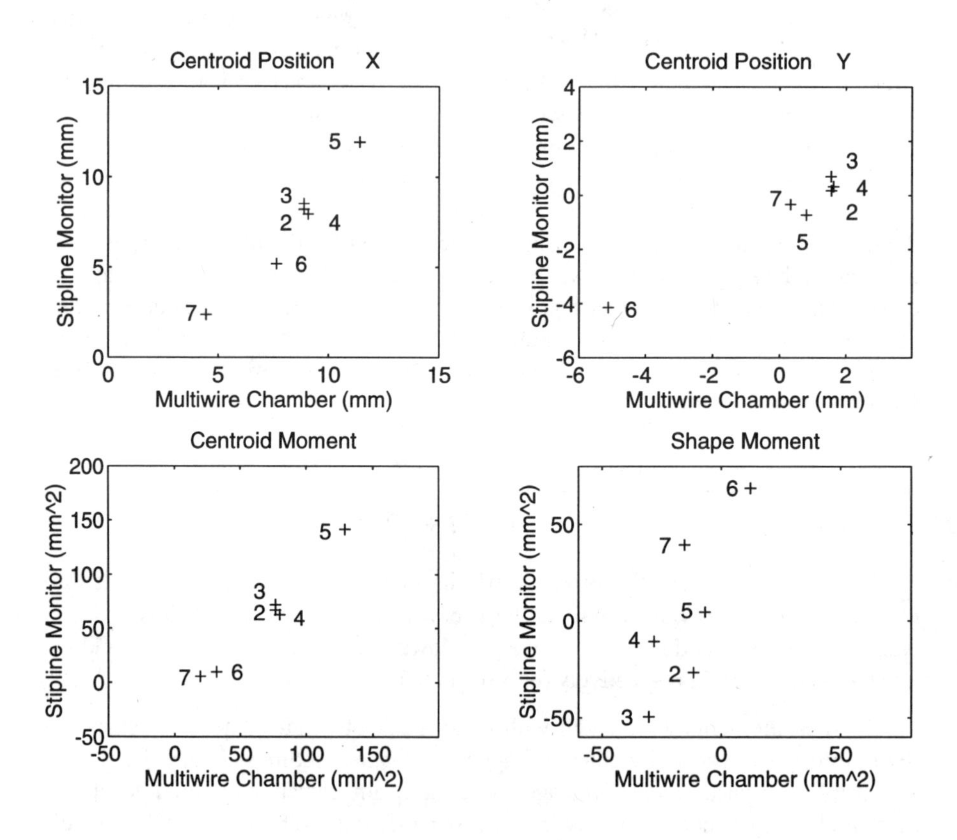

Figure 12. Scatter plots comparing the beam parameters as measured with the stripline monitor to those measured with a nearby multiwire chamber.

The results demonstrate the difficulty of the measurement. The monitor signals are small, about 0.5 mV peak, and compensation must be made for amplifier gain variation and cable delays. The measurements are made using averaging and beam off background subtraction. The areas under the peaks are used for the calculation. The overall results indicate that the trend is correct, i.e. widening the beam increases the measured quadrupole moment due to the beam shape. The reproducibility however, is comparable to a significant change (double) in width as seen by comparing runs 3 and 4 which were taken under similar conditions but 2 hours apart.

The signals from runs 6 and 7 were also processed point by point along the bunch. It was found that the x centroid position moved about 3 mm from head to tail and the y about 5 mm implying that, at this point, the beam is moving crab-wise down the pipe. The quadrupole shape moment had no measurable change.

Beam Phase Monitors

The experimenters wish to have a timing signal that is phase locked to the beam bunches for opening counter gates, etc. The cyclotron RF system is driven by a frequency synthesizer, but the system does not presently attempt to lock the phase on the dees to that of the synthesizer. The beam phase can also vary w.r.t. that of the dees.

A simple capacitive probe consisting of a 7.5 in. long tube around the beam has been used on a high current beamline to provide a timing signal. The signal into a 50 Ω load is a bipolar pulse; a differentiated bunch shape with 25 mV positive and negative peaks and a total duration of 10 ns for a 141 μA CW beam. The signal passes through an integrator to form a unipolar pulse and then an AGC amplifier and is finally discriminated. The signals from any pair of the capacitive probe, the dees, or the synthesizer could be compared using a phase detector circuit. It was found that at 500 MeV the dees had a 5.4° variation w.r.t. the synthesizer at about 0.125 Hz due to the opening and closing of a resonator water cooling valve. The beam phase w.r.t. the synthesizer had a jitter of a few degrees to 10° in a band centred on 5 Hz due to vibration of the resonator cooling panels caused by the cooling water flow. The beam had a jitter of 6.6° at 60 Hz and 3.3° at 180 Hz w.r.t. the dees, probably due to ripple in the cyclotron magnetic field originating in trim coil power supplies. Beam loading can also create phase jitter or shift. A 1 kHz jitter of 4° was introduced in the phase of the beam w.r.t. the dees when the duty cycle of a 144 μA CW beam was reduced to 48 %. Some experiments are envisioned which will require a reduction of the phase jitter by an order of magnitude.

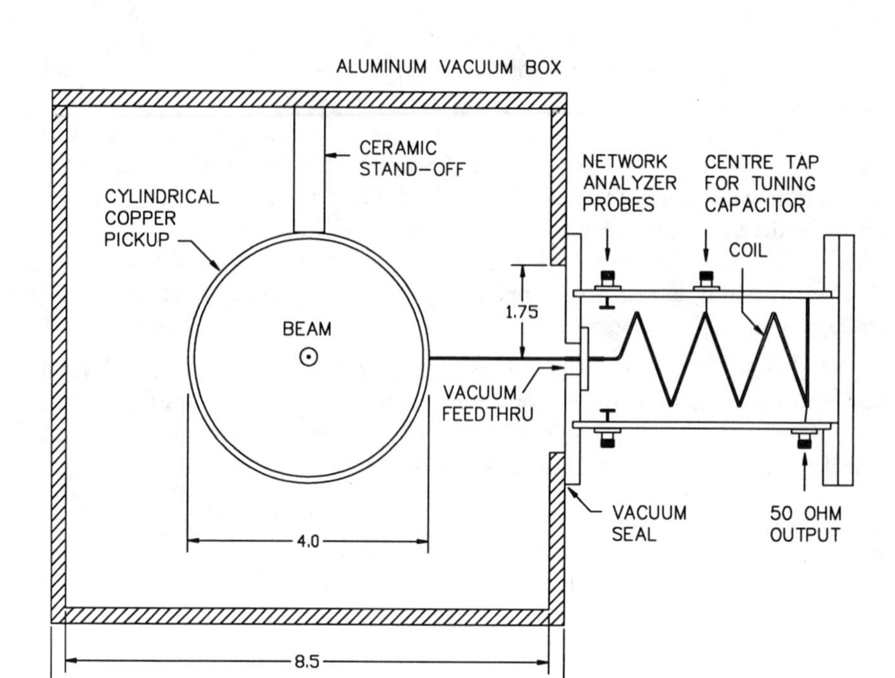

Figure 13. The low current phase monitor consists of a capacitive pickup and an external inductor. The dimensions are inches.

Recently a resonant beam phase monitor, incorporating ideas from an IUCF design (13), has been installed in a low current beamline. The monitor consists of a capacitive pickup with about 20 pF of stray capacitance to ground and an external inductor, fig. 13. The coil is tapped at its 50 Ω point to provide a signal output. The signal is amplified and the passes through 170 ft of heliax to the phase locked loop (PLL) electronics. The device resonates at the second harmonic of the bunch frequency where the RF background from the dees is small. The Q of the loaded system, 144, raises the impedance of the loaded pickup to 25 kΩ, greatly increasing the signal power for a given beam current. The filling time (the loaded Q over the resonant frequency) is about 3 μs.

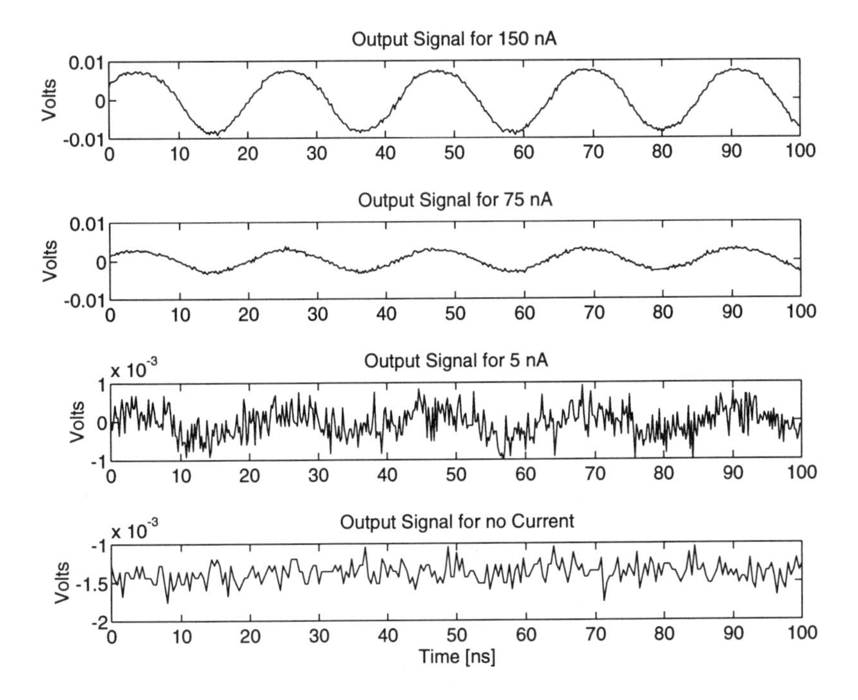

Figure 14. The amplified and averaged signal from the monitor for various beam currents.

The output signal from the monitor, following a 20 dB amplifier at the device and another in the electronics racks, is shown in fig. 14. The signals were recorded using a digital oscilloscope set to average over 256 samples. The monitor sensitivity is 0.33 μV/nA. The background RF is equivalent to only a few nA of beam.

The PLL electronics, fig. 15, are used to reduce the bandwidth of the system in order to reduce the effects of thermal and amplifier noise. The signal from the dees is used as a source of the fundamental frequency. The phase comparison is performed at the second harmonic so a frequency doubler circuit is required. The time constant of the PLL can be controlled by varying the size of the integrator feedback capacitor. Increasing the time constant reduces the beam current required to obtain a lock, but also decreases the frequency of the highest component of the phase jitter that can be tracked.

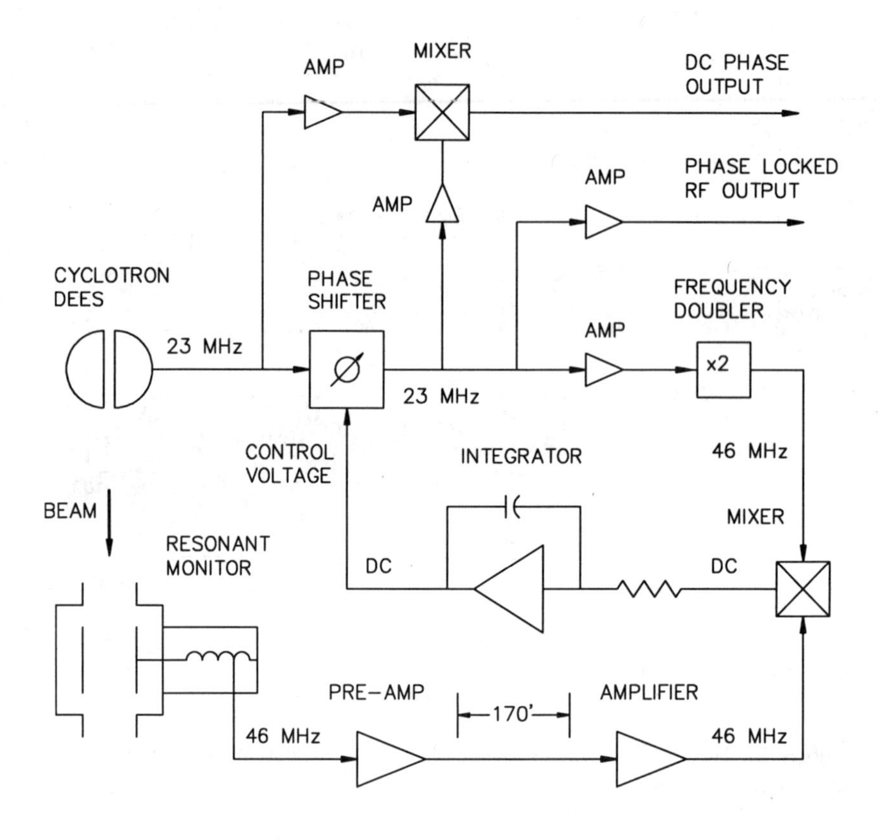

Figure 15. A block diagram of the phase locked loop circuitry.

Radiation Dosimetry Using Solid State Diodes

A proton therapy facility is in the testing stages. A 1 to 10 nA beam with energy between 65 to 110 MeV passes through a mylar window into air. On the vacuum side a secondary emission monitor consisting of 5 0.0003 in. aluminum signal foils and 5 bias foils at +100 V is used to detect beam bursts exceeding 10 nA and is part of the safety system. The monitors on the air side consist of a 16 x 16 wire multiwire ion chamber with a 3 mm wire spacing using a -300 V collection field, two total current ion chambers and a quadrant ion chamber all operating in dry air. The ion chamber foils are 0.001 in. Ni-Cu coated kapton spaced 0.25 in. and biased at +300 V. In addition, an insulated copper beam stop on a fast shutter provides signal when inserted.

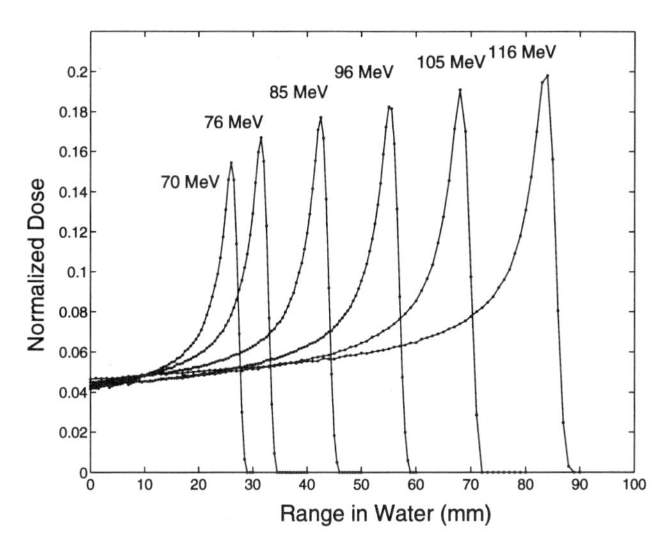

Figure 16. The location of the Bragg peak measured using silicon diodes is a function of the beam energy.

Common solid state diodes have been found to be useful for dosimetry.(14) Measurements at TRIUMF use a 1N4005 silicon diode, preconditioned by a dose of 0.1 Mrad. It has a disadvantage; its epoxy casing degrades the energy of the protons before they reach the silicon junction and offsets the measurements. In future, a planar junction which can be exposed directly will be tested. Care must be taken to prevent light from reaching the junction and causing interference.

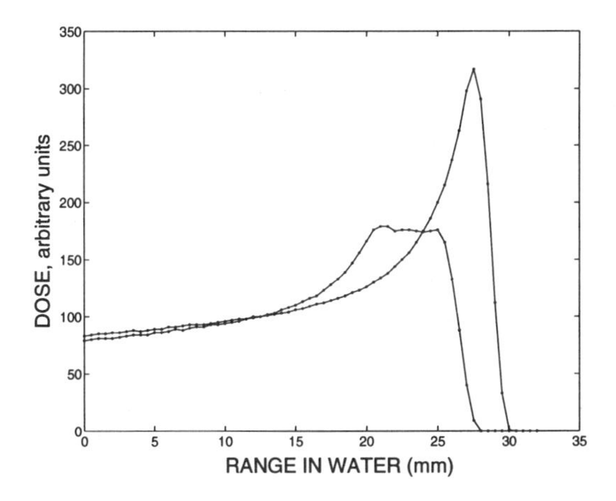

Figure 17. A modulator can be used to spread the Bragg peak along the beam path.

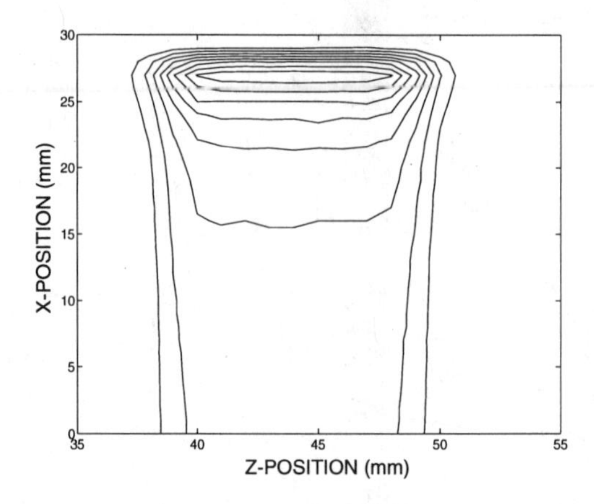

Figure 18. The Bragg peak can be spread laterally by using a diffuser.

The current from the diodes is measured with zero bias voltage. The diode current is converted to a NIM pulse train using either an Ortec 439 current integrator or a current amplifier and a V/F converter.

Figure 16 was produced by moving a diode in a water tank along the beam direction at about 1 nA of beam current. The location of the Bragg peak clearly varies with beam energy. In fig. 17 a plastic wheel of varying thickness was rotated in the beam to spread the Bragg peak along the beam path. The diode was raster scanned in a horizontal plane containing the beam path (which is parallel to the x axis) to produce a contour plot showing the uniformity of a diffused beam and the lack of edge effects which can be caused by collimation, fig. 18.

SCANNING WIRE AND BLADE MONITORS

Two types of wire scanners have been used in the primary proton beamlines. The first design, fig. 19, used a cam and follower to convert rotary motion from an electrical motor to a linear motion at the wires.(15) In a low current beamline aluminum blades 0.375 in. long in the beam direction are used. Since secondary electrons are emitted from a thin surface layer only microns thick, the signal current increases in proportion to the blade length in the beam direction. Good profiles of a CW 15 nA beam have been obtained. Sub nanoamp profiles have been measured by detecting particles scattered from the blade.

The blades are replaced by 0.005 in. dia. Au plated Mo wires in high current beamlines. A monitor at a location that images the stripping foil has been used with beams from 1 μA (at reduced duty cycle) to 150 μA CW. A profile recorded while using a stripping foil which had been damaged by beam during normal

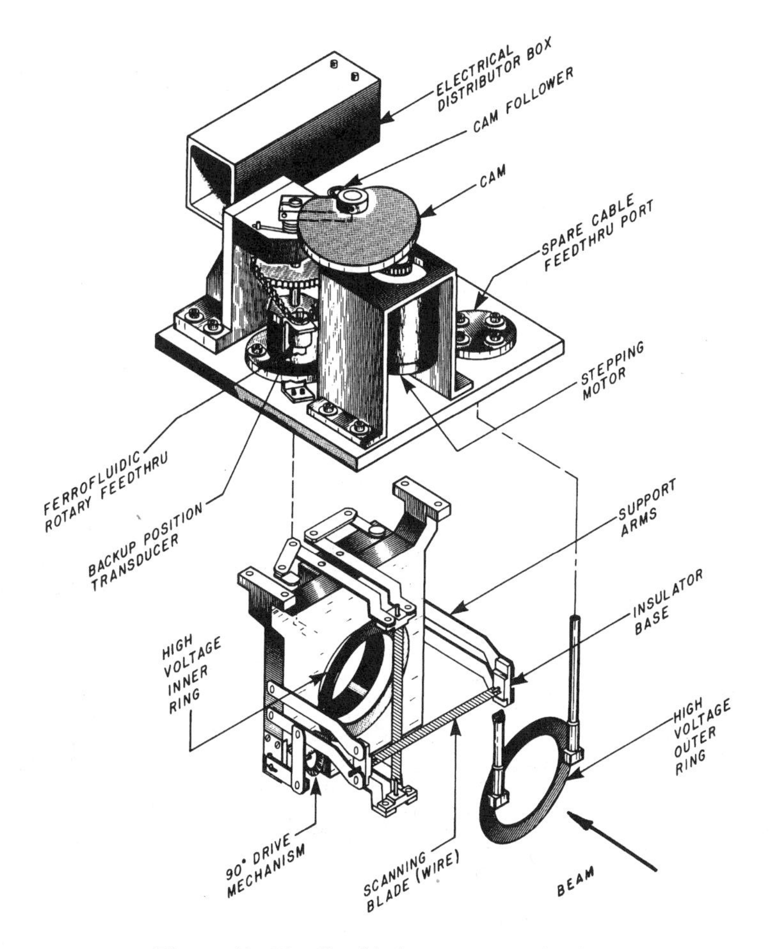

Figure 19. The first blade scanner mechanism.

running is shown in fig. 20. The Y profile has a distinct notch rather than its usual approximately Gaussian shape.

A periscope and lighting system allows visual inspection inside the cyclotron tank without breaking the vacuum. Figure 21 shows a picture of the damaged foil taken through the periscope telescope using a CCD camera and a digital image capture system. To the left of the round object in the centre (a structural feature on the tank wall) is the foil. The left and top sides of the foil are mounted in an "L" shaped frame hanging from the extraction probe's azimuthal arm. The pyrolytic graphite of the foil, normally a perfect rectangle, has cracked and the upper shard has bent out of the normal plane of the foil. In another case, foil damage has presented as a notch in the X profile. Profiles are attached to each shift log to track the foil condition.

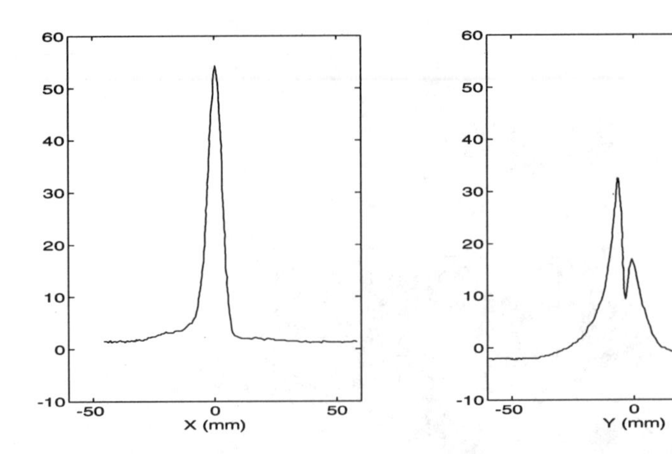

Figure 20. The X and Y profiles of the damaged extraction foil.

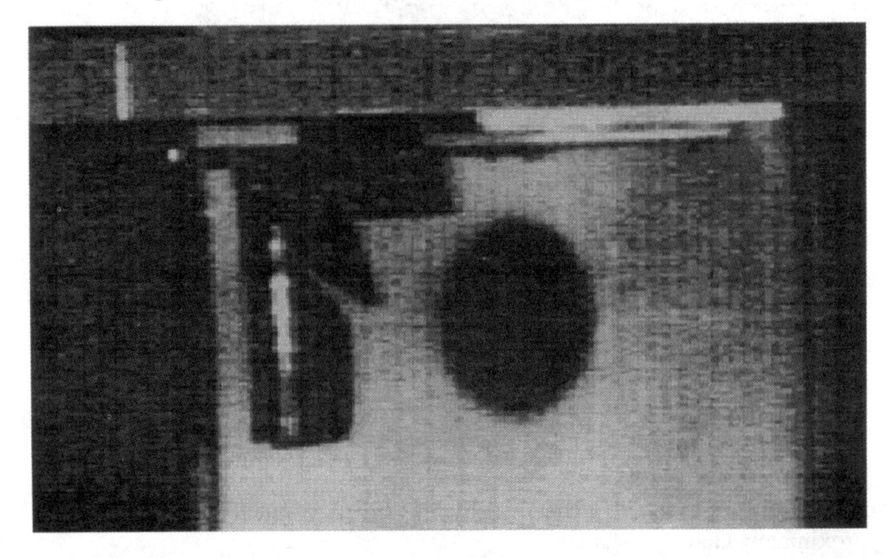

Figure 21. The damaged foil viewed through the periscope. The view is in a vertical–approximately azimuthal plane. The foil is rectangular with a tear on the right hand side.

A newer profile monitor design uses an air cylinder to drive the wires at 45° through the beam, fig. 22. A prototype of this device, fitted with 0.25 in. long Be-Cu blades, has been in use for several years in a low energy beamline to provide profiles of 65 to 110 MeV beams at 1 μA to 60 μA CW. One of the new monitors, fitted with wires, has been constructed and has been initially tested with beam. Four more of these monitors are under construction and three of them will be used to measure the emittance in our high current beamline.

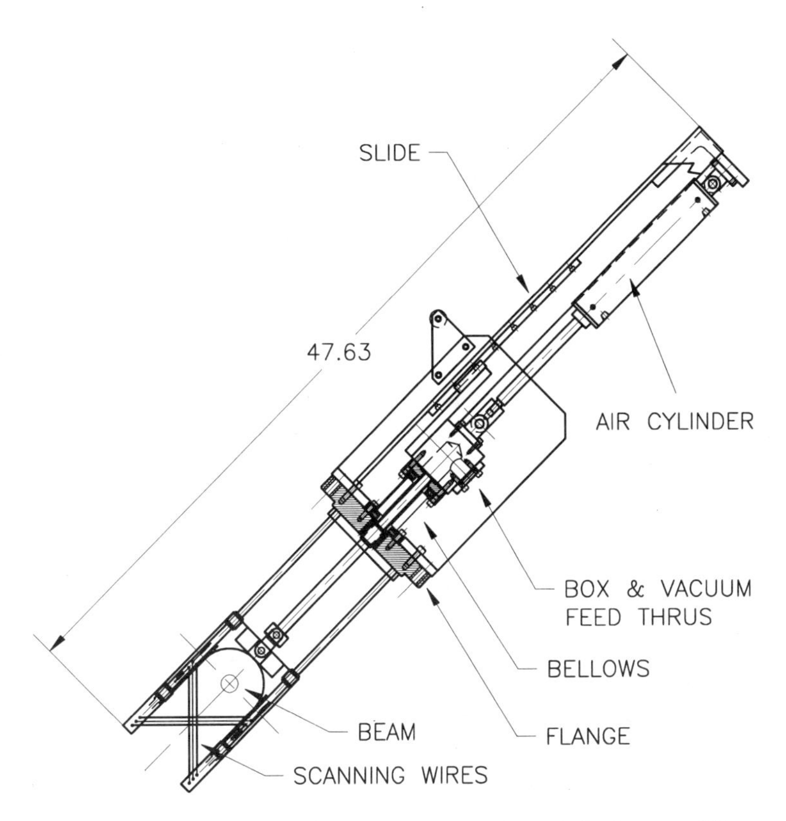

Figure 22. The air driven wire scanner passes through the beam at 45°.

CONCLUSION

Though TRIUMF has been in operation for almost 20 years, beam diagnostics are still being improved and replaced. These changes are driven by the needs of experimenters for improved beam availability, tighter tolerances on beam properties and new beamline configurations. New beam applications such as isotope production and proton therapy require new types of monitors.

ACKNOWLEDGMENTS

Luigi Rezzonico (PSI) designed the crossed wire head and Richard Lee adapted the tomography algorithm. Mike Mouat wrote the PSU program. Ewart Blackmore, John Vincent and the proton therapy group measured the Bragg peaks. Tom Ries, John Stewart, Derick To and Siggy Turke contributed to the wire scanner designs. Andy Hurst recorded the foil image. Tim Ellison (IUCF) adviced on the of the low current phase monitor design and Shane Parfitt constructed the device. George Mackenzie instigated and encouraged much of this work.

REFERENCES

1. Mackenzie, G.H., "Beam Diagnostic Techniques for Cyclotrons and Beam Lines", 8th International Conf. on Cyclotrons and Their Applications, 1978
2. Mackenzie, G.H., "Beam Instrumentation for an H$^-$ Cyclotron Meson Factory", Proc. of the Workshop on Advanced Beam Instrumentation", Vol. **2**, (Tsukuba, 1991), pp. 443-452
3. Ries, T.C., "Mechanical Design Control and Implementation of a New Movable Diagnostic Probe for the TRIUMF Cyclotron", Beam Instrumentation Workshop, AIP Conf. Proc. 319, (Sante Fe, 1993), p. 265
4. Rawnsley, W.R., "Timing and Current Measurement from the TRIUMF Cyclotron Probes", Proc. of the Workshop on Advanced Beam Instrumentation, Vol. **1**, (Tsukuba, Japan, 1991), pp. 339-347
5. Rezzonico, L., Adam, S., Humbel, M., "Diagnostics for High Intensity Beams", this workshop
6. TRIUMF Annual Report of Scientific Activities, p. 153, 1992
7. Yokoyama, I., Kase, M., "A Beam Phase Meter for RIKEN Ring Cyclotron", this workshop
8. Rawnsley, W.R., Howard, G.E., "Improving the Efficiency of the Design of Cylindrical Stripline Beam Monitors", Beam Instrumentation Workshop, AIP Conf. Proc. 319, (Sante Fe, 1993), pp. 255-264
9. Shafer, R. E., "Characteristics of Directional Coupler Beam Position Monitors", IEEE Particle Accelerator Conference, Vol. **NS-32**, No. 5, (Vancouver, 1985), pp. 1933-1937
10. Jachim, S.P., Webber, R.C., Shafer, R.E., "An RF Beam Position Measurement Module for the Fermilab Energy Doubler", IEEE Trans. on Nuclear Science, Vol. **NS-28**, No. 3, June 1981, pp. 2323-2325
11. Yin, Y., Rawnsley, W.R., Mackenzie, G.H., "Measurement of Transverse Geometric Moments of the TRIUMF Beam with a Multistrip Monitor", Proc. of the European Particle Accelerator Conf., 1994
12. Miller, R.H., Clendenin, J.E., James, M.B., Sheppard, J.C., "Nonintercepting Emittance Monitor", 12th Int. Conf. on High Energy Acc., 1983, pp. 602-605
13. Ellison, T.J.P., "Nondestructive Diagnostics for Measuring the Phase, Position, and Intensity of 15 enA Beams from the IUCF Cyclotron", Proc. 11th Int. Conf. on Cyclotrons and their Applications, 1987.
14. Koehler, A.M., "Dosimetry of Proton Beams Using Small Silicon Diodes", Radiation Research Supplement 7, 1967, pp. 53-63,
15. Rawnsley, W.R., Ries, T.C., and Mackenzie, G.H., "A Scanning Secondary Emission Monitor", IEEE Particle Accelerator Conf., 1987, pp. 553-555

Radiation Effects in Accelerator Components

M.J. Borden
Los Alamos National Laboratory, MS H838, Los Alamos, NM, USA 87545

Abstract

A review of basic radiation effects is presented. The fundamental definitions of radioactivity are given for alpha, beta, positron decay, gamma-ray emission and electron capture. The interaction of neutrons with material is covered including: absorption through radiative capture, neutron-proton interaction, alpha particle emission, neutron-multi-neutron reactions and fission. Basic equations defining inelastic and elastic scattering are presented with examples of neutron energy loss per collision for several elements.

Photon interactions are considered for gamma-rays and x-rays. Photoelectric collisions, the Compton effect and pair production are reviewed. Electron-proton interactions are discussed with emphasis placed on defect production. Basic displacement damage mechanisms for photon and particle interaction are presented. Several examples of radiation effects to plastics, electronics and ceramics are presented. Extended references are given for each example.

Optical and X-ray Imaging of Electron Beams Using Synchrotron Emission *

Mark Wilke
Los Alamos National Laboratory, Los Alamos, NM 87545

Abstract

In the case of very low emittance electron and positron storage ring beams, it is impossible to make intrusive measurements of beam properties without increasing the emittance and possibly disrupting the beam. In cases where electron or positron beams have high average power densities (such as free electron laser linacs), intrusive probes such as wires and optical transition radiation screens or Cherenkov emitting screens can be easily damaged or destroyed. The optical and x-ray emissions from the bends in the storage rings and often from linac bending magnets can be used to image the beam profile to obtain emittance information about the beam. The techniques, advantages and limitations of using both optical and x-ray synchrotron emission to measure beam properties are discussed and the possibility of single bunch imaging is considered. The properties of suitable imagers and converters such as phosphors are described. Examples of previous, existing and planned applications are given where available, including a pinhole imaging system currently being designed for the Advanced Photon Source at Argonne National Laboratory.

INTRODUCTION

The imaging of synchrotron radiation provides a useful alternative to invasive techniques for determining electron and positron beam profiles. In the case of storage ring beams, the use of wires, or Cherenkov or optical transition screens will disrupt the beam. High-intensity electron beams in high duty factor electron linacs, such as those used for free electron laser applications, will quickly damage the screens or wires. The synchrotron imaging technique also provides a direct measure of the electron beam profile, as opposed to methods such as beam collision techniques (which requires a second beam or a probe beam) or estimates of beam diameters from beam lifetime calculations.

Synchrotron light has been considered for use as a beam diagnostic since the first observation in 1947(1). As higher energy light sources have been constructed, diagnostic methods have been developed using visible through x-ray radiation. Techniques tend either to measure the distribution and direction of the angular distribution of the synchrotron radiation to infer a beam emittance or to directly

* Work performed under the auspices of the Department of Energy

image the beam bunch using an appropriate optical technique for the type of emission being imaged(2−7).

This paper discusses the imaging technique in both the visible-ultraviolet and the x-ray portions of the spectrum for measuring beam profiles. Included are descriptions of the methods, advantages, and limitations of the technique.

OPTICAL SYNCHROTRON RADIATION IMAGING

Analytical Optical Resolution Estimates

The following derivations of resolution use the definitions of Jackson for critical energy E_c

$$E_C = \frac{h\omega_C}{2\pi} = \frac{3h\gamma^3}{2\pi}\left(\frac{c}{\rho}\right) \tag{1}$$

or, equivalently, the critical wavelength λ_c given by

$$\lambda_c = \frac{2\pi\rho}{3\gamma^3} \tag{2}$$

where ω_c is the critical wavelength, $\gamma=1/(1-(v/c)^2)^{1/2}$, ρ is the radius of the bend defined in Fig. 1, c is the speed of light, and h is Planck's constant(8). It should be noted that Jackson's value of E_c is two times the value used by many other authors.

Figure 1 is a schematic of an optical imaging system. An optical element of aperture diameter D images the synchrotron radiation emitted by particles turning the bend of radius ρ at approximately a distance l_s from the bend through an angle of $\theta \sim D/l_s$. Particles emit radiation of a given wavelength λ or equivalently energy E into a cone of angle θ_c given by

$$\theta_c \cong \frac{1}{\gamma}\left(\frac{E_c}{E}\right)^{1/3} = \frac{1}{\gamma}\left(\frac{\lambda}{\lambda_c}\right)^{1/3} \qquad E_c \gg E, \tag{3a}$$

$$\theta_c \cong \frac{1}{\gamma} \qquad E_c \cong E, \tag{3b}$$

$$\theta_c \cong \frac{1}{\gamma}\left(\frac{E_c}{3E}\right)^{1/2} = \frac{1}{\gamma}\left(\frac{\lambda}{3\lambda_c}\right)^{1/2} \qquad E_c < E. \tag{3c}$$

As shown in Fig. 1, for large D, the particle appears to have a horizontal width of $\Delta\rho \cong \rho\theta^2/8$ as it turns the bend. The particle is observed for a time $t_D = \rho\theta/c$. When

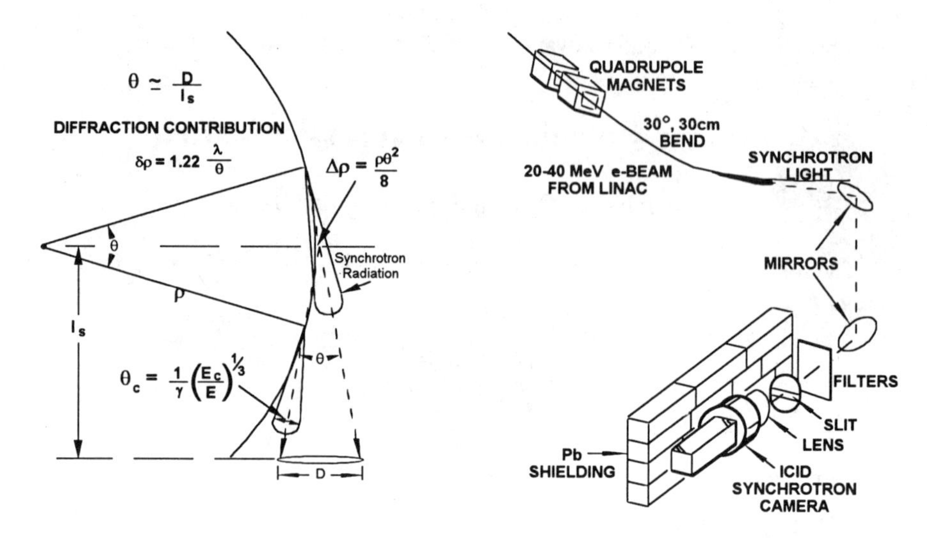

Figure 1. Schematic of optical system for imaging synchrotron radiation.

Figure 2. Schematic of the APEX linac visible synchrotron emission imaging system.

D is small and the wavelength of observation λ is long, diffraction becomes important and the particle appears to have an effective width $\delta\rho = 1.22\lambda/\theta$. The resolution of the imaging system resulting from the combined effects of the particle motion and diffraction is then given approximately by

$$R = \sqrt{(\Delta\rho)^2 + (\delta\rho)^2} = \sqrt{\left(\frac{\rho\theta^2}{8}\right)^2 + \left(1.22\frac{\lambda}{\theta}\right)^2}\,. \tag{4}$$

The optimum value of θ defined as θ_o, and therefore the optimum value of D given by $D_o = l_s\theta_o$, is found by minimizing R with respect to θ, and is given by

$$\theta_o = 1.9\left(\frac{\lambda}{\rho}\right)^{1/3}. \tag{5}$$

Therefore, the optimized resolution, R_o, is given by

$$R_o = 0.78\left(\rho\lambda^2\right)^{1/3} = 0.78\,\rho^{1/3}\left(\frac{hc}{E}\right)^{2/3}. \tag{6}$$

In order to compare θ_o with θ_c, we can note that the optics system will image at wavelengths in the visible so that $\lambda \gg \lambda_c$ ($E \ll E_c$). Substituting Eq. 2 for λ_c in Eq. 3a gives

$$\theta_c = \left(\frac{3}{2\pi}\frac{\lambda}{\rho}\right)^{1/3} = 0.78\left(\frac{\lambda}{\rho}\right)^{1/3}, \tag{7}$$

which by Eq. 4 gives the resolution R_c of an optical system when the aperture is $D = l_s\theta_c$

$$R_c = 1.56\left(\rho\lambda^2\right)^{1/3} = 1.56\,\rho^{1/3}\left(\frac{hc}{E}\right)^{2/3}. \tag{8}$$

The resolution therefore appears to be worse than R_o by a factor of two. The instantaneous cone implies a resolution given by Eq. 8. However, analogous to the enhancement of resolution in synthetic aperture radar, the horizontal resolution of an image produced by the synchrotron radiation beam scanning the whole aperture is given by Eq. 6 and not Eq. 8. The approximate vertical resolution is given by Eq. 8.

Above E_c, the emitted synchrotron power per unit frequency falls rapidly. We may define an approximate limit to the achievable resolution of an optics system, which is determined by the usable signal level. When $E \cong E_c$, so that from Eq. 3b $\theta \sim 1/\gamma$, the major contribution to R is from the diffraction term $\delta\rho$. If one substitutes $\Delta\rho$ and $\delta\rho$ in terms of θ into Eq. 4 and sets $\theta = 1/\gamma$ and $\lambda = \lambda_c$ as given by Eq. 2, then the defined limiting resolution R_l of an optics system that images a synchrotron beam at the critical energy is given by

$$R_l = 2.6\frac{\rho}{\gamma^2}. \tag{9}$$

There is also a technological limit to the achievable resolution, because as λ becomes shorter, it becomes more difficult to produce diffraction-limited optics.

Table 1 lists values of several of the parameters defined above, including the resolution of an optical system imaging the beam at $\lambda = 500$ nm for two cases: a 30 cm bend in a linac where the beam energy $E_b = 40$ MeV and a 3900 cm bend in a high energy synchrotron where $E_b = 7000$ MeV.

In the first case, $\lambda_c > \lambda = 500$ nm, and Eq. 3c was used to calculate θ_c. This value of θ_c was then used to derive an equation for R_c in a way similar to that used in deriving Eq. 8. In the second case, where $\lambda \gg \lambda_c$, Eq. 3a was used to calculate θ_c and Eq. 8 was used to calculate R_c. From Table 1 it can be seen that for the first case, θ_o is 4.6 times larger than θ_c. One can therefore expect an improvement in the horizontal resolution of the beam by using a lens of aperture $D \approx l_s \theta_o$,but the vertical beam resolution will be diffraction limited by θ_c to R_c=133 μm.

Table 1. Calculated parameters of a system imaging an electron beam at λ=500 nm.

γ	λ_c (nm)	E_c (eV)	θ_o (rad)	θ_c (rad)	t_D (ps)	R_o (μm)	R_c (μm)	$R_l \approx 2.6\rho/\gamma^2$ (μm)
			E_b =40 MeV		ρ = 30 cm			
79	1261	0.984	0.023	0 .005	23	30	133	95
			E_b=7000 MeV		ρ = 3900 cm			
13700	0.032	39058	0.004	0 .002	578	151	302	0.42

As defined here, θ_c is conservative. The angular distribution of the synchrotron radiation has a large portion of the total power in the wings. Also in the first case (where $\lambda_c > \lambda$), the value of θ_c more nearly represents the value from the peak to the 1/*e* value of the distribution, and not the full-width at half-maximum (FWHM) value. The value of θ_o is only twice as great as that of θ_c in the second case.

There are several relevant time scales. Imaging systems such as cameras typically have interrogation times on the order of 10^{-2} to 10 seconds, and are therefore time integrating devices. Linac macro-pulse lengths and synchrotron orbit times are on the order of microseconds while the micro-pulse separations may be 10^{-8} to 10^{-7} seconds. Micro-pulse lengths may be in the picosecond regime, but the use of synchrotron light to observe micro pulse is limited by the time the synchrotron light illuminates the imaging system (t_D), which is typically greater than micro-pulse lengths, as seen in Table 1.

Finally, we note in the second case that the resolution of 151 μm at 500 nm can be seen to be considerably worse than the 0.42 μm that would be achievable if diffraction-limited optics were available at the critical energy of 30 keV.

Optical Imaging of a Linac Beam

Figure 2 shows a system used to image an electron beam in a bend of the APLE Protoype Experiment (APEX) via optical synchrotron emission. APEX was a free electron laser driven by a linac which could be run at 20 to 40 Mev(9). The beam format typically consisted of 10 to 15 psec, 1 to 5 nC micro pulses from a photo injector spaced that were typically 47 nsec apart. The macro pulse could be varied from a single micro pulse to a pulse 100 μs long containing a maximum of 10 μC.

The beam was directed to the laser wiggler by a 60 deg achromatic, isochronous bend composed of two 30 deg bending magnets. The beam was matched to the bend by focusing the beam to a horizontal waist halfway through each of the 30 cm radius, 30 deg bending magnets.

Glass windows were located at each bend in order to determine the beam profile and trajectory via the synchrotron emission. A periscope configuration of mirrors relayed the light to the shielded intensified CID camera (ICID). The sensitivity of the S20-response ICID at full intensifier gain was 1.2×10^9 V/J/cm^2 at 500 nm with a camera noise level of 10 mV. The view was limited by the lens aperture to 48 mrad in the vertical direction and to 5 mrad in the horizontal direction by the 0.5 cm slit placed in front of the lens. Therefore, the horizontal resolution was diffraction limited to 133 μm. The slit was originally selected to match θ_c. As Table 1 indicates, matching θ_o would have given 30 μm resolution. The camera recorded images on video tape, and the images were later digitized for analysis.

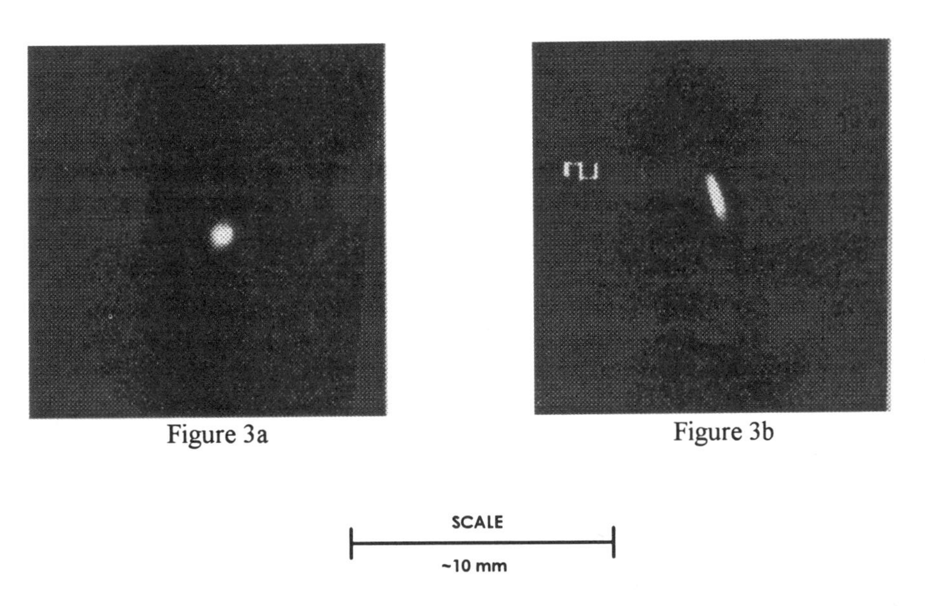

Figure 3a

Figure 3b

Figure 3. Two synchrotron emission images of the Linac beam. a) Beam focused in both directions. b) Beam with an optimal horizontal focus.

Two images from the camera are shown in Fig. 3. In Fig. 3a, the 38 MeV beam was focused in both the horizontal and vertical directions. The FWHM was 765 μm in both directions. In Fig. 3b, the beam was optimally focused in the horizontal direction, which resulted in a horizontal FWHM of 382 μm and generally produced better lasing. Note that the beam is slightly tipped. The major axis of the elliptical beam in Fig. 3b was 1.6 mm.

To determine how low a beam energy this diagnostic would be useful for, images were taken as E_b was varied. Figure 4 is a plot of the camera peak signal normalized to the total macro-pulse charge. For this camera, the noise floor was about 4×10^3 V/C. The signal was also calculated from the synchrotron energy radiated per unit frequency per solid angle, knowing the geometry and response of the imaging system. The calculated response is given as the dashed curve in Fig. 4. The signal decreases abruptly at low E_b, showing that imaging synchrotron radiation in the visible range is no longer practical for these beam conditions for beam energies below about 23 MeV.

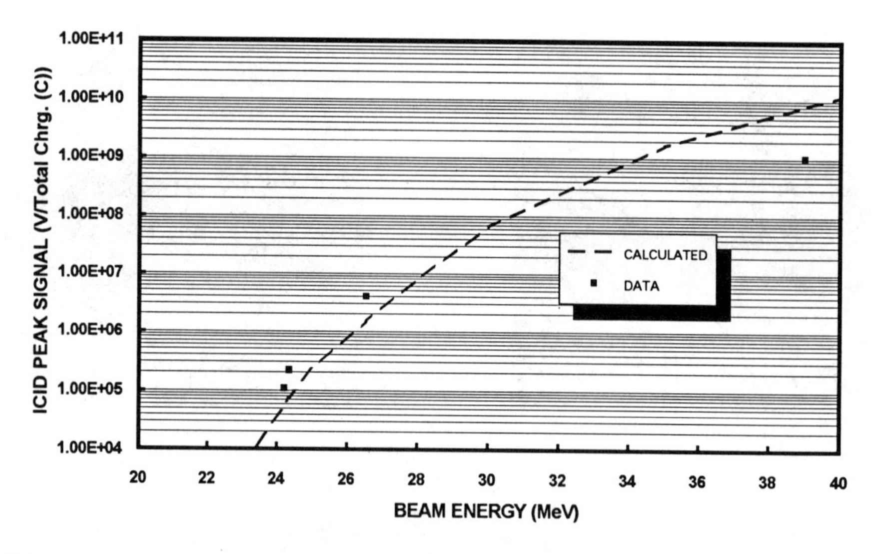

Figure 4. Plot of charge-normalized synchrotron image signal vs. beam energy.

X-RAY SYNCHROTRON RADIATION IMAGING

Although it is difficult to construct diffraction-limited optical elements that are useful at high x-ray energies, it is possible to build pinholes for imaging. In the case of pinhole imaging, there is a tradeoff between geometrical blurring due to the finite size of the pinhole and loss of resolution due to diffraction. By imaging at high energies, diffraction is minimized. This permits the use of pinholes small enough to overcome geometrical blurring. Therefore, in order to have adequate signal, the use of pinhole imaging of high-brightness beams is limited to high-energy, high-current synchrotrons.

Analytical X-Ray Pinhole Resolution Estimates

Figure 5 illustrates the geometrical blurring of the image due to the finite size of the pinhole. Two points within the beam, separated by ΔR, generate two circular images of radii r_i at the image plane when projected through a pinhole of diameter D. The two images are defined as resolved when separated by a distance $\Delta R_i \geq C r_i$ where C is a constant between 1 and 2 defined by some resolution criterion. Here we define ΔR_i as the distance between the two circles when the secant of the intercept region of the two circles is equal to r_i, as shown in Fig. 5. Therefore $C = \sqrt{3}$. It can be seen from Fig. 5 that the various parameters can be related by

$$\Delta R = \frac{l_o \Delta R_i}{l_i} = \frac{C r_i}{M} = \frac{1+M}{M} CD \ , \tag{10}$$

where $M = l_i / l_s$ is the magnification of the pinhole.

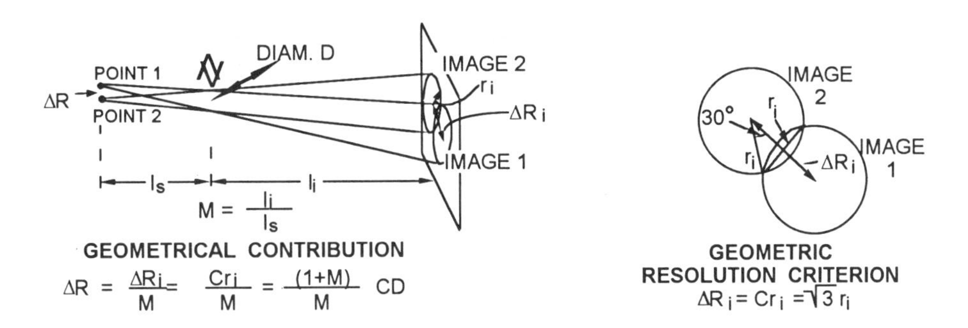

Figure 5. Schematic defining various pinhole imaging parameters.

The diffraction contribution to the blurring of the image is again $\delta\rho = 1.22\lambda/\theta = 1.22\lambda l_s /D = 1.22hc/ED$. There is also a contribution equivalent to $\Delta\rho$ from apparent motion of the particle as it rounds the bend while the synchrotron radiation sweeps over the pinhole. For large γ, $\Delta\rho$ is fractions of a micron and will be ignored. Therefore, the resolution R that is possible using a pinhole of aperture D is given by

$$R = \sqrt{\left(\frac{1+M}{M}CD\right)^2 + \left(1.22\frac{hc}{E}\frac{l_s}{D}\right)^2} \quad . \tag{11}$$

The optimized resolution, R_o, for the optimum pinhole diameter D_o, at a given photon energy E, is found by solving for D_o after differentiating Eq. 11 with respect to D_o and setting the result equal to zero. The optimum values are then

$$D_o = \sqrt{1.22\frac{M l_s}{1+M}\frac{hc}{CE}} \quad \text{and} \tag{12}$$

$$R_o = \sqrt{2}\frac{1+M}{M}CD_o = \sqrt{2.44 l_s C \frac{1+M}{M}\frac{hc}{E}} \quad . \tag{13}$$

For reasonable values of M in the range of 1 to 4, there is little M dependence. Both D_o and R_o vary as $E^{-1/2}$, so that imaging at higher energies yields a corresponding improvement in resolution. Table 2 lists values of D_o and R_o for different values of E and M for l_s=1400 cm. Comparison with Table 1 shows that a pinhole imaging system that images a high-energy synchrotron will have improved resolution compared to a system imaging at 500 nm.

Table 2. Values of D_o and R_o for different values of E and M for l_s=1400 cm.

	E=39058 eV		E=80000 eV	
	M=1	M=4	M=1	M=4
D_o	12.5	15.8	8.7	11.0
R_o	61.2	48.4	42.8	33.8

Experimental considerations

Figure 6 is a schematic of a synchrotron pinhole imaging system showing the various elements involved.

Pinholes with the optimum pinhole diameters given in Table 2 are technologically challenging to fabricate. Pinhole imaging, at the higher x-rays energies that are required to achieve better than 100 μm resolution, requires high-Z pinholes of several millimeters thickness. Crossed slits may also be used in place of a pinhole. A crossed-slit beam diagnostic is currently under test at the Brookhaven National Light Source.(5)

High aspect Au pinholes 50 μm in diameter and several inches long for neutron pinhole imaging have been produced at Los Alamos National Laboratory by plating Au on Cu wire and acid etching away the wire mandrel(10). It is difficult to obtain round wires with diameters on the order of 10 μm for use as mandrels, but some manufactures will supply short lengths suitable for pinhole fabrication. Plasma etching techniques can be used for reducing the diameter of thicker wires. Carbon wires with 8 μm diameters are commercially available but tend not to be round. Carbon wires require a sputtered metal coating before Au can be chemically plated. A 2.8 mm thick, 10 μm diameter Au pinhole is currently being designed for the Advanced Photon Source (APS). The APS pinhole will be fabricated using the mandrel technique by plating the Au on a 10 μm Cu wire.

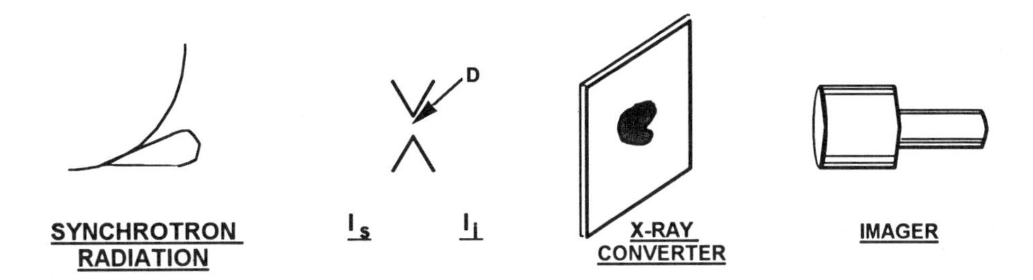

Figure 6. A schematic of a synchrotron pinhole imaging system.

Characterization of a small diameter pinhole is also difficult. Continuous x-ray tube sources exist that produce source sizes of several microns. Energies are on the order of 150 keV, so diffraction is minimized. However, neither this energy nor source size is sufficient to eliminate consideration of the calibration source dimensions and beam diffraction from the pinhole characterization.

As with any object in a high-energy synchrotron x-ray beam, pinhole heating is a concern. Electroplated Au alloys are typically more sensitive to heat than other Au alloys. The Au begins to distort at temperatures as low as 300 to 400 deg C because of trace impurities. Standard procedures that pre-collimate and pre-filter the x-ray beam and cool the pinhole can all be used to mitigate heating. As an example, finite-element heat transport modeling of pinhole designs was done for the APS pinhole. At 1400 cm, the peak design flux for the APS pinhole is 20 W/mm ^{2}. The x-ray beam was filtered with 5 mm of Be and 2 mm of Al so that the spectrum peaked at 30 keV, and the x-rays were collimated through a 0.5 cm diameter aperture. Calculations gave temperatures below 150 deg C in both the filter stack and pinhole assembly when the assembly was cooled to 23 deg C at a radius of 2 cm. The x-ray beam could be collimated further to reduce the total incident power to less than 1 watt at the pinhole, making it possible to avoid having to liquid cool the pinhole assembly. Extreme collimation of the x-rays, however, makes alignment of the system difficult.

Several possibilities exist for visualizing the pinhole image. Figure 6 shows an image converter such as a phosphor recorded with an electrooptic imager. In all cases, a major concern is the image size. Reasonable magnifications still produce beam images with dimensions of less than 1 mm. Therefore, the inherent resolution of any converter must be considered. A second major concern is the effect of the radiation on the converter. The pinhole geometry and spectral filtering reduces the dose, but the integrated dose can be substantial if the diagnostic is used for long periods. Use of a shutter can help.

Film provides the simplest method of visualization. Due to development time, film is not a real-time method suitable for monitoring beam conditions. A somewhat faster method is to scan the pinhole image plane with a detector. If a slit is used, tomographic views of the beam profile may be obtained. By rotating the slit to obtain several beam profiles, a 2-D image can be generated in almost real time.

Direct electrooptic imaging of the beam is difficult because of the duty factor and energies of the beam. Direct x-ray imaging with charge-coupled-device (CCD) cameras has proven useful at low x-ray energies, and low-cost cameras have now been evaluated(11). Front side CCD illumination is possible above fractions of a keV. Above about 10 keV the CCD becomes increasingly transparent to the x-rays and quantum efficiency decreases rapidly. Further difficulties arise from finite CCD well sizes and from radiation damage over long periods of use. A CCD charge well will accommodate $\gtrsim 10^5$ electrons. One electron-hole pair is generated per 3.65 eV. An 80 keV x-ray can generate $>2\times10^4$ electrons per well and several hits will saturate the well. Therefore, dynamic range is limited. When integrated doses approach several krad, permanent damage becomes apparent even for CCDs using multipin phasing. The CCD noise level begins to increase, further limiting dynamic range(12).

Microchannel plate image intensifiers (MCPIIs) can also be used to detect the x-rays. Generally, the input side of the MCPII is coated with an $\sim 1\ \mu$m thick, high-Z photocathode such as Au. The MCPII can then be turned off and on by controlling the voltage across the MCP. Above about 30 keV, the direct excitation of the phosphor by x-rays penetrating the Au and MCP gives a signal comparable to that from the Au photocathode.

Phosphors imaged with an electrooptic device provide a simple method for observing the x-ray image, but for very small images, phosphor grain sizes typically a few microns may complicate resolution. Organic scintillators also may be used, but in general the organics become nonlinear and sustain radiation damage at lower dose levels than inorganics. Organic scintillators also have a lower Z and therefore, although they are typically fast and more efficient at converting radiation to light, they are more transparent to high-energy x-rays. High-Z loaded liquid scintillators have shown some promise for use as fast, bright, x-ray−to−light converters(13). Because of self-absorption and internal scatter, phosphor screens must be thin. Therefore, phosphors containing high-Z elements such as Gd, Y, and Eu, and scintillators such as CsI are often used. These phosphors generally show long decay times of tens to hundreds of microseconds and often have long lag components on the order of milliseconds(14,15). Some of the inorganics have also been evaluated at high x-ray doses(16). Ruby (Al_2O_3:Cr^{3+}) screens used for direct accelerator beam monitoring are also useful for x-ray applications.

Table 3. Properties of some common phosphors and two fast phosphors.

PHOSPHOR TYPE	PEAK WAVELENGTH	DECAY TIME to 10% (μsec)	POWER EFFICIENCY (W/W)	COLOR
P-11 (BE)	460	70 - 80	0.089	Purple/Blue
P-15 (GG)	390/530	2.6 - 2.8	0.044	Green
P-16 (M)	380	0.12	0.05	UV
P-20 (KA)	560	500[a]	0.14	Yellow/Green
P-22 R	680	1000	0.14	Red
P-RED	680	(See b)	0.063[c]	Red
P-24 (GE)	510	1.5	0.024	Green
P-31 (BH)	520	30 - 35	0.060[c]	Green
P-36 (KF)	550	0.25	(See b)	Yellow/Green
P-46 (KG)	540	0.16	0.010	Yellow/Green
P-47 (BH)	420	0.08	0.07	Purple/Blue
P-48 (KH)	420/540	0.12	0.06	Yellow/Green
WL-1201	395	0.0008	0.011	Purple/Blue
WL-1198	520	0.00013	0.001	Green

NOTES: a - Varies with current density
b - Data not available
c - Data accuracy is questionable

Information regarding phosphors with P-designations can be found in various handbooks(17). Most of the available information is pertinent to kilovolt electron beam excitation, because of its importance to cathode ray tubes. Even so, the information is difficult to correlate. Phosphors often emit at several wavelengths. The various bands of a given phosphor may have different time responses and brightness for different types and levels of excitation. Activator doping levels also influence behavior and may vary, depending on the manufacturer. Phosphors are almost always produced as powders. Methods for forming screens include combining the phosphor with organic binders, settling, plasma spraying, and *in situ* growth. The various methods cause wide fluctuations in a given phosphor's properties, especially for x-ray excitation. Publications typically give relative phosphor properties, especially conversion-to-light efficiencies, for a given type of excitation, screen thickness and geometry. Therefore, it is best to test a phosphor, particularly its efficiency, specifically for a given application. With these caveats, Table 3 is provided, listing some properties of the more common phosphors(18).

Table 3 also contains the properties of two subnanosecond phosphors that have been extensively evaluated for pulsed x-ray spectroscopy applications. The efficiencies listed for the P-designation phosphors are for electron excitation at optimal thickness and beam energy. The efficiencies for the WL-designation phosphors are absolute values of total emitted power into 4π per absorbed x-ray power at 8 keV measured at the Cornell synchrotron. Phosphor grain sizes, which are important indicator of a phosphor's resolution, are not listed. Grain sizes are typically on the order of one to several microns. Screen resolutions are also a function of the coating method used, because some techniques lead to clumps in the deposition.

Table 4. Properties of WL-1201 and WL-1198

	WL-1201	WL-1198
COMPOSITION	ZnO:Ga (.003 mole Ga)	CdS:In (.001 mole In)
DENSITY	5.61 g/cm^3	4.82 g/cm^3
EMISSION PEAK	390 nm	520 nm
FWHM	18 nm	20 nm
TIME RESPONSE:		
FWHM	730 ps	150 ps
1/e DECAY	800 ps	130 ps
X-RAY CROSS SECTIONS:		
10 keV	127 cm^2/g	102 cm^2/g
80 keV	0.6 cm^2/g	1.5 cm^2/g
OPTICAL ABSORB. LNG.	242 cm^2/g	150 cm^2/g
GRAIN SIZE	3.5 μm	3-25 μm
LINEARITY AND RAD HARDNESS	GOOD	?

Because there is little published literature on the properties of the WL-designation phosphors and because they are useful for pulse measurements, Table 4 is included which lists measured properties(19, 20). The properties of the WL phosphors can be modified by controlling the percentages of the activators. In particular, time response can be traded for efficiency.

Observation of the emission of the phosphor is a simpler problem than direct x-ray observation. Where time resolution is not an issue, slow, bright phosphors can be imaged with standard television cameras, either vidicons or CCDs. Fast and/or less efficient phosphors require more sensitive cameras for time resolved imaging. High-quality, cooled CCD cameras are now capable of imaging at the tens of photons per pixel level. To resolve individual bunches requires some method of gating the imager, such as a gated MCPII. The MCPII also provides gain control, so that imaging at the several photons per pixel level is possible.

It is possible to estimate signal levels for an imager using the data from Table 3. Figure 7 is an illustration of a phosphor screen being excited by x-rays incident with flux P_{XI} in W/cm^2 (or J/cm^2 for time-resolved single-bunch data). It is assumed that the phosphor grain size is small compared to the incremental excited volume. The intensity, I_o in W/cm^2/sr (or J/cm^2/sr), of the phosphor at the detector location is given by

$$I_o = \frac{P_{Ix}C}{4\pi}\sigma_x \cos\varphi \exp(-\sigma_o\tau/\cos\theta_o) \bullet$$
$$\left\{\frac{1}{\sigma_o\cos\varphi - \sigma_x\cos\theta_o}\left[\exp\left(\tau\frac{\sigma_o\cos\varphi - \sigma_x\cos\theta_o}{\cos\theta_o\cos\varphi}\right) - 1\right] + \right. \quad . \tag{14}$$
$$\left. \frac{-R}{\sigma_o\cos\varphi + \sigma_x\cos\theta_o}\left[\exp\left(-\tau\frac{\sigma_o\cos\varphi + \sigma_x\cos\theta_o}{\cos\theta_o\cos\varphi}\right) - 1\right]\right\}$$

Where R is the reflectivity of an x-ray transparent substrate or reflector, θ_o and ϕ are defined in Fig. 7, σ_o and σ_x are the optical and x-ray mass absorption coefficients in cm^2/g, and τ is the phosphor thickness in g/cm^2. C is the efficiency in W/W (or J/J). The values of σ_o and C listed for the WL-phosphors in Table 4 were determined by fitting intensity data for samples with different values of τ excited by x-rays from a monochrometer at CHESS. When $\theta_o=\phi=0$ degrees and the reflectivity is zero,

$$I_o = \frac{P_{XI}C}{4\pi}\frac{\sigma_x}{\sigma_o - \sigma_x}\left[\exp(-\sigma_x\tau) - \exp(-\sigma_o\tau)\right] \quad . \tag{15}$$

For completeness, when the detector is on the side of the phosphor illuminated by the x-rays so that $90<\theta_o\leqq 180$ deg, and the reflective substrate of reflectivity is on the opposite side, the brightness is given by

$$I_o = \frac{P_{Ix}C}{4\pi}\sigma_x \cos\varphi \bullet$$
$$\left\{\frac{1}{\sigma_o\cos\varphi - \sigma_x\cos\theta_o}\left[\exp\left(\tau\frac{\sigma_o\cos\varphi - \sigma_x\cos\theta_o}{\cos\theta_o\cos\varphi}\right) - 1\right] + \right. \quad . \tag{16}$$
$$\left. \frac{-R}{\sigma_o\cos\varphi + \sigma_x\cos\theta_o}\exp\left(\frac{2\theta_o\tau}{\cos\theta_o}\right)\left[\exp\left(-\tau\frac{\sigma_o\cos\varphi + \sigma_x\cos\theta_o}{\cos\theta_o\cos\varphi}\right) - 1\right]\right\}$$

When θ_o=180 deg, ϕ=0 deg and the reflectivity is zero,

$$I_o = \frac{P_{XI}C}{4\pi}\frac{\sigma_x}{\sigma_o + \sigma_x}\left\{1 - \exp\left[-\tau(\sigma_o + \sigma_x)\right]\right\} \quad . \tag{17}$$

Equation 15 gives the intensity at a detector located on the back side of a front-side, x-ray–exposed phosphor screen. Equation 15 would be applicable to a situation where the phosphor was coated on an optically transparent substrate with the phosphor toward the x-ray beam and the substrate toward the detector. Equation 17 gives the intensity at a detector observing the x-ray illuminated side of the phosphor at normal or near-normal incidence.

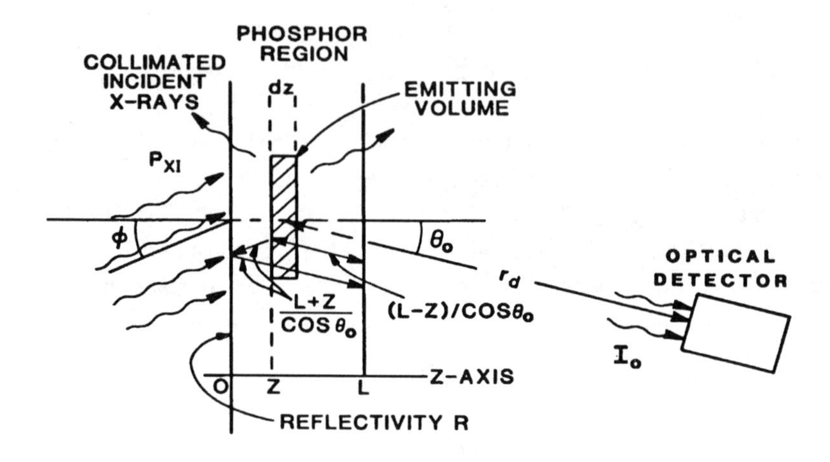

Figure 7. Geometry of x-ray–excited phosphor and detector.

If s is the sensitivity of an imager (such as a camera) to an optical flux F in, for example, V/W/cm^2 (or V/J/cm^2), the peak signal S in volts of a system imaging the phosphor with a lens of f-number f is given by

$$S = sF = s\frac{\pi}{4}\frac{1}{f^2}I_o \quad . \tag{18}$$

As an example, we can now estimate the peak camera signal for a system imaging the APS beam, where E_b=7000 MeV and ρ = 3900 cm. It is assumed that the camera has an f/2 lens viewing the back side of a WL-1201 screen which is imaging the beam at a magnification of 4 through a 10 μm pinhole located 1400 cm from the beam. The spectrum is filtered to peak at 80 keV through 0.5 cm of Be, 0.16 cm of Al, and 0.035 cm of Mo. The filters reduce the total beam power by a factor of 22.6. During the initial operation, APS will run a 100 mA beam of 20 bunches at the ring frequency of 270 kHz. The initial beam will be ~100 μm in diameter. The 5.4 $\times$ 10^6 pulses per second will produce 910 W/cm^2 at the pinhole. After passing through the pinhole and filters, the incident x-rays will produce an image at the phosphor with a peak x-ray flux P_{XI} = 2.5 $\times$ 10^{-2} W/cm^2.

Figure 8 is a plot of F/P_{XI}. F/P_{XI} is plotted both for a camera system imaging from the side opposite the x-ray illumination (I_o given by Eq. 15) and for a camera system imaging from the same side as the x-ray illumination (I_o given by Eq. 17). The optimal thickness for the back-side case is 26 mg/cm^2, where the value of F/P_{XI} is 4.37×10^{-7}. The value of F/P_{XI} for the front-side case asymptotically approaches a maximum value of 4.43×10^{-7}.

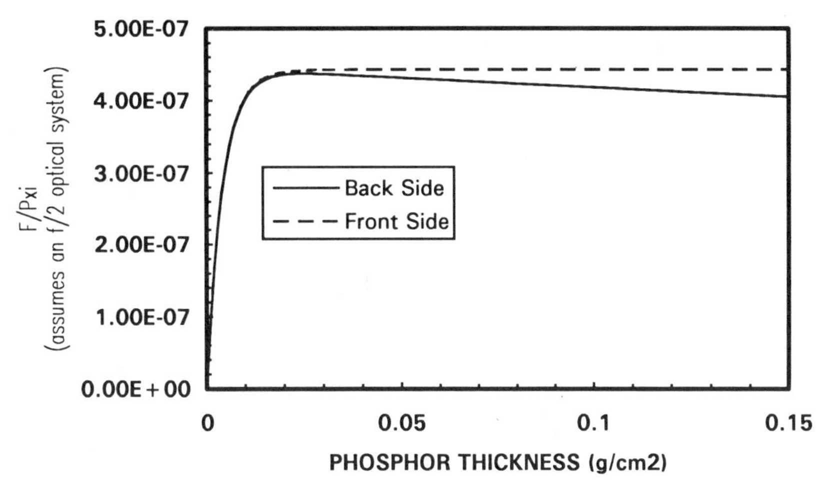

Figure 8. Optical flux at camera target normalized to x-ray flux at WL-1201 phosphor (F/P_{XI}) for a camera observing from the back side or front side relative to the x-ray illumination.

Assuming back-side observation, the optical flux F at the camera target using the optimum thickness of phosphor is 1.1×10^{-8} W/cm^2. Assuming a camera sensitivity of 10^9 V/W/cm^2 (similar to the camera used to image the optical synchrotron emission above), the signal would be 11 V if the camera did not saturate first.

The purpose of using a fast phosphor such as WL1201 is to image a single pulse out of several in a narrow macro pulse. In the case of some modes of operation at APS, there may be more than 20 bunches which would yield $>5.4 \times 10^6$ x-ray pulses per second. The bunches may not necessarily be evenly spaced about the ring, which would decrease the time between x-ray pulses and require nanosecond phosphor response. A single 18.5 nC bunch produces 2.0×10^{-15} J/cm^2 of optical flux at the camera for the conditions described above. This represents about 4000 photons/cm^2 at the camera target at the 400 nm emission wavelength of WL-1201. A camera with 40% quantum efficiency would record only 1600 events/cm^2/bunch. An image formed with these few points would be granular and hard to interpret. In order to do single-pulse imaging it is necessary to use a phosphor more efficient than WL-1201 but still faster than the shortest bunch spacing. One could also increase the solid angle of the x-ray imager.

The phosphor information in this paper is by no means comprehensive and more efficient phosphors may exist. There are several prospects for increasing the solid angle of the x-ray imaging system; these however, have the disadvantage of added complexity over simple pinhole imaging. Larger pinholes may be used if the pinhole transfer function can be accurately measured. The beam can then be imaged and the measured point response of the pinhole can be used to numerically recover the beam profile. This technique requires a small source to characterize the pinhole. Henke-like sources exist that produce source sizes on the order of a few microns at ~100 keV. The source is relatively weak and counting techniques must be used to do the characterization. Pinhole sizes greater than the beam profile can be used to do penumbral imaging, as is done with neutrons, thereby greatly increasing the solid angle(21). Other coded aperture techniques that have an increased solid angle are possible, including annular and zone plate imaging. These techniques require fabrication of a more complicated aperture, and the deconvolution methods often generate low-level image artifacts, which would complicate imaging of beam halos.

Equation 6 indicates that resolution of a system using optical elements is proportional to $\lambda^{2/3}$. Therefore a system imaging in x-rays should produce improved resolution and still have large solid-angle acceptance. The difficulty of fabricating x-ray optics often compromises the resolution a system obtains. The Kirkpatrick-Baez (K-B) microscope, which uses crossed grazing-incidence cylindrical mirrors, has successfully x-ray imaged inertial-confinement laser targets with 5 μm resolution at greater than 5 keV x-ray energies(22). The K-B microscope has achieved acceptance angles of $\Delta\Omega=3\times10^{-7}$ ster as compared to the 4×10^{-13} ster acceptance angle of the 10 μm pinhole at 1400 cm. Note that $(\Delta\Omega)^{1/2}=5.48\times10^{-4}>1/\gamma=7.3\times10^{-5}$ rad, so that the K-B microscope acceptance is approximately greater than θ_c. The geometry of the K-B microscope requires shorter optics-to-source distances for higher energy imaging where the required mirror angles are more glancing. The shorter distance increases difficulties from heat loading in the support structures, which may not be at a glancing angle. Imaging at lower energies would mitigate this problem.

Wölter x-ray microscope optics, consisting of consecutive hyperboloid and ellipsoid surfaces that are used to x-ray image inertial-confinement laser targets, have been shown to give $\lesssim$10 μm resolution, depending on the direction in the image plane relative to the optical axis(23). Unfortunately, Wölter optics are extremely difficult to fabricate.

CONCLUSIONS

Imaging of electron and positron beams using synchrotron emission is a useful diagnostic for determining beam position and profiles. Imaging in the visible range is a useful technique for low energy machines down to beam energies of about 20 MeV. Above this beam energy, there is ample light for single bunch imaging. The optical aperture and therefore optimal resolution are determined by balancing the blurring of the beam as it rounds a bend and diffraction of the optical aperture. For high-energy machines, because of the increased bending radius of the magnets, optical resolution in the visible range is $\gtrsim 150\ \mu m$. Resolution is proportional to $\lambda^{2/3}$ and therefore, imaging at shorter wavelengths is advantageous. Imaging at shorter wavelengths has the disadvantages of the difficulties associated with the fabrication of high-quality x-ray components.

Simple pinhole imaging of high-energy beams is possible. Optimal resolution necessitates a compromise between diffraction through the pinhole and geometrical blurring from the finite pinhole diameter. The resolution is improved by imaging at higher energies. The problems are fabrication of ~10 μm pinholes thick enough to work at high energy, decreasing signal levels above the critical energy, low imaging efficiency due to the small solid angle of the pinhole, and the converter and imager technologies required. Imaging at TV rates is relatively easy. To image a single bunch with ~30 μm resolution is difficult because of the small signal levels. Various more sophisticated x-ray imaging techniques discussed, such as coded imaging and x-ray optics, may increase signal levels but have the drawback of increased complexity.

ACKNOWLEDGMENTS

The author thanks David Bowman and Roddy Walton for engineering information and Peter Gobby for manufacturing information regarding the micro pinholes. I also thank S. Iverson and W. D. Turley for information regarding the WL-designation phosphors and Bingxin Yang and Alex Lumpkin for information regarding pinhole imaging at APS. Thanks to N. S. P. King for providing electrooptic imager information. Lastly, thanks to Vaikunth Stewart for the constructive editorial comment.

REFERENCES

1. Elder, F. R., Gurewitsch, A. M. and Langmuir, R. V., "Radiation from Electrons in a Synchrotron," Phys. Rev. **71**, 829 (1947).
2. Nawrocky, R. J., Galayda, J., Yu, L. H. and Shu, D. M., "A Beam Profile Monitor for the NSLS VUV Ring Employing Linear Photodiode Arrays," IEEE Trans. Nucl. Sci. **NS-32**, 1893 (1985).
3. Hofmann, A., "Electron and Proton Beam Diagnostics with Synchrotron Radiation," IEEE Trans. Nucl. Sci. **NS-28**, 2132 (1981).

4. Greegor, R. B. and Lumpkin, A. H., 'Synchrotron Radiation as a Low Gamma, Non-Intercepting Diagnostic Concept for the Average Power Laser Experiment (APLE)," Nucl. Inst. Meth. **A318**, 422 (1992).
5. Krinsky, S., Bittner, J., Fauchet, A. M., Johnson, E. D., Keane, J., Murphy, J., Nawrocky, R. J., Rogers, J., Singh, O. V. and Yu, L. H., "Storage Ring Development at the National Synchrotron Light Source," **BNL-46615**, (Brookhaven National Laboratory, Sept. 1991), Ch. 2, pp. 14–20.
6. Hofmann, A. and Robinson, K. W., "Measurement of Cross Section of a High-Energy Electron Beam by Means of the X-ray Portion of the Synchrotron Radiation," IEEE Trans. Nucl. Sci. **NS-18**, 937 (1971).
7. Akbari, H., Borer, J., Bovet, C., Delmere, Ch., Manarin, A., Rossa, E., Sillanoli, M. and Spanggaard, J., "Measurement of Vertical Emittance at LEP from Hard X-Rays," in *Proceedings of the 1993 Particle Accelerator Conference*, (IEEE Wash. D.C. 1993), pp. 2492–2494.
8. Jackson, J. D., *Classical Electrodynamics* (John Wiley & Sons, Inc., New York, 1962), Ch. 14, pp. 481–488.
9. Feledman, D. W., Bender, S. C., Byrd, D. A., Carlsten, B. E., Early, J. W., Feldman, R. B., Goldstein, J. C., Martineau, R. L., O'Shea, P. G., Pitcher, E. J., Schmitt, M. J., Stein, W. E., Wilke, M. D. and Zaugg, T. J., "Operation of the APEX Photoinjector Accelerator at 40 MeV," in *Proceedings of the 1992 Linear Accelerator Conference Proceedings*, (AECL Research, Ottawa, 1992) pp. 603–605.
10. Armstrong, S. V. and Gobby, P. L., "Computer Aided Electroplating of Tapered Copper Electroforming Mandrels for Pinhole Imaging Components," Plating Surf. Finish., **76**, 47 (1989).
11. Dunn, J., Young, B. K. F. and Shiromizu, S. J., "X-Ray Sensitive CCD Instrumentation for Short and Ultra-Short Pulse Laser-Produced Plasma Experiments," presented at the 10th Topical Conference on High-Temperature Plasma Diagnostics, May 8–12, 1994, Rochester, NY., Proc. to be published in Rev Sci. Instrum., (1994).
12. Janesick, J., Elliott, T. and Pool, F., "Radiation Damage in Scientific Charge-Coupled Devices," presented at the 1988 IEEE Nuclear Science Symposium, Orlando, Fl. Nov 11–19, (1988).
13. Ashford, C. B., Berlman, I. B., Flournoy, J. M., Franks, L. A., Iversen, S. G. and Lutz, S. S., "High-Z Liquid Scintillators Containing Tin," Nucl. Instr. Meth., **A243**, 131(1986).
14. Moy, J. P., Koch, A. and Nielsen, M. B., "Conversion Efficiency and Time Response of Phosphors for Fast X-Ray Imaging with Synchrotron Radiation," Nucl. Instr. Meth., **A326**, 581 (1993).
15. Zurro, B., Ibarra, A., Acuña, A. U., Sastre, R. and McCarthy, K. J., "A Comparison of Phosphors for Broadband Plasma Emission Detectors," presented at the 10th Topical Conference on High-Temperature Plasma

Diagnostics, May 8–12, 1994, Rochester, NY., Proc. to be published in Rev Sci. Instrum., (1994).
16. Shu, D., Warwick, T. and Johnson, E. D., "Diagnostic Phosphors for Photon Beams at the ALS and APS,", Rev. Sci. Instrum., **63**, 548 (1992).
17. Electronic Industries Association (EIA) Tube Engineering Advisory Council (TEPAC), "Optical Characteristics of Cathode Ray Tube Screens," **TEPAC Pub. No. 116**, Electronic Industries Association, Wash., D.C. (Dec. 1980).
18. From the ITT Electro-Optical Products Brochure: "Special Purpose Photosensitive Devices," ITT Corp., (Jan. 1989).
19. Wilke, M., Moy, K., Iverson, S., Hockaday, M. and Blake, R., unpublished data.
20. Gleason, J. K. and Turley, W. D., "Photophysics and Chemistry of Ultra-Fast ZnO and CdS Phosphors," Proc. 180th Meeting of the Electrochemical Society, Phoenix, AZ., Oct. 13–18, 1991.
21. Delage, O., Garconnet, J-P. and Schirmann, D., "Neutron Penumbral Imaging of Inertial Confinement Fusion Targets at Phébus," presented at the 10th Topical Conference on High-Temperature Plasma Diagnostics, May 8–12, 1994, Rochester, NY., Proc. to be published in Rev Sci. Instrum., (1994).
22. Marshall, F. J., Delettrez, J. A., Epstein, R. and Yaakobi, B., "Diagnosis of Laser-Target Implosions by Space-Resolved Continuum Absorption X-Ray Spectroscopy," Phys. Rev. E, **49**, 4381 (1994).
23. Remington, B. A. and Morales, R. I., "Laboratory Characterization of Wölter X-Ray Optics," presented at the 10th Topical Conference on High-Temperature Plasma Diagnostics, May 8–12, 1994, Rochester, NY., Proc. to be published in Rev Sci. Instrum., (1994).

REAL TIME SINGLE SHOT THREE-DIMENSIONAL MEASUREMENT OF PICOSECOND PHOTON BUNCHES

(1994 FARADAY CUP AWARD INVITED PAPER)

Edouard Rossa
European Organization for Nuclear Research (CERN)
CH-1211 Geneva 23, Switzerland

Abstract

A new method using a streak camera to monitor the particle density distribution in three dimensions in space was developed for the LEP (Large European Electrons Positrons collider). This paper provides a basic overview of the experiment. The synchrotron radiation emitted by the particles creates the image of the density distribution of particles in the bunches. The optical set-up allows us to see the front view or the top and side view of the photon bunch simultaneously at successive turns. The practical limits of our set up are discussed and some examples of bunch oscillations in three dimensions are given.

INTRODUCTION

Small dipole magnets produce synchrotron radiation to monitor the shape of both LEP beams [1] [2].The density distribution in space of the emitted photon bunches is proportional to the density of the particles in the e^+ and e^- bunches. The light is collected by two beryllium mirrors in the vacuum tube. Two achromatic lenses create the image of the source on the double sweep streak camera in an underground optical laboratory [3] The optical set-up allows observation of the side view and the top view of both photon bunches simultaneously, and up to 50 single successive bunch passages can be recorded on one image. The front view of the photon bunches can also be displayed. The system gives the image of instabilities in all three dimensions and extracts the bunch length and transverse dimensions [4] of the photon bunches. But due to the synchrotron radiation they are also images of the e+ and e- bunches.

STREAK CAMERA

A double sweep streak camera built by A.R.P. [5] with CERN specifications, has been used at LEP for five years with increasing sophistication. A brief description of this instrument follows.

a)Focus

Photons hit the photocathode of the vacuum tube where electrons are emitted and accelerated, and the internal electrical focusing field gives the image of the input photocathode onto a phosphor screen.

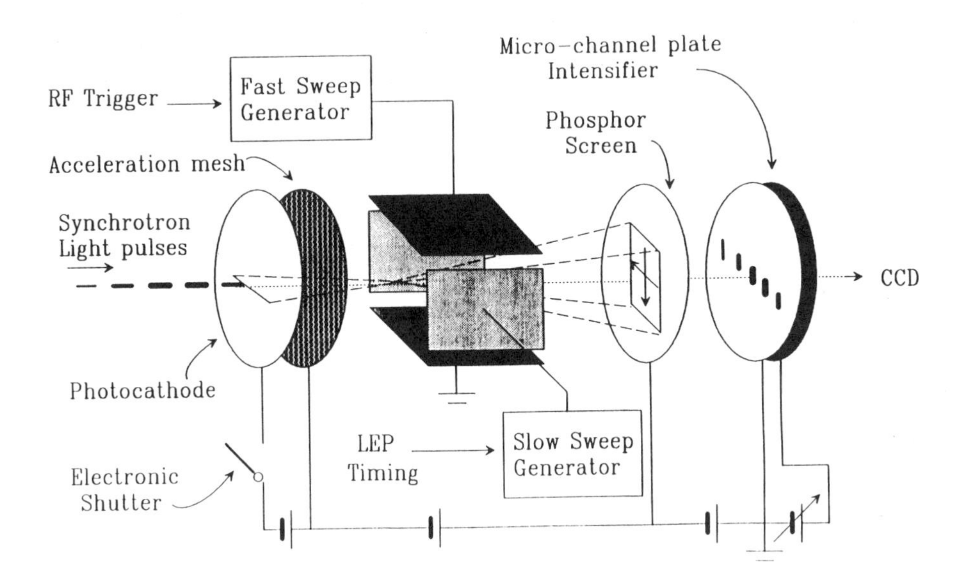

Figure 1: Set-up diagram of the double sweep streak camera.

b)Fast sweep

After acceleration the electrons emitted by the photocathode are quickly deflected by an electric field as they pass between two horizontal plates. This causes the bunch of electrons to rotate in space and strike the phosphor screen leaving an image of its length, as if we were observing it from above.

It is very important to note that we never diaphragm the incoming photon bunches with a slit as in usual streak camera set-up. Thus the image on the screen is the density projection of the three dimensional bunch and two dimensions are preserved (the longitudinal and one of the transverse). The trigger of the fast sweep is very precisely synchronized with the RF (Radio frequency) signal.

c)Double sweep

A second pair of plates, perpendicular to the first, allows successive streaks to be staggered. The composite image on the phosphor screen is recorded by a CCD camera and digitized before being processed by a computer. The software displays the bunch projected density with colours and the frame is rotated by 90° so that the streaks (time scale direction) always appear horizontal.

With the readout system provided by the firm A.R.P., the streak camera can be left running at a rate of 10 to 25Hz. The data analysis is done locally with an IBM PC which creates and refreshes a colour displays in RGB standard. These images have been transmitted to the LEP control room by means of a commercial TV set, in PAL standard which degrades the resolution significantly. All pictures can be recorded in digital form and kept on file for off-line analysis.

A new digitizing system made by the firm DATACUBE in VME standard has been implemented by CERN. The system allows a fast remote control and easy transmission of the data. It works in parallel with the local A.R.P. set-up. DATACUBE images are transmitted to the main control room via dedicated high resolution RGB fibre optics channels.

OPTICAL SET UP

The synoptic of the equipment is shown in figure 2.

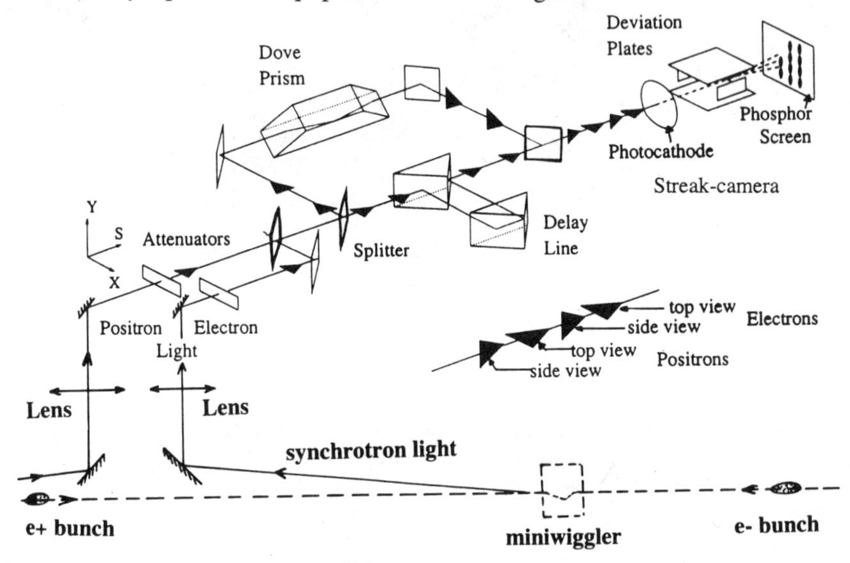

Figure 2: The Optical set-up provides side and top view at the same time.

The convention for the axes is defined in this figure 2: X, Y, S are respectively horizontal, vertical and longitudinal axes of the bunches in the machine. The synchrotron radiation of the e+ and e^- bunches arrives slightly separated in time ($\approx$ 500 ps). An automatic attenuation system consisting of motorized, continuously varying, neutral density filters and a photomultiplier keeps the intensity of the light constant. A semi-transparent mirror divides the light, and a dove prism rotates one photon-beam by 90^0 about the longitudinal axis. The achromatic lenses create the image of the light source, (e+ or e- bunch in the dipole magnet) on the surface of the photocathode. Hence the streak camera displays both top and side view which are the density projection of the beam in the horizontal and vertical plane respectively. The side- and top-view of both beams can be shown simultaneously in one streak of the camera. So one single streak gives the following views (figure 3)

The software is tracking the bunches in real time and extracts the length, transverse dimensions and the center of gravity of each photons-bunch.

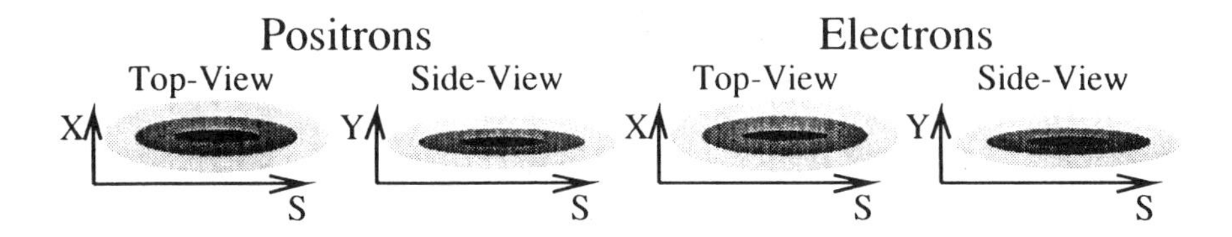

Figure 3: Example of image obtained in single streak for one passage of one bunch of positrons and one bunch of electrons.

RESOLUTION LIMITS OF THE ENTIRE SYSTEM

The limits of resolution, at the input photocathode plane, for the streak camera and the digitizing system are listed below:

Maximum sensitivity = 1 photon/count per pixel
Light intensity digitized by 8 bits A.D.C.

X (transverse)	< 15 μm
Y (transverse)	< 15 μm
S (length)	< 0.8 mm
Jitter (fast sweep)	< 2 ps (< 0.6 mm)

The transversal spatial resolution limit has been obtained by measuring the smallest possible spot size. A picosecond laser pulse measured with the S.C. has shown [3] a F.W.H.M. of 6 ps which corresponds to sigma = 0.8 mm. Which means that the resolution of the S.C. itself is better than this. We have measured the trigger jitter, with a laser diode pulse of sigma = 12 ps, by taking the standard deviation of the center of gravity on the S.C. The 2 ps measured hence also include the jitter of the laser diode.

INFLUENCE OF THE TRANSVERSE SPOT SIZE ON MEASURED LENGTH

The streak camera in this application is used, contrary to other applications, without a slit in front of the photocathode in order to get a complete image of bunch size with instabilities in all dimensions. Thus there is a finite spot size in the direction of the fast sweep which broadens the measured bunch length. For a Gaussian distribution, with a transverse dimension σ_x of the light spot on the phosphor screen of the streak camera, the broadening of the measured bunch length σ_{meas} is expected to be:

$$\sigma_{meas} = \sqrt{\sigma_\ell^2 + \sigma_x^2} \tag{1}$$

σ_{meas} is the length measured on the phosphor, and σ_x the finite spot size there. Hence the correction to achieve from the measured σ_{meas} the real σ_ℓ is

$$\sigma_\ell = \sqrt{\sigma_{meas}^2 - \sigma_x^2} \tag{2}$$

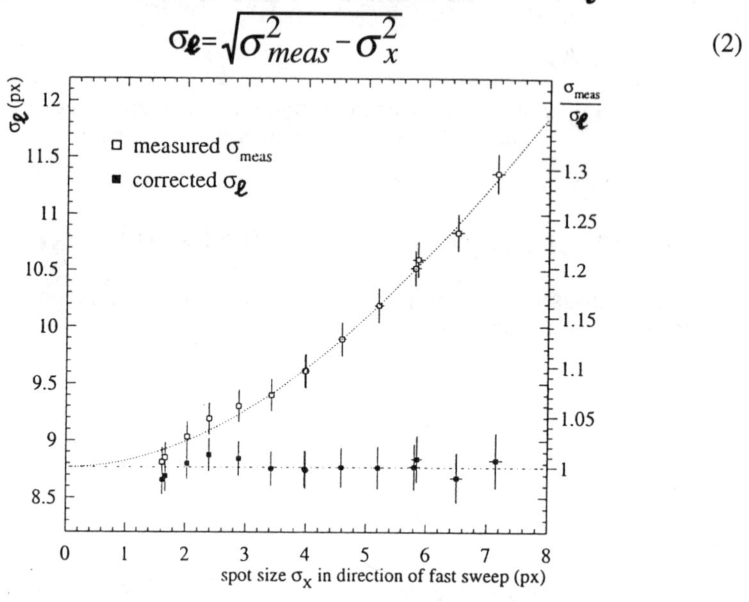

Figure 4: Influence of spot size on measured length.

To verify this effect, the light pulse of a laser diode was projected by a lens onto the S.C. The light was strongly attenuated to have no intensity effects as described later. The position of the lens was varied to achieve different spot sizes on the photocathode. The gain on the micro-channel-plate of the S.C. was adjusted to have a good signal/noise ratio. The spot size was measured in focus of the S.C. Then the fast sweep was set to the maximum streak speed and the length of the pulse was measured. Figure 4 shows the result obtained.

This proves that the correction according to equation 2 is valid for a wide range of spot sizes σ_x. Even when the spot dimension on the phosphor becomes comparable to the dimension of the longitudinal bunch profile the correction still yields the correct result.

INFLUENCE OF LIGHT INTENSITY IN A SINGLE SHOT MEASUREMENT

It is known that the measured pulse length increases with higher intensity of the incident light pulse especially for short pulses [7]. The influence of total intensity is plotted in figure 5.

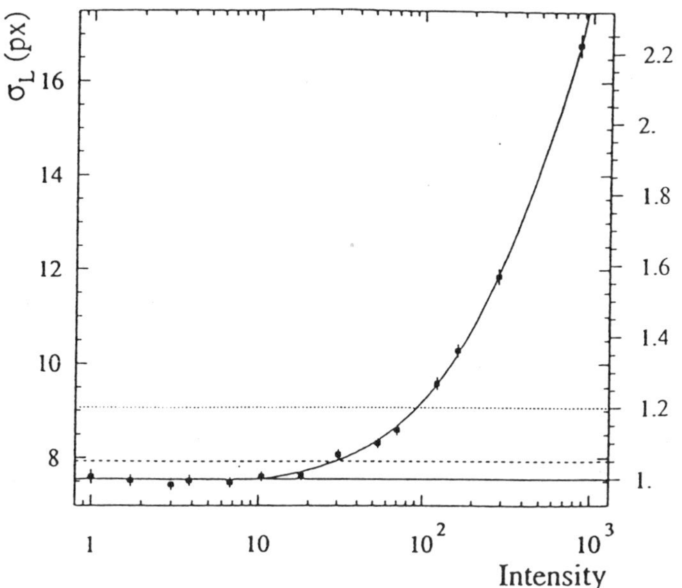

Figure 5: Influence of light intensity on measured length

The horizontal scale is 1 femptojoule per unit. The measurements taken in our set up [8] [10] of the influence of the total light intensity lead to the following conclusion:

1) When the light level is very low, on average smaller than 1 photon per pixel, then the noise starts to contribute to the tails of the distribution and the measured sigma increases by more than 5%.
2) When the photon flux reaches 30 femtojoules per pulse then the measured values increase by 5%.

Since the pixel's signals are digitized by 8 bits ADC, the useful range is large enough for LEP beam diagnostics, where we have one automatic level controller to keep the light at the right value for a continuous and safe operation 24h per day. So the dynamic range in this application does not impose any limits for maximum intensity.

SUMMARY OF THE MAIN PERFORMANCES

Three views of the density projection of single bunches are available:

- *front view (focus mode) :* *plane x, y respectively horizontal and vertical dimensions,*
- *side view:* *y, s respectively vertical and longitudinal dimensions*
- *top view:* *x, s respectively horizontal and longitudinal dimensions.*

Thus the three dimensions of the photons-bunches can be extracted in real time.

The presented way of correcting the measured bunch length for the spot size allows us to very precisely determine the bunch length in real time, without losing the possibility of observing all kinds of instabilities in the bunches of LEP [9] [11].

Nevertheless in some particular cases, it is also possible to cut the photon bunches into slices [4] and display the density distribution in each slice, at each bunch passage as in a real time tomography.

This is a non-intercepting detector which could be used continuously to monitor the behaviour of the bunches 24 h per day.

EXAMPLE OF RESULTS OBTAINED IN THE LEP MACHINE

The next figures gives some typical results obtained with our equipment for the same bunch on successive turns. The first turn is recorded always at the top of the screen and the last at the bottom. Figure 6 shows an example of six successive bunch passages of stable beams at 45 GeV.

Figure 7 displays a dipolar oscillation for both beams, electrons and positrons. Then figure 8 shows a quadrupolar oscillation and figure 9 gives a clear head to tail effect. Finally, figure 10 demonstrates a very complicated evolution of a bunch from turn to turn. The side view shows a head to tail oscillation in the vertical plane, but there is also a longitudinal oscillation of the center of gravity superposed to the elongation clearly displayed by the top view (horizontal plane).

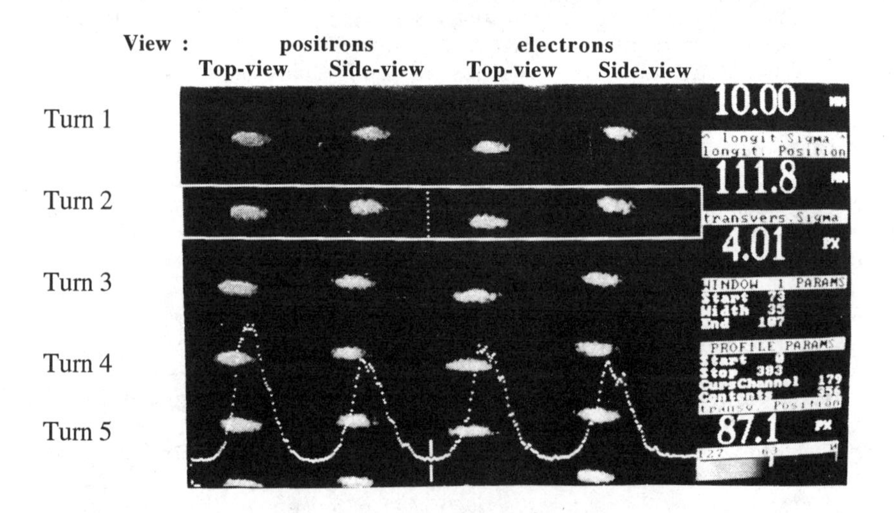

Figure 6: Stable beam at 45 GeV, with a plot of each longitudinal profile of the views selected by the rectangle.

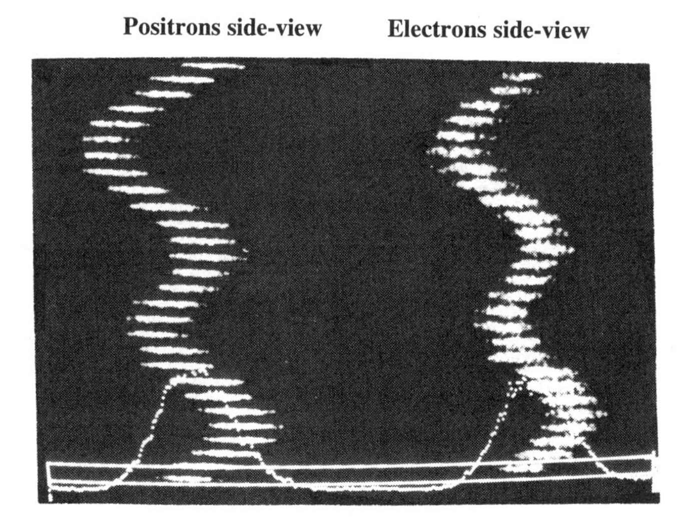

Figure 7: Longitudinal oscillations of beam of positrons and of electrons when the longitudinal feedback is off.

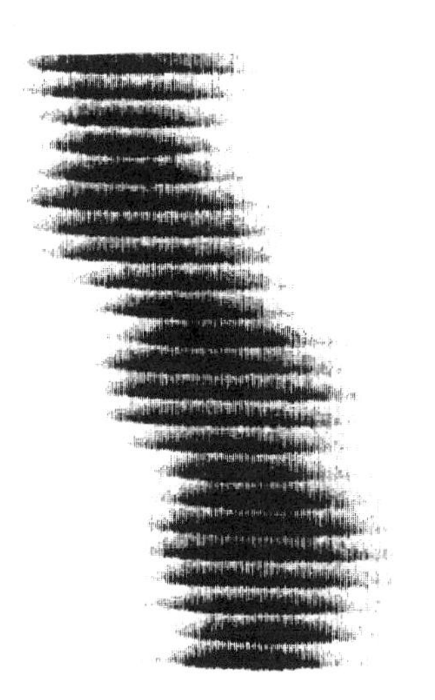

Figure 8: Quadrupolar oscillations.

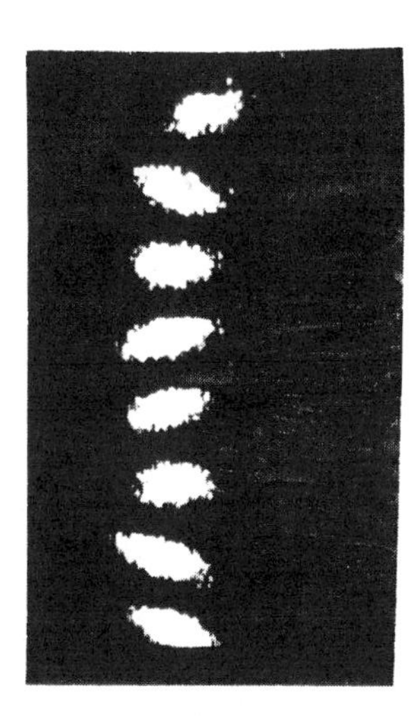

Figure 9: Head -to-tail in vertical plane oscillations

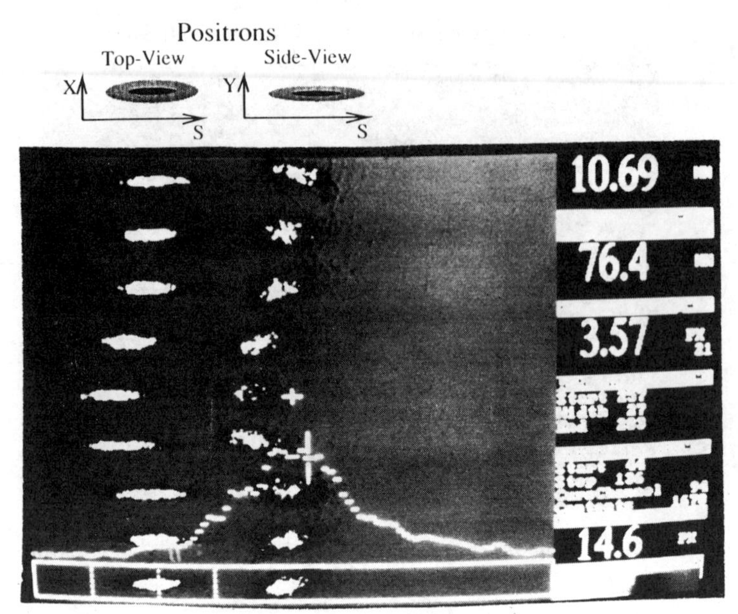

Figure 10: Simultaneous observation of the top view(left) and side view (right) of the same bunch on successive turns. This demonstrates a mixture of instabilities: elongation, oscillation and head to tail effects simultaneously.

NEW DEVELOPMENT

Remote control

A complete remote control system is being implemented. All the parameters of the experiment will be controlled from the main control room, 2 km away from the underground optic-laboratory.

Summary of the main parameters to be controlled:

a) Light source: The current in the small dipole magnet controls the direction and the intensity of the emitted light, but also a slight tuning of the angle of the beam in the vacuum chamber is useful.

b) Optic channel: All mirrors and achromatic lenses in the tunnel are motorized.

c) Parameters of the streak camera: All parameters of the streak camera must be remote controlled by the use of small stepping motors.

d) Parameters of the timing: The fine synchronization of the images on the CCD with the LEP turn clock and with the master RF source is done remotely.

Thus the bunch selection, number of turns, and bucket number could be done from the main control room. Moreover, a picosecond timing allows to continuously follow the phase change between the RF and the particle bunches. Although with the new bunch train project the time between bunches will be reduced, it will be possible to analyse each bunch in the train [12] separately.

Image processing.

A new software developed at CERN provides easy parameter control but also gives a very fast image analysis from the DATACUBE digitizer with automatic bunch tracking and calculation of all the dimensions of the bunches e+ and e-. Figure 11 shows a copy of the high resolution screen image obtained from the DATACUBE board. All the profiles, and the history of the dimensions, are displayed on the same screen and sent to the main control room via RGB transmission.

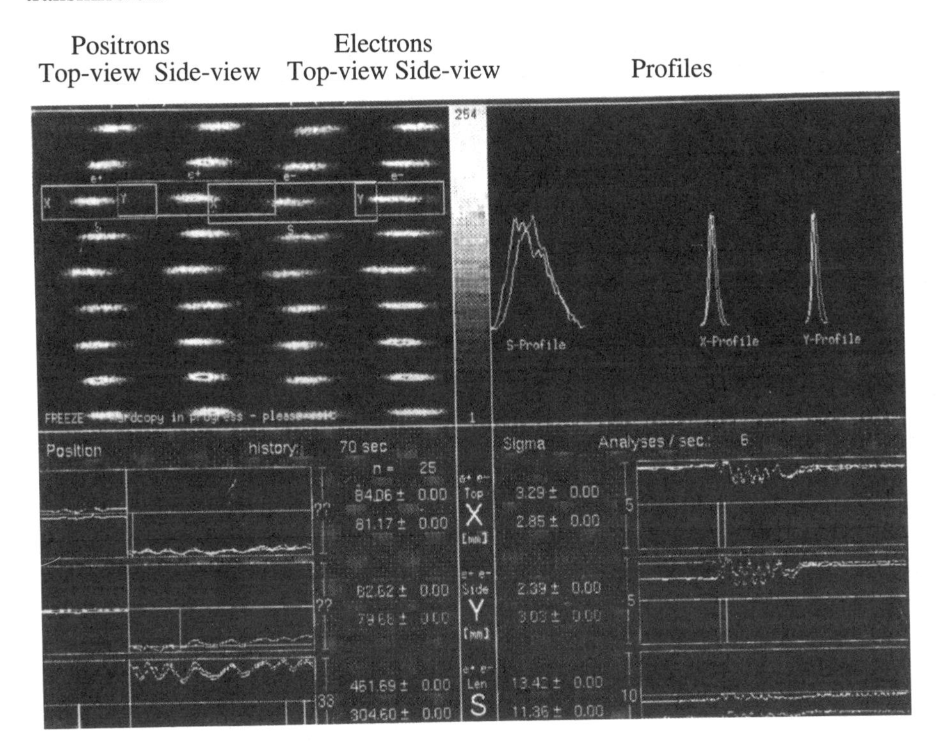

Figure 11: Display presentation in main control room.

ACKNOWLEDGMENTS

The development, on bunch diagnostics for LEP machine, could not have been done without the generous help of a number of people. CERN staff, but also students and fellows have contributed over the last eight years to the project. Special thanks are due to C. Bovet for his permanent support, to F. Tecker and K. Hanke for their very efficient work with the streak camera, to N. Adams who contributed to the picosecond timing. Many thanks are due to L. Disdier and F. Madeline who participated in the optical set-up.

The miniwiggler and light collection channels were designed by C. Bovet and M. Placidi and implemented by T. Verbeck and C. Grunhagel.

J.P. Corso, M. Sillanoli and R. Perret have fulfilled the mechanics of the underground optical laboratory.

The interlock controller was designed by G. Baribaud and J.J. Savioz and they also have contributed to the hardware of the remote control.

The software for remote control timing was made by A. Burns, D. Merel, G. Morpurgo and J. De Vries.

The early image analysis software for A.R.P. digitiser was done by J.C. Mathae and F. Tecker.

The DATACUBE image analysis software was implemented by Y. Solberg and M. Werner.

The new generation of streak camera remote control software and DATACUBE image analysis is being implemented by A. Burns and J. De Vries to provide a user friendly interface to the main control room.

Many thanks are due to all colleagues from the SL-BI and SL-AP Groups, F. Tomasini and J.M. Roth from the firm A.R.P.

REFERENCES

1. C. Bovet, M. Placidi ," A dedicated synchrotron radiation source for LEP diagnostics", LEP Note 532, April 1985, CERN, 1211 Geneva 23, Switzerland.
2. F. Meot, "Synchrotron Radiation Interferences at the LEP Mini-wiggler", CERN SL/94-22 (AP), CERN, 1211 Geneva 23, Switzerland.
3. E. Rossa, N. Adams, F. Tomasini, J-M. Roth, "Double sweep streak camera for LEP", Proc. of 2nd European Particle Accelerator Conference, Nice, June 1990, pp. 783-785, Edition Frontières, B.P. 33, 91192 Gif-sur-Yvette Cedex, France.
4. E. Rossa, C. Bovet, L. Disdier, F. Madeline, J.-J. Savioz, "Real time measurement of bunch instabilities in LEP in three dimensions using a streak camera", Proc. of the third European Particle Accelerator Conference, Berlin, 1992, Vol. 1, p.144-146.
5. A.R.P. (Applications de la Recherche en Photonique), 52, Avenue de l'Europe, 78 160 Marly-le-Roi, France - Fax (+33) 1 39 16 56 06 /// A.RP. Centre de Transfert de Technologies -Route de Hausbergen - 67309 Schiltigheim, Cedex - Fax (+33) 88 28 35 52.
6. G. Baribaud, C. Bovet, R. Jung, M. Placidi, E. Rossa, Y. Solberg, "Three dimensional bunch observation in LEP", Proc. XVth Int. Conf. on High Eng. Acc., Hamburg, Germany, 1992, Vol. 1, pp. 212-214.
7. D.J Bradley et al., "Intensity dependent time resolution and dynamic range of Photocron picosecond S.C., II Linear photoelectric recording", Rev. Sci. Instrument. 51 (6), June 1980.
8. E. Rossa, F. Tecker, J.C. Mathae, "Performance limits of a Streak Camera in Real Time three-dimensional measurement of Bunch Oscillation in LEP". Proc. of the 1993 IEEE Particle Accelerator Conference, Vol. 3, p. 2492, 2494.
9. E. Rossa, " Picosecond diagnostic", submitted to Optical Engineering .
10. F. Tecker, "Evaluation of the performance of a Streak Camera in beam diagnostics at LEP", diploma work PITHA 93/7, November 1993, Physikalische Institut RWTH Aachen, 52056 Aachen, Germany.
11. K. Hanke, "Measurement of the bunch length of LEP with a streak camera and comparison with results from LEP experiments", diploma work PITHA 94/1, May 1994, Physikalische Institut RWTH Aachen, 52056 Aachen, Germany.
12. C. Bovet , "LEP Instrumentation with Bunch Trains" This conference. CERN SL/94-77 (BI).

Instrumentation for Medical Beams

W. T. Chu

Lawrence Berkeley Laboratory, Berkeley, CA 94720, U.S.A.

Abstract

In recent years, accelerated heavy charged-particle (proton and light-ion) beams have been clinically used at an increasing number of accelerator facilities worldwide. Several hospital-based accelerator facilities dedicated to radiation therapy of cancer have been constructed, and their number is growing. Descriptions are presented of diverse instruments that have been developed in modifying extracted particle beams for clinical application, measuring the delivery of treatment beams, and controlling the treatment process to ensure patient safety.

INTRODUCTION

For almost a hundred years, electron and photon beams have been used for treatment of cancer. Electrons deposit their energy over a broad peak; the energy deposition by photons is characterized by an exponentially-decreasing absorption with penetration depth. Protons and light ions, being electrically charged and much heavier, have definite ranges with Bragg peaks followed by sharp distal dose falloffs and sharp penumbrae. Radiation oncologists can take advantage of these characteristics by depositing high Bragg-peak doses in irregularly shaped tumor volumes while sparing the surrounding healthy tissues and critical organs, expecting enhanced tumor control with reduced complications.

Based on the physical advantages of dose localization, as early as 1946 Robert R. Wilson proposed the rationale for using accelerated heavy charged-particle (proton and light-ion) beams for radiotherapy of human cancer and other diseases.(1) In 1954 at the Radiation Laboratory of the University of California, Berkeley, now the Lawrence Berkeley Laboratory

(LBL), Cornelius A. Tobias and John H. Lawrence performed the pioneering work in first therapeutic exposure of human to proton, deuteron, and helium-ion beams. Now, four decades later, an increasing number of heavy charged-particle radiation treatment beams is burgeoning worldwide.

In clinical trials underway now, several different particle beams are used: in addition to accelerated protons and light ions, negative pions and fast and slow neutrons are also utilized. In this paper, the term "medical beams" is limited to mean proton and light-ion beams with penetrating ranges of approximately 3–30 cm in water, specifically prepared for treatment of human cancer. Both proton and light-ion beams are considered, but the emphasis is on proton beams. Instrumentation for medical beams includes patient positioners and rotating gantries, but they are not discussed in this paper.

CLINICAL REQUIREMENTS OF MEDICAL BEAMS

Clinical requirements with many competing specifications drive the designs of medical beams. For example, clinicians may ask for a large treatment field, up to 40 cm x 40 cm, which may be provided using a scattering method, necessarily degrading the beam emittance. Another clinical requirement is the sharp lateral dose falloff (penumbra) required at the boundary of the treatment field. One may try to achieve it by increasing the apparent source-to-axis distance (SAD), which is not possible on a treatment beam line mounted on a rotating gantry. Achieving it using heavy collimation may not be allowed either because it may unacceptably reduce the beam utilization efficiency. Instrumentation must be constructed weighing the many pros and cons of competing designs and implementations, and finding the optimal solution that satisfies all of the clinical requirements.

The radiation oncologist's needs on medical beams may succinctly be stated as following: "Place a *therapeutically effective* dose distribution in the target volume while sparing the surrounding healthy tissues from unwanted radiation as much as possible. And do this reliably, economically, and, above all, safely." A list of clinical requirements of medical beams is discussed in a recent LBL report on the performance specifications for proton medical facility.(3) These clinical requirements of medical beams include specifications of such physical quantities as the residual range of the beam in patient, the extent of range modulation to cover the thickness of targets, the maximum dose rate, as well as the minimum dose rate that can be precisely controlled, the beam spill structure, the maximum attainable port size, the dose uniformity across the ports, the effective source-to-axis distance (SAD), the allowable degradation in distal dose falloffs and lateral penumbrae, which affect normal tissue sparing, and the attainable precision in delivered dose. This list was developed by LBL in collaboration with the Northeast Proton Therapy Center (NPTC) of the Massachusetts General Hospital in Boston. NPTC later used it in its request for proposal (RFP) as the basis of constructing a proton medical facility.(4)

PHYSICAL CHARACTERISTICS OF MEDICAL BEAMS

Upon entering an absorbing medium, energetic heavy charged particles are slowed down by losing their kinetic energy mainly through ionization of atoms and molecules in the medium. The specific energy loss of a particle, *i.e.*, the energy loss per unit path of the absorber traversed (usually expressed in keV/μ in water), increases with decreasing particle velocity, giving rise to a sharp Bragg peak near the end of its range as shown in Fig. 1, curve (a). When a beam of monoenergetic heavy

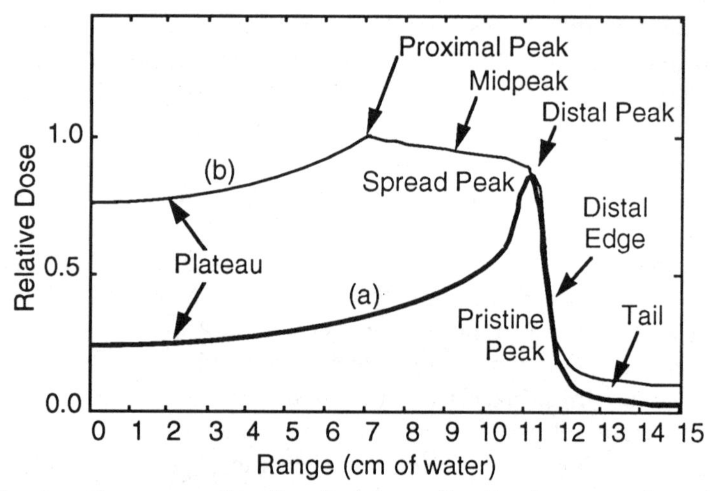

Fig. 1. Shown are the physical dose distributions as a function of penetrating depth of (a) a typical Bragg curve of a pristine beam, and (b) a typical specific ionization curve of a beam whose energy is modulated to widen the stopping region.

charged particles enters the patient body, the depth-dose distribution is characterized by a relatively low dose in the entrance region near the skin (plateau) and a sharply elevated dose in the Bragg peak. The Bragg peak is often spread out to cover an extended target by modulating the energy of the incident particles, and summing their ionization, as shown in Fig. 1, curve (b). The radiation dose abruptly decreases beyond the Bragg peak, sparing critical organs and normal tissues located downstream of the target volume from unwanted radiation.

SIMPLE BEAM LINES

A simple medical beam line (also called a treatment nozzle) is schematically shown in Fig. 2. Most of the instruments discussed below are described in length in a recent review article,(5) and their discussion in this paper is kept to minimum.

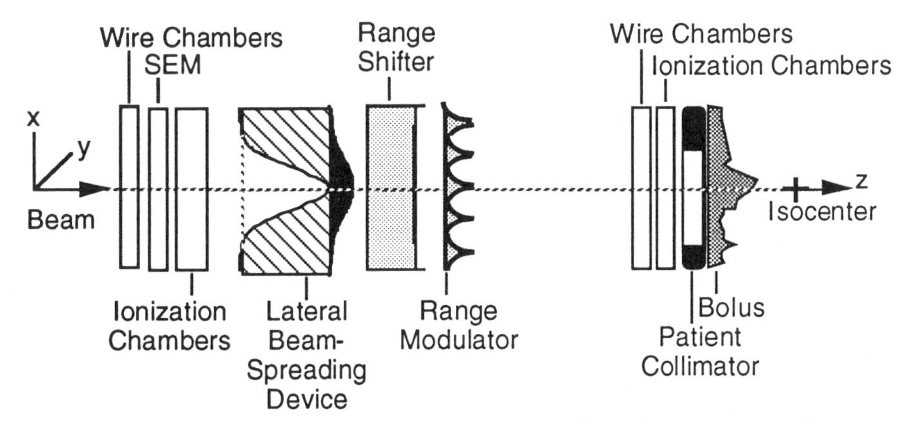

Fig. 2. A typical simple proton nozzle. As an example, a contoured scatterer is depicted as a lateral beam-spreading device. It is made of a low-Z material, such as Lucite (hatched), and a high-Z material, such as lead (black).

Range Adjustment

One of the clinical requirements is that the beam range in patient should be available for values between 3.5 g/cm^2 – 32 g/cm^2 for fields smaller than 22 cm x 22 cm. Here the field size is specified indicating that if the beam is laterally broadened using scatterers, the energy loss in them must be included in estimating the energy of an extracted beam. Another clinical requirement specifies that the beam range adjustment should be in steps of 0.1 g/cm^2, and in steps of 0.05 g/cm^2 for ranges < 5 g/cm^2. To provide various ranges, either a beam with an appropriated energy must be extracted (*e.g.*, from synchrotrons) or an extracted fixed-energy beam must be degraded by absorbers (*e.g.*, from cyclotrons or linacs). The resulting energy spread should then be reduced by magnetic analysis and collimators. When the beam energy is changed, the entire beam transport and beam delivery systems must track the beam energy.

The finer energy adjustments, called for in the latter clinical requirement above, may be achieved more readily by placing a mechanical range shifter in the treatment nozzle close to the patient. Placing it downstream of all beam-steering magnets eliminates the needs for retuning the beam between energy changes. The mechanical range shifter may be a *variable water column*, which places specified thickness of water in the beam path. The Bragg curve of a particle beam in a target may be measured by placing one ionization chamber upstream of the water column, and the second ionization chamber downstream of it and immediately upstream of the target. If a series of measurements at various water thickness settings is made, the dose measured by the second chamber (relative ionization at a given depth of water) normalized to the readings of the first chamber (the total number of the incident particles) produces the Bragg curve of the particle beam inside a water absorber. A *binary filter* or a *set of two wedges* may also be used as a range shifter. A binary filter adjusts the range of a beam by means of a set of metal or plastic plates of various thicknesses. In a double-wedge system, the two wedges may be placed in opposing directions and moved in such a way that the particles in the finite beam spot traverse a adjustable yet uniform-thickness absorbing material.

Lateral Broadening of Beams

The next clinical requirement to satisfy for medical beams is making large radiation fields, up to 40 cm x 40 cm, with a dose uniformity of better than ±2% over the entire treatment field. In the simplest method, it is accomplished using scatterers. A narrow pencil beam scattered by a thin scatterer produces an approximately 2-dimensional Gaussian dose distribution at isocenter. Here a scatterer is called thin when the kinetic energy of the particle does not change significantly by traversing it. The dose distribution as a function of the radial distance, r, from the central axis is $D(r) = \frac{1}{\pi \tilde{r}^2} e^{-\left(r^2/\tilde{r}^2\right)}$, where $\tilde{r}$ is the rms radius of multiple scattering, and is related to the scattering angle and the drift-space from the scatterer to the isocenter. The pencil beam may be also broadened using multipole magnets in the beam line.

The resulting Gaussian-like beam traverses a *contoured scatterer*,(6) which is shown in Fig. 2 as an example of lateral beam-spreading devices. For this contoured scatterer, the energy absorbing power is constant at all radial distances. However, the beam particles traversing it near the central axis encounter more high-z material, whereas those at larger radial distances encounter more low-Z material. Thus, the central rays are scattered out more than the peripheral rays, flattening the Gaussian-like dose profile, and making it into a larger uniform-dose field. Alternatively, the lateral spreading of beams may be accomplished using a *double scattering system with occluding rings*. These mechanical beam spreaders are passive devices and therefore simple to use. But the beam-tuning requirements are stringent as the beam must enter the device at its dead-center to produce the desired dose flattening effect.

A *wobbler* relies on magnets to flatten the dose profile without resorting to scattering materials. Its use relaxes the stringent beam-tuning requirements encountered with the scattering methods. A wobbler system consists of two dipole magnets placed in tandem with their magnetic field directions orthogonal to one another and to the beam direction. The magnets are energized sinusoidally with the same frequency but with a 90 degree phase shift between them. A beam entering a wobbler system along its axis emerges from it with the beam direction wobbling around the original beam direction, and "paints" an annular-shaped dose distribution. A large area of uniform dose is obtained by painting the treatment area in several concentric annuli with different diameters, each with a certain predetermined particle-number fraction.

Range Modulation

The beam emerging from the accelerator is not truly monoenergetic, but has some energy spread or dispersion. As the treatment is always achieved using many accelerator pulses, both the dispersion within a single pulse and that of several consecutive pulses of the extracted beam must be considered. Usually this dispersion is negligible for synchrotron beams, $\Delta E/E<10^{-4}$, with the pulse-to-pulse energy variation of $<10^{-3}$; but it can be significant in cyclotron beams, $\Delta E/E>10^{-2}$. This energy spread plus the range straggling of the incident particles in the absorbing material, including the part of patient body itself, located along the beam path up to the stopping region, contributes to the finite width of the Bragg peak. The resulting Bragg peak is still narrower than the clinical target thicknesses, typically 0.5 cm – 16 cm, and the beam range must be modulated in order to cover the extent of the target thickness with the Bragg-peak doses. The energy spread and range straggling also contribute to the deterioration of the steepness of the distal dose falloff of the spread-out Bragg peak.

The width of the Bragg peak may be spread out by a *range-modulating propeller*. The propeller, is a fan-shaped stepped absorber, which is made to rotate in the beam so that the appropriate thickness of the propeller "blades" intercept the beam. The profiles of the blades are designed in such a way that, when the beam traverses the propeller rotating at a predetermined rate, a constant *biological* dose is imparted across the entire width of the spread-out Bragg peak. As the clinical requirement on the range modulation specifies the steps of 0.5 g/cm^2 from 0 to 16 g/cm^2, thirty-two propellers are needed if a propeller is made for each step. The following solution may be considered to reduce the number of propellers. A propeller for a given width of spread-out Bragg peak can be used to produce a narrower width, if the particle beam is turned off while certain thick parts of blades are in the beam line. Using such a scheme, it is possible with only one propeller designed for the maximum width to produce all other spread-out Bragg peaks with narrower widths. However, if the beam-off periods get too large, the beam utilization efficiency may drop to an unacceptable level. As a compromise, say, eight propellers may be constructed instead of thirty-two propellers.

Measurements of Medical Beams

To correctly accomplish the beam modifications described above, it is necessary to measure the spatial parameters of the beam during an irradiation. For each treatment, accurate measurements of the delivered dose and dose distribution in the patient must be made without fail. In the example of a beam-line setup shown in Fig. 2, the beam is tuned using two sets of wire chambers, which measure x and y positions and dimensions of the beam spot at two different positions in the beam line. (Here, the beam axis is taken as +z direction, and the lateral directions x and y.) *Parallel-plane, segmented-element ionization chambers*(7) are used as dose detectors. The ionization, Q (measured in coulombs), produced by

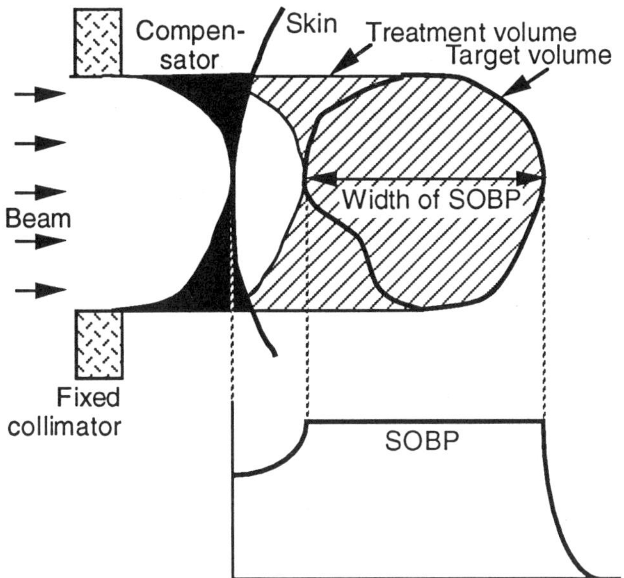

Fig. 3. The compensator adjusts the distal dose falloff edge to conform with the distal surface of the target. The width of the spread-out Bragg peak is made equal to the thickest part of the target. This fixed-modulation method produces a cylindrical treatment volume (hatched area), and the normal tissues upstream of the target are irradiated unnecessary.

the beam passing through such a chamber is proportional to the absorbed dose, D (measured in Gray), by what is known as the Bragg-Gray equation: $Q = \frac{D\rho V}{W}$, where ρ is the mass density (kg/m^3), V is the volume (m^3), and W is the ionization energy (eV) needed to produce an ion pair. Each ionization chamber, for example, may have two charge collecting planes, one of them divided into four quadrants to detect the position of the center of the beam, and the other divided into several concentric circles to measure the size of the beam spot if the Gaussian distribution of the beam profile is assumed and the beam is accurately centered. The ionization chambers are calibrated against a standard *thimble ionization chamber*, which is positioned at the center of the test target volume, and whose calibration is traceable to a standard source at the National Institute of Standards and Technology. For measuring ionization charges, the form of charge integrator that is particularly well suited to clinical applications is the *recycling integrator*. This circuit issues a pulse for every fixed increment of input charge, typically between 1 and 10 pC. When combined with digital electronics the recycling integrator has a large dynamic range, exceptional linearity, and excellent noise immunity.

A *secondary emission monitor* (SEM) is used as a backup to the ionization chambers. It has a lower dose sensitivity than the ionization chambers, but it serves well when the ionization chambers saturate because of a high dose rate.

Fixed-Modulation Beam Delivery Method

The beam-delivery method shown in Fig. 2 produces dose distributions as shown in Fig. 3. A *compensator (bolus)* is constructed in such a way that the dose falloff region conforms with the distal surface of the target volume. A fixed width of a spread-out Bragg peak, which is wide enough to cover the thickest part of the target, is used everywhere, thereby the name, *fixed-modulation beam delivery*. This method produces a cylindrical treatment volume (hatched area), and normal tissues upstream of the target are irradiated unnecessarily as indicated in the figure.

ADVANCED BEAM LINES

In realizing the full clinical advantage of heavy charged-particle beams, one of the important developments to achieve better dose localization is a beam delivery system that allows modulation of the width of the spread-out Bragg peak over the target volume according to the local target widths (*variable modulation*). The shape of the high-dose volume, in other words the *treatment volume*, can be made to conform more closely to that of the target volume, as shown in Fig. 4.

First, the radiation field for the most distal part of the target, slice M in the figure, is shaped by the aperture of a *multileaf collimator*, and a *raster scanner* deposits in it a uniform dose distribution. While the slice M is treated, a part of the slice N and all other slices upstream of slice M traversed by the beam receive the plateau dose, whereas the outside region receives none. Thus, to deliver a uniform dose to the slice N, a smaller dose should be delivered to the area upstream of M than to its periphery. One way of achieving this, for example, is to vary the speed of the raster scanning, faster in the areas upstream of M, and slower outside of it, while keeping the level of extracted beams constant. When combined with the already-deposited plateau dose, the resultant dose in the slice N becomes uniform. This process is repeated for all subsequent slices to produce a uniform dose distribution in the target volume while imparting minimal radiation outside of it.

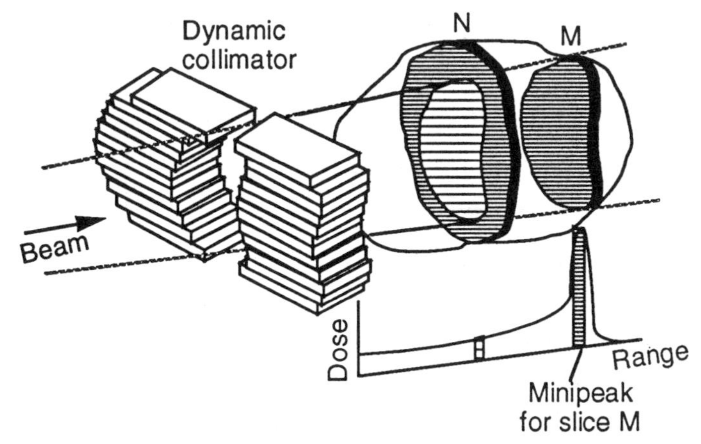

Fig 4. Schematic illustration of a variable modulation beam delivery method.

In principle, scan speed and/or beam intensity can be varied as a function of the spot location in the treatment volume to generate the desired dose distribution. In general, a scanner consists of one or two dipole magnets, one for the fast scan in the x direction and the other for the slow scan in the y direction. (Here the x and y represent arbitrary orthogonal directions.) Range modulation moves the stopping region of the beam spot, the Bragg peak, in the z direction.

A schematic representation of a raster scanner system developed at LBL(8) is shown in Fig. 5. Requirements for accuracy and stability of both the scanner magnetic field and the beam intensity are stringent. To ensure patient safety, the spatial distribution of the delivered dose is monitored in real-time using a *large-area high-resolution dose detector.* The system requires an *automated control* along with a means of very fast and fail-safe irradiation termination. Fast on-line measurements are required of various system parameters, such as scanning magnet currents, extraction levels of the particle beams, aperture of the multileaf collimator, widths of spread-out Bragg peaks and the residual ranges of the beams. A variable beam extraction level must be available over a large dynamic range.

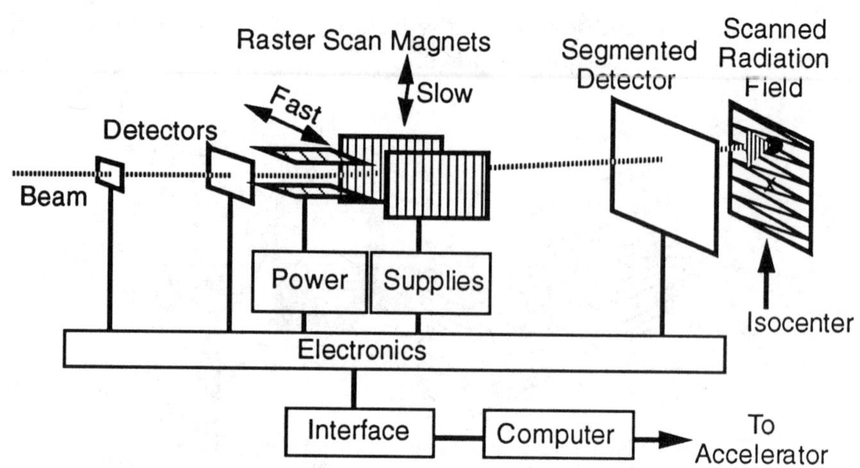

Fig. 5. Depicted is a general schematics of the LBL raster-scanning beam delivery system.

A large field of an arbitrarily specified dose distribution can also be obtained by moving a beam spot across the field in discrete steps. A predetermined amount of radiation is deposited after positioning a beam spot at a given location in the target volume. The spot is then move to the next position, and the process is repeated. This approach adopted at the Paul Scherrer Inst. (PSI) in Villigen, Switzerland to perform spot scanning of high-energy protons(9) is schematically shown in Fig. 6. The proton beam spot is moved in ±x direction by a sweeping magnet, and the movement of the spot in the y direction is accomplished by moving the patient. A position-sensitive ionization chamber measures the spot position as well as the delivered dose. A fast kicker magnet is placed upstream of the sweeping magnet, and when it is de-energized the proton beam is let through a narrow slit and the gap of the sweeping magnet. When the predetermined amount of particles are delivered, the fast kicker magnet is energized to deflect the beam away from the slit and into a slit plate, which acts as a beam stop. While the beam is blocked, the beam spot is moved to the next x position, and the irradiation is allowed to resume by de-energizing the kicker magnet. Upon finishing the irradiation of a line scan, the penetration depth (in the z direction) of the beam is shifted by the range shifter and the entire process is repeated to complete a x-z plane scan. Then the patient is moved in the y direction for the subsequent scans.

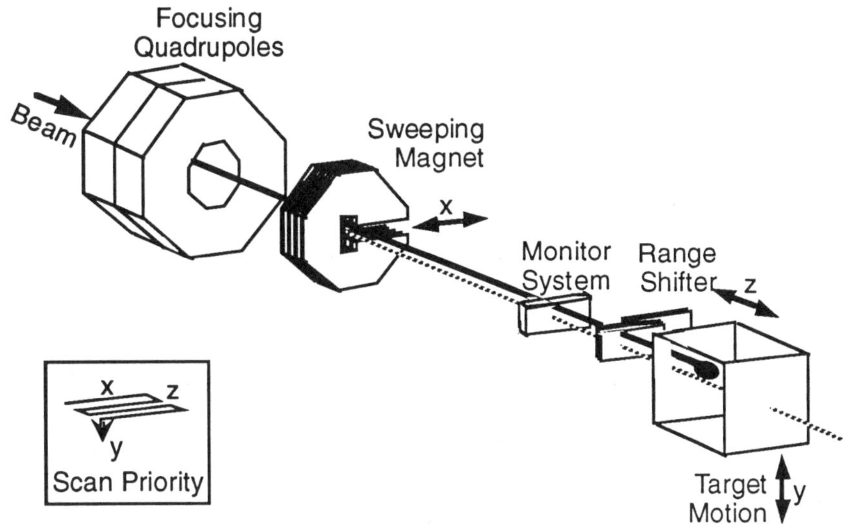

Fig. 6. Schematics of spot scanning developed at PSI.

A variable-modulation beam delivery may be achieved using a beam scanning system. Such a dynamic conformal therapy delivery must be controlled by a carefully-implemented treatment control system as described below.

CONTROL SYSTEMS FOR MEDICAL BEAMS

In the sections above the development of many instruments for *beam modification* and *beam measurement* have been described. However useful they might be, these instruments individually cannot reliably and safely operate the medical beams unless they are integrated into one coherent system operated under the command of a *treatment control system.* Based on the LBL particle therapy experience, it is evident that a much greater effort must be expended in the development of a control system than in the development of these individual hardware or systems of hardware. Recently a very thorough discussion on the rationale and structure of control systems for heavy charged-particle radiotherapy facilities have been presented(10), and the following summary is based on that reference.

A treatment control system must, above all, ensure *patient safety* by controlling the treatment procedure, measuring compliance of the actual treatment with the prescribed one, and providing feedback to the treatment planning process. As already stated, the major function of a control system is to control beam modification and beam measurements. For beam modification process, which tailors the radiation field for a prescribed patient irradiation, it controls such devices as collimators, range shifters, lateral beam-spreading devices, range modulators, and a gantry/patient positioner. The *fail-safe* control of these devices are required for patient safety. Before the beam undergoes any modification, the beam measurements ensure that the beam characteristics are correct before entering the beam delivery system. After the appropriate modification, additional measurements, as close to the patient as possible, ensure the treatment beam is correctly prepared and verifies the actual dose delivered to the patient. Beam measurements must dynamically determine radiation-field parameters for controlling the treatment and ensuring its correct delivery.

In addition, a treatment control system performs *accelerator and beam transport control* functions. This involves, besides an accelerator, the control of an extraction system for removing the beam from the accelerator, a beam transport system for channeling the particles to the desired treatment rooms, and a timing reference for process control. The important quantities controlled are the extracted beam energy, the beam intensity, and the intensity variations.

Structurally speaking, a control system may be divided into closely interlocking *software* and *hardware* beam delivery control systems. Software beam delivery control comprises the control algorithms and procedures that determine, set, and control the system parameters for a particular irradiation. Hardware beam delivery control comprises the hardware control functions necessary for the initiation, termination and control of an irradiation. Often, key control functions, such as starting and stopping of the beam, are implemented in both software and hardware for increased safety and reliability. *Monitoring* is an independent function during an irradiation for overseeing critical conditions and parameters of beam modification devices, radiation properties, or accelerator parameters. Monitoring is capable of vetoing an operation and terminating an irradiation, but is not designed for initiating an action or controlling a device.

A control system also performs *ancillary functions,* such as: the *human interface* providing video displays of pertinent information for the medical and operations personnel, *simulation* mimicking measurement information and status to allow testing of the system operation without the use of actual beams, *dispatching* which allocates the accelerator to a particular treatment room, the *prescription server* serves as the conduit between the treatment planning system and the treatment control system, and *archiving* which provides the nonvolatile recording of both raw and processed data sufficient to reconstruct a patient treatment, an accumulative history of each patient's treatments and create a data base for accelerator and treatment control system operation.

During an actual treatment procedure a control system typically performs the following tasks. First, it acquires dosimetry calibration information from the calibration procedure. Patient setup parameters are verified against the patient's prescription and hardware parameters set to desired values, *e.g.*, gantry/patient positioner parameters. When all treatment irradiation conditions are set and verified to be correct, the radiation is allowed to be brought into the patient. An abort of the irradiation is initiated upon normal completion of an irradiation. An abort can also occur upon detection of any fault condition. To guard against a software procedure failing without terminating the treatment, an acknowledgment within an appropriate time window of a procedure-specific *watchdog* occurs every dosimetry cycle. Failure of such acknowledgment leads to an abort of the treatment by the hardware beam delivery control function. To recover from an interruption caused by a computer failure, the state of the system is saved on disk every dosimetry cycle. Any abort of the treatment (by human intervention, interlock dropout, critical hardware failure, monitoring veto, or computer watchdog acknowledgment failure) should preserve the state of the system. Support of an immediate resumption or resumption at a later time provides the necessary clinical option of how to proceed with a treatment after an abort. The time to recover from an interruption needs to be prompt and the maximum possible dose uncertainty after recovery should not exceed the dose delivered in a dosimetry cycle. At the end of a patient treatment, the treatment data is archived and a summary to the appropriate medical personnel is provided.

SIMPLICITY AND COMPLEXITY

To be therapeutically useful, the instruments developed for medical beams must be simple in design and reliable in operation. However, in designing an instrument, not only its physical characteristics, but also the biological and therapeutic effects must be taken into a consideration. Often a clinical device, *e.g.*, a range-modulating propeller, seems deceptively simple, yet its "algorithmic information content" is large. In other words, its design is simple and complex at the same time. In the following, several simple medical beam instruments are examined, and the complexities of their algorithmic information contents are discussed.

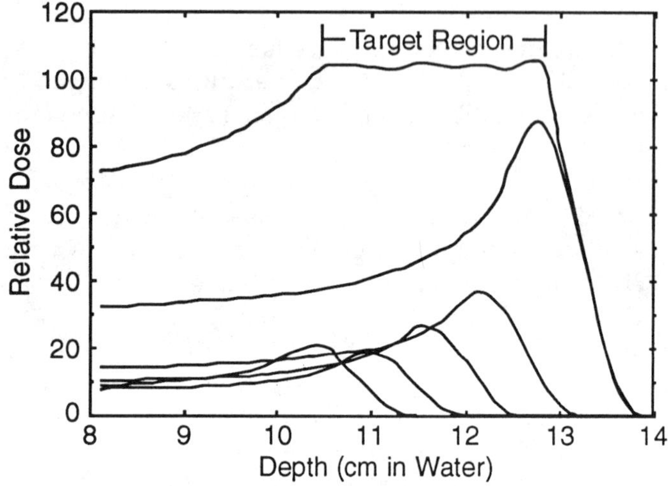

Fig. 7. To cover the target region with the Bragg-peak dose, several Bragg curves of the beams with different ranges are stacked to create an appropriate width of spread-out Bragg peak.

The range modulation may be accomplished using a mechanical device such as a range-modulating propeller. Or, as schematically shown in Fig. 7, different ranges may be stacked between the successive lateral spreading of beams. These two methods of spreading Bragg peaks out may be different, but the end results in achieved dose distributions are the same.

For clinical use, a *biologically* uniform dose distribution within the entire target volume is required. As can be observed in Fig. 7, the dose in the distal part of the spread-out Bragg peak is almost entirely made up by Bragg-peak doses; whereas, toward the proximal peak, the contribution of plateau doses increases and that of Bragg peaks diminishes. If the radiobiological effectiveness (RBE) varies across different parts of the Bragg ionization curve, the physical dose distribution must be adjusted to achieve a biologically uniform dose across the target region. The example shown in Fig. 7 is for proton beams, and the cell killing capability per unit dose remains approximately the same for the radiation over the entire width of the spread-out Bragg peak. For heavier-ion beams, to achieve a biologically uniform dose throughout the target region, the physical dose distribution is made to slope down as one moves from the proximal to the distal part of the spread-out Bragg peak. For low-LET radiation, such as a proton or helium-ion beam, the beam quality is less complicated than with heavier-ion beams. Here, LET stands for linear energy transfer. Nuclear fragmentation plays an insignificant role for protons and helium ions, and the variation in LETs is smaller for them than for heavier ions. For heavier ions, beam fragmentation must be taken into account since cell survivals depend not only on the dose-averaged LET but also on the charge and mass of the projectile particles.

The other point to observe here is that the clinical requirements are not only to provide a large uniform dose, but also a sharp dose falloff at the distal edge of the spread-out Bragg peak. The steep dose falloff is important in sparing the normal tissues located distal to the target. The treatment dose is often limited for consideration of the normal tissue complications. Therefore, the sharp dose falloffs are an important factor that influences the success and failure of treatments. As shown in Fig. 7, the distal part of the spread-out Bragg peak is largely composed of the Bragg peak of the most penetrating beam. This is deliberately done to steepen the slope of the distal edge.

The design of a proper spread-out Bragg peak involves the knowledge of not only the physical characteristics of ionization and multiple scattering of heavy charged particles traversing absorbing media, but also the cell responses to various radiation parameters. Furthermore, how the details of clinical requirements are satisfied by the instrument design influences therapeutic outcome of the treatments using the device. The mechanical design of a range-modulating propeller and its operation are simple, but its algorithmic information content is large and complex.

Next, the techniques to achieve sharp lateral dose falloffs (penumbrae) are examined. In order to take a full advantage of the dose localization properties of the heavy charged-particle beams, great care must be taken to minimize the width of the lateral dose falloff. The apparent source size of the radiation field causes a penumbra, which contributes to broadening of the width of the lateral dose falloff. In order to minimize the apparent source size, the diameter of the beam has to be small where the beam traverses material. This can be accomplished by placing any scattering and range-shifting material upstream of where the beam begins to spread transversely. However, any amount of material in the beam introduces an additional beam divergence due to multiple scattering. Positioning the material far upstream of isocenter may lead to an intolerable increase in beam spot size at the target, which, if collimated, results in a reduction in the dose rate. Also, often the beam-line length is limited on the treatment nozzle mounted on a rotating gantry. On the other hand, placing the scattering material close to the target volume reduces the scattering effect on the low-velocity particles and preserves the modulated Bragg peak shape. For this reason the patient compensator is usually placed as closely as possible to the patient body. The unavoidable multiple scattering inside the patient body and other material in the beam path, especially those near the patient such as the compensator, also contributes to the broadening of the width of lateral dose falloff.

The amount of material used for scattering or energy degradation also affects neutron production and projectile fragmentation, which in turn affects the peak-to-plateau ratio of the spread-out Bragg peak. The ratio is enhanced when the material is placed far upstream of the target as a large portion of the neutrons and fragments diverge out of the main beam.

This analysis shows that the design of the scatterers and their locations in the beam line affect the physics and biology of the beam that enters the patient body, and consequently influence therapeutic effects.

The designer of an instrument for medical beams must gather all physical, biological and therapeutic information pertinent to the instrument, and integrate them before designing and fabricating the device. It is the role of the designer to translate its large algorithmic information content into a reliable and therapeutically effective instrument.

CONCLUDING REMARKS

Clinicians demand in many ways the same requirements for medical beams as those demanded by physics experimenters for their accelerated particle beams. There are, however, some important differences between them. The major differences are in the stringent requirements of inflexibility and immediacy of the requirements of the clinical beams. For physics experiments, the set-up time of an experiment is long, usually weeks and months; and the accelerator and beam changes between the extended data taking periods are few. On the other hand, for each patient treatment, the setup time is short, usually 10–15 minutes; and the accelerator and beam setups are changed often between treatments. For beam scanning, the setups may be changed many times during a treatment exposure that lasts typically only 2 minutes. For physics experiments, delays of hours, or even days, in providing the accelerator beams for experiments are, of course, not desirable but tolerable. On the contrary, when a patient is scheduled to be treated at a certain time, or if the patient is prepared on a treatment positioner, the beam should be ready upon demand within minutes. Probably 10–15 minutes delays are tolerated once in a while, but certainly the delays longer than 30 minutes are not acceptable. Once a patient exposure is initiated, it should successfully complete the process all the time. Typically a patient receives 30 fractions of exposures for a course of treatments. If 1000 patients are treated per year at a given accelerator facility, that represents 30,000 fractions per year. Clinicians do not want to see more than one unscheduled interruption of treatment per month, because such an interruption forces changes in the prescription of subsequent treatment doses and schedules, or worse, such interruption may make the patient data ineligible for inclusion in the statistics of the clinical trial. The fact that fewer than 12 interruptions are allowed per year for 30,000 fractions represents a better than 99.96% reliability of the treatment facility; the desired reliability of therapy facility operations is indeed stringent. If a detector malefactions in a physics experiment, the measurement can be repeated, or the bad data discarded later during the analysis. However, in clinical situations, once a part of a treatment dose is deposited in a patient, it cannot be discarded for any reason, and the treatment must be finished with an accurate dosimetry provided.

Anyone embarking upon a patient treatment project at an accelerator facility should be prepared to invest a very large effort in developing a treatment control system. As discussed in the main text, a well-developed treatment control system must oversee each treatment by controlling the beam modification and beam measurement procedures for correct treatment delivery. There should be no room for compromising in ensuring patient safety.

ACKNOWLEDGMENT

The author wishes to express his gratitude to many LBL personnel who conceived and designed many of the instruments described in this paper. Especially he wishes to acknowledge the effort of B. Ludewigt, T. Renner, K. Marks, M. Nyman, R. P. Singh, and R. Stradtner. This work is supported by the Director, Office of Energy Research, Energy Research

Laboratory Technology Transfer Program, of the US. Department of Energy under Contract No. DE-AC03-76SF00098.

REFERENCES

1. Wilson, R. R., "Radiological Use of Fast Protons," Radiology **47**, 487-491 (1946).
2. Tobias, C. A., Lawrence, J. H., Born, J. L., McComb, R. K., Roberts, J. E., Anger, H. O., Low-Beer, B. V. A. and Huggins, C. B., "Pituitary irradiation with high energy proton beams. A preliminary report," Cancer Res. **18**, 121-134 (1958).
3. Chu, W. T., Staples, J. W., Ludewigt, B. A., Renner, T. R., Singh, R. P., Nyman, M. A., Collier, J. M., Daftari, I. K., Kubo, H., Petti, P. L., Verhey, L. J., Castro, J. R. and Alonso, J. R., "Performance Specifications for Proton Medical Facility," March 1993, LBL-33749, (1993).
4. "Request for Proposals to Design and Construct Proton Therapy Equipment for the Northeast Proton Therapy Center at the Massachusetts General Hospital," NPTC-10, MGH, Boston, MA 02114, U.S.A., December 18, 1992.
5. Chu, W. T., Ludewigt, B. A. and Renner, T. R., "Instrumentation for Treatment of Cancer Using Proton and Light-Ion Beams," Reviews of Scientific Instrument, **64**, 2055-2122 (1993).
6. Gottschalk, B. and Wagner, M. S., "Contoured scatterer for proton dose flattening," Harvard Cyclotron Laboratory, a preliminary report 3/29/89.
7. Renner, T. R., Chu, W. T., Ludewigt, B. A., Nyman, M. A. and Stradtner, R., "Multisegmented Ionization Chamber Dosimetry System for Light Ion Beams," Nucl. Instrum. Methods in Phys. Res. **A281**, 640-648 (1989).
8. Renner, T. R., Chu, W. T. and Ludewigt, B. A., "Advantages of Beam-Scanning and Requirements of Hadron-Therapy Facilities," Int. Symposium on Hadron Therapy, October 18-21, 1993, Como, Italy (in press).
9. Pedroni, E., Blattmann, H., Böhringer, T., Coray, A., Lin, S., Scheib, S. and Schneider, U., "Voxel Scanning for Proton Therapy," *Proc. of the NIRS International Workshop on Heavy Charged Particle Therapy and Related Subjects* , July 1991, Chiba, Japan, 94-109.
10. Renner, T., Nyman, M. and Singh, R. P., "Control systems for heavy-charged particle radiotherapy facilities," to be published (1995).

CONTRIBUTED PAPERS

Initial Diagnostics Commissioning Results for the APS Injector Subsystems*

A. Lumpkin, Y. Chung, E. Kahana, D. Patterson, W. Sellyey,
T. Smith, and X. Wang
Advanced Photon Source, Argonne National Laboratory, Argonne, IL 60439

ABSTRACT

In recent months the first beams have been introduced into the various injector subsystems of the Advanced Photon Source (APS). An overview will be given of the diagnostics results on beam profiling, beam position monitors (BPMs), loss rate monitors (LRMs), current monitors (CMs), and photon monitors on the low energy transport lines, positron accumulator ring (PAR), and injector synchrotron (IS). Initial measurements have been done with electron beams at energies from 250 to 450 MeV and 50 to 400 pC per macrobunch. Operations in single turn and stored beam conditions were diagnosed in the PAR and IS.

INTRODUCTION

Significant progress has occurred in the last year at the Advanced Photon Source (APS) project which includes the beginning of commissioning of all the major injector subsystems. This process has of course included the commissioning of the primary diagnostics systems for each injector subsystem. When completed in 1996, the APS will be a synchrotron radiation user facility with one of the world's brightest x-ray sources in the 10-keV to 100-keV regime.(1) Its 200-MeV electron linac, 450-MeV positron linac, positron accumulator ring (PAR), 7-GeV injector synchrotron (IS), 7-GeV storage ring, and undulator test line provide the opportunity for development and demonstration of key particle beam characterization techniques over a wide range of parameter space. A description of the overall status with an emphasis on the diagnostic systems or techniques is provided. More detailed descriptions are provided at EPAC '94(2,3) in these proceedings (4-8), and/or in the previous instrumentation workshops.(9,10) Initial measurements have been done with electrons at energies from 250 to 450 MeV and 50 to 400 pC per macrobunch. Operations in single-turn and stored-beam conditions were diagnosed on both the PAR and IS. To date energy ramping in the IS to 1 GeV has been attained.

---

*Work supported by the U.S. Department of Energy, Office of Basic Energy Sciences, under contract no. W-31-109-ENG-38.

BACKGROUND

Space precludes providing a complete description of the accelerator facilities for the APS but some background information is needed. The baseline electron source is a thermionic gun followed by a 200-MeV linac operating at an rf frequency of 2.8 GHz, and a maximum macropulse repetition rate of 60 Hz. The design goals include 14-ps-long micropulses, separated by 350 ps in a 30-ns macropulse with a total macropulse charge of 50 nC. The 200-MeV linac beam will be focused to a 3-mm spot at the positron-production target. The target yield is about 0.0083 positrons per incident electron with a solid angle of 0.15 sr and an energy range of 8 ±1.5 MeV. The positrons will then be focused by a pulsed solenoid and about 60% of them will be accelerated to 450 MeV. The 450-MeV positrons are injected into the horizontal phase space of the PAR at a 60-Hz rate. As many as 24 macropulses can be accumulated as a single bunch during each 0.5-s cycle of the PAR. The injector (or booster) synchrotron accelerates the positrons to 7 GeV at which energy they can be extracted and injected into the designated rf bucket of the storage ring. A schematic of the APS accelerators, which lists the number of diagnostic stations, is given in Ref. 2 and shown in Fig. 1. The solid circle by a system indicates it has been commissioned with beam.

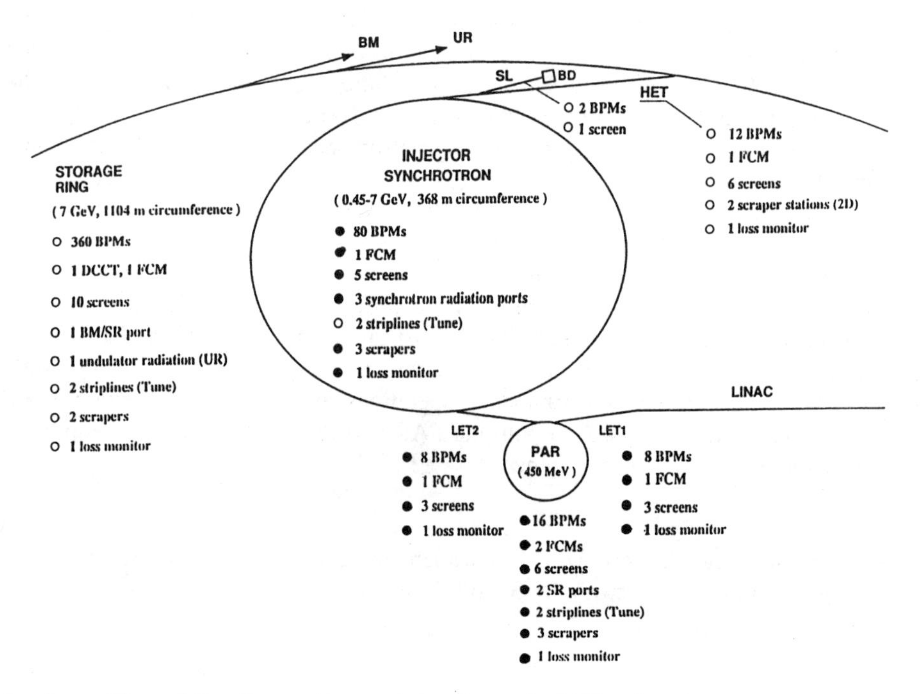

Fig. 1. An outline of the diagnostics systems for each APS subsystem.

The main design features of the subsystems are listed in Ref. 11. The 20-ns macropulse bunch length and the 200 to 400 pC per macropulse delivered by the linac into the low energy transport (LET) lines between the linac and PAR were measured by the fast current monitor. Macropulse repetition rates of 2 to 10 Hz have been provided with the ultimate design goal being 60 Hz. All experiments to date in the ring have used electrons. Recently a 360-MeV positron beam has been delivered into the LET1 line, and the 8-mA current was measured by the same fast current monitor.(12)

INITIAL DIAGNOSTIC RESULTS

The basic charged-particle beam parameters such as beam profile, position, current, beam loss, bunch length, and energy will be addressed. An initial report has been provided at the 1994 EPAC conference(2) which focused on the early LET1 and PAR results. In this report, results in the synchrotron for closing and correcting the orbit and ramping the beam energy to 1 GeV are reported. These tests are based on linac beam of 200- to 450-MeV energy and 50 to 400 pC per macrobunch.

Beam Profile Monitor

In the early stages of commissioning, one of the key diagnostic systems was the beam profile monitor based on an Al_2O_3 (Cr) screen material and a standard charge-coupled device (CCD) video camera. Three intercepting screens on pneumatic actuators are arrayed along the transport line z-axis between the linac and the accumulator ring. There are an additional six viewing screens/cameras in the PAR itself for single-turn tuning, three in the transport line to the IS, and five in the IS. The system images beam as low as 10 to 30 pC in a macrobunch from the linac. The images are displayed at standard 30-Hz rates on a monitor or digitized by a VME-architecture-based video digitizing system linked to a Sun workstation. Figure 2 of Ref. 2 shows the pseudo-3D representation of one of the first electron beam bunches injected into the LET1 line. Since the screen is at 45° to the vertical, the ellipticity of the beam is exaggerated in the uncorrected image. The beam size was a few mm (FWHM) in the vertical dimension. During commissioning, beam was readily transported through both the transport lines, the PAR, and even the 368-m-circumference IS using the observed image shapes and positions.

Beam Loss Rate Monitors

The loss rate monitors (LRMs), which will cover the entire extent of beamlines and accelerators, are operational on all but the HET and the storage ring now. A gas-filled coaxial cable acting as an ionization chamber was installed along the length of the transport lines and around the circumference of the PAR and IS. The gas mixture is 95% Ar and 5% CO_2, and the voltage across the center conductor to ground is 500 V. Figure 4 of Ref. 6 shows the clear effect in loss rate when one beam profile screen in the PAR is removed and its adjacent LRM cable is monitored in a strip-chart-mode on the workstation. Time flows to the left, so the drop from about 0.3 nA to 0.1 nA is clearly evident with <100 pC in the beam bunch. If the arrival time of the signals is viewed on a scope, few-meter axial resolution for losses can be determined. Reference 6 provides more details.

Beam Current Monitor

Monitoring of the current/charge in the transport lines and rings is based on the use of fast current transformers manufactured by Bergoz and in-house electronics. The electronics are described in detail by Wang elsewhere in these proceedings.(5) The current transformer signals are processed through a gated integrator and the output digitized to provide readouts on the workstation. During commissioning transported electron beams and positron beams from the linac were measured from 50 to 400 pC per macrobunch. In the PAR, both a fast current transformer (FCT) and integrating current transformer (ICT) were used to assess single-turn and stored-beam conditions. Kicker electrical noise interfered with using the FCT raw signal as a turns counter in the early commissioning so a photomultiplier tube (PMT) was used. Although the engineering screen clearly showed the stacking of charge in the PAR under 10 Hz injection rates (5 per 1/2 cycle), the digital oscilloscope trace is a more graphic presentation as shown in Fig. 2. The HP 54542 scope was set to cover 50 ms/div, and the envelope of five steps with increasing charge is clearly seen. About 200 pC per macrobunch was injected so we stacked to about 1 nC. This is not far from the nominal 250 pC per macrobunch of positrons needed. Only 24 bunches need to be accumulated for nominal operations.

Photon Monitors

In each of the three rings at APS at least one bending magnet's synchrotron radiation is viewed by photon detectors. Detection of this radiation can be used to count beam turns and provide a measurement for tune, beam size, and bunch length. A full complement of photon detectors and cameras was planned for radiation analysis. However, initial commissioning involved either a standard, CCD-based video camera and/or a photomultiplier tube (PMT). In the case of the PAR, an early commissioning issue was the large amount of electrical noise in the FCT signal from the nearby kicker magnet system. The PMT also showed noise, but the 300-mV signal from a single bunch/turn was much larger than the noise. The change from six turns in the PAR to thousands occurred in less than an hour

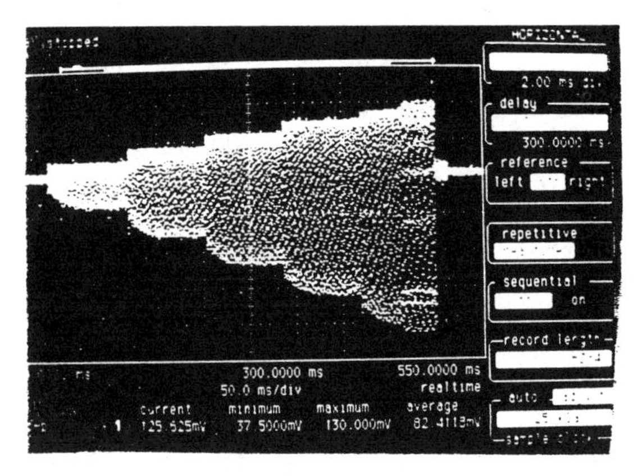

Fig. 2. A digital oscilloscope trace showing the accumulation of charge in the PAR as measured by the current monitor.

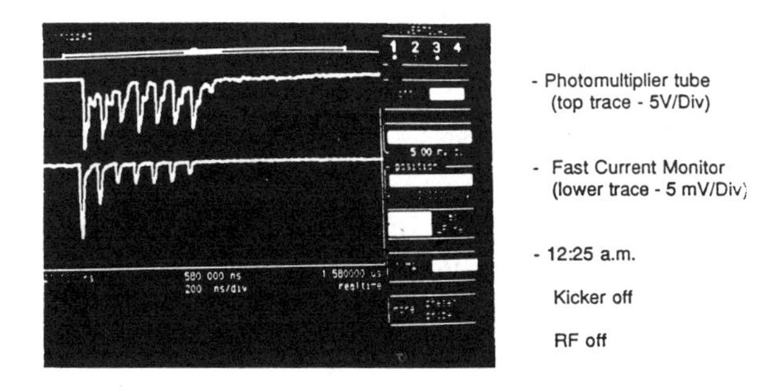

Time (200 ns/Div)

Fig. 3. Samples of PMT and FCT data showing early multiple turn detection in the PAR.

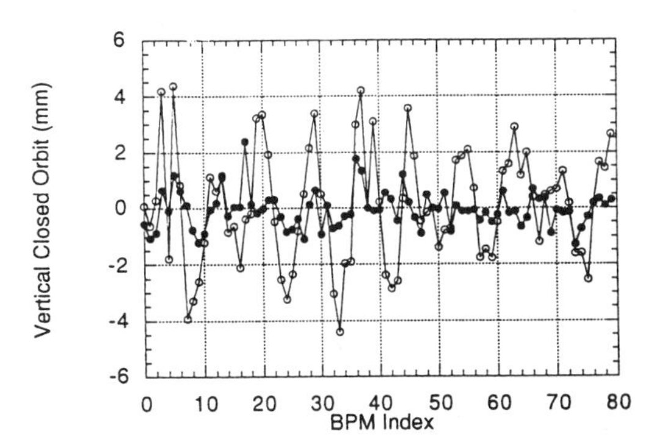

Fig. 4. Injector synchrotron rf BPM readings showed the closed orbit before (open circles) and after (closed circles) orbit correction. An rms-orbit error of ~0.5 mm resulted.

with the PMT diagnostic after several shifts of fighting the kicker noise. Figure 3 shows an example of multiple turn data with the PMT.

For PAR operations at 250 MeV, the lower energy resulted in a transverse damping time which corresponds to 80 to 100 ms. Our standard photon monitor camera easily tracked the progress of such damping in a series of six to eight frames taken 60 ms apart. This was reported at EPAC 94.(3) As a side note, the synchrotron radiation imaging has been very useful during early commissioning, even at low charge (~100 pC).

Beam Position Monitor

Beam position monitors (BPMs) utilize a series of stripline pickup devices: eight in the LET and sixteen in the PAR. The latter are described by Sellyey(7) in these proceedings. The processing electronics are designed to provide single macrobunch detection in the first transport line and require beam to be stored in the PAR. Position resolution of about 100 μm has been demonstrated.

In the injector synchrotron, 80 rf BPM stations are composed of 10-mm-diameter pickup buttons in sets of four and storage ring-like electronics(8). After initial triggering problems at the low injected beams of commissioning, the use of a timing signal to take data in the first few turns when charge was the highest produced useful results. For Fig. 4, the global orbit for 400-MeV electrons is shown to be closed and corrected. The initial rms orbit error was 2 mm and it was corrected to about 0.5 mm using the singular value decomposition technique (SVD). Corrector magnet supply jitter was an early limit on this process. A few weeks later, the beam energy was ramped from 400 MeV to 1 GeV.

Commissioning Experiences

Although machine commissioning stories have accumulated throughout the year, it still seems relevant to make a few comments on the February to August 1994 APS commissioning period.

- Don't trust the EPICS screen reading until the diagnostics have been timed with beam. Timing is critical.
- Be aware of commissioning beam parameter values versus operations parameter values. Timing scenarios don't always work, and peak intensities are not always as expected.
- Design for one to two orders of magnitude better performance than operations goals wherever possible (bunched beam, charge, size).

SUMMARY

In summary, key charged-particle beam parameter characterizations are well underway on the APS subsystems. Most of the diagnostic systems are now commissioned and are supporting the injector accelerator commissioning. The formidable task of instrumenting the third generation main storage ring has begun. We expect to commission the storage ring in 1995.

ACKNOWLEDGMENTS

The authors acknowledge the commissioning team members Mike Borland, Louis Emery, Nick Sereno, and Steve Milton for their extensive hours in the control room which included beam time for commissioning diagnostics. They also acknowledge the efforts of the Diagnostics Group technicians and staff who were instrumental in the testing, installation, and checkout activities.

REFERENCES

1. Moncton, D.E. , Crosbie, E., and Shenoy, G.K., "Overview of the Advanced Photon Source," Rev. Sci. Instrum., 60, 7 July 1989.
2. Lumpkin, A.H., et al., "Summary Test Results of the Particle Beam Diagnostics for the APS Subsystems," presented at the 1994 EPAC, London, England, June 26-July 1, 1994.
3. Lumpkin, A.H., Sellyey, W., Yang, B., "Proposed Time-Resolved Photon/Imaging Diagnostics for the APS," presented at the 1994 EPAC, London, England, June 26-July 1, 1994.
4. Wang, X., "Design and Commissioning of the APS Beam Charge and Current Monitors," these proceedings.
5. Wang, X., "Ultrafast, High-Precision Gated Integrator," these proceedings.
6. Patterson, D., "Design and Performance of the Beam Loss Monitor System for the Advanced Photon Source," these proceedings.
7. Sellyey, W., Barr, D., and Erwin, L., Design, Construction, and Wire Calibration of the PAR BPM Striplines," these proceedings.
8. Chung, Y., and Kahana, E., "Resolution and Drift Measurements on the APS Beam Position Monitor," these proceedings.
9. Proceedings of the Fourth Accelerator Instrumentation Workshop, Editors J. Hinkson, and G. Stover, Berkeley, CA, Oct. 27, 1992, AIP No. 281, (1993).
10. Proceedings of the Fifth Beam Instrumentation Workshop, Santa Fe, NM, Oct. 1993, AIP in press, Editors R. Shafer, and M. Plum.
11. Lumpkin, A.H., et al., "Overview of Charged-Particle Beam Diagnostics for the Advanced Photon Source (APS)," AIP No. 281, p. 150, 1993.
12. White, M., et al., "Status of the APS Linac," presented at the 1994 International Linac Conference, Tsukuba, Japan, Aug. 21-26, 1994.

Electron and Photon Beam Diagnosis in PLS Storage Ring

J.Y.Huang, M.K.Park, D.H.Jung, D.T.Kim, S.C.Won

Pohang Accelerator Laboratory, POSTECH,
San 31, Hyoja-dong, Pohang 790-784, Korea

Abstract

The beam diagnostic system in PLS storage ring, which is in commissioning now, is described. Components of the electron beam diagnostic systems are: 4 fluorescent screens for the first-turn beam detection, injection screen monitor, dc current transformer, beam scraper, betatron and synchrotron tune measurement system, and 108 beam position monitors composed of button style pickup electrodes. Among 108 BPMs, 96 have narrow band receivers and 12 have wideband receivers. With narrow BPMs, we achieved more than 65dB input dynamic range and better than $5\mu m_{rms}$ resolution. At every beamline frontend, there is an x-ray beam position monitor, composed of a pair of gold-plated diagonal electrodes, for the beamline beam steering and fast local orbit steering. Transverse beam sizes will be measured with two linear image sensor arrays and a $2d$ CCD array. A sampling oscilloscope and an optical-to-electrical converter module with $17ps$ risetime will be used for the measurement of short bunch length.

INTRODUCTION

The $2GeV$, low emittance PLS storage ring[1] has been completed and in commissioning now. We have developed, tested and installed various kind of beam diagnostic systems: 108 BPMs, 4 fluorescent screens, injection screen monitor, beam scraper, DCCT, tune measurement system, PBPM and beam profile measurement system. These are now all operational. Important points of the beam diagnosis for the 3rd generation accelerator will be accuracy, reliability, speed, computerization and intellegent diagnosis software. Special care has been taken for beam position measurement system and for commissioning instruments of the storage ring. As a consequense, we achieved better than $5\mu m$ resolution and more than $65dB$ input dynamic range of BPM. During the commissioning, we obtained clear beam orbit with only $0.1mA$ stored beam. For the first turn commissioning, we employed four fluorescent screens actuated by a simple vacuum pneumatic actuator. Stripline kickers were applied as the pickups, to get large input beam signal. Wideband BPMs were turned on to see the single bunch beam positions turn by turn. Even every BPM electrodes could be used to find the first turn beam trajectory, though we never needed to use them. All the monitors are interfaced to VME-bus local control computers and the local computers are connected to a central subcontrol computer via MIL1553B data communication network. Subcontrol computers and console computers are connected by Ethernet. Overview of these beam diagnostic instruments will be described.

BEAM POSITION MONITOR

Synchrotron radiation users demand that synchrotron light should be stable within a fraction of the beam size, typically less than $10\mu m$. To meet this requirement, BPM should have good resolution, long-time stability, reliability and high measurement speed for the fast and stable correction of closed orbit. Among the various existing methods of beam position measurement systems,[2,3,4] we adopted button style pickup electrodes and narrow-band heterodyne detector electronics.

Pickup Electrodes

Beam position is sensed by four button style pickup electrodes diagonally positioned. An achromat section of PLS vacuum chamber is formed by very long machined pieces of aluminun plates: $10m$ and $7m$ long respectively. To fit this chamber, we invented a pickup module composed of two pickup electrodes as shown in Fig.1. Modularization of pickup modules was very useful in accurate machining, fabrication, individual bench tests and in installation. A stainless steel-to-titanium bonded metal block was machined to form a plug-shaped BPM body. Then, two titanium button electrodes, ceramic insulators, titanium center conductor were vacuun brazed onto the BPM body to form a pickup module. Titanium was selected as the material of electrode and bottom part of the body, because its secondary electron yield coefficient is less than 1.0. This will avoid multiplication of secondary surface electrons yielded by the possible high frequency resonances in pickup electrode structure. Good electrical contact between pickup module and chamber wall is maintained by putting a spring rf contact around BPM body.

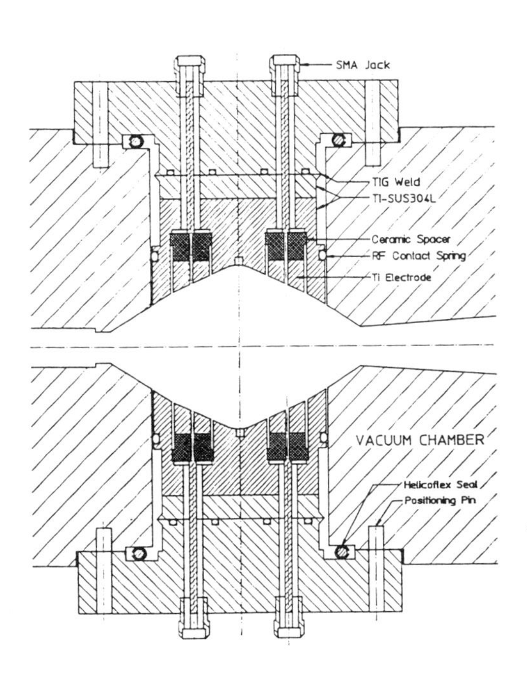

Figure 1. Layout of BPM components installed on vacuum chamber

Because of its simple structure, response of the button electrode to high frequency signal was excellent both in time and in frequency domain. In Fig.2 capacitance of the pickup electrode is less than $2pF$ and there is no significant resonances of the structure up to $10GHz$. However, low loss, low capacitance and fast rise time of pickup electrodes didn't seem to be the best, because of reflections of the signal between electrode and receiver input circuits. Instead, a kind of high frequency absorbtive material inside the structure of electrode assembly might be better for the processing electronics.

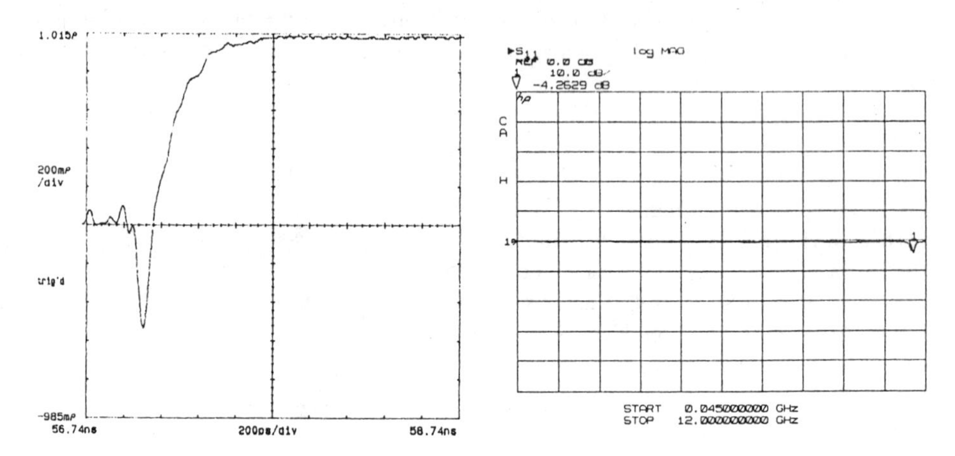

Figure 2. Responses of a pickup electrode a) in time domain reflectometry and b) in frequency domain reflection measurement.

Signal Processing Electronics

The short bunch signals picked up by four electrodes are delivered to the detector electronics through $24m$ Andrew Heliax 1/4" coaxial cables. Length of the cable is $35m$ at some places because of shielding wall structure of storage ring. However, cable length didn't affect BPM performance. Four pickup signals are multiplexed by fast rf switching module. To reduce channel-to-channel rf cross talk, we used low pass filters and three SPDP switches instead of a single SP4T switch. All the rf components and rf circuits are covered by machined aluminum cases. Giving special care in isolation, channel-to-channel isolation was improved to $40dB$ minimum when compared to $40dB$ maximum of a single SP4T switch. A block diagram of the signal processing electronics is shown in Fig.3. To widen dynamic range, we used two AGC amplifier/attenuators: one in the rf section and another one in the detector section of the receiver. AGC1 rf amplifier circuit has small input-current-dependent reflection. This current dependent effect was included in the receiver calibration data. Dynamic range is more than $65dB$ over two operation modes.

1. Mode I. Low current mode. When the beam current is between $0.1\sim30mA$. AGC1 is set to a fixed gain but AGC2 varies with input current. Mode I is used during the commissioning and single bunch fill operation.

2. Mode II. High current mode. When the beam current is between 3~500mA. AGC1 is set to a fixed attenuator. Most operation will be done in this mode.

In each operation mode, four pickup signals are detected in sequence and summed in the computer. If the sum is smaller than a reference value $20V$(or $10V$), AGC2 is fedback by a 12bit D/A converter to make the sum to be $20V$(or $10V$)$\pm 0.1V$. $20V$ sum reference voltage is the optimum value for the good linearity and large sensitivity, but the good measurement area is limitted to $\pm 7mm$. When the closed orbit distortion will be greater than $\pm 7mm$, possibly during commissioning, sum reference value will be set to $10V$ to make the measurement area wider. In this case, sensitivity of the system becomes smaller. From the experiences of beam position measurement during the early commissioning period, we found that the pickup signal is larger than we expected.[5] It seems that we don't need Mode I operation for the single bunch fill but will operate only in the commissining period. Then only Mode II operation with just a fixed attenuator, instead of AGC1, is enough. This will also reduce much of BPM calibration data.

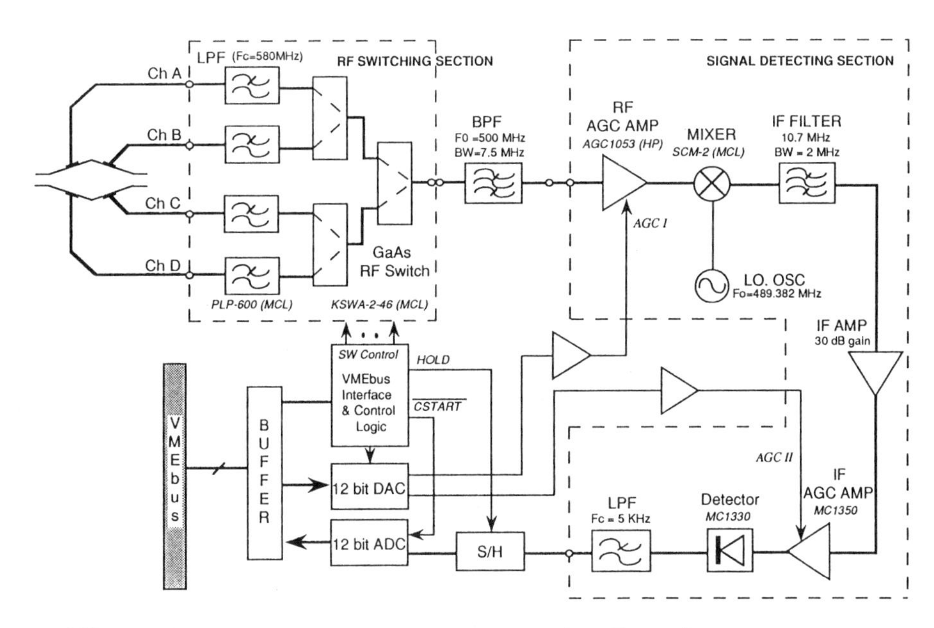

Figure 3. Block Diagram of Beam Position Monitor Electronics.

Sensitivity, resolution, and channel-to-channel gain differences were calibrated on the bench test setup. Resolution of BPM was better than $5\mu m$ typically. In Fig.4, input current vs resolution and calibration data of pickup button A are shown. All the pickup electrodes and cables were calibrated using Lambertson method.[6] We also measured sensitivities for all the BPM boards. An interesting result is that the sensitivity map of the whole system is linear in much wider range than the geometrical sensitivity of pickup electrode module. We guess this behavior as the logarithmic effect of detector electronics caused by coupling between AGC1 variable gain amplifier and AGC2 video detector. Geometric sensitivities S_x and S_y of pickup module were 6.18%/mm and 6.15%/mm respectively at the center of the chamber. Sensitivity of pickup and electronics system is 7~8%/mm.

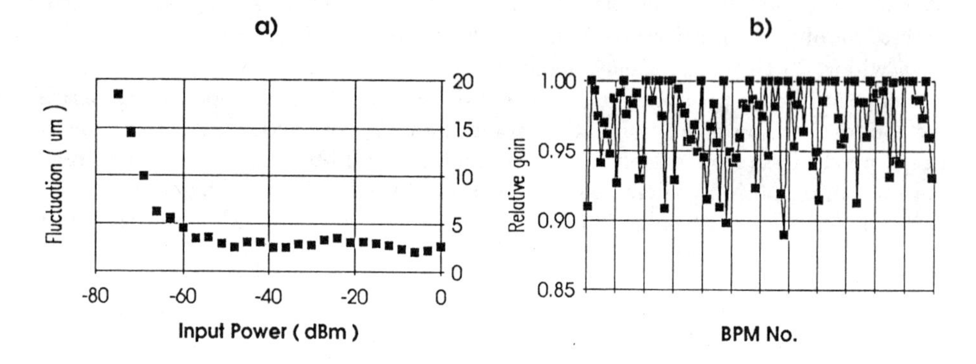

Figure 4. a) Resolution vs input beam current. In this graph, 0dBm correspond to 550mA, -55dBm to 1mA, -75dBm to 0.1mA respectively. Dynamic range is better than 65dB. b) An example of calibration data of pickup elcectrode and cable by Lambertson method. Only button A is plotted.

TUNE MEASUREMENT SYSTEM

Tune measurement system consists of rf tracking generator, rf power amplifier, stripline kickers, button style pickup electrodes, signal processing hybrid circuits and spectrum analyzer as shown in Fig.5. HP8560E spectrum analyzer equiped with built-in tracking generator was selected as the major instrument of the tune measurement system. With appropriate instrument parameter settings, the noise level has been lowered to -130dBm. Four stripline kickers, 17mm wide and 150mm long, are positioned diagonally inside the elliptical vacuum chamber wall. Starting from the empirical formula w/d~5, 50Ω longitudinal input impedance was matched by fine adjustment of d~3.5mm. Result of the computer simulation showed that 1A current through the stripline will kick the 2GeV beam by 1μrad/kick. Pickup electrodes are located upstream of a straight section of the ring where the betatron phase is advanced by $\pi/4$ from the kicker position. Although the longitudinal coupling impedance of the stripline kicker is maximum at 500MHz, strong harmonic peaks of the pickup signal will also appear at harmonics of 500MHz. Considering this and the frequency bandwidth of the rf amplifier(220~512MHz), we chose 321.6MHz for tune measurement. This is

300th harmonics of the revolution frequency $1.072MHz$. Using four coaxial line length adjusters, time delays between the pickup signals at input of hybrid network were adjusted within $10ps$. During the first-turn commissioning, we used striplines as the signal pickups by connecting the signal cable and the 50Ω load in opposite directions.

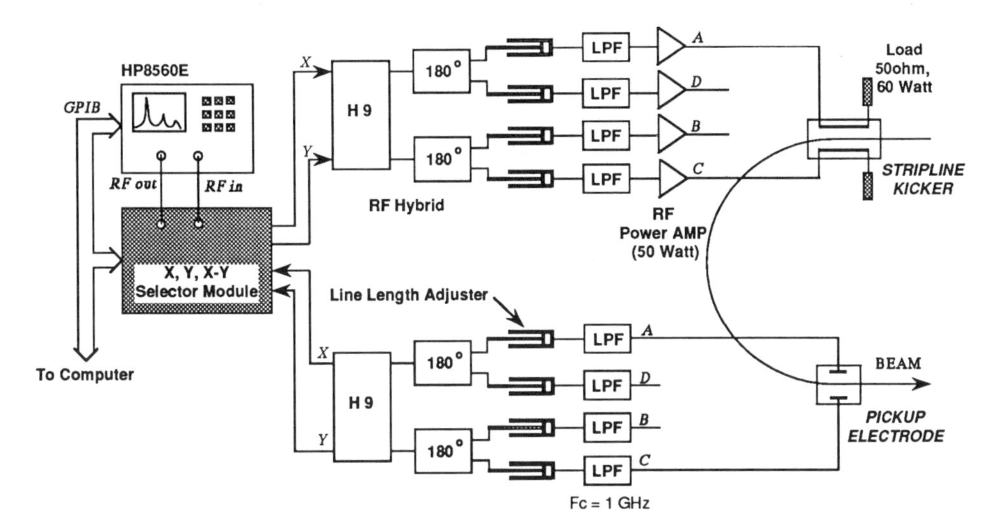

Figure 5. Layout of Tune Measurement System.

DCCT

To measure dc component of circulating beam accurately, we use a parametric current transformer supplied by Bergoz.[7] It is now widely used in many storage rings. A pair of magnetic cores are driven into saturation in opposite directions by an alternating current. Beam current will unbalance the core magnetizations by the amount corresponding to the beam current effect. A synchronous detector detects the second harmonic voltage induced on a third winding around two cores, then feedback a dc balancing current. This dc balancing current is proportional to the beam current. DCCT is assembled on the vacuum chamber as shown in Fig. 6. To avoid any beam impedance problem occurring from the chamber discontinuities or insulator gaps, we made inner wall of the DCCT chamber as exactly the same shape to the vacuum chamber wall. $1mm$ current-breaking gap was insulated by beryllium oxide gasket. Beryllium oxide gasket is also a good heat conductor, which is important for the bakeout of inner structure of DCCT assembly by conduction of heat. For the bakeout of ceramic break and bellows, a $200Watt$ carbon sheet heater is put around the ceramic break inside the core. During the bakeout, DCCT magnetic cores are water-cooled below $70°C$. Three pairs of thermocouples are connected to interlock system for the protection of DCCT: core temperature, bakeout heater temperature and cooling water temperature. Before operation we adjusted offset of the DCCT to zero. When the magnet power supplies were turned on in full current, however, DCCT read $20\mu A$. To reduce this kind of offset induced by the stray magnetic fields, we need to put a magnetic shield arround DCCT assembly.

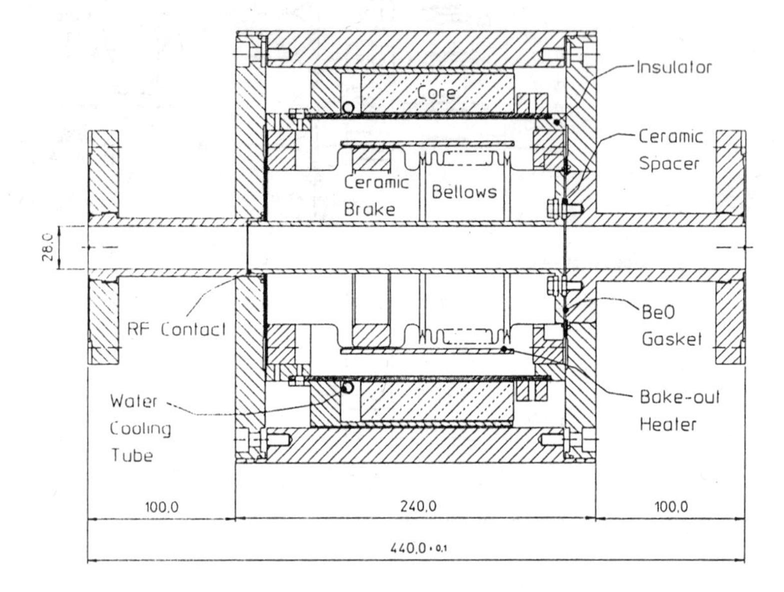

Figure 6. Layout of DCCT Assembly

FLUORESCENT SCREENS

Inspite of its primitive old-fashionedness, fluorescent screen is the most reliable beam finder during the first-turn commissioning. We used Chromox Type 6 fluorescent screen,[8] four-way cross vacuum chamber and vacuum air-actuator. A CCD camera sees fluorescent screen through a quartz viewport. Since the four-way cross chamber will induce large beam impedances when the beam is stored above $100mA$, screen monitors will be removed after the phase-I machine commissioning. The injection screen monitor will remain permanently. So its chamber wall profile is exactly the same to the neighborhood vacuum chamber. When the injection screen is off, beam sees no change of the chamber wall. Before the injection of beam to the storage ring, septum magnet and injection kicker magnet currents were adjusted by observing the displacement of injection point on the injection screen.

SYNCHROTRON RADIATION MONITORS

In each frontend of beamlines, x-ray beam position is monitored by a PBPM (photon beam position monitor). This will be used for beamline users to align their beamlines or for fast local beam steering system. $1.2\mu m$ gold is diagonally plated on the beryllium oxide plate to form a pair of photo-diodic electrodes. We used a $3mm$ thick beryllium oxide plate to pass only high energy x-rays over $10keV$ but filter out low energy lights for the good resolution of PBPM. The monitor will be calibrated by moving the detector assembly by stepper motor when the beam stored stably. Test of electronic circuit showed good linearity and wide dynamic range.

For the beam profile measurement system, we use Seya-Namioka beam port temporarily. A water-cooled molybdenium mirror reflects the beam in the vacuum chamber. This light is guided into the optical hutch which is located outside the shielding wall, $19m$ apart from the source. In vertical direction, small beam angle of $\theta_y \simeq (1/\gamma)(\lambda/\lambda_c)^{1/3} \simeq 2.4mrad$ cause large diffraction limit of imaging, $\delta_x \simeq \lambda/sin\theta_y \simeq 170\mu m$ with $400nm$ light. Transverse beam sizes σ_x and σ_y will be measured by two $25\mu m \times 512$ linear sensor arrays at $400nm$. A $2d$ CCD camera is also used for beam image processing. Beam emittances ϵ_x, ϵ_y are then calculated as $\sigma_x = \sqrt{\epsilon_x \beta_x + \sigma_\epsilon{}^2 \eta_x{}^2}$ and $\sigma_y = \sqrt{\epsilon_y \beta_y}$, where σ_ϵ and η_x are energy spread and dispersion function respectively. Bunch length will be measured with a fast optical-to-electrical converter, SD46, and a fast sampling oscilloscope, CSA803 Communication Analyzer having $17ps$ risetime and input spectral range of $1100 \sim 1600nm$. An x-ray diagnostic beamline is scheduled to be constructed for the accurate beam profile measurement.

REFERENCES

1. PLS, "PLS Design Report", 1991.
2. J.-C. Denard et al, "Beam Position Monitoring System for ELETTRA", in *Proc. of the EPAC*, (Nice, 1990), pp726-728.
3. R. Biscardi and J. Bittner, "Switched Detector for Beam Position Monitor", in *Proc. of the PAC, IEEE*, (Chicago, 1989), pp1465-1467.
4. J. Hinkson, "ALS Beam Position Monitor", in *AIP Proc. No. 252*, (Newport, 1991), pp21-42.
5. J.Y. Huang et al, "Analysis of the Electrical Signal Induced on a Pickup Electrode of PLS BPM", *Proc. of the 4th APPC*, (Seoul, 1990), pp1106-1109.
6. G.R. Lambertson, "Calibration of Position Electrodes Using External Measurements", ALS, LBL, LSAP Note-5, 1987.
7. K.B. Unser, "The Parametric Current Transformer, A Beam Current Monitor Developed for LEP", in *AIP Proc. No.252*, (Newport, 1991), pp266-276.
8. C.D. Johnson, "The Development and Use of Alumina Ceramic Fluorescent Screens", CERN/PS/90-42(AR).

The Beam Observation System of the ISOLDE Facility

G. J. Focker, F. Hoekemeijer, O.C. Jonsson, E. Kugler, H. L. Ravn
CERN, Geneva, Switzerland

During the reconstruction of the ISOLDE on-line mass separator facility at CERN one of the goals was the development and implementation of an efficient and user friendly beam observation system. The pick-up devices used on the two mass separators and the beam lines are mainly Faraday cups, wire scanners and wire grids. Most of these items were specifically designed at ISOLDE, in particular, a linearly moving wire scanner. The controls, data handling and display are all integrated in the PC based control system and can be accessed under Windows like any other element of the entire machine. The strong points of this measuring system as well as the limitations of beam observation in a low energy radioactive beam facility are discussed.

INTRODUCTION

The ISOLDE on-line mass separator facility at CERN (1) is mainly used for the production of pure samples of short-lived radioactive isotopes. It consists of two electromagnetic isotope separators, the General Purpose Separator and the High Resolution Separator, the latter now in its final phase of installation. The radioactive nuclides are produced in a suitable target bombarded with the high intensity beam of 1 GeV protons from the PS-Booster. After ionisation and acceleration to 60 keV, the ions are mass analysed. The selected beam from either of the two separators is then fed into a complex beam transport system and steered to the experiments.

The aim of the beam instrumentation is to obtain qualitative and quantitative information on intensity, distribution and position of these low intensity ($<10\mu A$) d.c. ion beams all along the beam path. In many positions the insertion of probes into the beam is impossible due to, for example, electrical potentials (as in the extraction gap) or difficult due to mechanical obstacles (as inside a magnet). In the design of the ISOLDE mass separators attention was paid to provide access for beam observation at the most critical regions such as the entrance to the magnets and the different focal points. However, if one wants to introduce several probes and a slit in a given position the available space along the beam axis may be limited, in particular if the focusing is strong. Therefore they need to be compact in the direction of the beam. In addition, emphasis was put on the probes to be easily interchangeable and radiation resistant because of the radioactive environment in which they have to operate.

The beam detection is in all cases based on the deposition of electrical charge from the beam on a probe. Unless suppressed, the emission of secondary electrons enhances the measured current by a factor depending on the surface conditions and the species and charge of ions collected.

PROBES

1. Moving wire scanners

The pick-up needle is mounted on a small chariot which also carries the pre-amplifier. It is guided on a rail and driven back and forth by a stepping motor via a 0.1mm thick piano wire. There is only one position-reference and only one direction is used for data-taking to avoid the effect of mechanical hysteresis. The development of this device was started more than 10 years ago by G. Sidenius and A. Lindahl at the Niels Bohr Institute in Copenhagen (2) and since then it has been further improved at CERN.

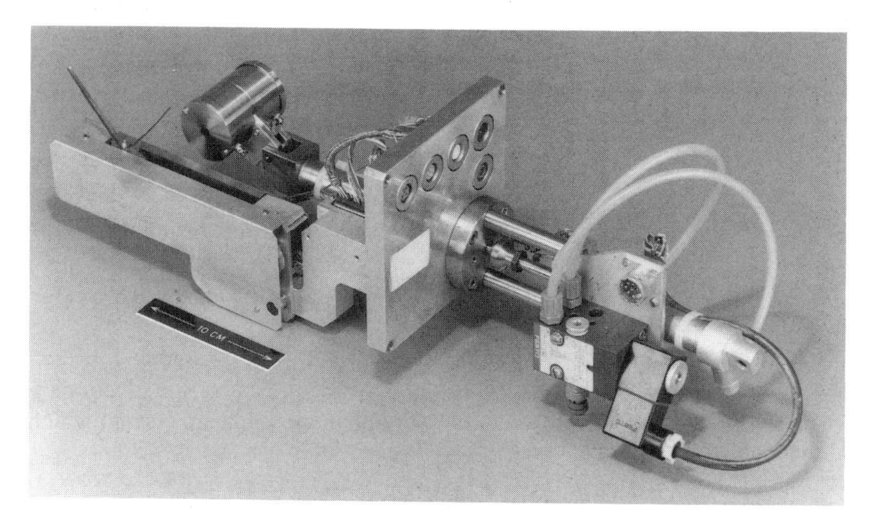

Figure 1. The standard beam observation unit.

In its standard version this beam scanner has a working range of 97 mm and is equipped either with a single needle of dia. 0.5 mm for scanning in one dimension or with a V-shaped pick-up bought from Danfysik. The former type is installed horizontally in the focal plane of the isotope separators to visualise the mass spectrum. The scanning range of this unit is limited to 3.5% of the mass spectrum. In order to cover the whole accessible mass range in the focal plane chamber of the General Purpose Separator, i.e. 13% on either side of the central mass, the scanner can be moved mechanically along the focal plane. Two scanners are mounted on the same support as backup for each other. A new version of this scanner unit with double stroke length is being developed.

In the configuration with a V-shaped pick-up (angle of 90°) the scanner chariot moves at 45° with respect to the horizontal plane thus producing both an x- and a y-scan of the beam profile. Mounted on a flange together with a Faraday cup they form the standard beam observation unit as shown in figure 1. Fifteen of these units are used in all the beamlines.

2. Wire grids

These robust devices are used where high spatial resolution is not required and where the main objectives are reliability and absence of semi-conductors because of high radiation levels near the target. Different versions with wire spacing ranging from 1,25 to 2,50mm that cover a surface of 75x75mm are used. Some of the grids were produced at the Niels Bohr Institute, others were bought from industry.

A special model was constructed at ISOLDE with only 20 horizontal wires (1mm spacing) with the aim of measuring the vertical profile of the central beam in the focal plane of the mass separators. This unit is fitted to the slit mechanism and can be moved into the beam together with one of the defining slits.

3. Faraday cups

The standard Faraday cup is of simple construction and has a diameter of 26mm. No additional cooling is used as the maximum beam current does not exceed 20μA at 60keV. Suppression of secondary electrons is achieved by applying a negative bias voltage of 130V to a repeller electrode inside the cup. The background current in the system is of the order of 10^{-12}A. Pneumatic drives are used for moving the cups in and out of the measuring position. These two positions are monitored by micro-switches.

In the "front ends" of the separators a special version of Faraday cup (active area 125mm dia.) is installed in the short gap between the extraction electrode and the first quadrupole lens This allows to measure the total ion beam current which is accelerated out from the ion source.

Complementary equipment

- Slits and apertures with well defined positions and dimensions, though being passive elements, are used at ISOLDE for defining sizes and positions of beams, in particular in the focal points of the ion optical system.
- Since the main purpose of the ISOLDE facility is the production of radioactive ions a dedicated tape transport is routinely used in particular for the identification of a selected isotopic beam and, combined with radiation detectors, for optimisation of proton beam, target and ion source parameters.
- In rare cases the following setups have been employed:
  i. high speed amplifiers for Faraday cup readings (3) when the ion production is achieved with a pulsed laser ion source (10μsec pulses),
  ii. a very thin (0.05mm diam.) fixed wire and electrostatic sweeping of the beam in the ISOLDE-3 high resolution separator.

DRIVERS AND AMPLIFIERS

The block diagram of the scanner controller system is shown in figure 2. The maximum electrical input current is 20μA full scale, the lowest current range actually used is 25pA. For high sensitivity both the bandwidth of the amplifier and the scanning speed are reduced. An automatic offset regulation is done which also compensates for parts of the influence of radioactive contamination left on the needle.

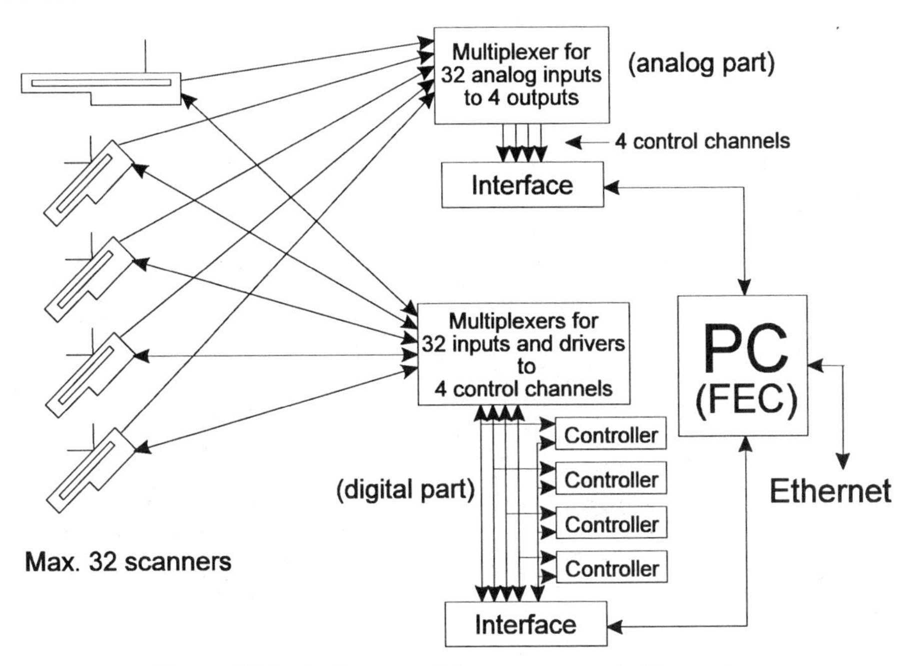

Figure 2 Block-diagram of the scanner controller system.

Linear micro-stepping motor-drivers are used. They don't produce the continuous electro-magnetic noise caused by chopping amplifiers and provide a very smooth movement even at lower speeds.

The scanner controller-system connects to the corresponding Front End Computer (FEC) via a special PC-card. The present card is based on an industrial board which has been thoroughly modified but it has technical limitations that cause the software to be slow and not reliable enough. A new PC-card is under development. For each control channel it will have an ADC and a FIFO-register (First In First Out). It will also have an improved digital interface.

The wire grid electronics is a compact unit which is mounted as close to the wire grid as radiation levels permit. It is designed to fit to the ISOLDE control system and memorises the data which can be read at any time via a relatively long cable and an interface-card inside the FEC.

The output from the Faraday cups is measured directly by a Keithley model 617 programmable electrometer. The switching between different cups is done by two Keithley model 706 multiplexers. A IEEE-488 controller, residing in the FEC, is in charge of the hardware-software communication with the ISOLDE control system.

CONTROLS AND DISPLAY

The most important feature of the ISOLDE instrumentation is its complete integration in the ISOLDE control system (4). It can be treated as any other part of the machine such as lenses or valves. All information is easily accessible even for a less skilled operator.

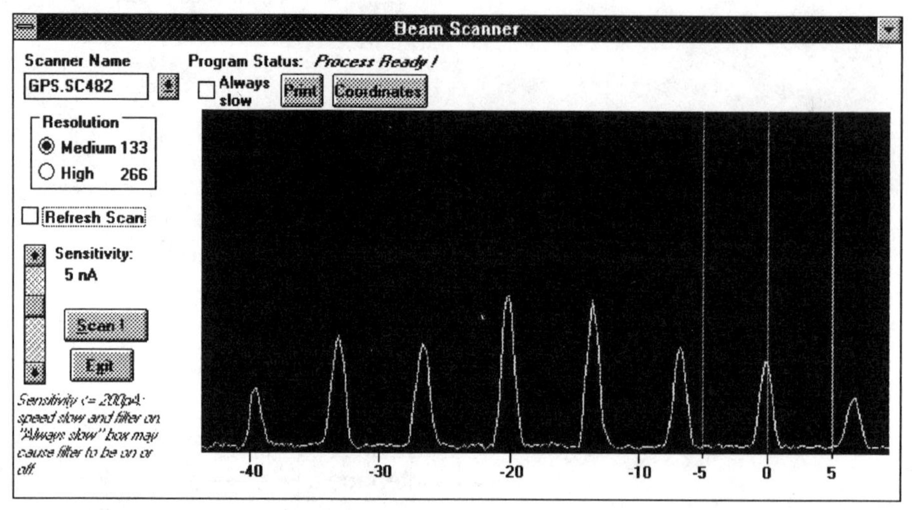

Figure 3. A focal plane scanner window showing the isotopic spectrum of radioactive Francium (mass numbers 220-227) as produced in a ThC_2 target. The intensity of the ^{226}Fr beam is 0.7nA.

The ISOLDE control system is entirely set up with PCs which are all connected to the CERN-wide Ethernet (Novell software). Most of the PCs are configured as FECs interfacing with the hardware and a few are configured as consoles. The FECs run C-written programs under DOS. The keyboard access to a FEC is in "NODAL", a widely used interpreter language at CERN. The consoles form the user interface via Microsoft Windows 3.1 and are programmed in C and Visual Basic. Any 386 PC or better on the CERN-wide office network may be used as a console. The database and other system-files are stored in a dedicated server.

The software in the FEC for the scanners may be complicated in comparison to that for a power supply, but for the consoles it makes very little difference whether they read data from a power supply or from a scanner. The only difference is that the approximately 270 datapoints from each scan are displayed graphically. The actual time for data handling and updating of the display is about 2 seconds. The new PC-interface will speed up data handling and allow to run four scanners simultaneously. Although the mechanical parts of the scanner would permit more than two cycles per second the updating frequency will be maintained in order to reduce the wear of critical components. For the same reason the scanners are switched off automatically after 10 minutes. Experience shows that higher updating speed is not really essential for setting up the beam. The spatial resolution depends on the number of datapoints and, of course, on the thickness of the pick-up needle (the apparent thickness may be higher due to possible mechanical vibrations). Figures 3 and 4 show typical scanner windows.

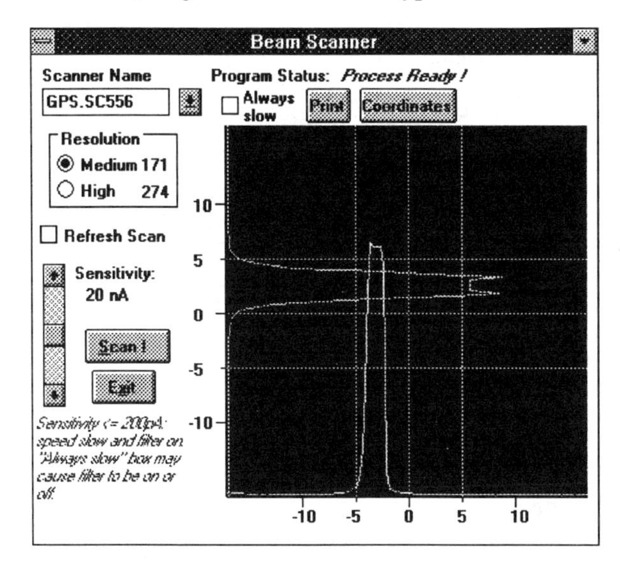

Figure 4. A typical beamline scanner window, ion current 9.3nA.

The readout of the wire grids with the ISOLDE control system is almost identical to that of the scanners. Although not as many datapoints are processed, the graphical display shows a fairly smooth picture similar to that of the scanners.

A window for a Faraday cup displays the measured current and the control buttons.

The Faraday cup reading of the ion current through the narrow slit in the focal plane of the separators, together with sweeping the field of the analysing magnet, is used to perform automatic mass scans. A program in the console allows to collect up to 3000 readings over the entire mass spectrum from mass 1 to 300.

SENSITIVITY

Although experiments at on-line isotope separators are occasionally done with ion beam intensities as low as 1 ion/sec., i.e. 10^{-19}A (5), it does not make sense to push the sensitivity of the beam observation equipment too far. In fact a continuous electrical background may be produced by the decay of radioactive nuclei which are implanted in the device itself or on nearby surfaces. This unwanted though decaying current, which may be as high as 10^{-9}A, depends on the preceding production of radioactive beams. In the read-out electronics of the scanners this effect is partly compensated by the offset regulation, but in general it sets a lower limit of some 10^{-11}A to the sensitivity of the beam observation for routine operation of an on-line mass separator facility.

To overcome this "blindness" when handling beams smaller than 10^{8} ions/sec we rely on a "mass calculator" program in the control system which, after calibration against an observable mass beam, calculates and adjusts the magnetic field (for a given acceleration voltage) for the selected mass. Thanks to the very high stability of the acceleration voltage and the excellent regulation of the magnetic field this program works accurately to 0.2 amu within the entire mass range.

ACKNOWLEDGEMENTS

The authors thank the technical staff at ISOLDE for their meticulous work in the assembly of the different probes and the CERN ECP-Division for the preparation of the many circuit boards. Jean-Luc Dautriat, French "cooperant" in 1988/89, made valuable contributions to the development of the scanners. Our thanks go to the ISOLDE Collaboration for helping with the production of some mechanical parts at university institutes. In particular we should mention the Isotope Separator Group at the Niels Bohr Institute in Copenhagen where the scanner development was started, but interrupted in 1985 when the group was dissolved.

REFERENCES

1. E. Kugler, D. Fiander, B. Jonson, A. Przewloka, H. L. Ravn, D. J. Simon, K. Zimmer and the ISOLDE Collaboration, "The new CERN-ISOLDE on-line mass-separator facility at the PS-Booster," Nucl. Instr. and Meth. **B70,** 41-49 (1992).
2. Private communication.
3. V. I. Mishin, V. N. Fedoseyev, H. J. Kluge, V. S. Letokhov, H. L. Ravn, F. Scheerer, Y. Shirakabe, S. Sundell, O. Tengblad, "Chemically selective laser ion-source for the CERN-ISOLDE on-line mass-separator facility," Nucl. Instr. and Meth. **B73,** 550-560 (1993).
4. O. C. Jonsson, O. Borch, A. Bret, R. Catherall, I. Deloose, G. J. Focker, D. Forkel, E. Kugler, G. Olesen, A. Pace, H. L. Ravn, C. Richard-Serre, G. Shering, O. Tengblad, H. J. Torgersen and the ISOLDE Collaboration, "The control system of the CERN-ISOLDE on-line mass-separator facility," Nucl. Instr. and Meth. **B70,** 541-545 (1992).
5. M. J. G. Borge, H. Gabelmann, L. Johannsen, B. Jonson, G. Nyman, K. Riisager, O. Tengblad, and the ISOLDE Collaboration, "The decay of ^{31}Ar," Nucl. Physics **A515,** 21-30 (1990).

Diagnostic Tools for the COSY-Juelich Synchrotron

J. Bojowald, K. Bongardt, J. Dietrich, H. Labus, H. Lawin, R. Maier,
R. Wagner *KFA Jülich*
and I. Mohos, *KFA Jülich/KFKI Budapest*

Abstract

The cooler synchrotron COSY, a synchrotron and storage ring for medium energy physics, is being used for internal and external target experiments since fall 1993. The cooler ring has delivered protons up to the design momentum of 3.3 GeV/c. In this paper some diagnostic tools in the COSY-ring are described. Almost all diagnostic components are installed and tested with beam. Special emphasis is given to the tools and methods, needed for optimizing the extraction process.

INTRODUCTION

COSY Jülich is a cooler synchrotron and storage ring with a proton momentum range from 270 to 3300 MeV/c. It has been conceived to deliver high precision beams for medium energy physics. Since its inauguration in April 1993 substantial progress in developing beams for the experiments has been achieved and the physics program has started with first measurements [1]. The diagnostic tools at COSY and results of measurements are described in ref. [2,3]. In this paper, the major diagnostic components for extraction are described.

MONITORING THE EXTRACTION PROCESS

Up to now substantial progress has been made for accumulation, acceleration and extraction of particles in COSY. At the end of the acceleration cycle, we have about $2 \cdot 10^9$ circulating particles at momentum of about 3.3 GeV/c, corresponding to an energy of 2.5 GeV. After reaching the flat top values, the particle number stays con2stant, before the resonant extraction process starts.

At flat top, the particles are slowly extracted over 1 s by exciting a third order resonance. The beam is monitored by Multiwire Proportional Chambers (MWPC) [2] in the extraction beam lines, see Fig. 1. Integrating over the 64 wires in each plane gives the beam intensity during the extraction time, shown in Fig. 2. Also included are the traces of the sextupole power supply, the quadrupole power supply with "Q-jump" and the average COSY-beam current measured by the Beam Current Transformer (BCT).

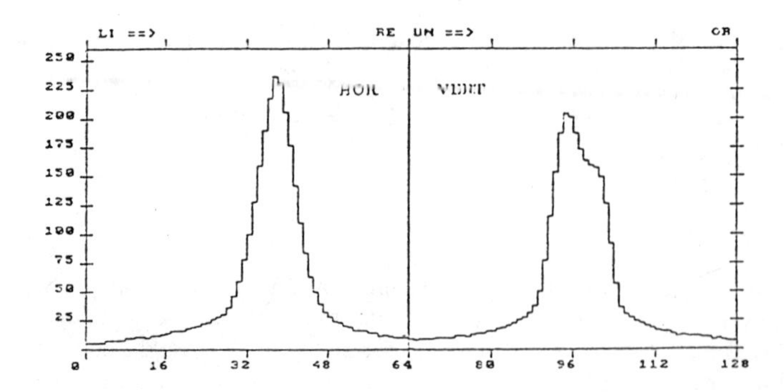

Figure 1. Beam profile of the extracted beam (1.1 GeV/c) measured by MWPC.

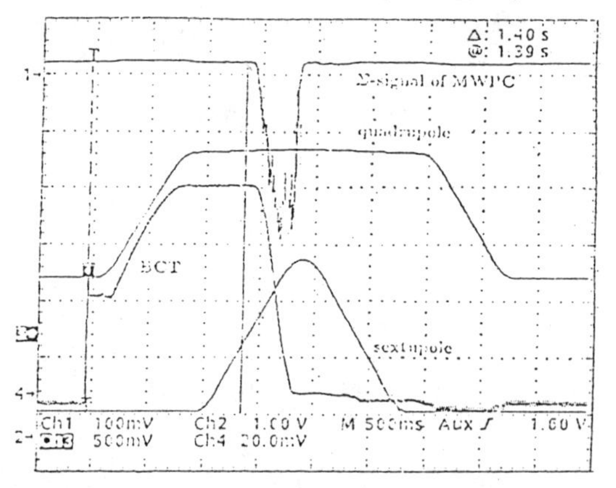

Figure 2. Extracted beam intensity (1.1 GeV/c) with the associated sextupole ramp, quadrupole ramp and average beam intensity in COSY measured with the BCT.

MEASUREMENT OF CHROMATICITY AND DISPERSION FUNCTION AT EXTRACTION ENERGY

In order to optimize the resonant extraction process, quite detailed measurements have been persued at extraction energy. Most of the experimental results are obtained by using the beam position monitors (BPM) either in broadband or in narrowband mode [4].

The broadband mode with 7 MHz bandwidth (time domain) reproduces the bunch shape and is used for the analysis of the beam turn by turn, e.g. for the analysis of the bunch shape (Σ-signal) itself or of beam oszillations (Δ-signal or

Δ/Σ-quotient) during time measurements with kicker excitation. The position sensitivity in this case is 0.4 mm. The narrow band mode (frequency domain) with remote presettable band pass filter bandwidths of 10, 100 and 300 kHz in the 0.2 - 60 MHz frequency range results in much higher resolution and in averaging over several turns due to the filter response time. It is used in diagnostic measurements during normal synchrotron operation, e.g. for closed orbit measurements. The position sensitivity is about 0.1 mm.

In Fig. 3, the measured horizontal dispersion curve along the complete COSY ring is shown for a momentum of 1.1 GeV/c. The data are obtained by shifting the closed orbit to an adiabatic frequency change. The measured static closed orbit is subtracted afterwards. The achievable accuracy of the dispersion curve is better than ± 1 m, obtained by averaging over different frequency changes.

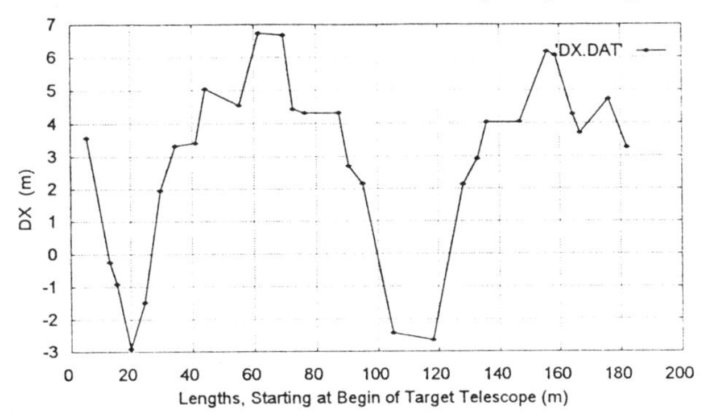

Figure 3. Measured horizontal dispersion curve at 1.1 GeV/c.

In Fig. 4, the theoretical dispersion curve from MAD calculations is plotted for some quadrupole settings. The two strong negative peaks at a position of 20 m resp. 110 m are the midpoints of the two long straight sections in COSY, 40 m in length each. Both straight sections are not dispersive free, indicating a shift of γ_T above the extraction energy [1]. The measured and calculated dispersion curves agree qualitatively, the large discrepancy at both negative peaks is not understood.

In Fig. 5 and 6, the measured horizontal and vertical chromaticities at 1.1 GeV/c are plotted, obtained with the same method as before. More datapoints are needed to determine the quadratic term in the tune dependence.

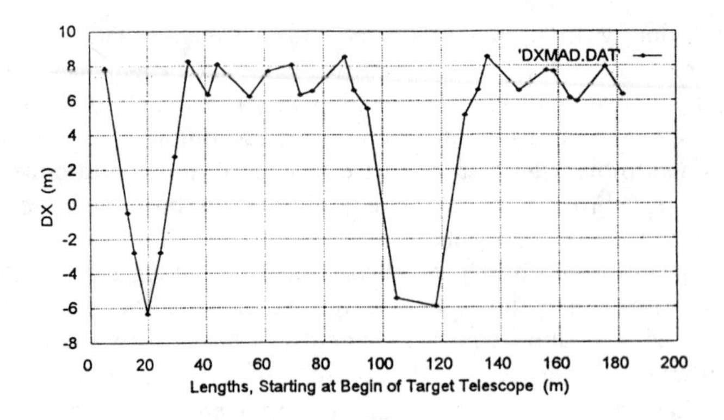

Figure 4. Theoretical horizontal dispersion curve.

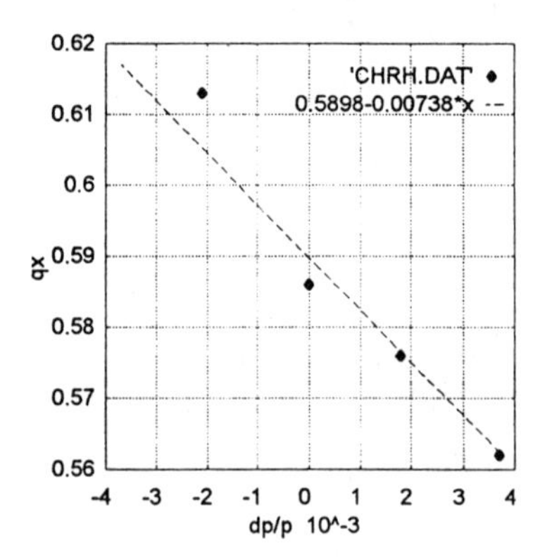

Figure 5. Measured horizontal tune dependence.

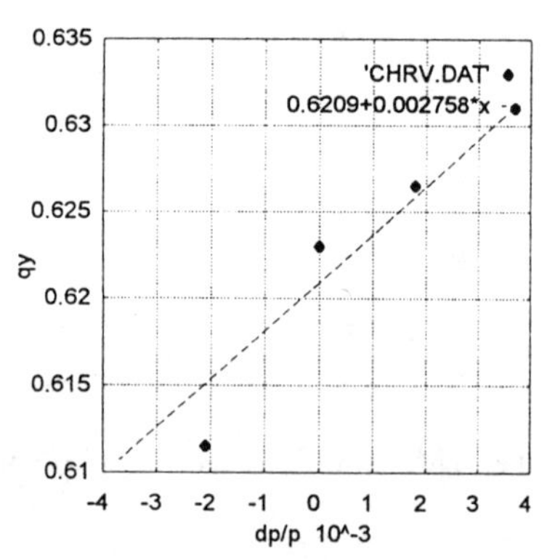

Figure 6. Measured vertical tune dependence.

STUDIES OF THE PHASE SPACE BOUNDARY AT THE ELECTROSTATIC SEPTUM

For the slow extraction process at COSY, the horizontal tune is moved towards a third order resonance and the particles are shifted towards the electrostatic septum by a local closed orbit bump. By exciting one or more sextupoles, in addition a phase space boundary (separatrix) is created. The correct orientation of this separatrix is essential for getting a good extraction efficiency.

Measurements are performed for studying the phase space boundary. The experiments are done by using two BPM's near the electrostatic septum, about 90°

apart in horizontal phase advance. A fast horizontal kicker, rise time less than 1 μs displaces the particles of the closed orbit. By operating the two BPM's in the broadband mode, the turn by turn oscillation around the closed orbit can be measured.

Figure 7 and 8 show sum and difference signal at both pickups, varying from turn to turn. The fast kicker is fired once after 5 turns. The obtained closed orbit values are about 7.5 mm at BPM 21 and about 0.5 mm at BPM 24. From this two values the divergence of the beam center at BPM 21 can be calculated.

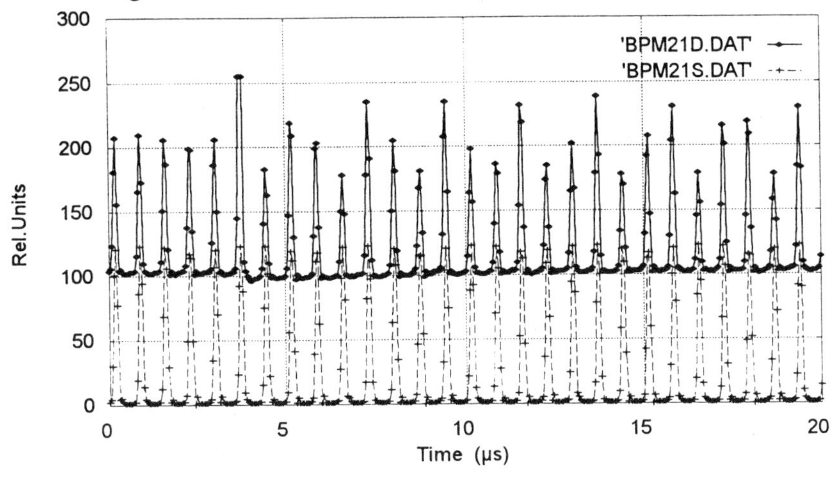

Figure 7. Turn by turn sum and difference signal at BPM 21.

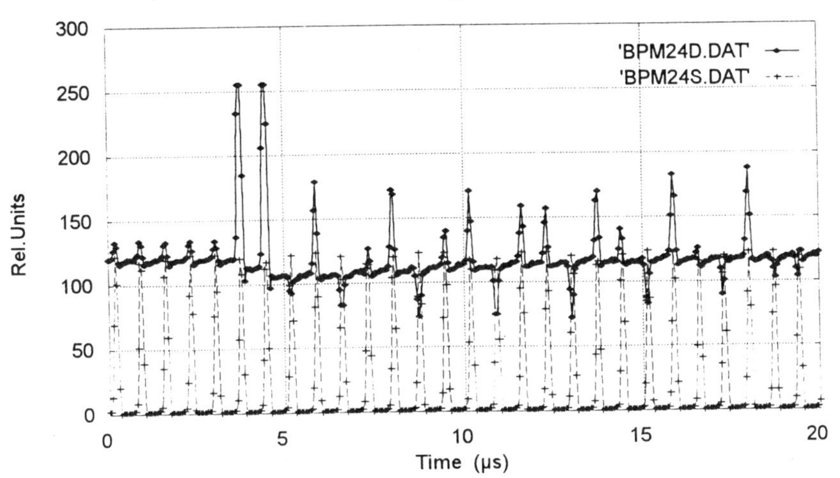

Figure 8. Turn by turn sum and difference signal at BPM 24.

Figure 9 show the FFT spectrum of BPM 21 from which a horizontal fractional tune of $q_x = 0.67$ is detected. This value is in good agreement with both, the theoretical value and the experimental value obtained by coherent beam excitation with a stripline unit.

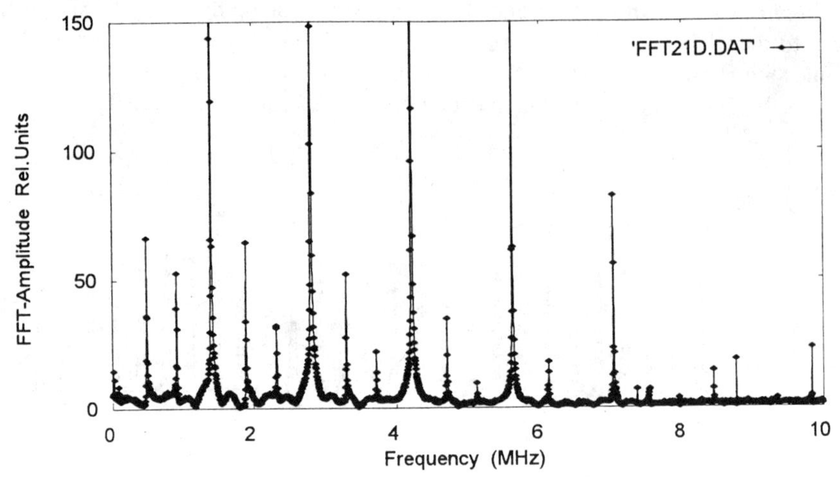

Figure 9. FFT spectrum of BPM 21.

Figure 10 shows the so obtained phase space oscillations of the displaced beam center at BPM 21, which is in front of the electrostatic septum.

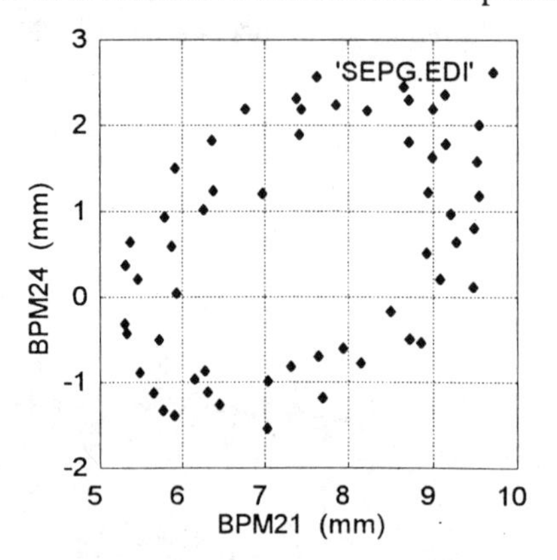

Figure 10. Forced horizontal betatron oscillations at BPM 21.

For detecting the separatrix, the phase space boundary created by sextupoles, the same measurements are performed, but now with switched on sextupoles. Due to the non-negligible beam size at the electrostatic septum, the interpretation of the experimental results is difficult, especially if the beam center is displaced nearby the separatrix. Part of the particles are stable here, parts are instable. Work is going on to get the separatrix in this case.

ACKNOWLEDGEMENT

We would like to thank all the colleagues of the COSY-team, especially the engineers and technicians of the diagnostic group, making these commissioning results possible.

REFERENCES

1. R. Maier, U. Bechstedt, J. Dietrich, U. Hacker, S. Martin, D. Prasuhn, P. v. Rossen, H. Stockhorst, R. Tölle, Status of COSY, *European Particle Accelerator Conference, EPAC'94*, London, 1994
2. J. Bojowald, K. Bongardt, J. Dietrich, H. Labus, H. Lawin, R. Maier and I. Mohos, Diagnostic Tools for the COSY-Jülich Synchrotron, *European Particle Accelerator Conference, EPAC'94*, London, 1994
3. J. Bojowald, K. Bongardt, J. Dietrich, U. Hacker, H. Labus, H. Lawin, R. Maier, I. Mohos, Diagnostic Tools for the Commissioning Period of COSY-Juelich, *Proc. First European Workshop on Beam Diagnostic and Instrumentations,* Montreux, May 1993, CERN PS 93-35, p. 171
4. J. Biri, M. Blasovszky, J. Gigler, I. Mohos, K. Somlai, Z. Zarandi, J. Bojowald, J. Dietrich, U. Hacker, R. Maier, Beam Position Monitor Electronics at the Cooler Synchrotron COSY-Jülich, *Real Time '93 Conference*, Vancouver 1993

The Use of LEP Beam Instrumentation with Bunch Trains

Claude Bovet
CERN, 1211 Genève 23, Switzerland

Abstract

Filling LEP with trains of bunches separated of only 247 ns instead of the present eight equidistant bunches (at 11 μs intervals) has important implications for the use of beam instrumentation. The BPM system will be limited in its measuring capabilities. Most other instruments will not be able to identify bunches unless further developments are made in their timing and acquisition electronics. A review of the situation is presented together with the planned actions to cope with the observation of individual bunches in trains.

INTRODUCTION

In order to reach the higher luminosity needed for $W^{\pm}$ physics, LEP is likely to run, in the future, with a newly proposed filling scheme of "Bunch Trains with Head-On Collisions" [1] to replace the present pretzel scheme (8 equidistant bunches).

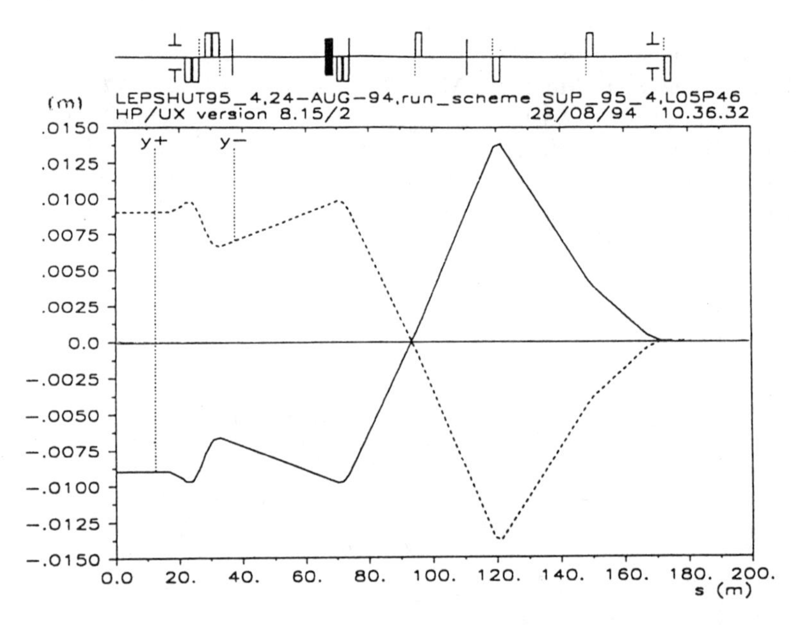

Figure 1. Vertical separation bumps for positrons (y^+ full) and electrons (y^- dashed) on the right side of the crossing point IP1 at 45 GeV.

The trains will have a maximum length of 220 m and if four trains are circulating in each beam, an adequate separation has to be provided over a length of ±110 m around the 8 IP's. The vertical separation foreseen around IP1 is shown in Fig. 1 and poses specific problems for some instruments layed down in this section of the machine: the tune-meters, the synchrotron light sources (mini-wigglers) and the polarimeters.

The short distances between bunches in the trains will influence in different ways the measurements that can be done with the various instruments. For someones the measurements will remain valid and be representative of all bunches in the train, for some others the measurements will concern only the first bunch or be completely corrupted. All instruments are reviewed here to see how best they can be used with bunch trains and what modifications should be foreseen for them to be more performing.

Front-end Electronics

Bunch trains impose severe constraints to the use of the existing electronic systems which have been conceived for bunches separated by 22 µs. Table 1 reviews the situation of the front-end electronics of the various instruments. The aim and the proposed action are explicitated in more details under the specific names of the instruments, in the following paragraphs.

Table 1: Front-end Electronics with Bunch Trains

Instrument	Possibility of 100ns-gating	Aim	Action
BCT	no	parallel acquisition	new front-end needed
BOM WB	yes	orbits for individual bunches	special gating of 2 BPMs
BOM NB	no	no hope !	
Tune-meter	yes	bunch selection	additional gating
BEUV	yes	bunch selection	faster intensifiers
BEXE	no	16 bunches in a row	new pulsed bias needed
Streak camera	yes	32 bunches in the picture	more picosecond gating
Luminosity det.	yes	parallel acquisition	additional electronics
Polarimeters	yes	bunch selection	none

Bunch Selection

None of the acquisition systems used in LEP beam instrumentation is capable of reading data at the rate of the bunches in a train (4 MHz) but they all can read at the frequency of the trains (44 kHz). Therefore, provided that the front end can cope with a gate of 100 ns, a triggering sequence like 1,2,3,4 in Fig. 2 can provide a full acquisition of the 16 bunches in 4 revolutions. Alternatively, in order to reveal the systematic behaviour of the different bunches in a train, a sequence of acquisition as illustrated in line 5 could be specially helpful.

Such sequences can be produced and computer controlled by means of a VME module which can yield any timing sequence over 16 LEP turns, with a resolution of 50 ns and a jitter of 1 ns. All programmed sequences are synchronised to the beams by means of a dedicated timing system[2].

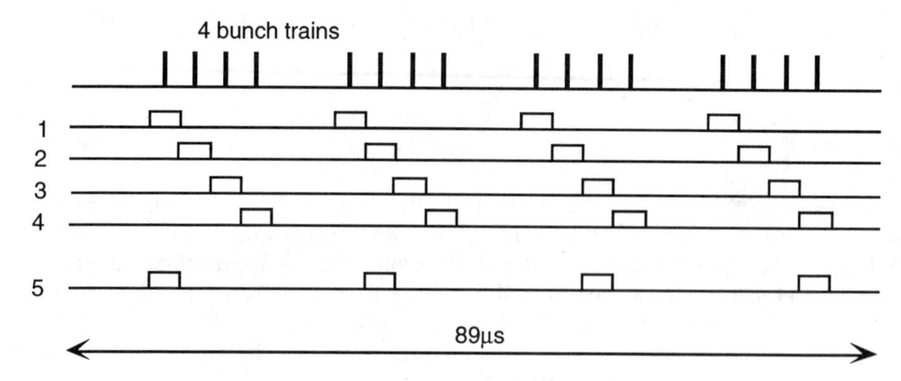

Fig. 2. Different gatings of individual bunches.

BUNCH CURRENT TRANSFORMERS[3]

BCTs in LEP are set up to measure the 16 bunches of the present pretzel scheme. For future bunch trains their integration window has been enlarged from 730 ns to 1100 ns, to accommodate trains of up to 750 ns, but the measurement cannot distinguish between bunches in a train. To do this, a new signal processing is needed. Software development is presently under way to get on line digital information from the bunch shape observation made with the help of a Lecroy 9350M oscilloscope. The scope sampling will be triggered by an external clock at a frequency of 1 GHz during the passages of the trains, i.e., for 1 μs every 22 μs. In its first version this should provide for bunch equalisation during filling. For more accurate measurements leading to lifetime determination for all individual bunches, more work is needed before next LEP start-up (March 1995).

BEAM ORBIT MEASURING SYSTEM[4]

BOM will measure only the first bunch in a train, and, for a fraction of the BPMs, this will only be possible for trains arriving at a collision point and not for those leaving.

BOM Wide-Band Electronics

This system, used for the 56 BPMs located close to the 8 crossing points, is auto-triggered by the first bunch of the selected polarity, and remains busy for 2 μs during the data acquisition sequence. For the measurement to be valid the signal should not be perturbed by another bunch signal arriving in the interval -20 ns/ +40 ns. This may occur for some BPMs close to the IP's when trains of bunches are crossing.

For incoming trains the first bunch will be measured correctly since the double time of flight from a BPM to the IP is always larger than 40 ns (see the distance to IP in Table 2). For outgoing trains the situation is more involved, because of the interferences between the two trains and Table 2 shows that two thirds of the pickups near the odd IP's, located at QL2B's and at QL4B's, will not work because of this odd coincidences.

Table 2. Wide band pickup excitation by trains with bunch separation of 87 λ_{RF}

Position	Distance to IP	Dt	dt(e–)	dt(e±)
	[m]	[ns]	[ns]	[ns]
Odd IP's				
QL1B R	+26.485	176.7	–176.7	70.3
QL2B R	+32.984	220.0	–220.0	**27.0**
QL4B R	+74.087	494.0	**–0.3**	246.7

Observation of Individual Bunches in Trains

Computer simulation has shown that all bunches will have different vertical closed orbits because of the different beam-beam kicks they experience at the various crossing points where the separation bumps have unequal amplitudes. This effect would be interesting to check at a point in the machine, with a suitable phase advance.

Contrary to NB electronics, the analogue part of the WB electronics can follow the sequence of bunches separated by 247 ns. But the digital encoding puts a dead time of 2 μs after each reading, which precludes the measurement of successive bunches in a train. In order to remove this constraint the following strategy can be employed. With some hardware modification an external trigger produced by the BST can be introduced to provide a gating of the wanted bunch in a train. This gating can be done of any bunch in a given train and be different on four successive revolutions so that all 16 bunches are eventually recorded. Such special gating could be introduced for a few WB pickups near one given pit.

BOM Narrow-Band Electronics

For BPMs with NB electronics a measurement can be validated only if the interval –600 ns to +90 ns is clear from any other signal. A detailed analysis (see Ref. 1) shows that a total of 80 NB pickups will not be able to measure outgoing trains with 4 bunches separated by 87 λ_{RF}. With trains of only 2 bunches this number reduces to 28 but, if the bunch spacing is increased to 150 λ_{RF}, then the number of deficient pickups rises again to 48. All of those pickups would work well with WB electronics but the conversion is a major operation which costs time and money. Due to the modularity of the components it would make sense to extend the WB system modulo 8 BPMs per pit. Converting 32 BPMs (for instance 8 per even IP) would cost 800 kSF and take a year to implement.

TUNE MEASUREMENT[5]

Individual bunch excitation for tune measurement is provided by four shakers located at 122 m and 152 m from IP1. They are fed with two types of pulses: the short ones are too long for exciting individual bunches in a train but the long ones can be used for exciting equally all bunches in a given train.

Bunch observation is presently done with a front end electronics similar to the wide band of BOM and therefore is auto-triggered on the first bunch of a train. A more sophisticated gating will have to be provided for the observation of the following bunches. Another possibility presently under scrutiny is to use signals obtained from two directional couplers located at ±957 m from IP1. The advantages are numerous : no displacement of the beams due to the vertical separation bump, no time interference due to crossing trains, higher sensitivity.

SYNCHROTRON RADIATION TELESCOPES[6]

The four BEUV telescopes are located in the machine tunnel around IP8. In their TV mode they integrate the light emitted by lepton bunches during 20 ms. The beam cross-section is therefore averaged over all bunches of a beam and this will not change with bunches grouped in trains. For single shot measurements the gated image intensifiers used so far could not separate bunches in a train, but this has become possible with the faster light intensifier (MCP), now under test, timed directly from the RF clock signal available at IP8.

HARD X-RAYS MONITORS (BEXE)[7]

The acquisition electronics integrates the signal during 2 μs and cannot separate bunches in a train. But the integrating front end shows a linear response for bunches separated by some 247 ns and the system, as it stands, will give profile measurements averaged over a whole train. For the selection of a particular bunch in a train a new development is required : of either a pulse generator for the polarisation voltage of the detector, or of a new front-end to the measuring system which would allow for a much shorter integration time.

In any case a sophisticated gating will have to be used in order to acquire all 16 bunches in a minimum of 4 revolutions (see Fig. 2).

STREAK CAMERA[8]

With the vertical bumps needed to separate the trains near IP1 the light of the mini-wigglers will have to compete with the radiation produced in the quadrupoles QL4. MD work and computer simulation are underway to see if these two sources can be disentangled, which would be desirable because the source in the QL4 is too long (3.70 m) to be focused properly and its characteristics are not stable since they depend on the bump amplitude.

Streaks can be fired at a maximum frequency of 45 kHz and can be triggered on any circulating bunch. The selection of different bunches in a train requires a fine tuning of the timing, modulo 247 ns, which can now be done remotely from the LEP control center. For the time being one given bunch can be seen at successive revolutions, or if the repetition time is set to a quarter revolution, the same bunch number can be seen from each train in succession, for a few revolutions.

In the future the timing module which divides the RF frequency to generate the streak camera triggers will have to be modified to allow for more sophisticated selections of different bunches (a to d) from each train (1 to 4), see Fig. 3.

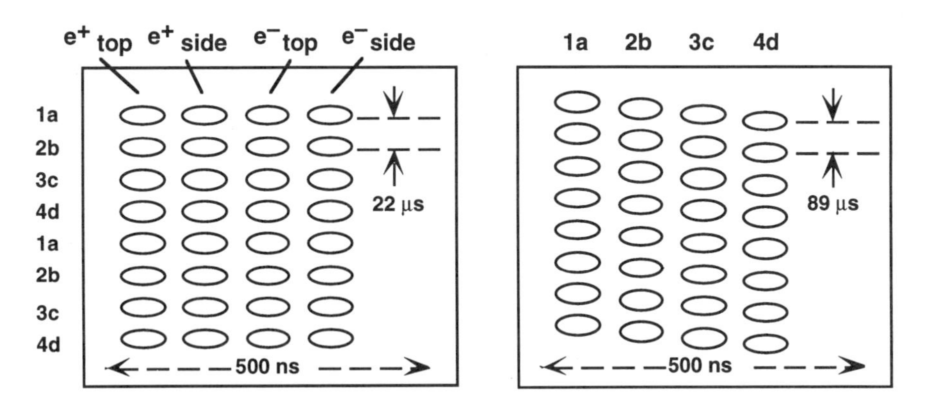

Fig. 3. Data taking sequences with the streak camera : 1,..4 trains, a,..d bunches

LUMINOSITY DETECTORS[9]

As it stands the system composed of 8 pairs of detectors is counting events in a time window of 120 ns centred on the crossing at the IP, every 11 μs. The digital recording takes several microseconds, therefore it is able to monitor the luminosity of only one bunch crossing per train, which is not adequate to represent the total luminosity and means a considerable loss in statistics. In order to count Bhabha events from all bunch crossings, additional electronic channels will have to be implemented with delayed gating corresponding to the different bunches in the trains. The software for the analysis of Bhabha events and the subtraction of background coincidences will have to be supplemented to take care of the added complexity.

POLARIMETERS[10]

Back scattered photons are obtained from the interaction of laser pulses with a given circulating bunch at a frequency of 100 Hz. Any bunch can be selected for this measurement and the grouping of bunches in trains is of no particular consequence.

The new machine optics foreseen for bunch trains and the vertical slope introduced by the separation bump near IP1 will prevent the use of the laser-beam interaction region (LIR) at its present position, at 73 m from IP1 (see Fig. 1). Plans are being made for moving the LIR to IP1, which will be more favourable from the optical point of view, acceptable for the bump (zero slope and a maximum displacement of 10 mm) and will give a perfect symmetry for e^+ and e^- measurements.

The multi-layered dielectric mirrors which are at the limit of standing the present level of radiation inside the LEP vacuum chamber will have to be replaced by all-metallic mirrors, in order to stand the additional radiation created by the vertical bump in the middle of LSS1.

REFERENCES

1. C. Bovet et al., "Preliminary report of the 1994 bunch train study group", CERN SL/94-72 (AP) , September 5, 1994.
2. G. Baribaud, D. Brahy, A. Cojan, F. Momal, M. Rabany, R. Saban, J.C. Wolles, "The beam synchronous timing system for the LEP instru mentation", Proc. of the ICALEPCS, Vancouver, 1989, pp. 192-197.
3. A. J. Burns, B. Halvarsson, D. Mathieson, I. Milstead, L. Vos, "Real time monitoring of LEP beam currents and lifetimes", presented at the EPAC, London, 27 June-1 July, 1994.
4. J. Borer, C. Bovet, D. Cocq, H. Kropf, A. Manarin, C. Paillard, M. Rabany, G. Vismara, "The LEP beam orbit measurement system", Proc. IEEE PAC, Washington, 1987, pp. 778-782.
   G. Baribaud, J. Borer, C. Bovet, D. Brahy, D. Cocq, H. Kropf, A. Manarin, F. Momal, C. Paillard, M. Rabany, R. Saban, G. Vismara, "The LEP beam orbit measurement system : status and running-in results", Proc. EPAC, Nice, France, 1990, **vol. 1**, pp. 137-9.
   G. Morpurgo, "The software for the CERN LEP beam orbit measurement system", Proc. ICALEPCS, Tsukuba, Japan, 1991, pp. 260-264.
   G. Vismara, "The new front-end Narrow-Band electronics for the LEP beam orbit measurement system", presented at the European Particle Accelerator Conference, London, 27 June -1 July, 1994.
5. K. D. Lohmann, M. Placidi, H. Schmickler, "Design and functionality of the LEP Q-meter", Proc. EPAC, Nice, France, 1990, **Vol.1**, pp. 771-776.
6. C. Bovet, G. Burtin, R.J. Colchester, B. Halvarsson, R. Jung, S. Levitt, J.M. Vouillot, "The LEP synchrotron light monitors", Proc. IEEE PAC, San Francisco, 1991, **Vol.2**, pp. 1160-2.
   P. Castro, R.J. Colchester, C. Fischer, J.J. Gras, R. Jung, J. Koopman, E. Rossa, H. Schmickler, J. Thomas, "Comparative precision emittance measurements in LEP", presented at the European Particle Accelerator Conference, London, 27 June -1 July, 1994.
7. H. Akbari, J. Borer, C. Bovet, Ch. Delmere, A. Manarin, E. Rossa, M. Sillanoli, J. Spanggaard, "Measurement of vertical emittance at LEP from hard X-rays", Proc. IEEE PAC, Washington, 1993, **Vol.3,** pp. 2492-5.
8. E. Rossa, "Real time single shot three-dimensional measurement of picosecond photon bunches", in the proceedings of this Workshop.
9. G.P. Ferri, M. Glaser, G. von Holtey, F. Lemeilleur, "Commissioning and operating experience with the interaction rate and background monitors of the LEP e^+e^- collider," Proc. EPAC, Nice, 1990, **Vol.1**, pp. 797-9.
   P. Castro, L. Knudsen, R. Schmidt, "The use of digital signal processors in LEP beam instrumentation," Third Annual Workshop on Accelerator Instrumentation, Newport News, 1991, pp. 207-216.
10. R. Schmidt, "Polarization measurements", Third Annual Workshop on Accelerator Instrumentation, Newport News, 1991, pp. 104-123.

Beam Line Instrumentation at the AGOR Cyclotron

J.M. Schippers, O.C. Dermois, K. Gerbens, H.H. Kiewiet,
P.A. Kroon and J. Zijlstra
Kernfysisch Versneller Instituut, 9747 AA Groningen, the Netherlands.

Abstract

Instrumentation for measurements of beam intensity, beam position and beam shape at the beam lines of the AGOR cyclotron is discussed. A non-intercepting beam current transformer and a capacitive pick-up can be used down to 1 nA beam current. In addition to wire grids (measurement of secondary emission current), a very sensitive non intercepting profile monitor, based on ionization measurements of the residual gas in the beam pipe, is presented.

INTRODUCTION

The superconducting cyclotron AGOR (1) at the Kernfysisch Versneller Instituut (KVI) in Groningen, the Netherlands, will be commissioned during the summer of 1995. The beams will be transported to different target area's in the existing KVI building. A new beam transport system has been designed and built (2) to accomplish the capabilities of the new cyclotron as well as the needs of the various users. Also new beam lines for the transport of the low-energy ion beams ($\simeq$10 keV/nucl) from the ion-sources towards the cyclotron have been built. In this paper an overview will be given of the beam-measurement equipment ("beam diagnostics") used in both the "low-energy" beam lines, as well as in the "high-energy" beam lines. Beams from the old Philips cyclotron at the KVI have been used to test all devices, so that their behavior was studied in a realistic environment.

The extracted beams of the AGOR cyclotron are characterized by the large variety of particles and energies to be handled: heavy ions between 5.6-100 MeV/nucleon as well as protons up to 200 MeV. The beam intensities for the different applications request a dynamic range of 10^4. Since uniformity of the different beam lines facilitates the operation and service, the capability of handling "all beams" has been an important design criterion.

The purposes of the beam-diagnostic instruments discussed here can be distinguished in measurements of: beam intensity, beam position and beam shape. Before discussing some of the devices, an overview of the signal processing in the control system is given.

SIGNAL PROCESSING AND CONTROL SYSTEM

The control of the cyclotron and all beam lines (3) will be performed via the central control system of AGOR. The control system has a three layered structure: equipment layer, control layer and operator interface.

In order to keep a flexible and accessible system, much attention is given to delegating tasks to local processors in the control layer. The local processors are linked as nodes in a network and messages are transferred with the Bitbus protocol (4). This protocol is a message oriented master/slave protocol optimized for high-speed transfer of short control messages between tasks in a hierarchical system. In our setup a Bitbus slave node usually corresponds to one piece of equipment or a set of strongly related devices.

Each controller can be equipped with various types of I/O-modules, such as ADC's, DAC's, modules for measurement of small currents, timer/counter modules, etc. After a dedicated preamplification stage, the signals from the beam diagnostic elements are sent to the appropriate I/O module(s).

The control layer uses several MicroVax systems. These systems are equipped with Bitbus interfaces, providing the connection to the beam line and accelerator equipment. The MicroVaxes are connected to each other and to the operator interface layer by Ethernet. The operator interface is based on X-Window workstations using a commercially available operator packages SL-GMS (5) and VSYSTEM (6).

A MAGNETIC PICK-UP FOR BEAM INTENSITY MEASUREMENT

The beam intensity measurements are performed by: magnetic pick-ups, (movable) faraday cups, ionization chambers and secondary-emission foils. Here only the magnetic pick-ups will be discussed.

The magnetic pick-up is essentially a beam current transformer, which offers a non-interceptive measurement of the beam current. It consists of a toroid which is mounted around the beam pipe, see fig. 1. The toroid is connected to an amplifier and the system acts as a LC circuit. If the particle beam is pulsed with a low frequency (200 Hz), the rise (or fall) of the beam current generates a damped sine-wave in the circuit. The magnitude of the signal is proportional to the beam current. A sample and hold unit detects the peak of the signal and is updated each pulse cycle. The signal output is available for Bitbus via a small bandwidth filter.

To minimize the influence of external magnetic fields, the toroid has been shielded with iron. Also the core material of the toroid should have a constant dB/dH over a wide range of B and the coercivity should be low. We have applied Vitrovac 6025F, having a coercivity less than 2 mA/cm and a permeability of 10^5. The toroid and the electronics are mounted on a heavy mechanical construction with mechanical shock absorbers to reduce microphonic noise. Also any mechanical contact with the vacuum chamber of the beam line has been avoided. The "mechanical isolation" has proven to be very effective, even with vacuum pumps in the near vicinity of the device. Presently eight current transformers are installed at the beam lines.

The pulsing of the beam will be performed in the low-energy beam line by applying a pulsed voltage to a pair of deflection plates. It is important that the rise (or fall) time of the beam current is much shorter than the time response of the LC circuit. The pulsing of the beam results in an effective decrease of the beam intensity. However, the beam duty cycle can be optimized to 50% or 90%. Therefore a beam current measurement does not interfere strongly with the beams.

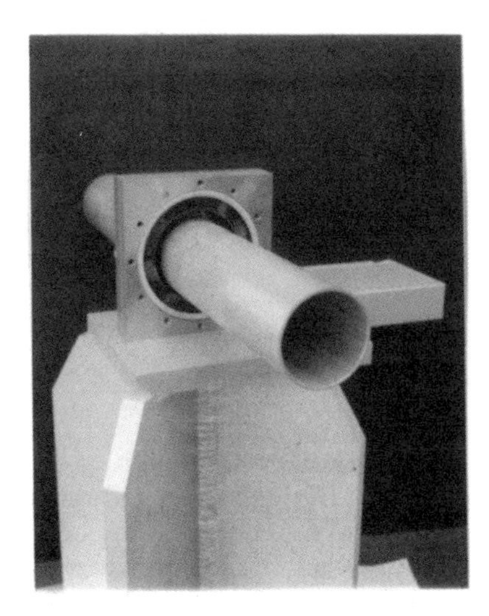

Figure 1. The beam current transformer for beam intensity measurements.

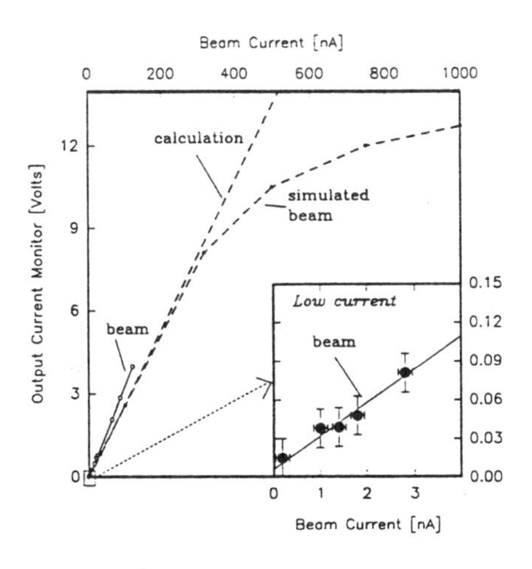

Figure 2. Response of the beam current transformer as function of beam intensity.

We have built a very sensitive current transformer: experiments at an existing beam line show that beam currents down to 0.5 nA can be measured with an accuracy of +/-0.5 nA, see fig. 2. With the present amplifier configuration the response is linear up to 300 nA. At higher beam currents clipping of the amplifier causes a non linear response, but current measurements up to 1 μA could be measured with sufficient accuracy.

A CAPACITIVE PICK-UP AS BEAM POSITION MONITOR

Accurate beam position measurements without intercepting the beam will be performed by capacitive pick-ups. Several of these diagonally-cut cylinders (length = 4 cm) are mounted in the beam lines to verify the beam centering

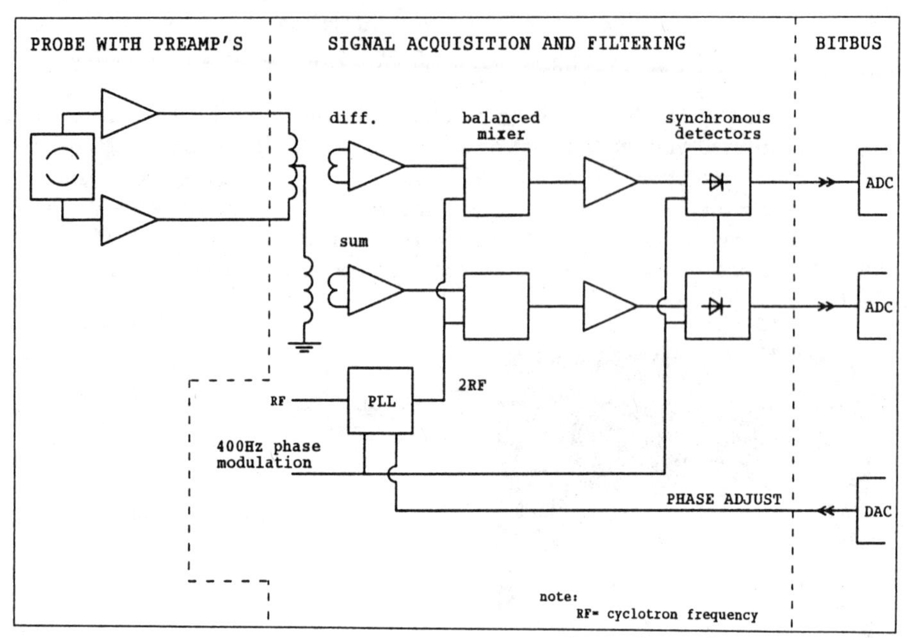

Figure 3. Signal processing of the capacitive pick-up electrodes.

behind dipole magnets. The AGOR beams are bunched with a repetition frequency equal to the cyclotron RF-frequency (24-62 MHz). In the longitudinal direction the beam bunches have a length of 5-20 cm, depending on the beam type. While passing a pick-up probe, each beam bunch induces a charge on both halves of the cylinder. The difference in charge is a measure of the beam position in one plane and the sum is a measure of the beam intensity.

Since the signals are very small (only a few μV per nA beam current) special signal processing electronics has been developed (see fig 3). After low-noise preamplification (FET input stage and 500 MHz bandwidth) the signals are sent into a combination of two transformers. There the sum signal and the difference signal are produced and sent to balanced mixers where the signals are multiplied with the second harmonic of the cyclotron RF-signal (2RF). The 2RF signal from the PLL is phase modulated with a frequency of 400 Hz, so that the 400 Hz signals at the output of the balanced mixers are proportional to the difference and sum signal.

Results of a bench test have shown that it is possible to detect a beam displacement of 1 mm while the beam current is only 1 nA. Presently these specifications are considered as sufficient for routine operation. For future applications the accuracy (or sensitivity) can be increased by optimizing the mechanical design of the probe, so that a larger fraction of the beam bunch is "seen" by the electrodes.

BEAM SHAPE MONITORS
Secondary Emission from Wire Grids

The "harps" are grids of 20 μm gold plated tungsten wires. Each harp has 48 wires, spaced 1 mm. The harp is mounted in a standard actuator and can be moved in or out of the beam pipe (see fig. 4). Much effort has been put in the mechanical design of the actuators and signal feed-throughs to minimize the dimensions. To allow quick service, the exchange of the wire grid can be done easily.

Figure 4.
The wire grid for profile measurements and its actuator. The aperture of the square wire frame is 50 mm.

If beam particles hit the wires, currents occur due to secondary emission of electrons from the wires. The distribution of the secondary-emission current measured at each wire, reflects the beam-intensity profile in one plane. The wire grid is surrounded by a ring on a positive voltage (50 V) to prevent that the electrons return to the wires. Using a beam of 24 MeV protons we measured an efficiency of 0.4 electrons per beam particle. For 200 MeV protons this would yield approximately 0.1 electron per proton, which is sufficient for beam currents larger than 10 nA.

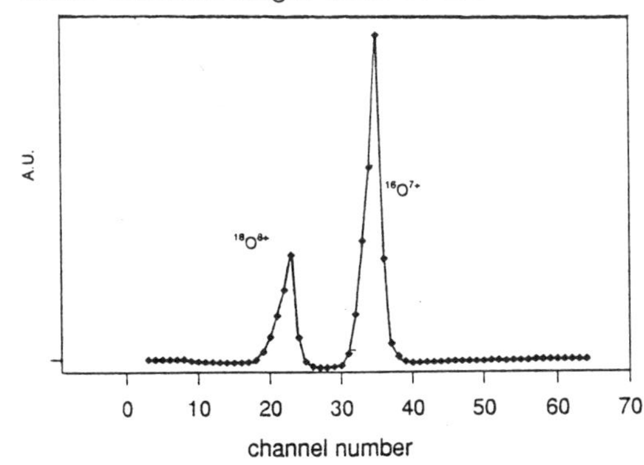

Figure 5.
The beam profile behind the analyzing magnet of a new ECR ion source shows the presence of two charge states in the ^{18}O beam.

Each harp has a dedicated box with electronics and a Bitbus node. First each signal is amplified in an I-to-V amplifier (50 MΩ feedback resistor in the high-energy beam lines). The amplified signals are multiplexed and sent to the control system via the Bitbus interface. Several status checks (e.g. on the performance of the actuator) are performed automatically. If the harp is in the beam, the profile is updated with a maximum speed of ten times per second. Offsets can be subtracted and for low beam intensity or noisy situations the profiles can be averaged in time.

It is expected that the harps will be the major diagnostic tools in the beam transport system. For cases where beam intensities are too low for accurate operation, special harps can be made: the wire frames could also be equipped with thicker wires or foils.

In the low-energy beam lines the harps are already used routinely for tuning the ECR ion source, see fig. 5.

Profiles from the Ionization of Residual Gas

At positions where a fully non-intercepting beam-profile monitor is required, residual gas monitors are mounted. These newly developed detectors (7) use the ions that are created by the beam particles while traversing the residual gas in the beam pipes (see fig. 6). The ions are extracted from the beam tube by means of a homogeneous electric field of 0.5 kV/cm and drifted towards a pair of micro channel plates (MCP's), which acts as a charge multiplier. The electrons generated in the MCP's are collected on strips parallel to the beam, so that a position sensitive read out with a 1 mm pitch, is obtained. The strips are connected to a similar electronics box as the harps are using. The signals from the strips represent the beam-intensity profile in one plane.

By choosing the amplification factor of the MCP's, one can operate within a wide range of beam intensities. In the residual gas of 10^{-7} mbar a 200 MeV proton beam of 1 nA will create about 135 ions $s^{-1}cm^{-1}$. In this case a moderate MCP-amplification of 10^4-10^5 is sufficient to collect the profile data.

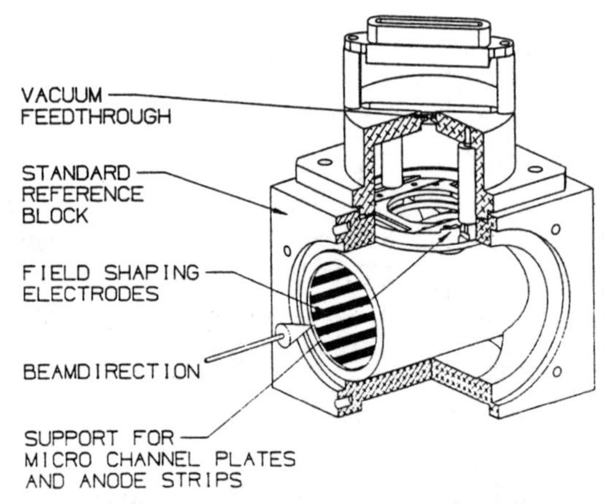

Figure 6.
The Residual gas beam-profile monitor mounted in an accurate standard position-reference block of the AGOR beam lines.

A large MCP-gain allows easy measurements even for very small beam currents. We observed that where viewing screens did not show any image (less than 0.1 nA beam current), the residual gas monitor could measure the profiles accurately.

Tests on a prototype as well as on a device in the final configuration have shown that the residual-gas beam profile monitor works very reliable and stable.

CONCLUSIONS

The beam measurement equipment at the AGOR beam lines is designed to be used for almost all extracted beams and also in the beam lines between the ion sources and the cyclotron. The designs offer a great flexibility in use and easy service. All beam-diagnostic elements are controlled and read-out by the central control system of AGOR and their signals can be analyzed online. Many devices do not intercept the beam, so that measurements can be performed without perturbing the experiments at the target stations. First results with beams from AGOR are expected in the summer of 1995, but most devices have already been tested successfully with beams from the old cyclotron at the KVI.

This work was performed as part of the research program of the "Stichting voor Fundamenteel Onderzoek der Materie" (FOM) with financial support from the "Nederlandse Organisatie voor Wetenschappelijk Onderzoek" (NWO).

REFERENCES

1. Schreuder, H.W., "AGOR: Recent achievements", in *Proceedings of the 13th Int. Conf. on Cyclotrons and their Appl.,*(World Scientific, Singapore, 1993)
2. Schippers, J.M., "The Beam Guiding System of AGOR", in *Proceedings of the 13th Int. Conf. on Cyclotrons and their Appl.,*(World Scientific, Singapore, 1993)
3. Kroon, P.A., "The AGOR Control System", *Annual Report 1991* , Kernfysisch Versneller Instituut, Groningen.
4. The Institute of Electrical and Electronic Engineers, Inc. *IEEE standard microcontroller system serial control bus,* IEEE Std 1118-1990.
5. SL Corporation, Corte Madera, California, USA, *SL-GMS Graphical Modeling System.*
6. Vista Control Systems Inc, Los Alamos, New Mexico, 87544, USA. *Vsystem.*
7. Schippers, J.M. and Kiewiet, H.H., "A Beam Profile Monitor Using the Ionization of Residual Gas", in *Proceedings of the 13th Int. Conf. on Cyclotrons and their Appl.,*(World Scientific, Singapore, 1993)

Experimental Diagnostics Using Optical Transition Radiation at CEBAF

J.-C. Denard, D. Rule†, R. Fiorito†, P. Adderley, K. Jordan, K. Capek
CEBAF, 12000 Jefferson Avenue, Newport News, VA 23606
†Naval Surface Warfare Center, Silver Spring, MD 20903-5000

Abstract

Optical Transition Radiation (OTR) devices have unique properties that allow them to complement the diagnostic tools more commonly used in particle accelerators. CEBAF is designed to produce a continuous electron beam accelerated up to 4 GeV by recirculating it five times through two 400 MeV superconducting linacs. We present two OTR applications that cannot be performed with standard fluorescent screens. The goal of the first one is to provide a multiturn "viewer" using the backward OTR emitted from a 0.8 μm thick aluminum foil. The foil must be thin enough to keep most of the beam in the machine after each passage. Looking at the successive turns in the linacs on the same screen will provide a new diagnostic device to help tune the machine. Replacing the ceramic of the present viewers with an Al foil is relatively simple and inexpensive. The preliminary results in single pass are encouraging. The goal of the second OTR application is to measure the emittance of high current continuous beams (≈ 200 μA) of low emittance ($5\ 10^{-9}$ m·rad) and size (≤ 50 μm rms). Standard fluorescent screens or wire scanners cannot withstand such an intense beam.

INTRODUCTION

The Optical Transition Radiation (OTR) phenomenon was observed as early as 1919 and its theory developed in 1945 (1); most of the experimental work using the OTR for beam diagnostics was done by Wartski (2)(3), Fiorito and Rule (4), but it is not yet commonly used for resident beam diagnostics. On the CEBAF accelerator, it is very easy to replace the standard ceramic disk of one of the beam viewers with an aluminum foil; the video camera is at the optimum angle to receive the most intense part of the backward OTR. We expect such a simple transformation to provide two new beam diagnostics to the machine, which in autumn 1994 is beginning multipass commissioning.

The first beam diagnostic is a multiturn viewer that will be useful very soon to tune up the successive recirculations of the machine. Thick screens disrupt the beam, but using a thin enough foil should provide a quasi-nondestructive diagnostic. During tune up, the machine yields pulsed beams at 60 Hz with a duty factor ≤ 0.6% to keep the average current below 1 μA. The current is kept low in order to limit the amount of radiation in the tunnel and avoid damaging the machine with the full CW beam at 4 GeV that carries a large amount of power.

---

* Supported by U.S. DOE contract DE-AC05-84ER40150

The second beam diagnostic is a profile monitor that is expected to give a good resolution for beam sizes down to 50 μm rms and currents up to 200 μA. The present diagnostics of the machine (wire scanners and fluorescent screens) do not stand more than a few μA average current when the beam is that small. A video camera observes the beam image and sends its video signal to be processed in order to retrieve the x and y rms dimensions. Three profile monitors at the proper locations will give the necessary data to compute the emittance. The fixed machine optics must assure a good transmission of the beam up to the dump, taking into account the emittance growth due to the foil. We concentrate here on the profile monitor; the rest of the process is straightforward. This diagnostic will be needed later and the experimental evaluation of the design will be done in several steps during the coming year.

Transition Radiation Properties

Transition radiation (TR) is generated by a charged particle beam at the boundary between two media of different dielectric constants. The visible part of its spectrum is called OTR. A beam outgoing from a foil surface generates the forward TR that has a symmetry of revolution about the beam direction. The spectrum of the radiation is constant in the visible frequency range. The backward TR is generated by the incoming beam on the surface of the foil; it presents a symmetry of revolution about the direction of specular reflection of the beam direction. The backward OTR spectral density is proportional to the reflectivity of the foil material.

The angular distribution of TR depends on the beam energy at small angles. With CEBAF beams whose energy range extends from 45 MeV up to 4 GeV, the distribution no longer depends on the beam energy for angles greater than 30 mrad. Although the peak of light density is at $1/\gamma$ (γ is the Lorentz factor), most of the radiated power is at large angles. This property is going to be important for the resolution limit of a profile monitor for high energy and low emittance beams.

A last property to emphasize is that a single electron radiates in a very short time as a point source; OTR is a surface phenomenon. One can find more details about OTR properties in references (1), (2), (3), and (4) and the references therein.

MULTITURN BEAM VIEWER

In last July 1994, for the first time, the beam reached the target of the first experimental hall after being accelerated by the two linacs. The commissioning phase aimed at recirculating the beam through the linacs is starting now and there is a high demand for diagnostics that can locate the transverse beam position of five turns. An OTR screen at the entrance and exit of each linac should provide the information. They yield enough light and we will see that, at least in theory, the beam does not experience too much emittance growth up to the last turn.

The first OTR test was done at 41 MeV in the injector with a CCD camera looking at a 1.5 μm thick Al foil. A beam about 3 mm in diameter remained visible down to 50 nA average current (5 μA pulses during 160 μs at a 60 Hz repetition rate). Since the light is abundant, the OTR viewers are now equipped with the standard vidicon cameras which are more resistant to radiation.

Effects of Energy Loss on the Beam

Due to collisions with the atoms of the foil material, the beam loses a small part of its energy, proportional to the track length in the foil. With 1 GeV incoming electrons, the EGS4 Monte Carlo code indicates that, after traversing an Al foil of 100 μm, 99.7% of the particles will lose less than 3% of their energy; since our foil is 100 times thinner, we expect the same percentage of electrons to lose less than $3\ 10^{-3}$ of their incoming energy. $3\ 10^{-3}$ is about equal to the acceptance of the machine. The relative average energy loss of the outgoing beam transported in the machine was computed from the EGS4 output distribution; it amounts to $1.1\ 10^{-6}$, a quantity that is negligible with respect to the beam energy spread ($2.5\ 10^{-5}$). We expect similar results for the successive energies in the machine, from 400 MeV to 4 GeV. We tried to measure the beam energy loss at 400 MeV and found it to be less than the resolution of our measurement: $\Delta E/E \leq 1\ 10^{-5}$. The orbit variation that the beam centroid experiences after such a small energy reduction is less than the beam size. In consequence, the foil has no significant effect on the beam energy and position.

Beam Scattering Through Very Thin Foils

Another effect of the collisions of the beam with the atoms of the foil is to scatter the outgoing particles, mostly at very small angles. The problems are the emittance growth and the amount of beam loss. E. Segré (5) and W. Scott (6) extensively reviewed the multiple and plural scattering theories; all of them agree in that the distribution width is inversely proportional to the momentum–and also to the energy for highly relativistic particles. Unfortunately, the relation with material thickness is not as easily found. Two theories seem to dominate the field: one based on Molière's work gives results if the path length in Al is greater than ≈ 1.5 μm, but the accuracy is questionable for thicknesses (in Al) smaller than 8 μm. The other theory, proposed by Keil et al., is an extension of Molière's for thinner foils; it is applicable between 0.08 and 8 μm of thickness for Al. This range corresponds to a mean collision number $0.2 \leq \Omega_0 \leq 20$, the regime of plural scattering, while the multiple scattering corresponds to more than ≈ 20 collisions. In the region of $\Omega_0 \approx 20$, the Keil and Molière approaches agree within a few per cent (6). Figure 1 shows the results obtained with those two theories. To our knowledge, there is very little experimental data in the plural scattering region.

We measured the projected distribution of a 41 MeV beam after a 1.5 μm aluminum foil inclined at 45° into the beam trajectory with a wire scanner installed downstream of the foil. The width of the vertical distribution, measured between the two 1/e points, increases from 3 to 7 mm upon insertion of the foil. Assuming both distributions to be gaussian for the sake of simplicity, we can remove the natural beam divergence contribution: $\sqrt{(7^2 - 3^2)} = 6.3$ mm. Taking into account the presence of a quadrupole between the foil and the wire scanner that reduces the measured beam size by 3% and knowing that the wire scanner is 6 m downstream from the foil leads to $\alpha_{1/e} \cdot E = 22$ µrad·GeV. One can see in figure 1 that it is notably smaller than the theory predictions, even if we consider the measurement uncertainty that amounts to 20%.

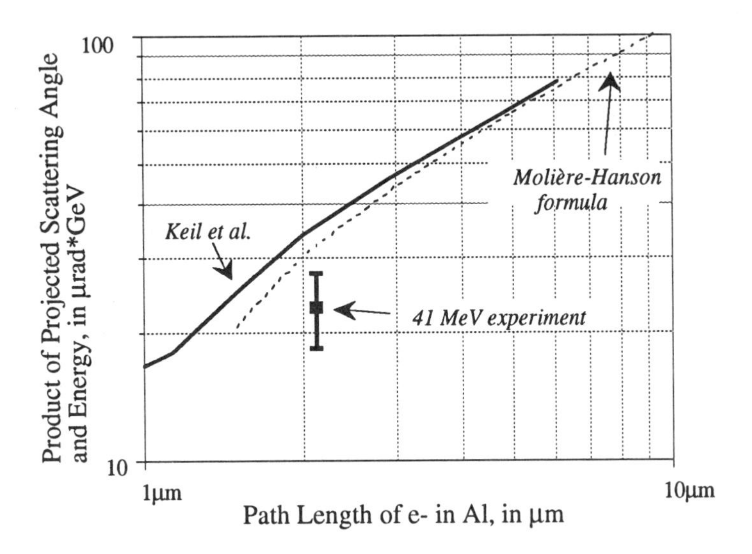

Figure 1: Projected scattering angle $\alpha_{1/e}$ of highly relativistic electrons in a thin Al foil, normalized at 1 GeV; $\alpha_{1/e}$ is the angle at which the distribution falls to 1/e of its value at 0°. The projected angle of a scattered particle is defined on any plane that contains the incoming trajectory; it is the angle between its incoming and outgoing trajectories projected on that plane. The dotted line curve is computed after Hanson's formula (5) which gives a very good approximation of Molière's first three terms. Both curves can be used for other materials than Al by calculating the equivalent thickness of Al that yields the same number of collisions.

Application to Multiturn Viewers

CEBAF is a low emittance machine, $2 \cdot 10^{-9}$ m·rad at 4σ, and the small beam divergence makes the beam scattering an important parameter. However, we limit our goal to seeing the five passes on a screen only to tune up the successive recirculations; then some beam losses (≈ 10%) and emittance degradation (up to ten times) can be tolerated in tune up conditions. We installed four OTR viewers, one at each end of each linac, equipped with a 0.8 µm Al foil (manufactured by Goodfellow) inclined at 45° with the beam direction. We made a rough estimate of the emittance growth with the scattering angles from Keil's theory (pessimistic); we estimated the emittance before and after each viewer, assuming only one is inserted at a time. We also assumed adiabatic damping in the linacs down to $2 \cdot 10^{-9}$ m·rad, a value that may be an intrinsic limitation of the machine. A last assumption (optimistic) is that the emittance grows linearly with the beam angle. In these conditions, the emittance after five passages of the beam through any one of the OTR screens will increase by a factor ranging from 1.4 to 2.5 with respect to the design value, depending on which viewer is used. This is acceptable.

During the commissioning of the first pass we had the opportunity to measure the beam loss after inserting the foil. The two viewers of the second linac gave respectively 8 and 6% loss, which is higher than expected but acceptable. We do not have reliable results for the first linac yet.

From the previous estimates and measurements, we are confident that the multipass viewers will see all passes and help in tuning up the machine. We are planning to do a few additional measurements to predict more accurately the beam scattering angular distribution.

PROFILE MEASUREMENT OF HIGH CURRENT CW BEAMS

Although 200 μA would be a modest intensity for a pulsed beam, it gives a 0.8 MW power to a continuous beam accelerated up to 4 GeV. In addition, the low emittance concentrates the power in small transverse beam dimensions (≈ 50 μm rms), and the consequence of a wrong steering would be melting the stainless steel vacuum chamber in a few tens of microsecond. There is presently no profile monitor available to measure the small emittance of the full current continuous beams. We plan to build one by replacing a standard ceramic viewer screen, oriented at 45° with respect to the beam, by a 0.8 μm Al foil. In a future upgrade, a CID camera will replace the regular vidicon to look at the backward OTR. The video signal is digitized and processed in order to display the beam profile. The CID camera presents a better linearity than vidicons and resists to higher radiation levels (500 krad) than usual CCD cameras. The issues we are going to discuss are the resolution of the profile measurement, the temperature of the foil, and the beam scattering.

Resolution

Similarly to synchrotron radiation, the highest power density of transition radiation is emitted within a small angle $1/\gamma$. But there is a fundamental difference for calculating the resolution limit due to diffraction: OTR fills the whole hemisphere, and furthermore, most of the total power is at angles greater than $1/\gamma$. The pupil factor is actually close to that of a spherical wave, so micron size resolution can be achieved with a wide aperture optical system (4). OTR is a surface phenomenon; there is no need for depth of field except for having the whole screen well focused. This can be achieved in conjunction with a wide aperture by orienting the camera about the normal to the foil.

Another issue related to resolution is the interpretation of the so-called formation length; it applies only to forward transition radiation, and therefore is not a limitation for the diagnostics presently discussed. It has never been observed; we think that if there is really a limitation due to the formation length, we should be able to measure it with the high energy and small size CEBAF beam.

Our 1.5 μm aluminum foil has a smooth surface with a uniform reflectivity; but it is not the case for the 0.8 μm foil that presents surface defects which may change locally the reflectivity. We are planning to take profile measurements of the same beam steered on several points of the surface to evaluate the error due to the imperfections. It would not be a problem with forward OTR.

Temperature

Excluding bremsstrahlung, almost all the collision energy ΔE lost by the beam goes into heating the foil. The EGS4 code yields $\Delta E = 0.41 \cdot 10^9$ eV/m for thicknesses much smaller than the radiation length in the material. The frame supporting the foil has a relatively high conductivity towards the vacuum chamber and will stay at room temperature as it receives powers of the order of 100 mW. Assuming the power deposited is evacuated only through radial conduction from center to edge of the circular foil, the temperature rise at the central point becomes independent of the thickness and amounts to:

$$\Delta T = \frac{\Delta E . I}{2 \pi K \sin \alpha} \left(\mathrm{Log}_e \left(\frac{r_{\mathrm{foil}} \sqrt{\sin \alpha}}{\sqrt{2 \sigma_x . \sigma_y}} \right) + 0.5 \right)$$

where I is the beam current in A; K the thermal conductivity of the foil material in W/m·K; α the angle of the beam direction with respect to the foil plane; r_{foil} the outer radius of the foil; and σ_x, σ_y the usual rms dimensions of the gaussian beam. For the sake of simplicity, the above formula has been established by considering an equivalent beam that has a homogeneous particle density equal to the maximum of the actual gaussian beam; the equivalent beam would illuminate the foil in a circle traversed by the same total current; the latter condition constrains the circle radius to be $r_{\mathrm{eb}} = \sqrt{2\ \sigma_x . \sigma_y / \sin \alpha}$. With I = 200 μA, K = 237 W/m.K, α = 45°, and r_{foil} = 12.5 mm, one obtains $\Delta T \approx$ 430°C, a temperature well below the melting point. The reflectivity of aluminum, an important parameter for backward OTR, does not change much at that temperature. The radiation from the hot central spot is negligible with respect to OTR in the visible spectrum.

Beam scattering

In order to limit the amount of radiation and its damaging effects on the machine components, we wish to keep the beam loss in the machine under 1% of the maximum current. The distribution of the scattering angle given by Keil et al. (6) has longer "tails" than the gaussian one: with our previous example of 0.8 μm foil inclined at 45° in the beam trajectory, the first pass beam at 845 MeV would have 99% of its particles contained within a 170 μrad spatial angle, almost 12 times more than the angle where the particle distribution falls to 1/e times its maximum; it is still within the acceptance angle of the machine at the future OTR monitor locations. However, we will feel more comfortable after conducting an experiment with the accelerator carefully tuned in order to be sure of the actual current loss.

Discussion

The above analysis shows the 0.8 μm Al foil to be adequate for measuring profiles of 200 μA beams down to 50 μm transverse dimensions. However, there is not much margin left if the beam has a higher current or smaller size. The OTR monitors now installed on the machine will allow us to measure resolution and scattering angle as well as verify that the foil withstand 200 μA CW beams. For the actual high current emittance measurement, three monitors will be installed later in each of the transport lines to the experimental halls. We are also planning to test another design that uses the forward OTR coming out of a thinner (0.15 μm) carbon foil in order to reduce the scattering angle and beam losses. Carbon also intercepts slightly less energy than Al, it withstands higher temperatures and we hope to find foils with a better thermal conductivity than Al. It would be a good solution for the few mA to be accelerated by the coming generation of superconducting FEL drivers.

CONCLUSION

We built a new beam viewer using the OTR from a foil thin enough to be almost transparent to the beam. Five recirculations in the linacs are expected to be seen on the same viewer. The transformation of a standard viewer into an OTR viewer was done simply and at low cost by replacing the fluorescent ceramic screen with a 0.8 μm Al foil. A similar device is expected to measure small beam sizes at high current (50 μm and 200 μA average). The maximum current capability should be increased by using forward OTR from a carbon foil. It would be useful for the milliampere range of continuous currents where superconducting FEL accelerators will operate. CEBAF, with its energy range and small emittance is the ideal machine to verify the plural scattering theory and the resolution limits of OTR at high energies.

ACKNOWLEDGMENTS

We would like to thank Y. Chao and D. Douglas for their machine optics calculations, P. Kloeppel for the many Monte Carlo analyses he did not hesitate to run, A. Hutton, B. Legg and the operations group for their support.

REFERENCES

1. Frank, I. and Ginsburg, V., J. Phys. USSR **9**, 363 (1945).
2. Wartski, L. et al., "Interference Phenomenon in Optical Transition Radiation and its Application to Particle Beam Diagnostics and Multiple Scattering Experiments," J. Appl. Phys. **46,** 3644 (1975).
3. Wartski, L., Doctoral Thesis, Université de Paris-Sud, Centre d'Orsay (1976).
4. Fiorito, R.B. and Rule, D.W. "Optical Transition Radiation Beam Emittance Diagnostics," in AIP Conference Proceedings 319 (Robert E. Shafer, ed., Santa Fe, NM 1993), pp. 21-37.
5. Segré, E., *Experimental Nuclear Physics* (John Wiley Inc., New York, 1953).
6. Scott, W., "The Theory of Small-Angle Multiple Scattering of Fast Charged Particles," Reviews of Modern Physics **35**, 231 (1963).

Measurement of 50-fs (rms) Electron Pulses

Hung-chi Lihn, Pamela Kung, Helmut Wiedemann
Applied Physics Department and SLAC, Stanford University, Stanford California 94305, U.S.A.

David Bocek
Physics Department and SLAC, Stanford University, Stanford California 94305, U.S.A.

Abstract

Electron pulses generated at the Stanford SUNSHINE facility with (2–4.6) $\times$ 10^8 electrons per microbunch have been measured as short as 50 fs (rms). The bunch length is determined by optical autocorrelation via a far-infrared Michelson interferometer using coherent transition radiation emitted at wavelengths longer than and equal to the bunch length. This frequency-resolved autocorrelation method demonstrates a better sub-picosecond resolving power than any existing time-resolved method. The experimental setup and the results will be described.

INTRODUCTION

In the development of particle accelerators, the measurement of the electron bunch length in the longitudinal phase space has played a very important role. Through bunch length measurements, it is possible to examine different bunch generation and compression techniques and, hence, find not only ways of optimizing accelerators' performance but also guidance to the design of future linear colliders, free-electron lasers, and high-intensity coherent light sources in the far-infrared regime.

To measure the bunch length, two approaches can be taken: the time-resolved and the frequency-resolved technique. The time-resolved technique uses fast electronics to resolve beam-generated light pulses in the time domain and derives the longitudinal bunch distribution. When the bunch length is reduced through the development of accelerators, faster processing speed and more advanced technology are required to achieve higher resolution. As a result, the highest time resolution achieved by the fastest time-resolved device, the streak camera, is on the order of a picosecond. However, this resolving power is not enough for facilities like SUNSHINE at Stanford University(1,2), where electron bunches can be compressed to the 100-fs range. To resolve such short electron bunches, new bunch length measuring technique must be used.

The frequency-resolved technique, on the other hand, resolves the frequency content of beam-generated light pulses in the frequency domain and deduces the particle distribution from the frequency information. Unlike time-resolved methods, this technique does not require fast processing speed, and the necessary broad bandwidth for short pulses can be achieved by optical methods.

Based on this technique, an optical autocorrelation method, similar to that used to characterize femtosecond laser pulses(3), has been proposed for sub-picosecond bunch length measurement(4). It utilizes the coherence property of the radiation emitted by short electron bunches to generate the optical autocorrelation via a far-infrared Michelson interferometer. By analyzing this autocorrelation measurement, it is possible to determine the electron bunch length down to sub-picosecond resolution.

In this paper, we will describe the experiment using this new autocorrelation method to measure the sub-picosecond electron bunches generated at SUNSHINE. We first describe the principle of bunch compression used at this facility and then discuss the experimental setup and the results.

CONCEPTUAL BACKGROUND

The bunch generation and compression system used at SUNSHINE, as shown in Fig. 1(a), consists of two major components: a 1½-cell thermionic RF gun and an alpha magnet with energy filters(1,2). The RF gun operating at 2856 MHz produces 2.5-MeV electron bunches in which the electrons are distributed along a thin line in the energy-time phase space with higher energy electrons located at earlier time and lower energy electrons, later time. These energy-time correlated bunches are then steered into the alpha magnet for compression. The magnet will guide the electrons in the magnet along α-shaped paths with higher energy electrons following longer paths and vice versa; hence, the earlier electrons in the bunch, which have higher energy, will spend more time in the magnet by following longer paths while the later electrons, less time. By correctly setting the magnet's strength, it is possible to compress part of the electron bunch into sub-picosecond duration. This optimally compressed part is then selected by energy filters and transported through a 30-MeV linear accelerator and a beam transport line to the radiation source point. When transporting the electron bunch, the velocity spread in the bunch can cause significant bunch lengthening; therefore, it is necessary to compensate for this effect by overcompressing the bunch so that the minimum bunch length is reached at the source point.

When these short electron bunches emit radiation, the total radiation is incoherent at wavelengths shorter than the bunch length and temporally coherent at wavelengths longer than and equal to the bunch length(5). The intensity of the coherent part exceeds that of the incoherent part by a factor of N_e, the number of electrons in each bunch. This coherent intensity enhancement makes it easier to detect the radiation from sub-picosecond electron bunches in the far-infrared regime. The radiation spectrum is the Fourier transform of the longitudinal bunch distribution; hence, the determination of the bunch length will become possible once this spectrum is measured by the frequency-resolved technique.

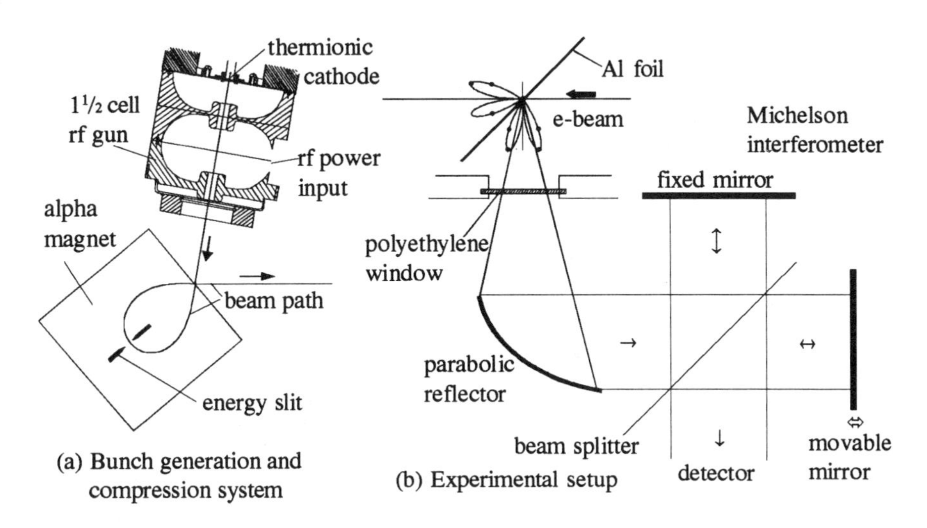

Figure 1. Schematic diagrams of (a)the bunch generation and compression system used at SUNSHINE and (b)the experimental setup.

EXPERIMENTAL SETUP

As stated in the previous section, 2.5-MeV electron bunches are produced by the RF gun every 350 ps for a macropulse duration of about 1 μs at a repetition rate of 10 Hz. These electron bunches, with a macropulse current of 700–900 mA at the gun exit, are guided into the alpha magnet. The magnet compresses part of each bunch into sub-picosecond duration and the undesirable part of the beam is filtered out. Finally, these sub-picosecond electron bunches, with (2–4.6) $\times$ 10^8 electrons each, are accelerated to 30 MeV and transported to the radiation source point.

At the radiation source point, as shown in Fig. 1(b), transition radiation is generated when the electrons pass through a 25.4-μm-thick Al foil. The foil supported by a copper ring is oriented at a 45° angle to the beam direction so that the backward transition radiation is emitted in the direction normal to the beam path and can easily be extracted from the vacuum chamber through a 1-mm-thick high-density polyethylene window of 19 mm diameter. Since the backward transition radiation is emitted at the Al surface, the focal point of an off-axis paraboloidal mirror is aligned with this surface to make the divergent radiation parallel without introducing optical path difference to the extracted light pulse. The parallel light then enters a far-infrared Michelson interferometer.

The interferometer consists of a 12.7-μm-thick Mylar beamsplitter, a fixed and a movable first surface mirror, and a room-temperature bolometer. The movable mirror is controlled by a Newport 850-B 25-mm actuator through a Newport PMC200-P 2-axis controller which in turn is controlled by a 486-based

PC via an RS-232 port. The bolometer consists of a Molectron P1-65 $LiTaO_3$ pyroelectric detector of 5 mm diameter and a pre-amplifier. This bolometer is designed to measure the integrated radiation energy from each 1-μs macropulse with a responsivity of *pre-amplification* $\times$ 1.21 $\times$ 10^3 V/J and is attached to a copper light cone, which will collect the parallel light into the bolometer. The bolometer signal is digitized into the computer through a National Instruments AT-MIO-16F-5 data acquisition board. With these computer interfaces, the autocorrelation measurement can be performed automatically through the program under the LabVIEW control environment implemented on the computer.

RESULTS AND DISCUSSION

By measuring the bolometer signal as a function of the position of the movable mirror through the computer program, typical 16-mm-long autocorrelation scans with 10-μm mirror step size were performed with a good signal-to-noise ratio. For example, the bolometer background noise was measured around 2 mV while the bolometer signal was measured about 200 mV after pre-amplification. However, in all the presented data, the pre-amplification was removed.

For an ideal beamsplitter which equally splits the intensity of the incoming radiation into two mirror arms, the intensity of the radiation combined at the detector, which will produce a proportional bolometer signal, can be expressed in the time domain as

$$\begin{aligned} I(\delta) &= \int_{-\infty}^{+\infty} |RT\{E(t) + E(t + \frac{\delta}{c})\}|^2 \, dt \\ &= 2\,|RT|^2\{\mathrm{Re}\int_{-\infty}^{+\infty} E(t)E^*(t + \frac{\delta}{c})\, dt + \int_{-\infty}^{+\infty} |E(t)|^2 \, dt\} \end{aligned} \tag{1}$$

or in the frequency domain as

$$\begin{aligned} I(\delta) &= \int_{-\infty}^{+\infty} |RT\,E(\omega)(1 + e^{i\omega\frac{\delta}{c}})|^2 \, d\omega \\ &= 2\{\mathrm{Re}\int_{-\infty}^{+\infty} |RT|^2\,|E(\omega)|^2 e^{i\omega\frac{\delta}{c}}\, d\omega + \int_{-\infty}^{+\infty} |RT|^2\,|E(\omega)|^2\, d\omega\} \end{aligned} \tag{2}$$

where δ is the optical path difference between both mirror arms, and R and T are the amplitude reflection and transmission coefficients with $|R|^2 = |T|^2 = 1/2$. Equation 1 and 2 are related by the Fourier transform of $E(t)$. Furthermore, the interferogram is defined as $I(\delta) - I_\infty$, where I_∞ is $I(\delta \to \pm\infty) = 2\,|RT|^2 \int_{-\infty}^{+\infty} |E(t)|^2 \, dt = 2\int_{-\infty}^{+\infty} |RT|^2\,|E(\omega)|^2\, d\omega$. The interferogram is, indeed, the autocorrelation of the light pulse (see Eq. 1), and its Fourier transform is the power spectrum of the pulse (see Eq. 2). $I(\delta)$ is equal to I_∞ for $|\delta|$ much larger than the pulse length and rises to the maximum of $2I_\infty$ at $\delta = 0$. Hence, the width of this peak around $\delta = 0$ in the interferogram can be used

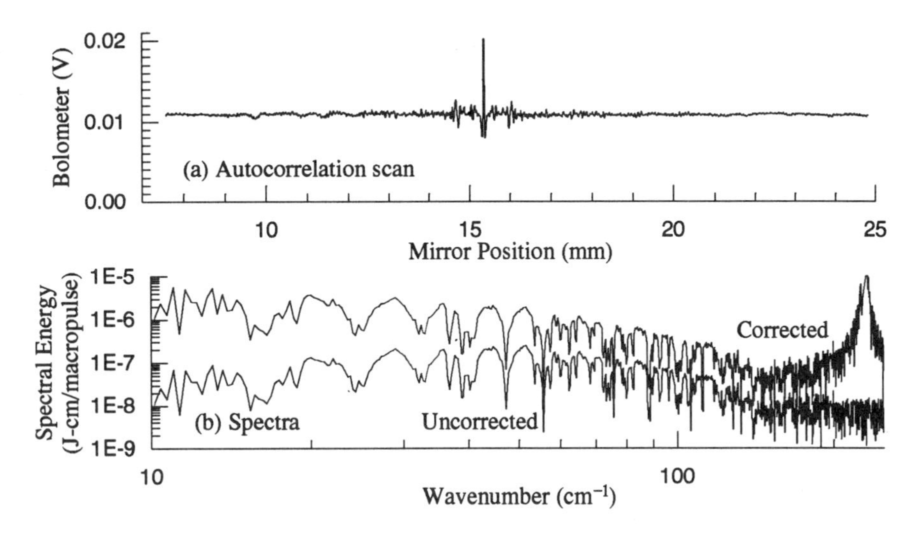

Figure 2. Typical 16-mm-long autocorrelation scan and derived spectra for radiation from sub-picosecond electron pulses. The autocorrelation scan is shown in (a). The raw spectrum and the one corrected for the Mylar beamsplitter efficiency are shown in (b).

to estimate the pulse length (which is equivalent to the bunch length). For example, the full width at half maximum (FWHM) of the interferogram is equal to the bunch length for a rectangular bunch distribution and is equal to $4\sqrt{\ln 2}\,\sigma_z$ for a Gaussian bunch distribution with an equivalent bunch length of $\sqrt{2\pi}\,\sigma_z$ (=0.75 FWHM). Most real bunch distributions are neither rectangular nor Gaussian, and the equivalent bunch length can be estimated as 75% to 100% of the FWHM of the interferogram. A typical autocorrelation measurement, as shown in Fig. 2(a), exhibits the sharp rise in signal to about twice of I_∞ when pulses from both mirror arms overlap.

However, a suitable beamsplitter used for the far-infrared regime does not provide constant ½:½ reflection and transmission at all frequencies but those varying as functions of frequency. This departure from the ideal beamsplitter is caused by the interference of light reflected from both surfaces of the Mylar beamsplitter, which is equivalent to thin film interference. Because of this interference, the efficiency of the 12.7-μm-thick Mylar beamsplitter, which is defined as $|RT|^2$, increases from zero at zero frequency to the maximum value of 0.17 at 115 cm^{-1} and then drops to zero again at 230 cm^{-1}, the first singularity of the beamsplitter, where light reflected from both surfaces of the beamsplitter forms destructive interference. Hence, the beamsplitter thickness of 12.7 μm is specially chosen to include the whole expected spectrum within the frequency range up to its first singularity.

The raw radiation spectrum of the autocorrelation measurement and the

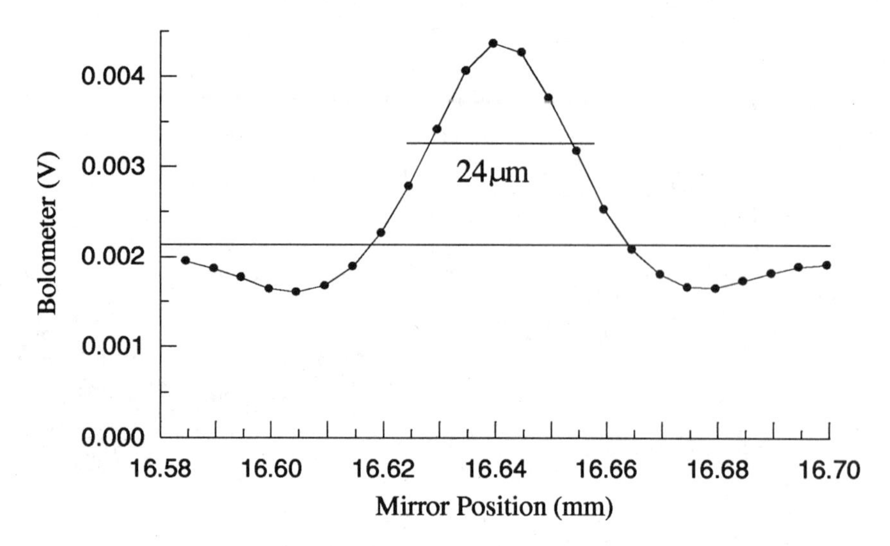

Figure 3. Fine autocorrelation scan of the main peak for the shortest electron bunch length obtained at SUNSHINE. The measurements are shown as • with mirror step size of 5 μm.

one corrected for the Mylar beamsplitter efficiency are shown in Fig. 2(b). The total 16-mm mirror movement corresponding to 32-mm optical path difference results in a spectral resolution of $0.3125\,\text{cm}^{-1}$. Below $10\,\text{cm}^{-1}$, the spectra are believed to be contaminated by slow drift of machine parameters during the half-hour measurement. The whole raw spectrum is well contained within the frequency range up to the first singularity of the beamsplitter. The spike around the first singularity of the beamsplitter in the corrected spectrum indicates a large over-correction due to the poor signal-to-noise ratio in this spectral range. The roll-off rate of the corrected spectrum at high frequency is closer to that of a rectangular distribution rather than that of a Gaussian distribution. The multitude of absorption lines in the spectrum are identified as water absorption lines(6) caused by the humidity in ambient air.

Utilizing the bunch information from the autocorrelation measurement, the performance of the accelerator was optimized to produce the shortest bunch length. With this optimization effort, the shortest bunch length was obtained and shown in Fig. 3. The FWHM of the peak in the interferogram is 24 μm of mirror movement which is equivalent to 48 μm of optical path difference. Hence, the equivalent bunch length is estimated as 36 μm (Gaussian) to 48 μm (rectangular) or, in the time scale, as 120 fs (Gaussian) to 160 fs (rectangular). The rms bunch length is calculated as 46 fs (rectangular) to 48 fs (Gaussian).

CONCLUSION

In summary, we have shown the results of using an optical autocorrelation method to measure the length of sub-picosecond electron bunches generated at SUNSHINE. The rms bunch length is measured about 50 fs. This new method has demonstrated a very good sub-picosecond resolving power which out-performs any existing time-resolved method; hence, it may be applied to any short bunch length measurement task. Furthermore, the Michelson interferometer used in this method fits on a $20 \times 30\,\text{cm}^2$ breadboard and is easily transportable. Combining this compact interferometer with the easily manageable transition radiation and the room-temperature bolometer, this new device would become a convenient diagnostic instrument for accelerators.

ACKNOWLEDGEMENT

The authors would like to thank for the technical support by M. Baltay, G. Husak, J. Sebek, R. Theobald, and J. Weaver in this experiment. We also appreciate the useful introduction on the techniques of far-infrared experimentation received from P. Richards and the help to use some critical equipment provided by A. Schwettman and T. Smith. This work was supported by the Department of Energy, Office of Basic Energy Sciences (contract No. DE-AC03-76SF00515) and the Office of Naval Research (contract No. N00014-91-C-0170).

REFERENCES

1. Kung, P. H., Lihn, H.-C., Bocek, D., and Wiedemann, H., "High-intensity coherent FIR radiation from sub-picosecond electron bunches," in *Gas, Metal Vapor, and Free-Electron Lasers and Applications*, edited by Similey, V. N. and Tittel, F. K., Proc. SPIE **2118**, 191(1994).
2. Kung, P. H., Lihn, H.-C., Bocek, D., and Wiedemann, H., "Generation and measurement of 50-fs (rms) electron pulses," Phys. Rev. Lett. **73**, 967(1994).
3. Fork, R.L., Greene, B. I., and Shank, C. V., "Generation of optical pulses shorter than 0.1 psec by colliding pulse mode locking," Appl. Phys. Lett. **38**, 671(1981).
4. Barry, W., "An autocorrelation technique for measuring sub-picosecond bunch length using coherent transition radiation," in *Proceedings of the Workshop on Advanced Beam Instrumentation* **1**, KEK, Tsukuba, Japan, April 22–24, 1991 (unpublished).
5. Nodvick, J. S. and Saxon, D. S., "Suppression of coherent radiation by electrons in a synchrotron," Phys. Rev. **96**, 180(1954).
6. Richards, P. L., "High-resolution Fourier transform spectroscopy in the far-infrared," J. Opt. Soc. Am. **54**, 1474(1964).

Synchrotron Radiation Monitor For DAΦNE

A.Ghigo, F.Sannibale, M.Serio
INFN Laboratori Nazionali di Frascati - 00044 Frascati (Roma) - Italy.

Abstract

The Synchrotron Radiation Monitor for both electrons and positron beams of DAΦNE, the LNF Φ-Factory, is described. The most important measurements the monitor needs to provide are: beam transverse dimensions, bunch length, beam transverse density, vertical and horizontal tunes. The source points are in two bending magnets, one of each rings, placed in a zero dispersion region. The collected radiation, in the visible range, is sent through two optical transfer lines to the same measurement bench placed in a dedicated laboratory outside the DAΦNE hall. The complete layout of the measurement bench and a detailed description of the detectors employed are presented.

INTRODUCTION

The DAΦNE accelerator complex (1), which is being built at LNF, consists of two storage rings and an injector for topping-up at 510 MeV. The stored positron and electrons beams circulate in opposite directions, intersecting in two interaction points. The first interaction point is dedicated to CP violation experiments, while the other one to hypernuclei experiments.

The synchrotron radiation monitors design for the DAΦNE storage rings is presented.

MEASUREMENT RESOLUTION OF A SYNCHROTRON RADIATION MONITOR

Horizontal Resolution and Source Length

In the hypothesis of gaussian beams, the horizontal resolution Δx can be calculated by the square root of the quadratic sum of three different errors, the *curvature error* Δx_C, the *diffraction limit error* Δx_D and the *depth of field error* Δx_{DF} (2).

Figure 1 shows the essential geometry of a light source in a constant magnetic field area (dipole magnet case). The curvature error is due to the finite size of the finite size and to the curvature of the particles trajectory within this length. It can be expressed by (3):

$$\Delta x_C = x - x_0 \tag{1}$$

with

$$x = \frac{l^2 R - a(R + x_0)^2 - \sqrt{l^2 (R + x_0)^2 \left(a^2 - 2aR - x_0^2 - 2Rx_0 + l^2\right)}}{(R + x_0)^2 - l^2} \tag{2}$$

where x_o is the half horizontal size of the beam, while the meaning of the others symbols may be derived from Fig.1. Equation 2 holds for small angles θ and ϕ.

The diffraction limit and the depth of field errors can be evaluated using (4,5):

$$\Delta x_D = \frac{\lambda}{2\,\upsilon} \qquad\qquad \Delta x_{DF} \approx \frac{\delta z}{2}\,\upsilon \tag{3}$$

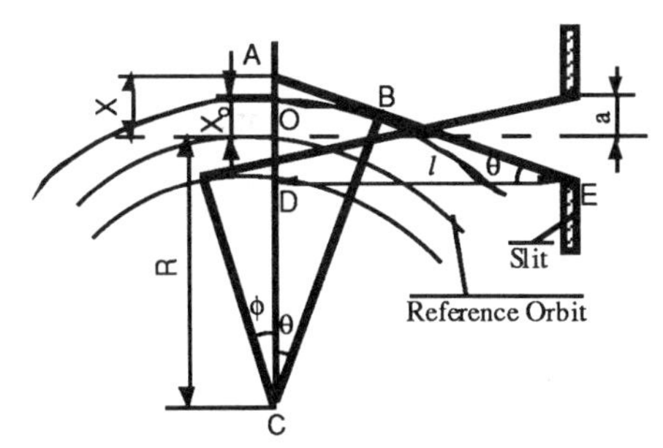

Figure 1. Typical Source Area Top View.

where λ is the wavelength measurement, $\upsilon = arctan\,(a\,/l\,) \approx a\,/l$ is the half aperture angle of the system source+slit and δz is the longitudinal extent (length) of the source (3):

$$\delta z = (\,\phi + \theta\,)\,R \approx 2\theta R \tag{4}$$

where

$$\theta \approx tan\,\theta = (x + a\,)\,/\,l \tag{5}$$

Vertical Resolution

In the vertical case we still have the *diffraction limit error* Δy_D and the *depth of field error* Δy_{DF}, but no more the curvature error:

$$\Delta y_D = \lambda\,/\,(\,2\vartheta\,) \qquad \Delta y_{DF} = \delta z\,\vartheta\,/\,2 \tag{6}$$

The quantity δz is given by Eq.4 where ϑ is the *total angular half-aperture of the photon beam* :

$$\vartheta = \left(\psi_{beam}^2 + \psi_{typ}^2\right)^{1/2} \tag{7}$$

with

$$\psi_{beam} = (\gamma_y\,\varepsilon_y\,)^{1/2} = \left(\frac{1+\alpha_y^2}{\beta_y}\,\varepsilon_y\right)^{1/2} \qquad \psi_{typ} \approx \frac{1}{\gamma}\left(\frac{\lambda}{\lambda_c}\right)^{1/3} \tag{8}$$

ψ_{beam} is the electron beam divergence and ψ_{typ} is the typical opening angle of the emitted photon beam. The quantities β_y , α_y , γ_y are the vertical Twiss parameters at the source point, ε_y is the beam vertical emittance (zero vertical dispersion has been assumed), λ_c is the critical wavelength (4) and γ the energy in rest mass units. The expression for ψ_{typ} holds if $\lambda \gg \lambda_c$.

A routine, called Synch1_2, which applies the theory previously described, has been developed (3). Its output file, with the DAΦNE parameters, is shown in Fig.2.

DAΦNE MAIN RINGS SYNCHROTRON RADIATION MONITOR

Source Area

Table 1 shows the relevant DAΦNE parameters and Table 2 the optical characteristics at the selected source point.

Two sources, one per beam, are foreseen. They are 18.5 deg inside the parallel face dipole magnets in the shorter DAΦNE ring half section (see Fig.4).

The geometry , which is the same for both of them, is shown in Fig.3. A water cooled Al with 35 mm diameter, placed 0.8 m downstream the source point, vertically deflects the photon beam, through a vacuum window, onto a slit 1.065 m far away from the source point. A window on the photon beam axis, upstream the source point, allows the alignment of the optical line and the calibration of the transverse dimension measurement using a laser in place of the synchrotron ligth.

Table 1. DAΦNE General Parameters.

Energy	510 MeV
Ring Length	97.69 m
Dipole Bending Radius	1.4 m
Natural Emittance	10^{-6} m rad
Natural Relative Energy Spread	3.97×10^{-4}
Particles/Bunch	9×10^{10}
Max Number of Bunches	120
r.ms. Natural Bunch Length	8.1×10^{-3} m
r.m.s. Anomalous Bunch Length	3.0×10^{-2} m

Table 2. Characteristics at Source Point (1% Coupling)

β_x	6.46 m
β_y	7.87 m
α_x	0.468
α_y	0.165
Dispersion	~ 0 m
Horizontal Dimension (rms)	2.5×10^{-3} m
Vertical Dimension (rms)	2.8×10^{-4} m

For what concerns resolution, an important choice is the working wavelength. We decided to work within the visible range (~ 400 ÷ 600 nm) in order to use the wide variety of commercial optical components and to have the optical channels in air. The vertical resolution, which is the more critical in DAΦNE, is still good enough with such a choice. The Synch1_2 output file in Fig. 2 shows that, with a maximum wavelength of 600 nm and a slit half aperture of 1 mm, the relative error on the vertical measurement is less than 3 %. In the same file all the others meaningful quantities of the DAΦNE monitors are given.

Optical Channel and Instrumentation Hall

Figure 4 shows the optical channels top and side views. The Synchrotron Radiation Monitor Instrumentation Hall is outside the concrete wall of the main rings hall, in a room at an higher level with respect to the machine plane. This feature makes the hall a radiation safe area, where it is possible to stay when the beams are stored. Each of the two optical channel starts from a beam expander downstream the slit, (see Fig.3), this is an achromatic system of lenses with focus in the source point. A set of matching mirrors transports the light from the source area to the optical table in the Instrumentation Hall. The first of these mirrors is visible in Fig.3, the other two will be placed in the Instrumentation Hall. The optical line, which is ~ 20 m long, has to be surrounded by an opaque black pipe to prevent distortions due to thermal effects in air and to avoid noise caused by environmental stray lights.

```
Synch 1.2  (400 - 600 nm) ; PARALLEL FACE DIPOLE ; 0.01 coupling
    LINE
        l0 (m)                                      : 0.5653
        d (m)                                       : 0.2656
        Slit Half Aperture (m)                      : 0.1000E-02
    RING PARAMETERS:
        Energy (MeV)                                : 510.0
        Relative Energy Spread                      : 0.3970E-03
        Natural Emittance (m rad)                   : 0.1000E-05
        Coupling                                    : 0.1000E-01
        Bending Radius (m)                          : 1.400
        Revolution Frequency (Hz)                   : 0.3068E+07
        Number of Bunches                           : 120.0
        Charge/Bunch (C)                            : 0.1442E-07
        Bunch Length (sigma) (m)                    : 0.3000E-01
    BEAM OPTICAL FUNCTIONS @ SOURCE:
        Beta x (horizontal) (m)                     : 6.457
        Alpha x (horizontal)                        : 0.4679
        Eta x (horizontal) (m)                      : 0.0000E+00
        Deta x (horizontal)                         : 0.0000E+00
        Beta y (vertical) (m)                       : 7.872
        Alpha y (vertical)                          : 0.1648
    INTEGRATION PARAMETERS:
        Lambda Start (nm)                           : 400.1
        Lambda Stop (nm)                            : 600.0
        Lambda Number Int. Steps                    : 2000
        Ksi Number Int. Steps                       : 50

    CALCULATION RESULTS:
        Total Average Current                       : 5.309 (A)
        Cut Off Lambda                              : 5.881 (nm)
        Source Magnet Edge Angle                    : 0.3230 (rad)
        l: Source-Slit Distance                     : 1.065 (m)
        Single e- Emitted Power                     : 0.3030E-11 (watt)
        Normal/Parallel Power Ratio                 : 0.3142
        Single Bunch Radiated Power                 : 0.2727 (watt)
        Total Radiated Power                        : 32.73 (watt)
        Source Length                               : 0.9299E-02 (m)
        Single Bunch Accepted Power                 : 0.2883E-03 (watt)
        Total Accepted Power                        : 0.3460E-01 (watt)
        One Turn One Bunch Ac. Ener.                : 0.9397E-10 (joule)
        Single Bunch Peak Power                     : 0.3746 (watt)
        # of Accepted Photons/bunch                 : 0.2365E+09
        Photon Average Lambda                       : 500.0 (nm)
        Electron Beam x dim (sigma) (m)             : 0.2529E-02
        x Measurement Error (sigma) (m)             : 0.3196E-03
        x Measurement Relative Error                : 0.7955E-02
        x Curvature Error (m)                       : 0.7735E-05
        x Depth of Field Error (m)                  : 0.4367E-05
    REMARK: In the following part the largest lambda has been used
        x Diffraction Limit Error (m)               : 0.3194E-03
        Electron Beam y dim (sigma) (m)             : 0.2792E-03
        y Measurement Error (sigma) (m)             : 0.6773E-04
        y Measurement Relative Error                : 0.2901E-01
        Diffract./Depth of Field Error Ratio        : 2.949
        Photon Beam Divergence (rad)                : 0.4677E-02
```

Figure 2. Synch 1_2 Output File: DAΦNE Light Source.

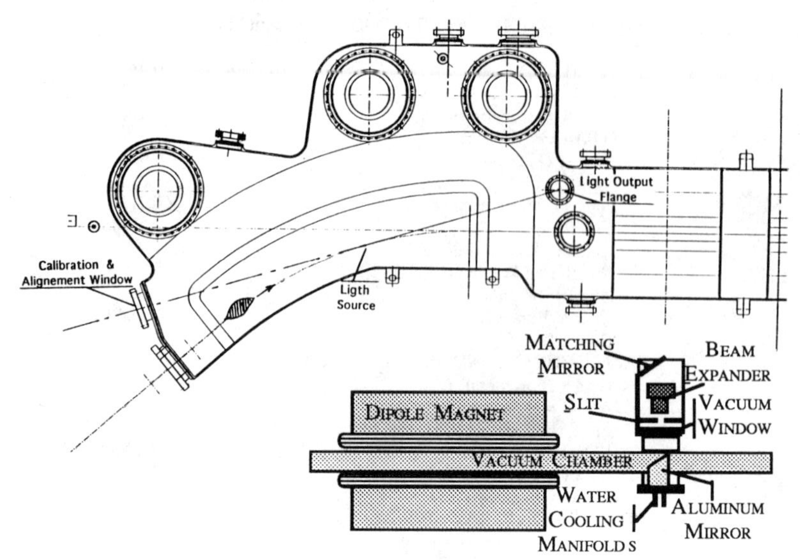

Figure 3. Source Area Layout.

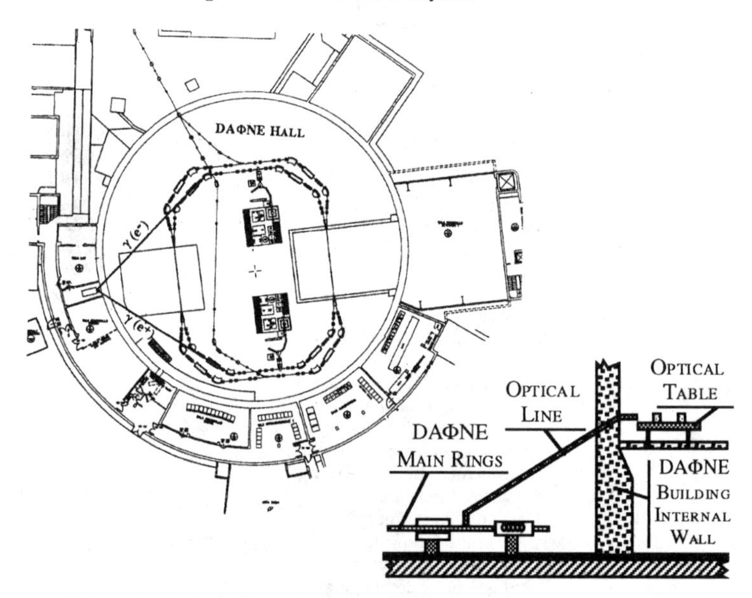

Figure 4. Optical Channels Lay-out. Side and Top View.

MEASUREMENTS

Figure 5 shows the optical table layout. By using splitters and two independent lines, all the measurements will be independently and simultaneously available for both beams.

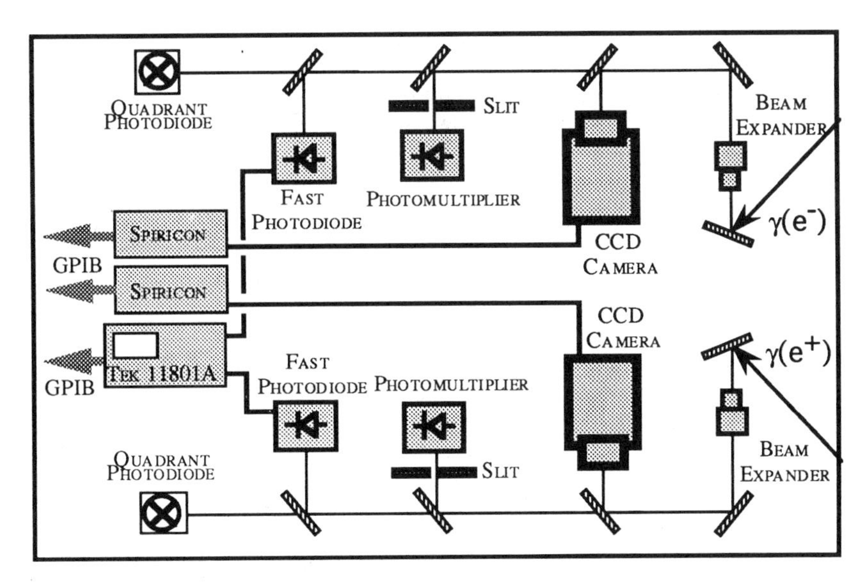

Figure 5. Optical Table Lay-out.

Transverse Dimensions

The measurement system is composed by a CCD camera and an image analyzer. The camera, a PULNIX TM 6 has a 752×582 video matrix with a pixel dimension of $8.6 \times 8.3\ \mu m^2$. A zoom objective in front of the camera permits to vary the overall magnification (optimum magnification ~ 0.5).

The image analyzer is the SPIRICON LBA 100A which is able to capture, display and analyze the camera image with repetition rates up to 15 Hz.

It is worth to mention that, because of the zero dispersion, the beam emittances can be measured (this measurement implies the knowledge of the ring optical functions at the source point).

Bunch Length

The bunch length is measured by a fast detector+large bandwidth oscilloscope system. The New Focus 143-4 photodiode ensures DC-25 GHz bandwidth and 17 psec rise time. The coupling between radiation and the photodetector active area (25 μm diameter) is provided by a single mode fiber optic with a GRIN lens collimator. The detector output is directly connected to the digital oscilloscope TEKTRONIX 11801A (50 GHz bandwidth and 200 KHz sampling rate).

Assuming gaussian beams with σ=100 psec, the estimated measurement error is ~ 0.3%. The oscilloscope sampling rate and record length imply that about 8000 turns are needed to sample the whole beam pulse.

The system sensitivity makes the measurement possible with stored currents as low as 2 mA. This feature should permit the study of the turbulent bunch lengthening in DAΦNE.

Beam Transverse Density and Tunes

Figure 5 also shows a slit+photomultiplier system for the beam transverse density measurement, whose set-up is shown in Figure 6. The beam is transversely excited by a sweeping oscillator+kicker system. The slit, horizontal for the vertical measurement and viceversa, selects the photons around the peak of the light distribution allowing a measurement of the density around this position. The effect of incoherent oscillations on the beam is to decrease this density (7). In this way the existence of non-linear phenomena (read beam-beam interaction, lattice non-linearities, ion trapping, etc) leading to incoherent oscillations, can be detected and measured (8).

Replacing the slit-photomultiplier assembly with a quadrant photodiode, the previous set-up can be used to simultaneously measure the horizontal and vertical machine tunes. The quadrant photodiode, which is sensitive to the photon beam center of mass position, is also used for the alignment of the line on the optical table (see figure 5).

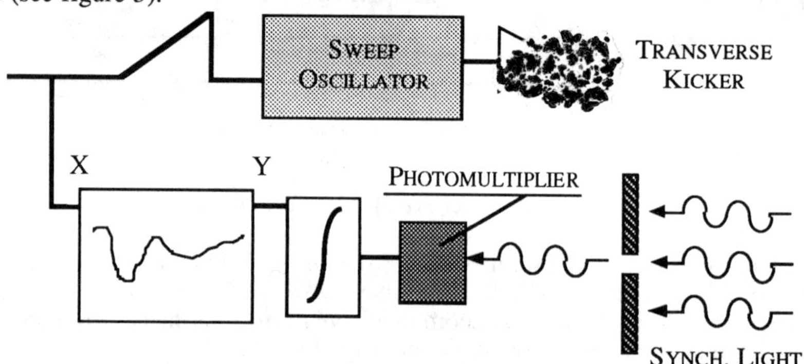

Figure 6. Density Measurement Set-up.

REFERENCES

1. The DAΦNE Project Team, "Overview of DAΦNE, the Frascati Φ-factory", Proceedings of the VI International Conference on High Energy Accelerators, Hamburg, Germany, 20-24 July 1992.
2. Bovet C., Placidi M., "A dedicated synchrotron radiation source for LEP beam diagnostics" LEP Note 532, April 11, 1985.
3. Ghigo A. and Sannibale F., "DAΦNE Main Rings Synchrotron Radiation Monitors Design Criteria" DAΦNE Technical Note CD-3, September 16, 1994.
4. Hofmann A. and Meot F., "Optical Resolution of Beam Cross-Section Measurements by means of Synchrotron Radiation" Nucl. Inst. Meth., **203** (1982) pp 483-493.
5. Sabersky A. P., "The geometry and optics of synchrotron radiation" Particle Accelerator, Vol. **5** (1973), p. 199.
6. Hofmann A., "Electron and proton beam diagnostics with synchrotron radiation" IEEE Transaction on Nuclear Science, Vol. **NS-28**, No 3, June 1981.
7. Serio M., Zobov M., "Measurement of transverse and longitudinal spectra", Proceedings of the First European Workshop on Beam Diagnostics and Instrumentation. May 3-5 1993, Montreux, Switzerland, p. 47.
8. Biagini M. et al.,"Observation of ion trapping at Adone", Proceedings of the XI-th International Conference on High Energy Accelerators, CERN -Geneva July 1980, p.687.

Experimental study of the E.L.S.A. electron-beam halo

G. Haouat, N.Pichoff, C. Couillaud,
J. Di Crescenzo, S. Joly, S. Seguin, S. Striby
Commissariat à l'Energie Atomique, Centre d'Etudes de Bruyères-le-Châtel, SPTN. BP 12 - 91680 Bruyères-le-Châtel. FRANCE.

Abstract

An experimental study of the transverse spatial distribution of the ELSA-linac electron beam has been initiated to investigate halo-producing mechanisms in the generation, acceleration and transport of high-intensity, high-brightness charged-particle beams. In an experiment presented here the beam profile has been analysed over four decades using an imaging technique. This technique consists in putting an optical transition radiation screen on the beam path. The emitted light is transported through conventional optics to an intensified video camera. The beam-profile behaviour is shown to be dependent on the current, the initial size and shape of the beam and on the transport configuration.

1 INTRODUCTION

The technolgy of high-intensity linear accelerators has made so much progress during the past decade that we can see an increased interest for their utilisation in number of applications such as isotopic separation, plasma heating with high-power free electron lasers fed by electron linacs, radioactive actinide and fission-product waste management, production of tritium, and transmutation of defense and commercial plutonium stocks with intense spallation-neutron sources driven by large proton or light-ion linear accelerators. However experimental observations and multiparticle simulations indicate that high-intensity beams in accelerators develop a low-density halo at the periphery of the central core of the beam. This outer part of the distribution leads to particle losses in the accelerator which, for a high-energy beam, may induce enough radioactivity or radiation damage in the structures to complicate considerably maintenance and operation of the machine. The major challenge is to control the beam profile through a better knowledge of the phenomena involved in halo formation.

A number of theoretical papers and numerical simulation studies are devoted to understanding the mechanisms of beam halo formation in high-intensity linear accelerators, and to learning how this halo develops in accelerator structures. However, predictions of the simulations may be quantitatively incorrect since, because of computer limitations, halo properties are described by only a few particles. Therefore experimental study of beam halo is necessary to check theoretical predictions and complement simulations.

Few experiments to date have been dedicated to precise determination of the beam structure extending in the very low density regions constituting the halo (1-4).We have undertaken, at Bruyères-le-Châtel, an experimental study of the transverse spatial distribution of the high-intensity, high-brightness electron beam delivered by the ELSA linac. The objective of this study is to analyze the beam profile over a very large dynamic range under varying conditions in the generation, acceleration and transport of the beam. An experiment is presented in which the beam profile intensity has been observed over four decades. The experimental setup and the measurement technique used for this work are described.

2 EXPERIMENTAL ARRANGEMENT

The ELSA linac, which has been designed for free-electron laser applications, is described in detail in Ref. 5. It is composed of a 2 MeV photoinjector cavity followed by three accelerating cavities. The photocathode preparation chamber is attached to the photoinjector. The accelerator delivers a pulsed electron beam of 18-20 MeV maximum energy. The temporal structure consists of macropulses, 20 to 150 μs duration, at a repetition rate of 0.1 to 10 Hz. The macropulse train consists of $\sim$20 ps long micropulses spaced 69.2 ns apart. Micropulse charge of 1 to 5 nC, corresponding to peak currents of 50 to 250 A, are used currently. Normalized transverse emittance is typically 10-30 π mm.mrad. This high-brightness, space-charge dominated beam is suitable for experimental study of the halo.

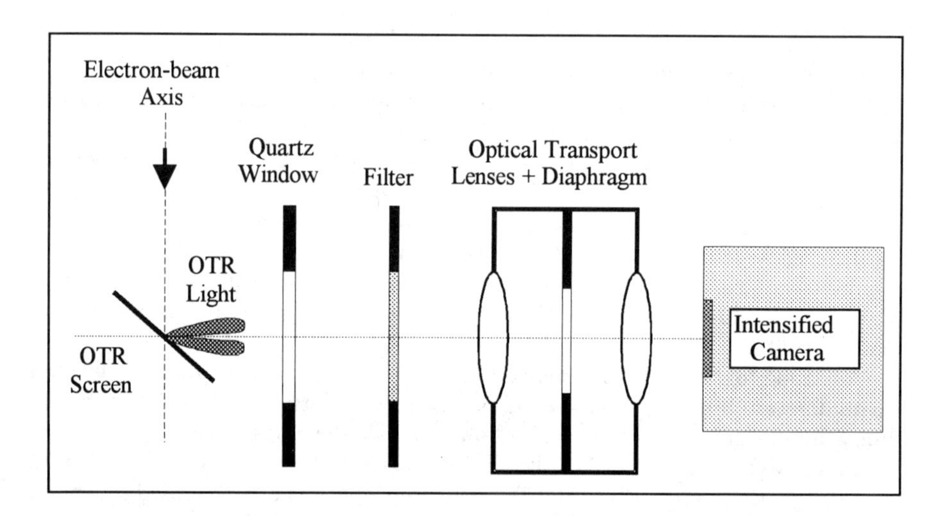

Figure 1: Schematic of the experimental setup for halo measurements.

The experimental apparatus developped for the halo measurements is a modified version of the devices used to diagnose transverse profile on the ELSA facility (6). It consists in a conversion screen, moved in the beam path by an actuator, image-transport optics, and a variable-gain video camera coupled to a data-acquisition system (Fig.1). This apparatus is located at the end of the linac.

The electron/photon conversion screen must fulfill requirements of accurate halo measurements; these are: good sensitivity, excellent response linearity over a wide dynamic range, no saturation effects, good spatial resolution, resistance to considerable energy deposition, and no image smearing induced by thermal effects. Optical transition radiation (OTR) converters fulfill these requirements; moreover, they are widely used for the ELSA-linac beam characterization because of the unique properties of the transition radiation process (7). Measurements were performed with two different kinds of OTR screens. We used converters made of a 8 μm thick aluminized (3000 $\AA$ layer) Kapton film stretched between two rings of 50 mm diameter. They constitute excellent screens with a perfectly reflecting surface, good mechanical properties, weak x-ray production and good behaviour to high-power beams. We used also 1-mm thick stainless-steel disks polished to optical standards. These rather thick screens do not suffer much damage at high currents.

The screen plane is oriented at 45° with respect to the beam axis so that the backward lobe of radiation is observed at 90° to the beam through a quartz window (Fig.1). The image of the beam transverse distribution formed on the screen is transported to the video camera by an optical system which has been designed for an observation field of 16 mm on the screen, corresponding to roughly 10 times the beam-core diameter. This optical system is capable of collecting the light originating from any point in the observation field and emitted in a cone of aperture angle $4/\gamma$, γ being the electron reduced-energy. Most of OTR light is assumed to be emitted in this cone. This ensures good response linearity as a function of the emission point. A low-pass filter with a cut-off wave length of 620 nm is placed in the optical-beam path to block the 532 nm radiation from the photocathode drive-laser, reflected or scattered in direction of the video camera. This filter reduces the background light in the profile images. For the same reason vacuum-gauge filaments are switched off during the measurements.

Images of the beam profile are analysed by an intensified camera whose video signal is transported to image-processing hardware composed of a Sofretec PITER500 8 bit frame grabber, a 486 DX PC clone and a TV monitor. The intensified camera consists in a voltage-controlled multichannel-plate light intensifier coupled to a Nocticon tube. Images are digitized and displayed for every macropulse (1 Hz cycle) on the real-time TV monitor. This allows beam adjustment such as positioning and focusing. Images can be stored on a Bernouilli box 90 PRO, with 90 Mb removable disks, for further off-line precise and detailed image processing.

3 MEASUREMENT METHODS AND RESULTS

The results of the first halo measurements, undertaken using the saturated-core experimental method, were not satisfactory. Later, measurements using the hole-bored screen method were made yielding better results.

3.1 The saturated-core method

The OTR screen is a thin aluminized Kapton film from which the beam core is first observed by adjusting the light amplification of the intensified camera below saturation. The light-intensifier voltage is then gradually increased so as to obtain a more observable and exploitable image of the halo. The core image is saturated during the measurements. In off-line data processing the profile images are normalized one to the other, by using the precise calibration of the light intensifier, to finally yield the beam-density distribution on an extended range. Measurement results are presented in Fig.2 for an electron beam of 16.5 MeV energy and 0.8 nC/pulse charge. This figure shows that the core and the halo exhibit very distinct shapes. The core, enclosing 99 % of the particles, is 2 mm diameter while the halo, which represents ~1 % of the total beam, extends over a diameter of at least 8 mm. This last value is not very significant since in this measurement the observation field was small (~10 mm).

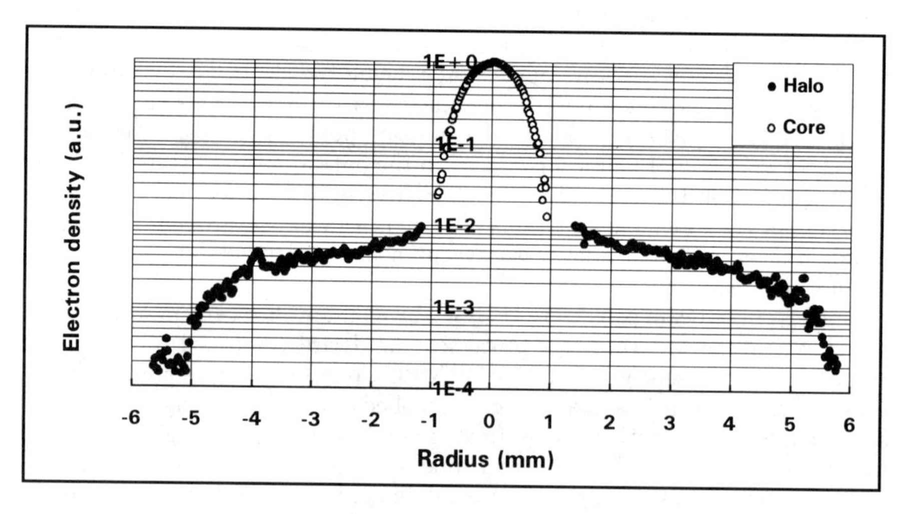

Figure 2: Beam-halo measurement using the saturated-core method.

In this experimental technique some undesirable phenomena can spoil observation of the halo such as multiple reflection of photons originating from the core on the various elements of the optical transport to the camera, scattering of these photons on the dust present in the transport system, or generation of secondary-emission electrons by the intense part of the photon beam which extend in the low-intensity region in the light intensifier. This led us to use the following method.

3.2 The hole-bored screen method

To reduce background light in the halo distribution we use an experimental procedure in which the OTR screen is a 1-mm thick, polished, stainless-steel disk with a hole bored in the center. The experimental equipment is the same as that of the previous experiment but the observation field has been increased to 16 mm. The beam core is first observed with its image below saturation by appropriate setting of the light intensifier. The electron beam is then steered towards the hole through which most of the core passes. This eliminates the intense source inducing background light, and by increasing amplification of the light intensifier the halo can be observed properly with considerably reduced background.

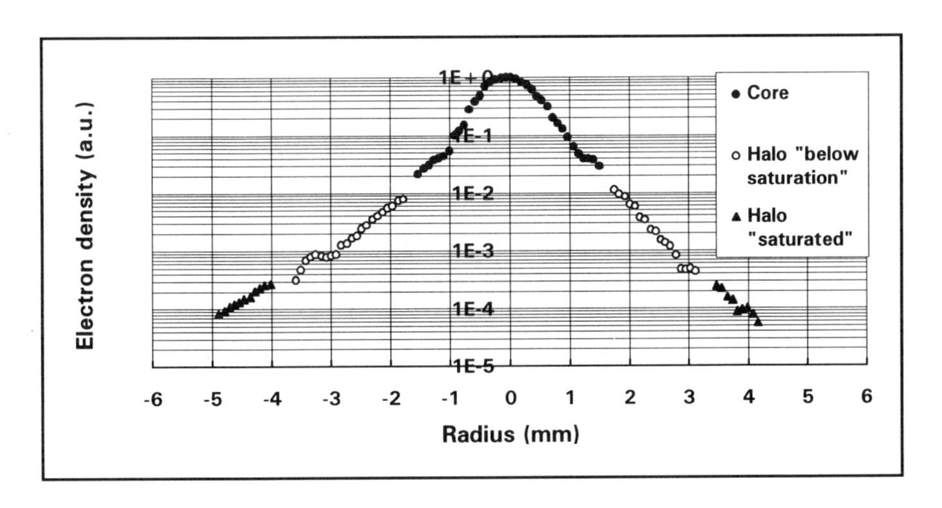

Figure 3: Beam-halo measurement using the hole-bored screen method.

An example of beam profile obtained using this method is displayed in Fig.3. The measurement was made at an electron-beam energy of 16.5 MeV and a micropulse charge of 0.8 nC. The diameter of the hole in the OTR screen was 2.5 mm. The beam transverse distribution was established with three settings of the light intensifier corresponding to *i)* the core below saturation outside the hole, *ii)* the core through the hole and the halo below saturation and *iii)* the core through the hole and the halo slightly saturated. Figure 3 shows that the beam profile has a dynamic range of at least four decades and that the density distribution presents an exponential shape in the conditions of this experiment.

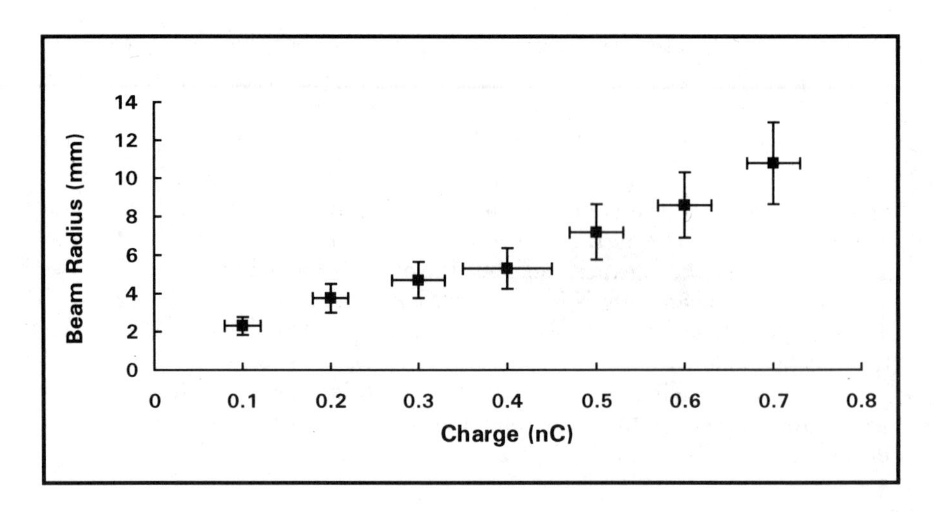

Figure 4: Variation with micropulse charge of the beam radius enclosing 99.9% of the particles.

Space-charge effect on the beam profile has been analysed at 16.5 MeV energy by measuring the transverse density distribution for micropulse charges ranging from 0.1 to 0.7 nC. This is displayed in Fig.4 which shows a large and rapid spot broadening with increasing charge. Recent measurements, not yet processed, extend the micropulse-charge range up to 5 nC.

4 SUMMARY AND CONCLUSIONS

The hole-bored screen technique seems to be promising though the background light in profile images is still important at low-density levels. Measurements are in progress to scan a larger illumination range by using screens with holes of different diameters, to correlate halo data to transverse-emittance values for the core, to analyse the dependence of beam-transport configuration on the halo and to study the influence of initial size and shape of the electron beam by varying the spatial distribution of the photocathode drive-laser beam.

To overcome problems encountered with the saturated-core method and to extend the capabilities of the hole-bored screen method, we are building a more sophisticated experimental setup. Its main features are: a thin OTR screen without hole, an optical transport system under vacuum with anti-reflection coating and multi-coated lenses, variable attenuation of light from the core through the use of circular neutral density filters mounted on field lenses, a gated intensified CCD camera with 12 bit resolution and a very high performance frame grabber. We expect to cover a dynamic range of seven decades with this apparatus.

REFERENCES

1. D. Kehne, M. Reiser and H. Rudd, "High brightness beams for advanced accelerator applications", AIP Conf. Proc. **No. 253**, 47 (1991).
2. G. Riehl, J. Pozimski and W. Barth, "Investigation of beam aberrations and beam halo by 3-dimensional emittance measurements", Proc. of the 1990 Lin. Acc. Conf., LA-12004-C, UC-910/UC-414 (1991) 755.
3. J. Palkovic, "Emittance growth and halo formation in a low energy proton beam", SSCL-Preprint-223 (Apr.1993).
4. A.H. Lumpkin and M.D. Wilke, Nucl. Inst. Meth. **A331**, 803 (1993).
5. R. Dei-Cas et al., "Status report on the low-frequency photo-injector and on the infrared FEL experiment (ELSA)", Nucl. Inst. Meth. **A296**, 209 (1990).
6. G. Haouat, C. Couillaud, J. Di Crescenzo, J. Raimbourg and S. Seguin, "Some electron-beam diagnostics on the ELSA linac", Proc. First European Workshop on Beam Diag. and Inst. for Part. Acc., Montreux, Switzerland. CERN PS/93-35 (BD) - CERN SL/93-35 (BI) (1993) 180.
7. L. Wartski, S. Roland, J. Lasalle, M. Bolore and G. Filippi, "Interference phenomenon in optical transition radiation and its application to particle beam diagnostics and multiple-scattering measurements", J. Appl. Phys. **46**, 3644 (1975).

The Planned Photon Diagnostics Beamlines at the Advanced Photon Source*

Bingxin Yang and Alex H. Lumpkin
Advanced Photon Source, Argonne National Laboratory, Argonne, IL 60439

Abstract

We present the planned photon diagnostics beamlines at the Advanced Photon Source. The photon diagnostics beamlines of the storage ring include two bending magnet sources and a dedicated diagnostic undulator. The bending magnet lines will employ the conventional UV/visible imaging techniques (resolution $\sigma \cong 40$ μm) and the x-ray pinhole camera (resolution $\sigma \cong 15$ μm) for the measurement of the positron beam size (design value: $\sigma \cong 100$ μm). The opening angle of the undulator radiation will be around $\sigma \cong 3$ μrad for its first harmonic (23.2–25.8 keV), and $\sigma \cong 1.7$ μrad for its third harmonic (70–72 keV), providing a good resolution for measuring the positron beam divergence size (design values: $\sigma \cong 9$ μrad for 10% vertical coupling and 3 μrad for 1% coupling). The undulator and its x-ray optics are specifically optimized for full emittance measurement of the positron beam. A major developmental effort will be in the area of detecting very fast phenomena (nanosecond and sub-nanosecond) in particle dynamics.

INTRODUCTION

The Advanced Photon Source (APS) will be a highly bright x-ray source for scientific and industrial users. With the present design, the contribution of the particle beam dominates the undulator photon beam brightness above the photon energy of several keV. The characterization and improvement of the particle beam emittance is a key technical task for the APS. Serving this objective, the goals of the APS photon diagnostics beamlines are twofold: (1) To provide on-line information with the positron beam data to the Operations Group and users on a 24-hour basis; (2) To provide information necessary for determining the origin of any observed emittance growth, transverse as well as longitudinal.

The positron beam size and divergence parameters for the APS storage ring under standard operating conditions are shown in Table 1. The following quantities of the positron beam will be of primary interest:

(1) The first moments of the positron population in the transverse phase space (x, x', y, y'): beam position and direction.
(2) The second moments of the positron population in the four-dimensional transverse phase space (beam transverse emittance): the beam size (x, y), beam divergence (x', y'), and tilt of the emittance ellipse.

* Work supported by the U.S. Department of Energy, Office of Basic Energy Sciences, under Contract No. W-31-109-ENG-38

(3) Longitudinal emittance: the beam energy spread (σ_E) and bunch length (σ_z).
(4) Single bunch characteristics: One or more parameters in (1) through (3) measured as a function of time within a single positron bunch, taken over a single pass or averaged over multipasses.
(5) RF bucket purity: The residual charge in the nominally empty buckets adjacent to a filled bucket.

Table 1 The APS Storage Ring Positron Beam Parameters (1)

(ε=8.2 nm)	Bending Magnet A (1/50 point nominal)		Bending Magnet B (1/8 point nominal)		ID Straight Section (center)	
	x	y	x	y	x	y
β (m)	4.6	14.6	1.8	18.4	14.2	10.1
α	1.5	-0.6	0.4	0.8	0	0
$1/\gamma$ (m)	1.4	10.7	1.5	11.2	14.2	10.1
$dq/d(\Delta E/E)$ (μm/0.1%)	0.05	0.0	92	0.0	0.0	0.0
	10% vertical coupling (ε_x=7.5 nm, ε_y=0.75 nm)					
σ_{eq} (μm)	185	104	114	117	325	87
σ'_{eq} (μrad)	73	8.3	70	8.2	23	8.6
	1% vertical coupling (ε_x=8.1 nm, ε_y=0.08 nm)					
σ_{eq} (μm)	193	34	119	39	340	29
σ'_{eq} (μrad)	76	2.8	73	2.7	24	2.8
	0.1% vertical coupling (ε_x=8.2 nm, ε_y=0.008 nm)					
σ_{eq} (μm)	194	11	120	12	341	9
σ'_{eq} (μrad)	76	0.87	74	0.85	24	0.90

The photon diagnostics beamlines at the storage ring include a bending magnet line and an insertion device line. They will employ synchrotron radiation from visible to x-ray wavelength region with the following techniques:

(1) imaging the positron beam with UV/visible radiation of the bending magnet.
(2) imaging the positron beam with x-rays with pinhole cameras.
(3) deducing all first and second moments of the positron beam with undulator photon beam measurement, with the final objective of single bunch, single pass measurement.

UV AND VISIBLE IMAGING TECHNIQUES

There are two fundamental limitations for high spatial resolution imaging of the positron beam:

(1) The diffraction limit: The synchrotron radiation concentrates strongly in the

forward direction, with a half cone angle defined by

$$\sigma'_{py} = \frac{1.07}{\gamma}\left(\frac{\lambda}{\lambda_c}\right)^{1/3} = 1.22\left(\frac{\lambda}{\rho}\right)^{1/3}, \qquad (1)$$

Hence the spatial resolution is limited to

$$\sigma_{py}[\mu m] \approx \frac{\lambda[\mu m]/2}{\sigma'_{py}[rad]} = 0.41\left(\rho\lambda^2\right)^{1/3}. \qquad (2)$$

(2) The finite source depth and curved positron orbit: The finite source depth and the curved positron orbit (radius ρ=39.8 m) contribute further to the transverse image size.

The minimization of the combined image size due to the above effect leads to the optimal acceptance angle of the order of (2)

$$\sigma'_{yR} \approx \left(\lambda / \rho\right)^{1/3}, \qquad (3)$$

and the optimal resolution

$$\sigma_{yR} \approx \left(\rho\lambda^2\right)^{1/3}, \qquad (4)$$

for λ=0.25 μm, $\sigma_{yR} \cong 40$ μm. Since the distance from the bending magnet source to the first optical element is limited by heat load and other practical considerations to be about 13 m or larger, the angular resolution required of the primary optical system is 3 μrad or better. In this resolution regime, we face technical challenges similar to those for the best ground-based astronomy telescopes. The instrument resolution can be limited by the third factor:

(3) Aberration of the camera system: Refractive optical elements are not suitable since the chromatic aberration alone will exceed the tolerance, all optical surfaces have to be stringently figured and polished, and the fluctuations due to air currents are significant.

We plan to use two stage imaging optics: A primary optical system, consisting of only mirrors, relays the image to the Optics Lab at about one-to-one magnification. The image will then be viewed in parallel by a number of cameras via beam splitters. Since the image of the primary system will be about 2 m from these cameras, the angular resolution requirement for the camera optics is relaxed to about 20 μrad. High quality commercial optics, in conjunction with bandpass filters, may used.

A key optical component is the first mirror which separates the UV and visible radiation from the x-ray beam. To achieve the objective of less than 1 μrad wavefront distortion, the rms temperature difference over the entire mirror should be less than 1°C. We plan to use a water-cooled Glidcop mirror with a central slot (Fig. 1). Under normal operating conditions (300 mA@7 GeV), the mirror absorbs 3 W of power out of a total of 1500 W coming down the beamline, and the highest temperature rise is 0.03°C as shown by a finite element analysis.

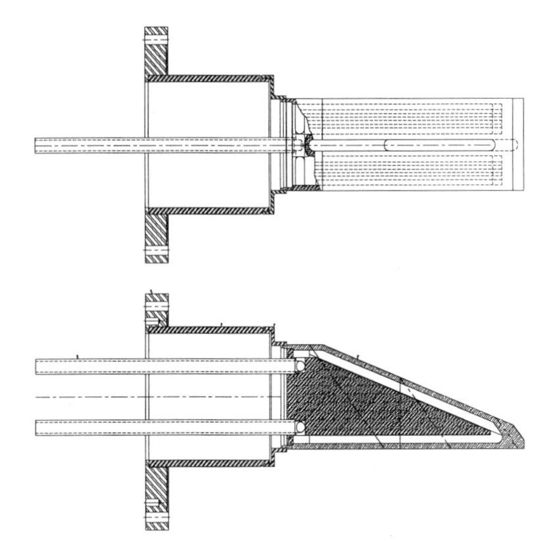

Fig. 1 M0 mirror design. The mirror is made of Glidcop and coated with rhodium. It is braze-joined to the 200-mm Conflat flange and is water cooled. The mirror has a 10-mm central slot to let the x-ray beam pass.

X-RAY PINHOLE IMAGING

The x-ray pinhole imaging possesses a few technical advantages over the UV/visible imaging:

(1) A pinhole camera provides the absolute position information when the locations of both the pinhole and the detector are precisely known.
(2) A pinhole camera is aberration free due to its simplicity. Even the deformation of the pinhole would not affect the image location or size but only flux throughput. Hence it is compatible with high power loads.
(3) A pinhole camera is achromatic: it is a broad-band optical instrument. Hence the efficiency of utilizing photons is high.

Table 2 Spatial Resolution of the X-ray Pinhole Camera

Name	Location (m)	Resolution (μm)	Function
35-BM-A	25.0	30	First Optical Enclosure (FOE)
35-BM-C	44.5	24	Medium Resolution Pinhole Camera Hutch
35-BM-D	68.5	15	High Resolution Pinhole Camera Hutch
35-ID-A	30	30	First Optical Enclosure (ID-FOE)
35-ID-B	50	16	High Resolution Pinhole Camera Hutch

Our x-ray pinholes are located about 12–14 m from the bending magnet sources.

The camera magnification ratio depends on the location of the screen detector which reads out the image and the wavelength of the photon used. Table 2 lists the expected location and spatial resolution for the bending magnet lines.

At its full magnification, the angular resolution of the camera will be about 15 μm, assuming a 10 μm pinhole can be made to operate at 100-keV photon energy. This will be sufficient for the beam of 1% vertical coupling. To measure beam sizes to a higher resolution the development of alternative x-ray optics will be required.

PARTICLE BEAM EMITTANCE MEASUREMENT WITH A DIAGNOSTICS UNDULATOR

At the APS, the positron beam vertical divergence is an order of magnitude lower than that of the bending magnet radiation, $1/\gamma$ being 73 μrad, and the angular information of the positron beam is practically lost. The well-collimated undulator beam, on the other hand, carries all useful information about the first and second moments of the particle beam. The diagnostic undulator is a low-power, low-K device with a short period. Its first harmonic energy is adjustable within 23.2 to 25.8 keV. At the expense of low flux and low tunability, we obtain high magnetic field uniformity due to weak field, and high collimation of the x-ray beam due to a large number of periods (N). Its basic parameter can be summarized in the following table.

Table 3 Operation Parameters of the APS Diagnostic Undulator

- Total length (L) : 3.0 m
- Magnetic period length (λ_u): 1.8 cm
- Number of period (N) : 158

(For 7 Gev / 100 mA positron beam)

Gap (mm)	11.5	15.5	38.0
B_0 (T)	0.24	0.13	0.002
K	0.40	0.22	0.004
Power (W)	530	150	0.04
First harmonic			
Energy (keV)	23.9	25.3	25.9
Flux (phs/s/0.1%BW)	3×10^{14}	1×10^{14}	3×10^{10}
ID / BM flux ratio	1300	450	0.1
Third harmonic			
Energy (keV)	71.8	75.9	77.6
Flux (phs/s/0.1%BW)	3×10^{12}	9×10^{10}	---
ID / BM flux ratio	58	2	

We plan to perform the following measurements with the undulator beam:

(1) *Pinhole Array (3)*
This is an extension of the scanning slits/pinhole method used in the accelerator physics. In principle, only three pinholes are needed for each dimension, x or y. The location of the photon beam waist can be obtained from the spacing of the pinhole images, the size of the waist from the on-axis pinhole image, and finally, with the size and the location of the waist known, the divergence of the photon beam can be obtained from the envelope of the pinhole images.

(2 Two-*Screen + Pinhole (Fig. 2)*
This is a modified version of the three-screen method used in accelerator physics. In this mode, two Laue crystal monochromators (C_1 and C_2) are in the undulator beam, followed by two x-ray screen detectors recording the beam cross section at these locations. A pinhole camera arranged in series with these crystals measures the source size (beam waist). From these recorded images, we can calculate the positron beam position and angle.

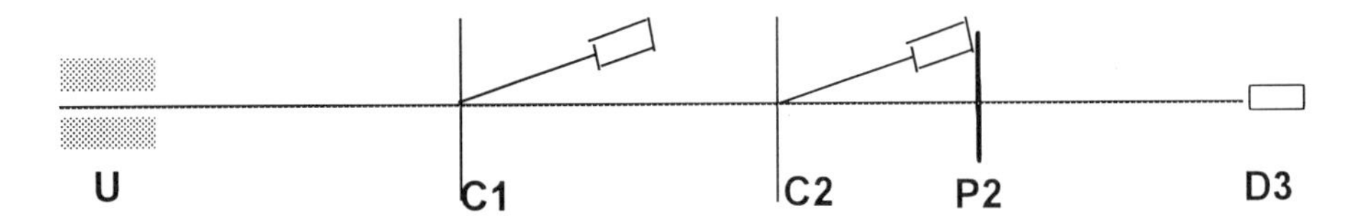

Fig. 2 Schematic diagram showing the experimental setup of the two screen and one pinhole method. The components are listed in the following table:

Symbol	Component	Location (m)
U	Diagnostic Undulator	0
C1	Monochromator/Screen 1	26
C2	Monochromator/Screen 2	37
P2	Pinhole	39
D3	Screen detector	58

While the second method is more promising in fast measurements (single bunch / single pass) due to its efficient use of photons, the first method has a significant advantage on the experimental side: its set up and data treatment are simple. Since all signals appear in one screen, it is compatible with streak camera measurements. Table 4 shows the expected performance of the diagnostic undulator beamline.

Table 4 Operation Plan for the Diagnostic Undulator Beamline

Positron Beam Vertical Coupling	10%	1%	0.1%
Undulator Gap (mm)	15.5	12.3	12.3
X-ray energy ω_n (keV)	25.3	73.1	73.1
Harmonic number n	1	3	3
Ideal Undulator Photon Beam			
size (μm)	1.4	2.4	2.4
divergence (μrad)	2.9	1.7	1.7
Trajectory Random Walk			
size (μm)	0.9	0.9	0.9
divergence (μrad)	0.6	0.6	0.6
Optical Instrument Resolution			
size (μm)	30	9	1.5
divergence (μrad)	< 1.0	< 0.8	< 0.5
Total Instrument Resolution			
size (μm)	30	10	3
divergence (μrad)	< 3.1	< 2	< 1.9
Positron Beam			
vertical size (μm)	87	29	8.7
divergence (μrad)	8.6	2.8	0.9
Resolution / Beam Size			
vertical size	0.35	0.35	0.35
vertical divergence	0.36	0.71	2.1

ACKNOWLEDGMENTS

We would like to thank Drs. Z. H. Cai, Y. Chung, E. Gluskin, W. Sellyey, P. J. Viccaro, and W. B. Yun for stimulating discussions. Dr. E. R. Moog has been especially helpful in our technical specification of the undulator. E. Rotela, S. Sharma, and I. C. Sheng designed the M0 mirror and other bending magnet front-end components. Efforts by Professor George S. Brown have been especially helpful to us all through this project.

REFERENCES

1. Glenn Decker, MAD calculations
2. R. C. Nawrocky, J. Galayda, L. H. Yu, D. M. Shu, IEEE Trans. on Nucl. Sci., NS, No. 5, (1893) 1985.
3. Z. Cai, et. al., *Phase-space measurement of undulator photon beam using a slit array and a CCD detector*, reported in SRI94.

Beam Instrumentation at the NSLS Electron Storage Rings

R.J. Nawrocky
Brookhaven National Laboratory, Upton, NY 11973

Abstract

The National Synchrotron Light Source at Brookhaven National Laboratory operates two electron storage rings, one at 750 MeV and another at 2.5 GeV, for the production of synchrotron radiation which is used in material science research, x-ray lithography, etc. The paper describes the instrumentation installed on the rings to measure and to monitor various parameters of the stored beams using electromagnetic pick-ups and synchrotron light. The acquisition and processing of beam signals, the interfacing of the processing electronics to the computer system and the display of measured results is discussed.

Ultrafast, High Precision Gated Integrator[1]

Xucheng Wang
Argonne National Laboratory, Argonne, Illinois 60439

Abstract

An ultrafast, high precision gated integrator has been developed by introducing new design approaches that overcome the problems associated with earlier gated integrator circuits. The very high speed is evidenced by the output settling time of less than 50 ns and 20 MHz input pulse rate. The very high precision is demonstrated by the total output offset error of less than 0.2mV and the output droop rate of less than $10\mu V/\mu s$. This paper describes the theory of this new gated integrator circuit operation. The completed circuit test results are presented.

INTRODUCTION

The gated integrator described here can accurately integrate a high-speed signal and hold the output DC level with an extremely small droop rate until a reset pulse returns its output to zero. The integrator is totally bipolar. Its output is independent of input pulse repetition rate and the "on" resistance of the gate switch. The relationship between the input and output of the integrator is clearly defined. The integrator is also capable of subtracting the input signal baseline offset within just one timing window. Its output offset errors introduced mainly by the charge injections from gate switches, opamp input bias current, and input signal baseline offset are greatly minimized if not eliminated.

BACKGROUND INFORMATION

A simplified schematic of the conventional gated integrator circuit is shown in Fig. 1, and Fig. 2 is a timing diagram of general gated integrator operation. The integrator has three operation modes: 1) integrating mode, 2) hold mode, and 3) reset mode. In the integrating mode, gate switch S_1 is closed and reset switch S_2 is open; integration of the input signal starts. In the hold mode, both S_1 and S_2 are open; integration is complete. The integrator holds the output DC level that is proportional to the integral of the input signal. In the reset mode, S_1 is open and S_2 is closed; the output returns to zero.

The output voltage of an ideal gated integrator is given by:

$$V_o(t) = \frac{1}{RC}\int_{t_1}^{t_2} V_i(t)\,dt = \frac{1}{RC}\int_{t_1}^{t_2} V_i'(t)\,dt + \frac{1}{RC}\int_{t_1}^{t_2} V_{bo}\,dt \quad , \tag{1}$$

---

[1] Work supported by U.S. Department of Energy, Office of Basic Energy Sciences, under contract no. W-31-109-ENG-38.

where $V_o(t)$ is the integrator output, $V_i(t)$ is the input signal containing the real signal $V_i'(t)$ and the signal baseline offset V_{bo} introduced by the signal detector and previous signal conditioning circuitry, C is the integrating capacitor, and R is the integrating resistor.

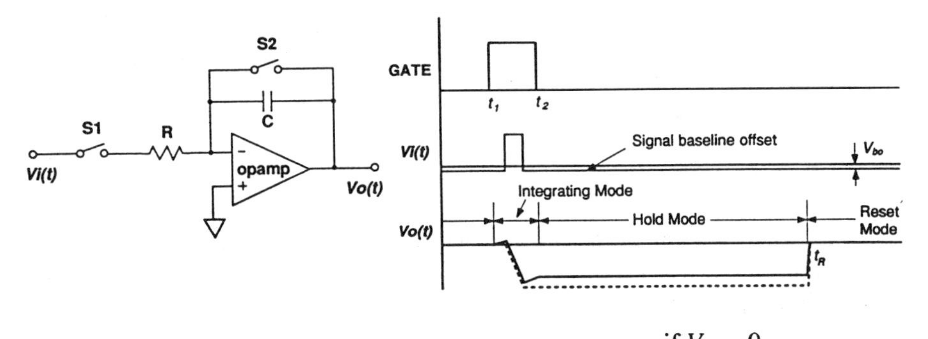

---- if $V_{bo} = 0$

Fig. 1 Basic conventional gate integrator

Fig. 2 Timing diagram of an ideal basic gated integrator

The first item in Eq. (1) is the desired output, and the second is the error output which is from the signal source. After considering the nonideal characteristics of components in the circuit, the output voltage of a practical gated integrator can be written as:

$$V_o(t) \doteq \frac{1}{(R+R_{on})C}\int_{t_1}^{t_2} V_i(t)\,dt + \frac{1}{C}\int_{t_1}^{t_R} I_b\,dt + V_{qs} + V_{os}$$

$$= \frac{1}{(R+R_{on})C}\int_{t_1}^{t_2} V_i'(t)\,dt + \frac{1}{(R+R_{on})C}\int_{t_1}^{t_2} V_{bo}\,dt + \frac{1}{C}\int_{t_1}^{t_R} I_b\,dt + V_{qs} + V_{os} \quad , \tag{2}$$

where V_{qs} is the total charge injection error from analog switches, I_b is the input bias current of the opamp, R_{on} is the "on" resistance of gate switch S_1, and V_{os} is the output offset voltage of the opamp.

The basic circuit is reasonably accurate for slow speed applications, provided that the input baseline offset is very low. The problems of error output become much more severe in a high-speed gated integrator because of the small integrator time constant and the limitations of nonideal components. The desired output is often buried in the error output. The output offset voltage can sometimes be as high as the saturation level of the opamp and is very unstable. Several modified gated integrator circuit configurations have been developed, each suited for different accuracy and speed requirements. Although some of these circuits are potentially very accurate, they are not fast, or vice versa.

In almost all these earlier designs, the solution to reducing the switch charge injection errors is to add compensation networks to each switch. The problem with this approach is that it is extremely difficult to trim all the network's components to obtain the accurate compensations, especially when a large quantity of integrators are produced.

In a relatively new design, the switch "on" resistance effect is minimized by placing a voltage-to-current converter (sometimes called a current pump) in front of the gate switch. Unfortunately, the current source becomes very unstable and highly dependent on temperature when it employs a high-speed opamp.

A typical approach to reducing the error caused by signal baseline in the earlier designs is to have two gated integrators in parallel with two gate windows open in sequence. In the first window, the channel 1 gated integrator integrates the total input signal. In the second window, only the signal baseline offset is integrated by the channel 2 gated integrator. Then the two integrators' outputs are subtracted to restore the desired output. The problem is that the circuit becomes so complicated that it is almost impossible to match the characteristics of the two gated integrators, and it often makes it more difficult to compensate other errors, such as charge injections.

The high-speed opamp usually has high input bias current. The solution to minimizing the errors caused by input bias current lies in adding the current source to cancel the bias current. The problem is that the input bias current is very sensitive to temperature, which makes accurate compensation difficult to achieve. The other solution is to have a large integrating capacitor value; however this is often undesirable because the large C value limits the gain and speed of the integrator.

THEORY OF OPERATION OF NEW DESIGN

The gated integrator described here provides an exceptional combination of high speed and high precision by introducing new design approaches. This gated integrator differs from other known circuits by creating a totally symmetrical architecture with separate AC and DC input signal paths. As shown in Fig. 3, the circuit consists of three sub-circuit elements: 1) gated integrator (around opamp U_2), 2) DC offset detector (around opamp U_3), and 3) input buffer amplifier (around opamp U_1).

The gated integrator includes a pair of input gate switches S_1 and S_2, a pair of reset switches S_3 and S_4, and a differential integrator. Two matched gate switches driven by two switch drive signals of the same pulse polarity are differentially applied to the integrator's inputs. The charge injection errors introduced by dynamic switching transitions are effectively cancelled because of the common mode rejection of the opamp. The charge injection error introduced by the reset switches is minimized in the same way.

To further take advantage of the symmetrical architecture, a DC offset detector is added with its input connected to the buffer amplifier's output and its output connected to the S_2 gate switch's input side. The DC offset detector is actually a unity gain sample and hold amplifier (S/H). The sample switch S_5 has a complementary drive signal relative to those for gate switches S_1 and S_2. In the hold and reset modes of integrator operation, S_1 and S_2 are open, S_5 is closed. The S/H is in the sample mode. C_s is charged to the voltage potential that equals the signal baseline offset (V_{bo}) plus the output offset of U_1 (V_{os}). Since U_3 is a unity gain buffer, its output V_{osi} equals the voltage potential of C_s. In the integrating mode, S_1 and S_2 are closed, S_5 is open. The S/H changes to hold mode. The output of U_1 feeds to the integrator's inverting input via S_1. The output of U_3 feeds to the integrator's non-inverting input via S_2. The signal seen by the differential integrator is,

$$V_{osi}-V_b(t)=[V_{bo}+V_{os}]-[V_i'(t)+V_{bo}+V_{os}]=-V_i'(t) \quad , \tag{3}$$

where V_{osi} is the output of the DC offset detector, $V_b(t)$ is the output of the input buffer amplifier, V_{bo} is the input signal baseline offset, V_{os} is the opamp output offset of U_1, and $V_i'(t)$ is the real signal.

As a result, only the real signal $V_i'(t)$ gets integrated. The input signal baseline offset subtraction is performed by only one gated integrator in one timing window instead of using two gated integrators and two timing windows as in the earlier designs.

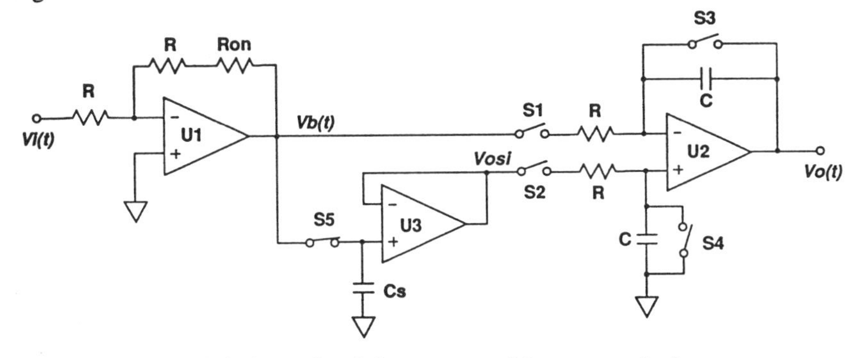

Fig. 3 Schematic of the new gated integrator design

Symmetrical architecture also helps to minimize the errors caused by the opamp input bias current by making the error accumulation rate proportional to the opamp input offset current rather than the input bias current. Hence, the drift is reduced by a factor of more than 10. The error can be further reduced by balancing the two integrating capacitors. Since the input offset current is much lower than the input bias current even for very high-speed opamps, the error item introduced by the input bias current is practically negligible. The extremely low output droop rate is also a result of the symmetrical architecture, since all discharging paths for the integrating capacitors, except very low offset and leakage currents, are cut off when the integrator is in hold mode. For most applications, the output droop before the completion of analog-to-digital conversion is virtually nonexistent. The cancellation of bias current related drift and very low output droop rate make many high-speed opamps suitable for integrator use and also allow the use of a small value capacitor for the integrating capacitor C, which directly results in high gain and speed.

The input buffer amplifier has three important functions. First, it buffers the input signal and provides low output impedance for the integrator's input. Second, it inverts the input signal's polarity so that the final output of the gated integrator has the same polarity as the input. Third and most important, the resistors in the buffer amplifier are arranged so that the gate switch "on" resistance is cancelled in the overall circuit transfer function.

The output voltage of the new gated integrator can be derived as follows:

$$V_o(t) \doteq \frac{1}{(R+R_{on})C}\int_{t_1}^{t_2}[V_{osi}-V_b(t)]\,dt + \frac{1}{C}\int_{t_1}^{t_2} I_{os}\,dt + V_{qs} + V_{oso} \quad , \tag{4}$$

where I_{os} is the input offset current of U_3, V_{qs} is total charge injections, and V_{oso} is the opamp output offset of U_3. Since the second and third items in Eq. (4) are virtually eliminated in the new design, and V_{oso} is usually much less than 1mV, the last three items in Eq. (4) are negligible. Thus the integrator output voltage can be written as:

$$V_o(t) \doteq \frac{1}{(R+R_{on})C}\int_{t_1}^{t_2}[V_{osi}-V_b(t)]\,dt$$

$$= \frac{1}{(R+R_{on})C}\int_{t_1}^{t_2}\left[\left(-\frac{R+R_{on}}{R}V_{bo}+V_{os}\right)-\left\{-\frac{R+R_{on}}{R}(V_i'(t)+V_{bo})+V_{os}\right\}\right]dt$$

$$= \frac{1}{(R+R_{on})C}\int_{t_1}^{t_2}\frac{R+R_{on}}{R}V_i'(t)\,dt = \frac{1}{RC}\int_{t_1}^{t_2}V_i'(t)\,dt \quad . \tag{5}$$

The input buffer amplifier can also be changed to an input gain amplifier if the amplification of the input signal is desired. This is accomplished by substituting the resistance value $G(R+R_{on})$ for $R+R_{on}$ in the input buffer amplifier's feedback, where G is the desired gain of the amplifier. The integrator output is then written as:

$$V_o(t) \doteq \frac{G}{RC}\int_{t_1}^{t_2}V_i'(t)\,dt \quad . \tag{6}$$

As a result of the cancellation of the switch "on" resistance in Eq. (5) and the elimination of both the errors introduced by nonideal components in the gated integrator circuit and the error carried with the input signal, the relationship between the integrator's input and output becomes clearly defined. Consequently, exceptionally high precision, fast response, and many other desirable features are realized in one gated integrator. The added baseline subtraction capability relaxes the

DC accuracy requirements on the previous signal conditioning circuitry and signal detector, which in many signal processing systems represents significant savings in the cost and time.

TEST DATA

Extensive data from both bench tests and units in use at the Advanced Photon Source (APS) have been gathered since the circuit was developed in February, 1992. The performance of the actual circuit constructed from inexpensive discrete components, even without carefully matching the analog gate switches and integrating resistors and capacitors, agrees extremely well with theoretical results. The circuit has the following tested performance:

Input pulse repetition rate: from 20 MHz down to single pulse

Input pulse duration: < 2ns — Output settling time: < 50ns

Full scale output: ± 10V — Dynamic range: 80 dB

Noise: < 0.1mV (rms) — Linearity error: < 0.1%

Total output offset error: < 0.2 mV — Output droop rate: < 10 μV/ μs

The long-term stability of the circuit has been proven excellent. The circuit is also very easy to use, requiring only minimum adjustments.

All test results presented here were obtained without any kind of filtering and averaging. Figures 4 and 5 show that the gated integrator is totally bipolar and has automatic reset capability. Fig. 6 shows that the integrator is used to calculate an arbitrary waveform area; the fast settling time is indicated. Figure 7 demonstrates that the integrator has extremely low output droop rate and noise level. The automatic input signal baseline capability within just one timing window is verified in Figs. 8 and 9. The -400mV baseline offset (middle trace in Fig. 8) is subtracted without causing the error output (bottom trace in Fig. 8). In contrast to Fig. 8 the output is severely distorted in Fig. 9 when the automatic baseline subtraction is disabled. More data from the applications in the APS systems can be found in the references (1,2,3).

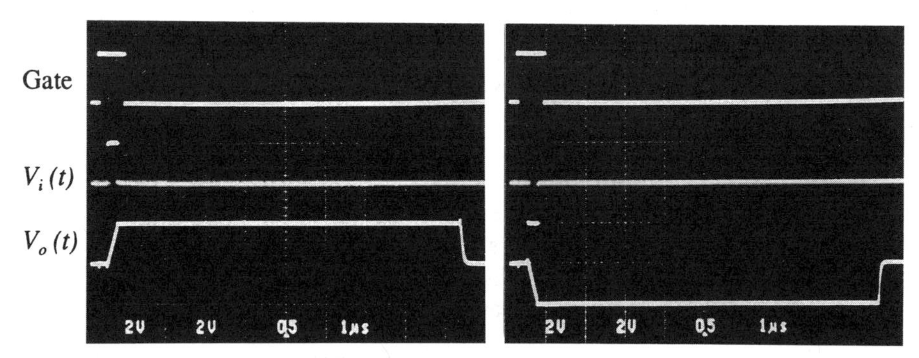

1 μs/div, 2V/div except top traces

Fig. 4 Integrate positive input pulse — Fig. 5 Integrate negative input pulse

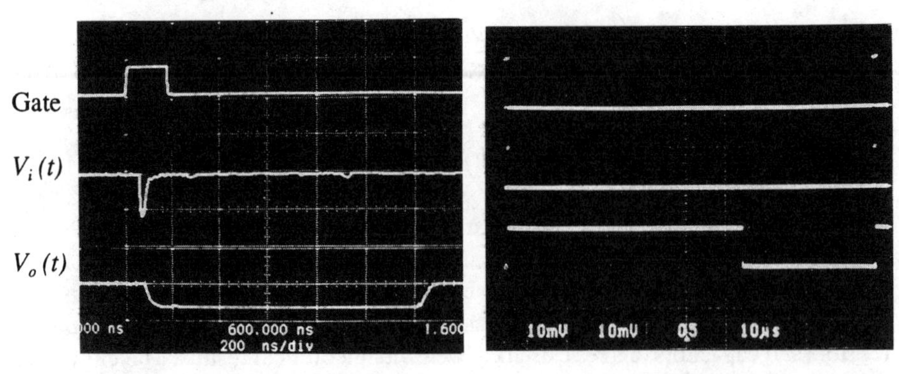

200ns/div, 200mV/div except top trace
Fig. 6 High-speed waveform area calculation

10μs/div, 10mV/div except top trace
Fig. 7 Very small output droop rate and noise level

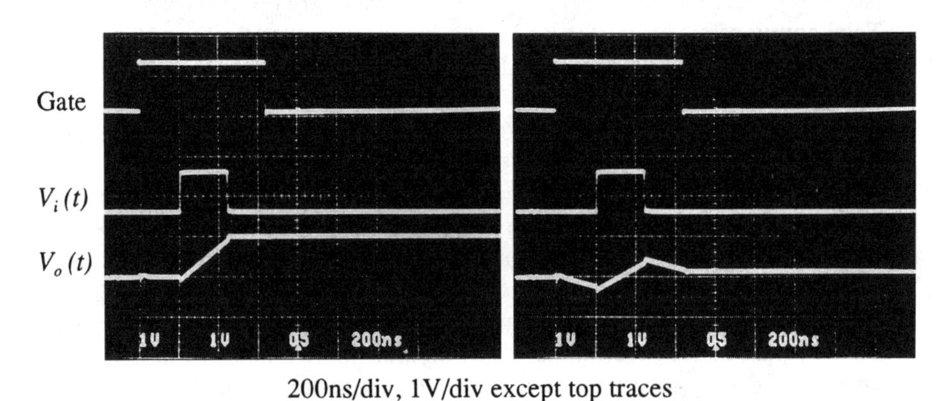

200ns/div, 1V/div except top traces

Fig. 8 Automatic baseline offset subtraction enabled

Fig. 9 Automatic baseline offset subtraction disabled

REFERENCES

1. X. Wang, "Design and Initial Tests of Beam Current Monitoring Systems for the APS Transport Lines," AIP Conference Proceedings No. 281, p. 234-241, 1992.
2. X. Wang, F. Lenkszus, and E. Rotela, " Design and Commissioning of the APS Beam Charge and Current Monitors," these proceedings.
3. X. Wang, M. Knott, and A. Lumpkin, " High Beam Current Shut-Off Systems in the APS Linac and Low Energy Transport Line," these proceedings.

Design and Commissioning of the APS Beam Charge and Current Monitors[1]

X. Wang, F. Lenkszus, and E. Rotela
Argonne National Laboratory, Argonne, Illinois 60439

Abstract

The non-intercepting charge and current monitors suitable for a wide range of beam parameters have been developed and installed in the Advanced Photon Source (APS) low energy transport lines, positron accumulator ring (PAR), and injector synchrotron. The positron or electron beam pulse in the APS has charge ranging from 100pC to 10nC with pulse width varying from 30ps to 30ns. The beam charge and current are measured with a current transformer and subsequent current monitoring electronics based on an ultrafast, high precision gated integrator. The signal processing electronics, data acquisition, and communication with the control system are managed by a VME-based system. This paper summarizes the hardware and software features of the systems. The results of recent operations are presented.

INTRODUCTION

Non-intercepting high accuracy beam current monitoring systems are required for the measurement of total charge, peak current, lifetime, and absolute beam loss. Figure 1 shows the layout of the charge and current monitors in the APS. All monitors shown in Fig. 1 have been installed and commissioned except those in the storage ring. The charge and current monitors described here feature new generation beam current transformers (1), a VME current monitoring electronics module based on an ultrafast, high precision gated integrator described in (2), graphical display and operator interface, and the current transformer housings with emphasis on noise shielding and grounding and use of standard commercial components.

DESIGN

A block diagram of the system and associated timing of a completed cycle are shown in Fig. 2 and Fig. 3, respectively.

The passage of a beam pulse through the current transformer induces a pulse siganl at the transformer's secondary winding. After suitable preamplification the signal is sent to the instrument room. The beam current monitoring electronics unit consists of a 6U VMEbus board that occupies a single slot in the card cage. The main features of the unit include a fast gated integrator with automatic baseline subtraction, completely programmable timing circuitry over the VMEbus, and an on-board 12-bit

[1] Work supported by U.S. Department of Energy, Office of Basic Energy Sciences, under contract no. W-31-109-ENG-38.

A/D converter. A beam pretrigger derived from the linac timing system enters a programmable delay generator that triggers a programmable width generator. This produces the pulses used to control the linear gate of the gated integrator. When the gate is opened by the gate pulse, the signal passes to the fast integrator. The integrator holds an output DC level proportional to the beam charge and current after the gate window is closed. This DC level is digitized by a 12-bit A/D converter linked to the VMEbus. A reset pulse returns the gated integrator's output to zero after the data conversion is done. Associated operating programs are of the Experimental Physics and Industrial Control System (EPICS) platform. The resulting digital data is converted to beam charge and current information and displayed on the workstation screen. The programs provide mouse-controlled operation for system setup and control.

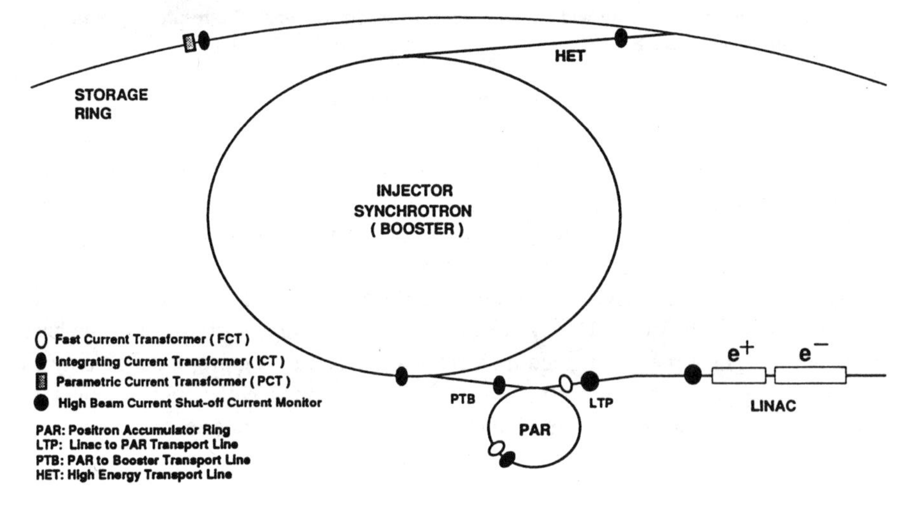

Fig. 1 Locations of beam current monitors in the APS

Beam Current Sensors

The beam current sensors installed, except those in the linac, consist of two types of current transformers: fast current transformer (FCT) and integrating current transformer (ICT), both manufactured by Bergoz. The FCT is a passive AC transformer with a 1-ns rise time. The FCT is installed in the Positron Accumulator Ring (PAR) for observing the bunch length down to several nanoseconds and in the linac to PAR transport line (LTP) for measuring the beam charge and peak current. The FCTs output can be expressed as:

$$U_{out}(t)=\frac{R_o}{N}i_b(t); \quad \textit{then the beam current } \; i_b(t)=\frac{N}{R_o}U_{out}(t) \quad ,$$

where $U_{out}(t)$ is FCT output voltage, $i_b(t)$ is beam current, R_o is FCT output termination resistance, and N is the number of turns.

The ICT is a passive AC transformer designed to measure the total charge in a very short beam pulse with high accuracy. Its output can be expressed as:

$$\int U_{out}(t)\,dt = \frac{R_o}{N}\int i_b(t)\,dt; \quad \textit{then the total charge} \quad Q = \int i_b(t)\,dt = \frac{N}{R_o}\int U_{out}(t)\,dt \quad ,$$

where $U_{out}(t)$ is ICT output voltage, $i_b(t)$ is beam current, R_o is ICT output termination resistance, N is the number of turns, and Q is the total charge.

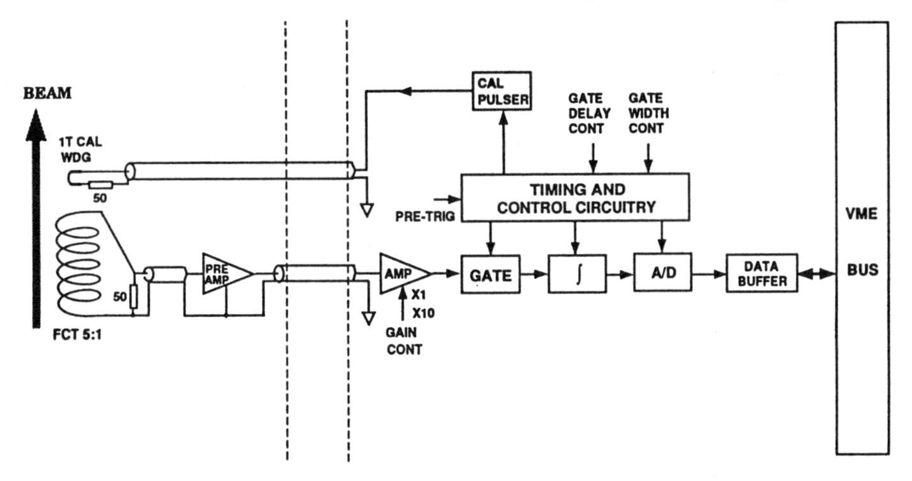

Fig. 2 Block diagram of the VME current monitoring electronics module

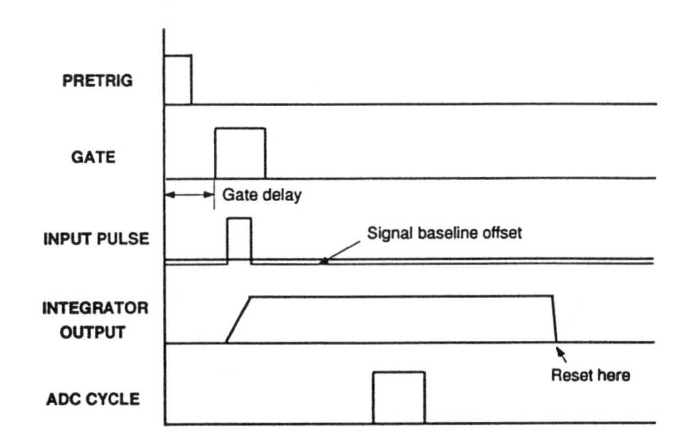

Fig. 3 Timing diagram

The Measurements of Total Charge, Peak Current, and Average Current

An ultrafast, high precision gated integrator has been developed to accurately calculate the transformer's output waveform area that is proportional to the total beam charge, and to hold its output DC level for digitizing. This gated integrator provides fast response and high precision by introducing new design approaches. The various circuit errors usually associated with a high speed gated integrator are virtually eliminated. Consequently, the transfer function of an ideal integrator can be used in the measurements. If the input signal falls within the gate window of t_2-t_1, then

$$V_o=\frac{1}{RC}\int_{t_1}^{t_2} V_i(t)\,dt \ ,$$

where V_o is the gated integrator's output DC level, $V_i(t)$ is input signal voltage, and RC is the integrator time constant.

The total charge of a beam pulse can be calculated directly from the gated integrator output voltage V_o . Let the ICT's output be connected to the gated integrator's input, then

$$V_o=\frac{1}{RC}\int_{t_1}^{t_2} V_i(t)\,dt=\frac{1}{RC}\int_{t_1}^{t_2} U_{out}(t)\,dt=\frac{1}{RC}[\frac{R_o}{N}\int_{t_1}^{t_2} i_b(t)\,dt]=\frac{R_o}{RCN}Q$$

$$\text{then} \quad Q=\frac{RCN}{R_o}V_o=kV_o \ , \qquad k=\frac{RCN}{R_o} \ ,$$

where k is the conversion constant.

The gated integrator output DC level V_o is first digitized, and then multiplied by the conversion constant to obtain the total charge data for the display.

To make the system integration easier, the peak current measurement in the LTP is also made by the gated integrator instead of another type of electronics, such as peak detector. The peak current readout is obtained by integrating the FCT output and dividing the total charge data by the linac pulse width. That is,

$$I_b=\frac{Q}{t_w} \ ,$$

where t_w is the pulse width of the linac macropulse.

Since there is only one beam bunch in the PAR or Injector Synchrotron at any time, the average current in the rings can be simply obtained by dividing the total charge of the single bunch by the ring revolution time:

$$\bar{I}=I_b\frac{\sigma}{T}=\frac{Q}{\sigma}\times\frac{\sigma}{T}=\frac{Q}{T} \ ,$$

where I_b is the peak current, σ is the bunch length, and T is the ring revolution time.

Software and Control Interface

The APS control system is based on EPICS, a distributed database-driven control system in use at several accelerator facilities. The charge and current monitor is interfaced to EPICS at the device layer. The device layer provides a standard environment between the driver layer, which accesses hardware, and the database record layer. EPICS provides an extensive set of record types from which a control database may be constructed. The device layer for the current monitor supports three record types: 1) an Analog In record to read the ADC values, 2) a Pulse Delay record to control the monitor's internal gate delay and width timing, and 3) Long In record to provide readback of the delay and gate values. In addition to these three record types, the current monitor database uses Calculation records to convert ADC input values to charge and current, Buffer records to accumulate successive measurements, and Sequence records to control the Buffer records.

The APS timing system provides triggers to the monitor electronics to time the gate-to-bunch arrival. Since the current monitoring electronics does not provide an interrupt to signal a data conversion, the APS event system, part of the timing system, is used in lieu of a monitor interrupt. The event system is capable of triggering processing of specific records. An appropriately timed event is used to trigger processing of the Analog In record that fetches a value from the current monitor. The processing of this Analog In record triggers a cascade of record processing through a link mechanism which converts the input value to charge and current through Calculation records and stores the converted values in Buffer records.

The data collection rate for the LTP and PAR is 60 Hz. Rather than transfer converted values at the collection rate, the samples are accumulated in the Buffer records and transferred to an X-windows based operator interface as a single data set at a 2 Hz rate. This greatly reduces the packet rate on the Ethernet local area network (LAN). The data set is displayed graphically as an xy plot of charge and current vs. bunch number.

In addition to the xy display of charge and current vs. bunch number, the operator interface for the monitors displays the ADC voltage, charge, and current numerically. Slider controls are provided for the monitors' internal gate delay and gate width timing.

Mechanical Design

Figure 4 illustrates the mechanical design for the synchrotron current transformer housing. The transformer is external to vacuum and encased in a copper and Mu-metal multi-layer shell to shield the transformer from external noise. The vacuum chamber is formed by a commercial ceramic break and a welded bellows which provides the protection for the ceramic break. A stainless steel tube with the same aperture as the synchrotron vacuum chamber is suspended inside the ceramic break and a bellows and is attached to the downstream end flange. A gap of 2mm is kept between the tube and the upstream end flange to limit the bandwidth of the cavity formed by the housing to prevent it from ringing with the beam frequency components. The transformer is supported by three G10 sectors attached to the inner wall of the housing, and is held axially by a G10 ring.

OPERATION AND PERFORMANCE

The charge and current monitors described have been installed and used extensively to support the commissioning and operation of the APS injector subsystems including the PAR, Injector Synchrotron, and beam transfer lines of the APS since February, 1994. The VME gated integrator modules are also installed in the linac to provide beam current measurements and readouts for all Faraday cups and current monitors which use the wall current monitor instead of the current transformer as beam current sensors.

Each system was calibrated prior to operation by injecting the test pulses to a single-turn calibration winding of the transformer to simulate the beam pulses of fixed charges. The timing variables for the gated integrator must also be set properly. The gate delay and width are adjusted so that the beam pulse falls well within the gate window. This control requires adjusting only when the system is installed since the trigger to the beam pulse delay is a constant for a given location. However, if the IOC where the VME electronics card is seated needs rebooting, then both gate delay and width have to be readjusted to the original settings after the rebooting.

Figure 5 shows the typical oscilloscope traces of the LTP current monitor. The electron beam pulse picked by an FCT (middle trace) inside the gate window (top trace) is integrated by the gated integrator to produce an output DC level (bottom trace) proportional to the beam current for the digitizing. The output is returned to zero by a reset pulse after the analog-to-digital conversion is completed so that the gated integrator is ready for next cycle beam pulse.

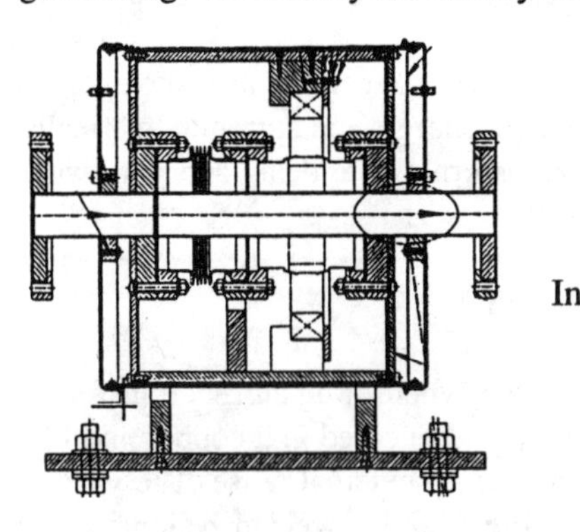

Fig. 4 Mechanical design

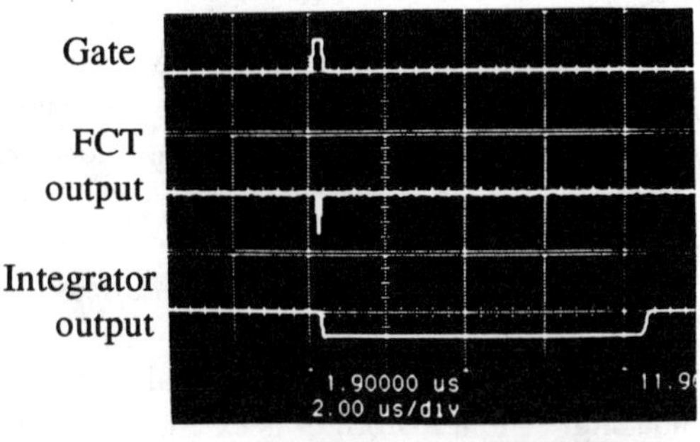

Middle trace: 80mV/div
Bottom trace: 400mV/div
Fig. 5 LTP current monitor signals

The PAR current monitor data is shown in Fig. 6. The small physical size of the PAR determines its high beam revolution frequency. The beam pulse separation seen by the ICT is only 102 ns . Because of this, the ICT output signal (middle trace) shows a large signal baseline offset relative to the signal amplitude. The offset level changes as beam intensity changes. The high performance of the fast gated integrator is well demonstrated here in the single bunch charge measurement. As shown in Fig. 5, the signal baseline is subtracted and only one beam bunch signal is integrated within the 100ns gate window.

Figure 7 shows that the PAR current monitor data is displayed on the workstation both numerically and graphically as an xy plot of charge and average current vs. bunch number. Three linac bunches were injected into the PAR, and the current monitor tracked the stacking of the bunches in the PAR.

To optimize the system resolution with low beam current signal, a programmable AGC amplifier is being added to automatically scale the beam signal so that the gated integrator's output DC level is always close to the full range of the A/D converter.

The system grounding method discussed in (3) was applied to all current monitors installed in the machine and has been proven very effective in reducing the noise.

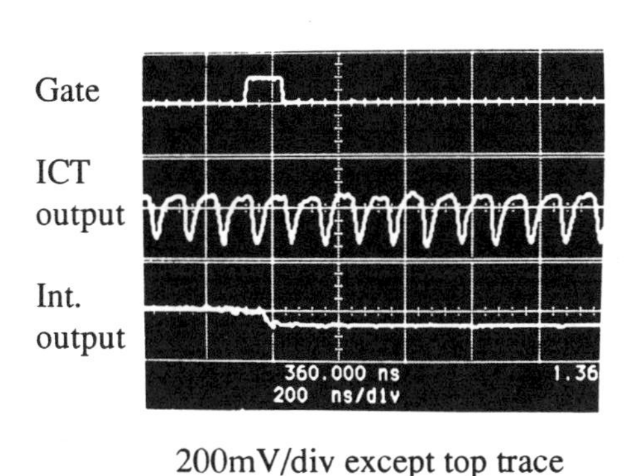

Fig. 6 PAR current monitor signals

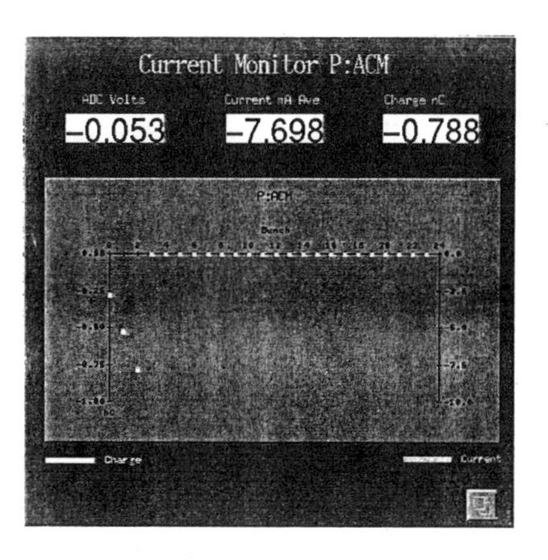

Fig. 7 PAR Current monitor display

REFERENCES

1. K. Unser, "Design and Preliminary Tests of a Beam Intensity Monitor for LEP," Proceedings of the 1989 IEEE Particle Accelerator Conference, Volume 1, p. 71.
2. X. Wang, " Ultrafast, High Precision Gated Integrator," these proceedings.
3. X. Wang, "Design and Initial Tests of Beam Current Monitoring Systems for the APS Transport Lines," AIP Conference Proceedings No. 281, p. 234-241, 1992.

High Beam Current Shut-off Systems in the APS Linac and Low Energy Transfer Line[1]

X. Wang, M. Knott, and A. Lumpkin
Argonne National Laboratory, Argonne, Illinois 60439

Abstract

Two independent high beam current shut-off current monitoring systems (BESOCM) have been installed in the APS linac and the low energy transport line to provide personnel safety protection in the event of acceleration of excessive beam currents. Beam current is monitored by a fast current transformer (FCT) and fully redundant supervisory circuits connected to the Access Control Interlock System (ACIS) for beam intensity related shutdowns of the linac. One FCT is located at the end of the positron linac and the other in the low energy transport line, which directs beam to the positron accumulator ring (PAR). To ensure a high degree of reliability, both systems employ a continuous self-checking function, which injects a test pulse to a single-turn test winding after each "real" beam pulse to verify that the system is fully functional. The system is designed to be fail-safe for all possible system faults, such as loss of power, open or shorted signal or test cables, loss of external trigger, malfunction of gated integrator, etc. The system has been successfully commissioned and is now a reliable part of the total ACIS.

INTRODUCTION

The BESOCM system is designed to prevent the unintentional acceleration of high beam currents that could result in unexpected and hazardous radiation fields. Two BESOCM systems have been installed at locations shown in Fig. 1. The current design is based on detecting beam current of a single 30ns macropulse in the positron linac and linac-to-PAR transport line (LTP)(1). Should measured beam current exceed the preset trip levels, accelerator operation is immediately shut down through hard-wired trip circuits (i.e. no software links) interfaced to the ACIS.

The trip levels of two systems specified by the APS Accelerator System Division Radiation Safety Policy Committee are (2):

Positron linac BESOCM	25 nC / macropulse	for electrons only
LTP BESOCM	0.67 nC /macropulse	for electrons or positrons

The main purpose of the positron linac BESOCM is to limit the radiation level for all downstream locations in case of the positron target failure. It is understood that the positron current level at this location will never reach a level high enough to cause safety concerns. Therefore, it is not necessary to set a positron trip level for the system.

A BESOCM trip will occur if the total charge in one macropulse exceeds the trip level. The safety envelopes are, however, based on averaging the current over some

[1] Work supported by U.S. Department of Energy, Office of Basic Energy Sciences, under contract no.W-31-109-ENG-38.

time period. Because of this, a BESOCM trip does not indicate a violation of the safety envelope, but rather that a condition has been detected, which might have resulted in a violation of the safety envelope if it had been allowed to persist.

The high reliability of the system is ensured by using two completely redundant channels in each system starting from the FCT all the way through to the ACIS connectors. Furthermore, each system employs a continuous self-checking function to test for various failure conditions. The accelerated beam will be inhibited if a fault condition is detected.

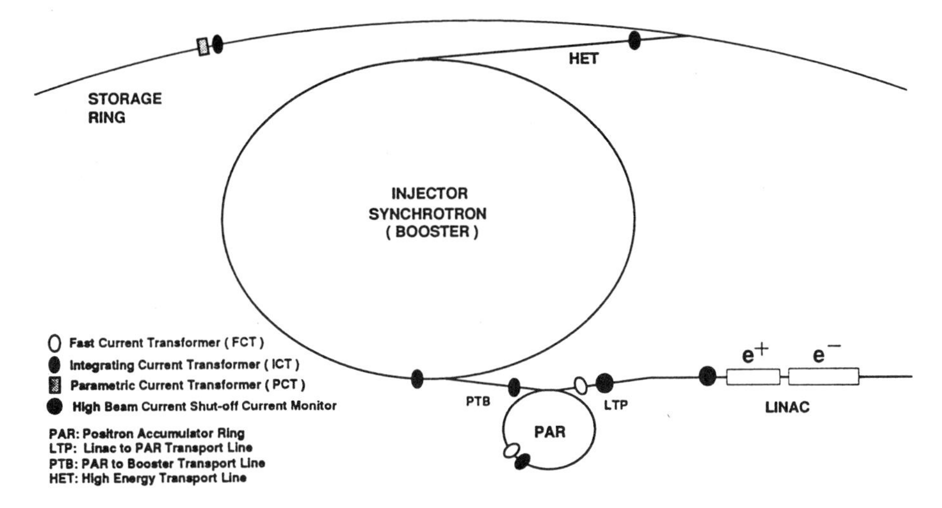

Fig. 1 Locations of the beam current monitors in the APS

SYSTEM DESCRIPTION

The design and basic operation of the systems presently installed will be described by referring to the block diagram in Fig. 2 and the system timing diagram in Fig. 3.

The beam total charge measurement for pulsed beam of charged particles is performed by an FCT and a fast gated integrator described in (3) and (4). This gated integrator, with its fast response, well-defined output, and very small droop rate, makes it possible to accurately measure the total charge of the single 30ns linac macropulse and to provide a stable DC level proportional to the charge for limit-checkings.

The output signal from the preamplifier is integrated by the gated integrator during a gate interval that is controlled by the timing and control unit. The external trigger signal to the timing and control unit is derived from the linac modulator #1 trigger so that the gate pulse is synchronized to the 2 to 60 Hz linac firing. It was determined that there is a significant delay plus rise time associated with the turn-on of the modulators. Based on this, a 1.5-μs delay is added between the modulator #1 trigger and the leading edge of the gate pulse for beam signal as shown in Fig. 3. In

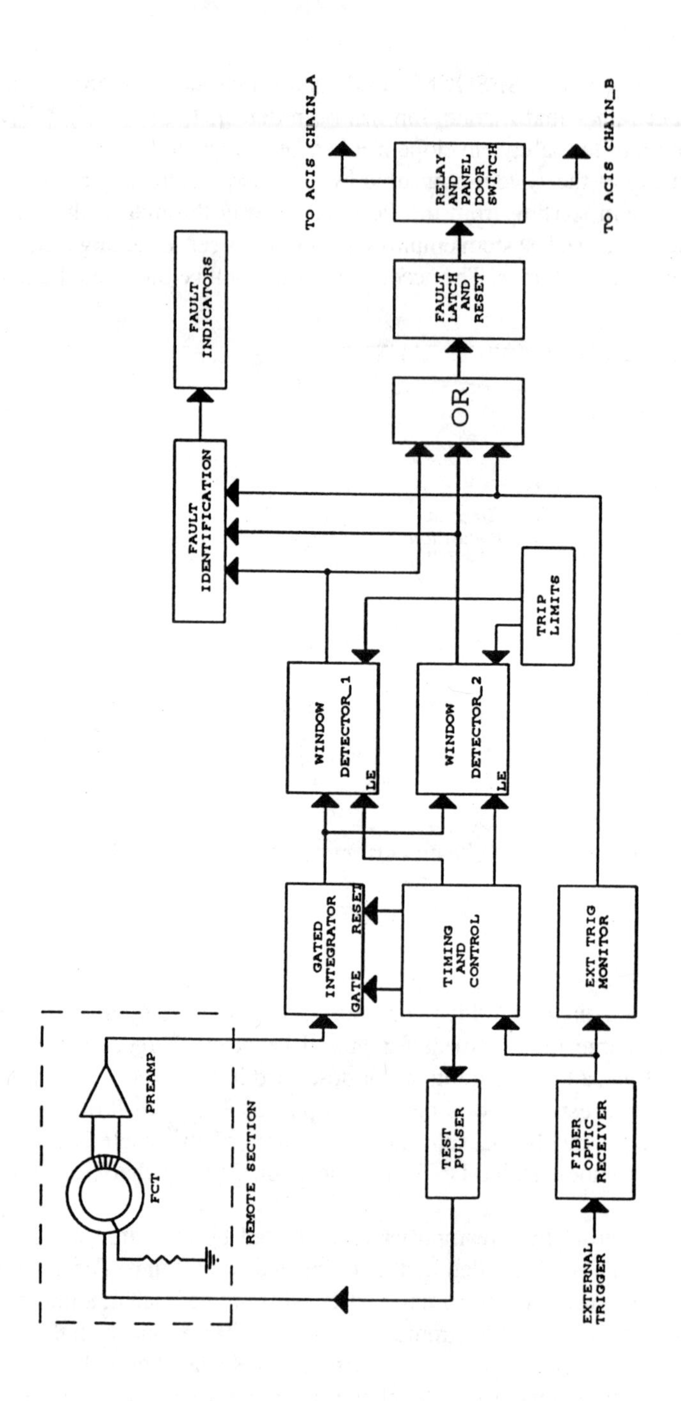

Fig. 2 Block diagram of the one channel BESOCM system

addition, the gate is set to be wide enough (6.5μs) to cover the time period when beam could be injected from the gun and overlap the rf power pulse in the accelerators (5). The system also employs an external trigger monitor designed to check for the presence of the external trigger signal. If the external trigger is absent for longer than one second, a fault will be generated to disable linac operation.

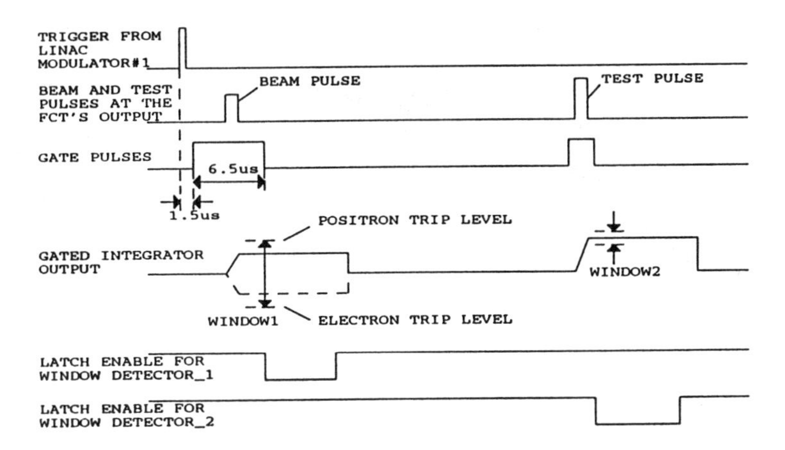

Fig. 3 Timing diagram of the BESOCM system

As illustrated in Fig. 3, the bipolar gated integrator has a positive or negative output DC level proportional to the positron or electron beam total charge, respectively. The integrator's output DC level is monitored by both window detector 1(WD1) and window detector 2 (WD2), each enabled in a different time slot. The WD1 has an upper limit corresponding to the positron trip level and a lower limit corresponding to the electron trip level. It is used to detect an out-of-range beam signal, when enabled after the gate window for the beam signal is closed and before the gated integrator's output is reset to zero. Approximately 30μs after the beam pulse, a test pulse also synchronized to the linac beam is sent to a single-turn test winding in the current transformer. The detected signal from the test winding is integrated by the same gated integrator and compared to closely-set high and low limits of the WD2. When a failure condition occurs, such as malfunction of the gated integrator, open or shorted signal or test cables, etc., the gated integrator's output DC level for the test pulse will be outside the preset limits. It is important to note that the system is always under continuous test even without beam.

If any one of the four limits of the two window detectors is exceeded during the operation, the interlock relay is opened and the ACIS will shut down all linac systems.

All system faults are latched, identified, and indicated on the electronics chassis' front panel. The critical checkpoints in the system are buffered for access from the front panel. The fault can be reset locally or remotely via the control system. Photographs of the system's front and rear panels are shown in Fig. 4 and Fig. 5. The system is fully redundant with two identical channels in each chassis. Two independent power supplies are provided for the separate channels. The front panel door shown in Fig.4 has to be locked to close the interlock loop. The controls behind the door are for setting the trip levels and adjusting the timing variables. The adjustments can only be made by authorized personnel.

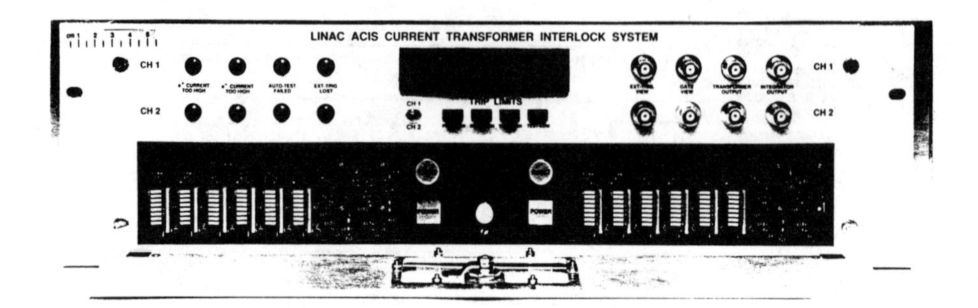

Fig. 4 The front panel of the BESOCM electronics chassis

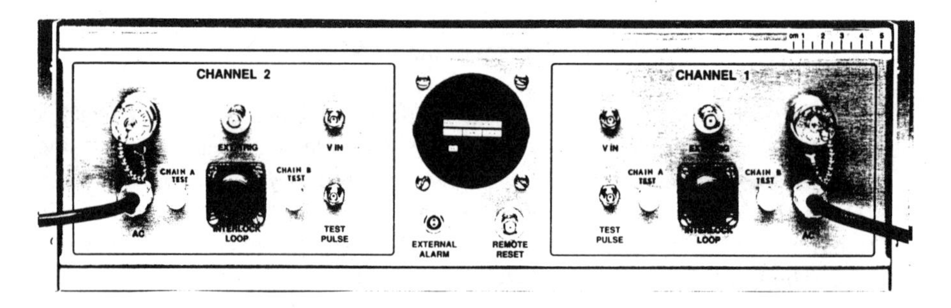

Fig. 5 The rear panel of the BESOCM electronics chassis

OPERATIONAL EXPERIENCE

The positron linac and LTP BESOCM systems have been installed and operating since February, 1994. All system functions were tested as part of the ACIS validation test procedure before operations started. Overall system reliability is excellent. Both systems have performed the job as required and aborted in the fail-safe mode. A typical operational cycle with real linac electron beam (the APS injector sub-systems are currently running on the electron beam only) is shown in Fig. 6. A 1.5-μs gate delay and 6.5-μs gate width are indicated.

The only problem encountered which remains to be resolved is the noise-related false trips generated by the LTP BESOCM system especially during the early commissioning stage. The positron linac BESOCM system was almost never tripped because of high trip levels. As noted in a previous section, the gate window of the gated integrator is set for a long period (6.5μs) to ensure that a linac beam triggered at any time during the linac rf pulse active time can be monitored by the gated integrator, and the trip level set for the LTP BESOCM is only 670pC. This means that the gated integrator has to integrate the noise over a 6.5-μs window while it has only 30ns to integrate real beam signal. It is clear that noise can sometimes dominate the integral and produce a false trip. The high voltage PAR kicker pulse power supplies near the LTP BESOCM current transformer housing were identified as the main noise source. The intermittent kicker noise picked up by the FCT has a noise pattern shown in Fig. 7. It consists of a high frequency burst which is integrated with no effect, and a clean low frequency sinewave following the initial high frequency transient, which results in a DC offset and sometimes a trip. To solve the problem, the shielding and grounding of the kicker power supplies were modified and improved. This reduced the number of false trips. Unfortunately, the results have been unstable due to the complicated kicker grounding structures. Some upgrades from the BESOCM system side have been discussed and will be implemented soon, including using heavier and more thorough shielding for the LTP FCT and triaxial cable to transmit the signal to the instrument room, shortening the gate window slightly to eliminate the majority of noise effect, and equipping the system with a firmware-based method of averaging the beam current over some time period within the safety envelope.

CONCLUSION

The high beam current shut-off systems were developed to eliminate the possibility of accidental radiation exposure due to unintentional acceleration of high beam currents. This capability allows management to define a safety envelope of operation that contributes to significantly improved protection at the accelerator. The system's high relialiability is ensured by employing hardwired design, redundant monitoring channels, built-in self-checking functions, and many other fail-safe features. The system described satisfies the present operational requirements. Some system upgrades are to be implemented.

ACKNOWLEDGMENTS

The authors wish to thank Bob Hettel at SSRL for the helpful information and discussions; John Galayda, Michael Borland, and Glenn Decker for their support during the commissioning; and Richard Voogd, Peter Nemenyi and others for their outstanding work during construction of the systems.

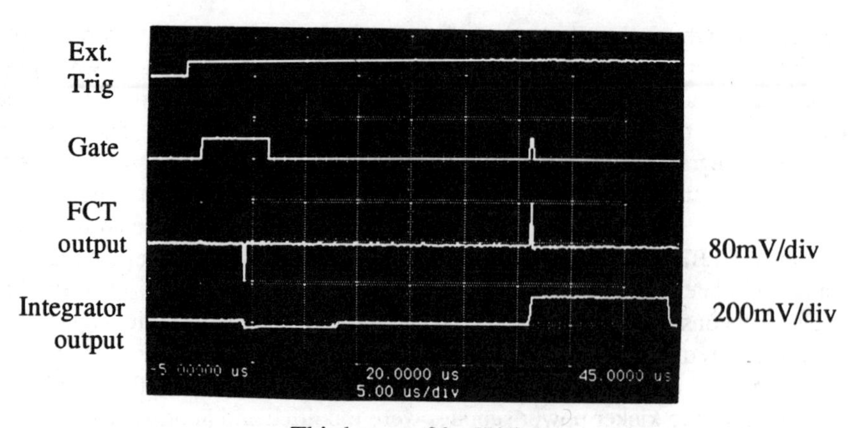

Third trace: 80mV/div
Bottom trace: 200mV/dig

Fig. 6 BESOCM signal traces

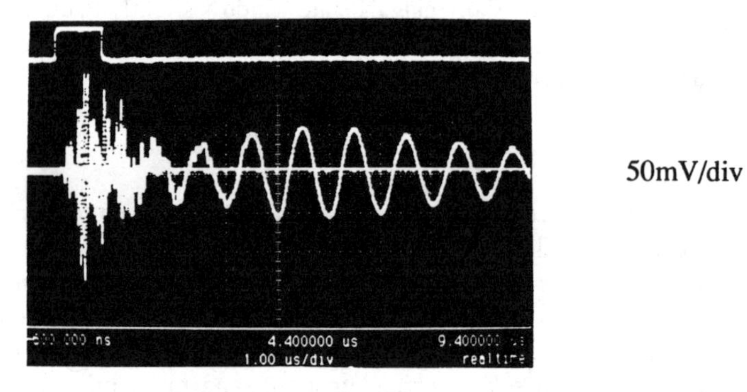

Fig. 7 The PAR kicker noise

REFERENCES

1. R. Hettel, private communication.
2. J. Galayda, M. Borland, private communication.
3. X. Wang, F. Lenkszus, and E. Rotela, "Design and Commissioning of the APS Beam Charge and Current Monitors," these proceedings.
4. X. Wang, " Ultrafast, High Precision Gated Integrator," these proceedings.
5. J. Galayda, A. Grelick, private communication.

CURRENT MONITORS FOR INTENSITY MODULATED BEAMS

Mark Ball, Brett Hamilton
Indiana University Cyclotron Facility, Bloomington, Indiana 47408

Abstract

A beam intensity modulation system (BIMPS), that works in conjunction with the beam splitting system to allow beams of different intensities to be sequentially delivered to two different areas, has already been in use for many years. The operators could not, however, tune the cyclotrons with the BIMPS in operation using the existing beam instrumentation systems in the cyclotron beamlines which consisted mostly of non-electron-suppressed stops. Since the BIMPS duty factor (e.g. as low as 1/100 when operating with a 100 μs high intensity (HI) pulse at 10 Hz) usually exceed the ratio of the HI to LOW beam intensities (varying in the range from 10 to 100), the stops would, to first order, merely read out the LOW beam intensity. Thus there existed no way to monitor the HI beam intensity and transmission efficiency unless operating continuously in the HI beam mode. To allow BIMPS operation at all times, a new system of intercepting and nonintercepting beam current monitors have been added to the cyclotron beamlines. The system consists of electron suppressed stops and nonintercepting beam pickups with high output bandwidth of (10 kHz) signal processors to allow accurate sampling of the short duration HI beam pulses. The electronics for the stops are straightforward; there are, however, important technical trade-off in the design of the nonintercepting system design. The amplifier input voltage noise and relatively low coupling impedance of the nonintercepting pickups cause the minimum detectable HI current to decrease with the square root of the HI beam pulse length; as the pulse length is shortened, the system timing constraints also become more critical. Although the BIMPS is capable of providing beam pulse durations as short at 10 μs, the minimum pulse length for operation was chosen to be 100 μs. The electronics have time constants of 200 μs allowing measurement accuracies of better than a percent. Since the most rapid modulation frequency used for filling the Cooler is 20 Hz (corresponding to 1 transverse cooling time constant) the HI pulse does not begin to significantly contribute to the radiation levels or to the total beam losses in the cyclotrons until the modulation ratio of HI to LOW beam exceeds 50.

NON-INTERCEPTING BEAM INTENSITY MONITOR SYSTEM

A non-intercepting beam intensity monitor system (1), (NIP) has been developed for BL1 common, BL2, and BL3, that will provide the operator with a real time, non-destructive, beam current measurement during the time the BIMPS is being used. This report discusses the design of the BL3 system. Because of the poor signal-to-noise, (s/n), level, it will have the worst dynamic range, and present the biggest challenge to the basic design. The electronics for the other beamlines will be similar to BL3, but with different gains and front end configurations. It should be noted that the BL1 monitor will not read the same as Stop 5-1. The reason for this is there is still unbunched beam in BL1, which cannot be measured by an RF pickup. This unbunched beam will be lost in the injector cyclotron so there is no point in measuring it. Instead, the BL1 monitor will be calibrated against the closest BPM in the beamline.

Electronics

The electronics were designed to operate within the following parameters.

I_{High} 50 μA maximum to as low as possible (< 50 nA)

I_{Low} 2 μA maximum to as low as possible (1 nA)

The minimum pulse width in which the system can operate is 1ms, and the intermediate frequency, (IF), is 50 kHz for I_{High}, and 2.777 kHz for I_{Low}.

Pickups for the system are stripline electrodes (these were used because they were in the beamline). On the front end, 4 rf amps are used. The output to the amps are summed together, doubling the signal while contributing nothing to the noise. After the combiner, the signals are shipped to the control room. Cable losses weren't taken into account in the s/n equations, but attenuator values can be changed to make up for any losses. In the control room a high gain, high output amplifier is used. After the amp, the signal is split and sent to mixers, used in both the high beam, I_{High}, and low beam, I_{Low}, systems. The local oscillator frequencies are 3/2f+IF, or 5/3f+IF.

Following the mixer in the I_{High} IF the signal passes through a 50 kHz lowpass filter, is amplified, and rectified, using an Analog Device, AD637 (2), RMS-to-DC converter. A series of 10 kHz filters are needed on the output to minimize the AC ripple. The rectified signal is sent to a two S/H circuits. The first S/H is used for fast acquisition of the rectified signal, a second S/H is in series with the first to hold the signal during Lo BIMP time. The timing for the first S/H comes from the Cyclotron timing system, (Jorway), which also goes to a one shot to delay the trigger on the second S/H. The acquisition time of the first S/H is 6 μs, with a droop rate of 20 mV/s. The acquisition time of the second S/H is 300 μs, with a droop rate of 5 mV/min. The held signal goes to a 6 Hz lowpass filter and then to a multiplier which is used to compensate for the different energies used in the cyclotron. After the multiplier, the signal goes to an ADC.

The I_{Low} signal uses an IF of 2.777 kHz, and a Burr Brown, UAF40,(3) active band pass filter to set the BW to 100 Hz. The signal is rectified and filtered then sent to a S/H, used to hold the signal during Hi BIMP time. As in the I_{High} system, a multiplier is used to compensate for the different energies used in the cyclotron. The signal then goes through a 5 Hz filter and to an ADC.

The gain in the I_{High} system (G_{sys}) is set so that a 40 MeV, 50 μA beam will produce 10 V at the input to the S/H, V_{sys}. The voltage V_{sig} at the pickup and G_{sys} were found from the following equations:

$$V_{sig} = \frac{K_h \, \omega \, L_{pu} \, R \, I_{DC}}{\beta \, c \, 4}$$

$$G_{sys} = \frac{10V}{V_{sig}}$$

where the frequency, f, is 45 MHz, $\omega = 2 \cdot \pi \cdot f$, the length of the pickup (L_{pu}) is 0.305, β is 0.283 (40 MeV), c is the speed of light, the beam current, I_{DC}, is 50 μA, the resistance (R) is 50Ω, and K_h is 2. The result for G_{sys} is 10,600.

For the I_{Low}, the same equations can be used. With 2 μA as the maximum beam current and β of 0.283, the G_{sys} will be 266,000.

Signal-to-Noise

To find the s/n ratio, the minimum V_{sys}, must be calculated, using a β of 0.566 (200 MeV) and I_{DC} of 1 nA. This results in a V_{sys} of 100 μV. To calculate the noise voltage, V_n, the Johnson noise of a 50Ω resistor, .63 $n\mathrm{V}/\sqrt{\mathrm{Hz}}$, is multiplied by the noise figure of the first amplifier, 1.6, yielding a V_n of 1 $n\mathrm{V}/\sqrt{\mathrm{Hz}}$. The system V_n, $V_{n,sys}$, is found by:

$$V_{n,sys} = \frac{1nV}{\sqrt{Hz}} \times \sqrt{BW} \times G_{n,sys}$$

The noise gain of the system, $G_{n,sys}$, is different than G_{sys}, because the effect combiners and mixers have on noise is different than the effect they have on rf signals. $G_{n,sys}$ is 7665, so the $V_{n,sys}$ is 1.7 mV. With 50 nA of beam, the s/n ratio is 5.9. The sensitivity with which signals can be measured is determined by low pass filters in the final stages of the IF section. There are two 6 Hz low pass filters in the final stages of the IF section which improves the s/n to 532. The precision of the system is 0.1 nA.

For the s/n ratio of I_{Low}, I_{DC} of 1 nA will be used and β of 0.566, resulting in a V_{sys} of 5 mV. For $V_{n,sys}$, the same equation as before can be used, but with $G_{n,sys}$ of 192,000, and BW of 100 Hz. The $V_{n,sys}$ is 1.9 mV. Using the results of V_{sys}, and $V_{n,sys}$, the s/n ratio is 2.6. This system has the same sensitivity as the I_{High} system.

STOP ELECTRONICS

Beam stop electronics were developed in order to better monitor beam currents after the BIMP's. These electronics are connected to the following stops which have electron suppression in BL1-5, BL2-4, BL3-4 and to the A-Probe and South-Probe which do not have electron suppression. Five more stops can be added in the future. This system was designed to operate within the following beam parameters:

I_{High} 50 μA maximum to as low as possible (hopefully < 5 nA)

I_{Low} 2 μA maximum to as low as possible (hopefully < 100 pA)

The minimum pulse width for high beam current is > 100 μs with an cycle time > 10 Hz.

The first stage amplifier is located at the pickup in the beamline. The stops have a typical capacitance of, C=100 pF. Using t=10 μs we calculate the input resistance, R=25.5 kΩ of the amplifier from the following equation:

$$V_{output}=V_{input}(1-e^{-t/RC}) \qquad where\frac{V_{output}}{V_{input}}=0.98 \quad (\approx 4\tau)$$

The gain of this amplifier is set such that 10 V out is equal to I_{High} maximum (50 μA),

$$G_1=\frac{V_{Amp\,output}}{V_{Amp\,input}} = \frac{10V}{50\mu A\ 25.5k\Omega} =7.8$$

The output which is low pass filtered at 6.4 kHz is than sent to the control room for further processing.

The I_{Low} signal requires further amplification at the pickup in order to reduce the effect of interference (mostly 60 Hz) during transmission to the control room. This gain is calculated so the I_{Low} maximum (2 μA) equal 10 V out of amplifier #2.

$$G_2=\frac{V_{Amp\,output}}{V_{Amp\,input}\ G_1} = \frac{10V}{2\mu A\ 25.5k\Omega\ 7.8} =25.1$$

This signal is low pass filtered at 6.4 Hz than sent to the control room.

In order to minimize the effect of 60 Hz ground loops during transmission back to the control room a differential receiver is used. These devices offer high common mode rejection and greatly enhance the system's dynamic range.

The I_{High} signal than passes through a two stage sample and hold (S/H), the first with a acquisition time of 5 μs and a droop rate of 20 mV/s and the second a acquisition time of 300 μs and a droop rate of 5 mV/min. By operating in this configuration the signal of interest can be acquired accurately within the 100 μs minimum pulse time and also held for long periods of time with a low output droop rate. The acquisition time for the I_{Low} need not be so fast and a single stage S/H with an acquisition time of 300 μs will suffice.

The outputs of the S/H circuits than pass through several stages of low pass filters, a selector switch and are than displayed on an analog meter.

The input amplifier noise is 20 nV/√Hz. The noise voltage, V_{noise}, can be calculated as follows:

$$V_{noise} = 20\frac{nV}{\sqrt{Hz}}\sqrt{\frac{BW_{AMP}\,BW_{Filter}}{N_{Samples/Second}}}$$

where $N_{Samples}$, which is the hi pulse rate of 10 Hz, and bandwidths, BW_{Filter} of 1 Hz, BW_{AMP} of 6.4 kHz, V_{noise} = 500 nV.

For I_{High} which has a gain, G, of 7.8, the output noise is 3.9 μV. V_{Signal}, which is simply, $I \cdot R \cdot G$, for 5 nA of beam current is 1 mV giving us a signal-to-noise (s/n) ratio of 255. It is clear that other factors such as component dynamic range (S/H resolution), rfi and D.C. offsets will determine our low level measurement threshold.

For I_{Low} which has a gain of 196, BW_{Filter} of 1 Hz, and BW_{AMP} of 6.4 Hz which is less than $N_{Samples}$, the output noise is simply:

$$V_{Noise} \approx 20\frac{nV}{\sqrt{Hz}}\sqrt{BW_{Filter}}$$

V_{Signal} for 100 pA of beam current will be 0.5 mV giving us a signal-to-noise ratio of 130.

System Displays and Controls

Stops and probes are selected and beam currents are observed on a new analog meters chassis located in the operators console. The NIP's outputs are digitized and observable on the computer. All NIP's can be monitored simultaneously while only one stop can be seen at a time (the stop must be put in).

The computer uses the NIP values to calculate the transmission efficiency through the injector and the main stage cyclotrons. This information is displayed in a continuous fashion (when using NIP's and no stops are in) on the control console.

System Timing

In order to capture the signal of interest, in this case I_{Low} and I_{High}, (for stops and NIP's), timing signals must be determined to a resolution of $< 10\ \mu s$. These timing signals are used to trigger the S/H's and are dependent upon the cyclotron operating parameters. Three separate Jorway timing signals are needed to accomplish this. One each for BL-1, BL-2, and BL-3.

The number of turns must be known in order to derive these timing signals. This can be calculated as follows:

$$Number\ of\ Turns = \frac{q\,(K_{out} - K_{in})}{4\,V_{rf}}$$

where K_{in} is the beams kinetic energy entering the cyclotron, K_{out} the exiting energy and q the charge of the particles being accelerated ($q=1$ for protons). Once this is known the time delays through each machine can than be calculated as follows:

$$t_{delay} = \frac{Number\ of\ Turns}{rf_{freq}}$$

The BIMP high current Jorway trigger, $BIMP_{trig\ Hi}$ time is first set, then all beamline timing triggers are derived from that along with the cyclotron delays. This procedure must be done for both cyclotron control the BIMP's as well as cooler control of the BIMP's. The Jorway timing signal for each of the three beamlines can than be determined as follows:

$$BL-1_{trig} = BIMP_{trig\ HI} - 5\mu s$$
$$BL-2_{trig} = BL-1_{trig} + t_{delay_{inj\ cyc}}$$
$$BL-3_{trig} = BL-2_{trig} + t_{delay_{main\ cyc}}$$

REFERENCES

1. Mark Ball, Tim Ellison, Brett Hamilton, *Indiana University Cyclotron Facility Scientific and Technical Report*, May 1992 to April 1993, Ind. Univ., Bloomington, IN., p204
2. Analog Devices, *Linear Products Databook*, 1990/91
3. BURR-BROWN, *Integrated Circuits Data Book*, volume 33

Intensity Measurements of slowly extracted Heavy Ion Beams from the SIS

P. Heeg, A. Peters, P. Strehl

Gesellschaft für Schwerionenforschung mbH, D-64291 Darmstadt

Abstract

The paper reports about performance tests of newly designed Secondary Electron Monitors (SEM), Ionization Chambers (IC) and Multi Diode Counters (MDC). Especially the linearity of the detectors with respect to the specific energy loss will be discussed. Calibration has been performed by means of scintillation particle counters at the lower end of the intensity region. The status of the Cryogenic Current Comparator (CCC), which is provided for absolute measurements and calibration of detectors above some nA of beam current is reported, too.

DETECTORS FOR INTENSITY MEASUREMENTS

Since absolut particle fluxes have to be known to determine cross sections of measured nuclear reactions as well as to optimize the efficiency of particle extraction and beam transportation a great variety of detector devices are in use around the laboratories. Table 1 gives a selection of commonly used principles for particle detection.

Table 1. Principles of intensity measurements

Detector Principle	on-line	nondestructive	radiation resistant	intrinsic calibration	Vacuum compatible	Output signal
Track Counting	-	-	-	+	-	$= N$
Faraday Cup	+	-	+	+	+	$= N\zeta e$
Ionization Chamber	+	-	+	-	-	$= Ne\Delta E/W$
Scintillation Pulse Counter	+	-	-	+	-	$= N$
Scintillation Current Monitor	+	-	-	-	-	$\sim N\Delta E$
Secondary Electron Monitor	+	o	+	-	+	$\sim NdE/dx$
Residual Gas Monitor	+	+	+	-	+	$\sim Np\Delta E$
Nuclear Reaction Monitor	+	o	+	-		
Beam Transformer	+	+	+	+	+	$\sim N\zeta e$

+ = yes; o = only under favorable conditions; - = no; N = Number of ions; ζ = ionic charge; e = elementary charge; W = the effective average energy to produce one ion pair; p = pressure

Fig. 1 shows the SIS intensities achieved at present. They will be improved in the next years to reach the incoherent space charge limit of the SIS. Most of high energy experiments with the SIS use the slow extraction mode. In this mode the ratio between the revolution time of the particles in the SIS and the extraction time is in the order of 1μs to 1-10 s which means that typical SIS currents of some mA correspond to external electrical currents below 1 nA.

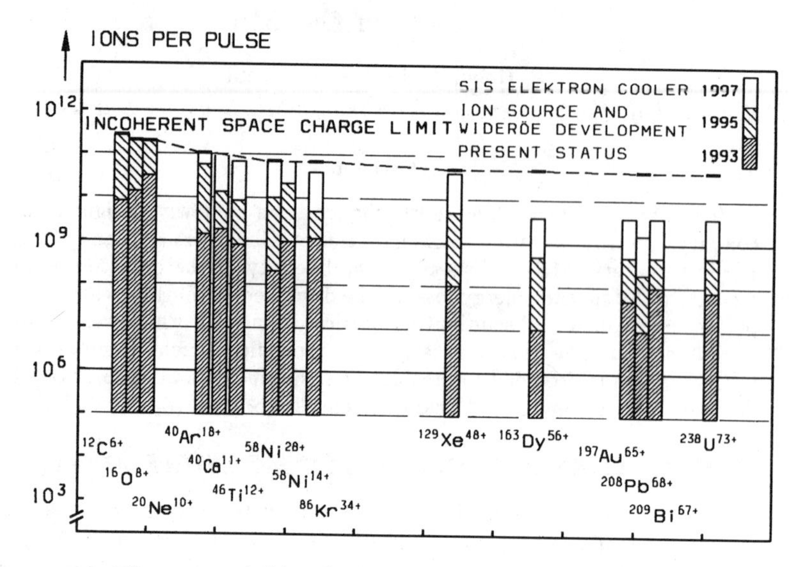

Figure. 1. Achieved and planned SIS intensities (1)

Considering the intensity ranges as well as the ion species which have to be covered it becomes evident that various types of detectors have to be used to determine the extracted particle fluxes. Furthermore, at present there is no detector available which can measure absolutely particle fluxes in the current range between about 10^{-12} - 10^{-6} A for all kinds of ion species.

Obviously, at the low intensity end up to about 10^6 particles per second (pps) calibration can be performed by reference to particle counters, while at the high intensity end a current of about 1μA ($10^{11} \ldots 10^{13}$ pps depending on ζ) is the lower limit where beam transformers of fluxgate type can be used for calibration and nondestructive absolute flux measurements. To extend the region below 1μA down to about 1 nA a new type of beam transformer using the principle of a Cryogenic Current Comparator (CCC) has been developed at GSI. The monitors which have been investigated for the determination of flux rates in the medium intensity region are: The scintillation current monitor, the ionization chamber and the secondary electron monitor.

Scintillation Current Monitor (MDC)

To extend the range of scintillation particle counters up to about 10^9 pps an attempt has been made to read out the scintillation light in current mode by photodiodes (2),(3),(4). In this mode the dependence of the light output on the deposited energy becomes important, in contrast to the pulse counting mode. Therefore, supposing the energy loss of the heavy ions can be taken from tables of stopping power, a linear relation between the light output and the energy loss will simplify the calibration of such a detector very much since only one calibration is necessary for all ion species and different energies.

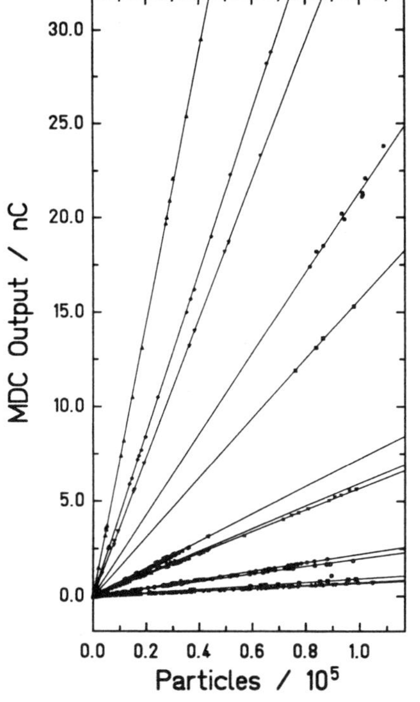

Figure. 2. MDC output as a function of the beam intensity.

This has been studied with a detector consisting of a round plastic scintillator sheet of NE108(5) surrounded by 15 photodiodes SFH100(6), a geometry which minimizes the position dependence of the output evaluating the sum signal.

A series of measurements (4) has been performed in the energy range between 200 and 1800 MeV/u for ions of C, Ne, Ar, Kr, Xe and U. Figure 2 shows the detector output as a function of the beam intensity determined by a particle counter. In the Figure the elements and energies (MeV/u) are from bottom to top Ne (1800), Ne (1200), Ne (700), Ne (300), C (270), Ar (800), Kr (800), Kr (500), Kr (300), Xe (1095), Xe (200), U (900), U (600), U (300). The relation between the slope of the fitted straight lines and the corresponding energy loss will be discussed in Fig. 4.

Ionization Chamber (IC)

The output of an ionization chamber should be proportional to the energy loss of the heavy ions, too. An ionization chamber of only 5 mm length, two 1.5 μm Mylar windows, filled with a mixture of 90% Ar and 10% CO_2 at 1 bar has been tested (7). Since the "target thickness" corresponds to about 1 mg/cm^2, the number of created electrons in the gas is in the order of 10^4 electrons/particle which means that secondary electrons from the foils can be neglected. Furthermore, assuming that losses by recombination and escaping electrons are also small, the chamber output should not only scale with the energy loss but can even be calculated using the well known W-values (8),(9) of gases. Figure 3 gives the results, extracted from a series of calibration measurements using again the particle counters. A straight line with a slope of 1.0 has been drawn to demonstrate the excellent agreement between calculated and measured values. In the near future further measurements will be performed to study the dependence on the beam energy and higher intensities more in detail as well as to test the dependence on the beam spot size.

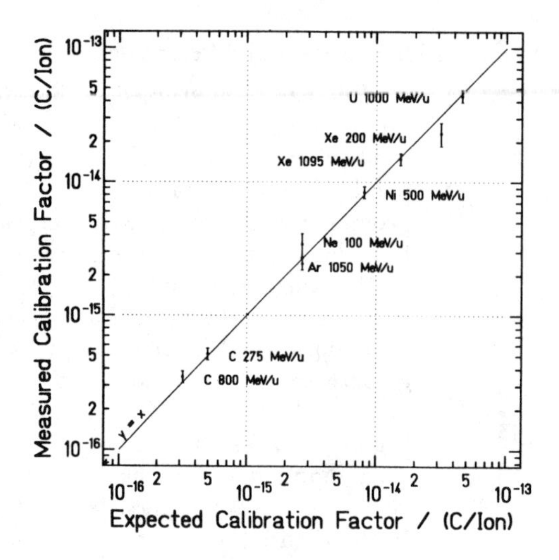

Figure. 3. Measured calibration factor for the ionization chamber versus calculated values

Secondary Electron Monitor (SEM)

Probably the SEM (10),(11) is the most used monitor to measure intensities with a relative method. A prototype consisting of nine 25 μm Aluminum foils has been built. The foils are slightly curved to increase the mechanical strength which reduces microphonic noise signals due to vibrations. With a spacing of 10 mm between the foils the overall lengths of the monitor is below 120 mm. The dependence of the output signal on the collecting voltage has been tested and no change beyond 20 Volts could be detected. In routine operation a collecting voltage of 100 Volts is applied. Since the detector consists only of metal and Al_2O_3-ceramics, it is even bakeable up to 300°C. Again there should be a proportionality between the output signal and the specific energy loss of the heavy ions. Since only about 30 electrons / (MeV/ mg/cm^2) can be expected (12) the gain of the SEM is low taking into consideration that only highly charged ions are accelerated in the SIS. To com-

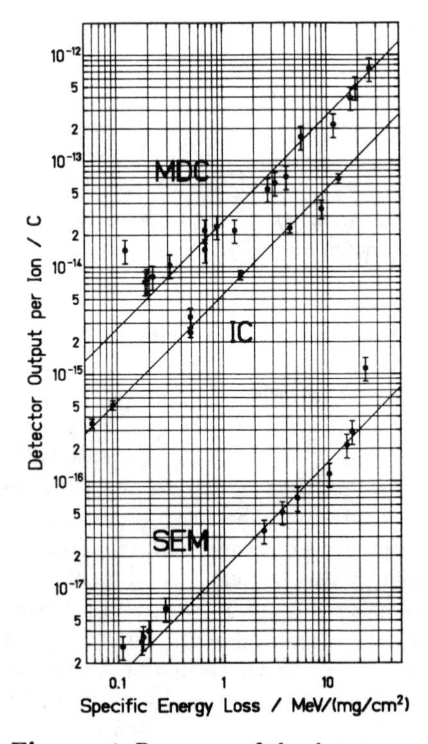

Figure. 4. Response of the detectors

pare the SEM with the monitors discussed above and to test the linearity with respect to the specific energy loss a similar series of calibration measurements as described above has been performed. Defining the detector response as the output charge per incoming ion (for the MDC it is the slope in Fig. 2), Fig. 4 shows the response of the SEM and the other detectors as a function of the specific energy loss calculated according to (13). The straight line through the data of the SEM results from a fit and corresponds to 23 electrons/(MeV/mg/cm^2).

Discussion of the Results for the MDC, IC and SEM

The significant differences of the three detector principles consist in the dependence of the response from ion species and their energy: while the IC output *equals* NeΔE/W, the MDC output is only *proportional* to NΔE, the energy deposited into the detector. The SEM output scales with the *specific* energy loss which is independent from the target thickness.

Obviously, the MDC gives the highest output signal due to the high target thickness. But, keeping in mind table 1 an essential shortcoming of the MDC is the low radiation resistance which will result in a very short lifetime at the planned high intensities. While the ionization chamber delivers only a slightly smaller output signal the linearity is excellent in the considered intensity region and, as discussed above there is no need for an experimental calibration. The SEM has the lowest output, but due to the high radiation resistance and the excellent vacuum compatibility this monitor and the IC will be further developed for their use in routine operation.

THE CRYOGENIC CURRENT COMPARATOR (CCC)

Obviously, for further detector developments based on relative methods and their calibration it is essential to decrease the gap in the intensity region where no absolute methods are available. Therefore at GSI a new type of beam transformer has been developed (14),(15), using the principle of the CCC(16), with the attempt to cover the current region above some nA with a non-destructive absolutely calibratable current measurement. The main components of the device are shown in Fig. 5.

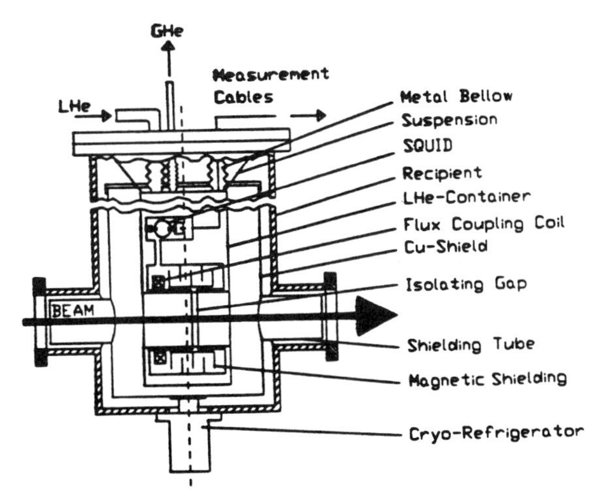

Figure. 5. Principle of the GSI Cryogenic Current Comparator

A current of some nA produces a magnetic field in the order of 10^{-14} T at a distance of 10 cm. The superconducting flux transducer shields this small azimuthal field from stray fields of bending magnets or the earth's magnetic field. A flux coupling coil with a SQUID is able to detect this small magnetic field. In the meantime the cryostat has been extensively tested and optimized leading to a Helium boil-off rate of 5.6 l LHe/d which corresponds to a heat loading of 170 mW and is in good agreement with the calculated value.

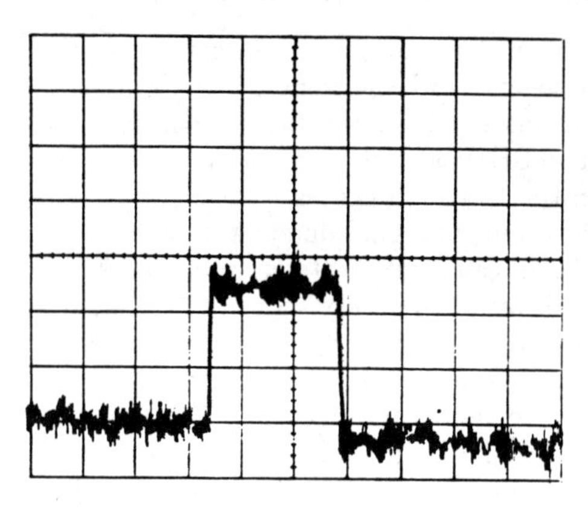

Figure. 6. Output of the CCC for a 10 nA test pulse, taken with a bandwidth of 100 Hz (x-axis 1s/div, y-axis 50 mV/div)

Using a one wire loop around the flux transformer to simulate the ion beam, a sensitivity of 175 nA/ϕ_0 has been achieved, where $\phi_0 = h/2e \cong 2 \cdot 10^{-11}\, T \cdot cm^2$ is the flux quantum and 1 ϕ_0 corresponds to 2.5 V in the most sensitive range of the SQUID system.

Figure 6 shows the output signal for a 10 nA test pulse. The slight zero drift of about 0.1 nA/s is probably caused by imperfect superconducting contacts. In the early spring of 1995 the whole CCC equipment will be installed in a test beam line behind the SIS, together with various other detectors for comparative measurements.

REFERENCES

1. Blasche, K., et al., "Status Report on SIS-ESR", Proc. of the Fourth European Particle Accelerator Conference, EPAC, (London, June 1994), to be published.
2. Keller, O.,"Untersuchungen zum Betrieb eines Szintillationsdetektors im Strom-Mode", Diplomarbeit, (FH Darmstadt 1994).
3. Heeg, P.,"Intensity Measurements of Energetic Heavy Ion Beams between 10^5 and 10^9 pps", Proceedings of the First European Workshop on Beam Diagnostics and Instrumentation for Particle Accelerators, (Montreux, May 3-5, (1993), CERN PS/93-35 (BD), CERN SL/93-35 (BI)), pp. 96-99
4. Heeg, P., Keller, O.,"A Scintillator - Photodiode - Beam Intensity Monitor", Proc. of the Fourth European Particle Accelerator Conference,

EPAC, (London, June 1994), to be published.
5. Nuclear Enterprices Technology, Ltd.
6. Siemens AG, Germany
7. Heeg, P., Junk, H., Stelzer, H., "An Ionization Chamber for Intensity Measurements in the High Energy Beam Lines", GSI Scientific Report 1993, (GSI 94-1, March 1994, ISSN 0174-0814), p. 302.
8. Sauli, F.,"Principles of Operation of Multiwire Proportional and Drift Chambers", (CERN 77-09, May 1977).
9. Knoll, G. F.,"Radiation Detection and Measurement", 2nd ed., (Wiley & Sons, New York 1989).
10. Tautfest, G. W. and Fechter, "A Nonsaturable High-Energy Beam Monitor", H. W., Rev. Sci. Inst., 26, 2, (1955), p. 229 ff.
11. Agoritsas, V.,"Secondary Emission Chambers for Monitoring the CPS ejected Beams", Symposium on Beam Intensity Measurement, (Daresbury 1968), pp. 117-151.
12. Ziegler, C. et al., "Performance of the Secondary-Electron Transmission Monitor at the FRS", GSI Scientific Report 1993, (GSI 94-1, March 1994, ISSN 0174-0814), p. 291.
13. Ziegler, J. F., Handbook of stopping cross-sections for energetic ions in all elements, Vol. 5, "The stopping and ranges of ions in matter", (Pergamon Press, 1980).
14. Peters, A., Vodel, W., Dürr, V., Koch, H., Reeg, H., Schroeder, C. H., "A cryogenic current comparator for low intensity ion beams", Proceedings of the First European Workshop on Beam Diagnostics and Instrumentation for Particle Accelerators, (Montreux, May 3-5, 1993, CERN PS/93-35 (BD), CERN SL/93-35 (BI)), pp. 100-104
15. Peters, A. et al., "A Cryogenic Current Comparator for Nondestructive Beam Intensity Measurements", Proc. of the Fourth European Particle Accelerator Conference, EPAC, (London, May 1994), to be published.
16. Harvey, I. K., "A precise low Temperature DC Ratio Transformer", Rev. Sci. Instrum., Vol. 43, (1972), p. 1626.

Ripple Measurements on Synchrotron Spill-Signals in the Time- and Frequency-Domain

P. Moritz
Gesellschaft für Schwerionenforschung mbH, D-64291 Darmstadt, Germany

Abstract

The analysis of ripple on spill-signals from the heavy-ion synchrotron SIS is presented. The signals are obtained either in time- or frequency domain, depending on the particle detector. An active ripple-injction method for determination of power supply ripple and extraction time constants for a spill regulator is shown.

INTRODUCTION

The slow-speed resonance extraction of particles at GSI´s heavy ion synchrotron SIS always shows some fluctuations. A primary point of interest is the identification and quantification of the various contributions of the magnets and their power supplies to these fluctuations. The spill fluctuations can be split into *non-coherent contributions* like energy spread of the particles, noise in the magnet power supply systems and their regulators, and into *coherent contributions* like power supply ripple on harmonics of the main power system and harmonics of regulator system clock frequencies.
There are many different ways to look for ripple signals (1). A passive detection method and an active ripple excitation method have been applied to the SIS beam for ripple investigations.

DETECTION OF RIPPLE SIGNALS

Measurements of the spill-signals need a signal from a target detector. Usually, there is a scintillator type detector signal available. These signals have a shaped pulse form and constant pulse width as well as a definite amplitude due to frontend signal processing. The particle count is coded in the pulse repetition frequency. The measurement goal is to demodulate the frequency modulation. An integrator or low pass fed with the signal will give an analogue voltage that is proportional to the particle count. A spectrum analyzer, usually of the FFT-type, is able to give a frequency domain display of the spill signal. Time resolution and error correction depend on the various circuit parameters. A somewhat different equipment setup to get the necessary information is shown in Fig. 1.

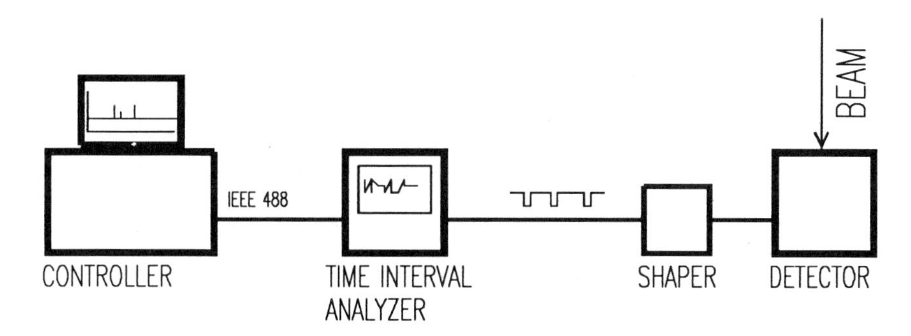

Figure 1. Equipment setup for passive detection of spill-signal spectrum

Here, a time-intervall-analyzer counts the number of pulses of the detector-signal during fixed and contiguous time-intervalls. The result is read out via IEEE-488 bus, giving a data stream of:

$$D(n) = F(ti(n)) \qquad \textbf{Eq. 1}$$

where,

n	- number of measurement
D	- Data value
ti	- time-intervall
F	- counts

Thus, a plot of all D(n) over n gives spill intensity over time. Another display feature is the histogram plot H(D(n)) over D(n), where H(D(n)) are the summed events of counts D(n).
With the aid of the controlling computer, an FFT-transformation of the D(n) can easily be performed. The procedure gives a view in the frequency domain. This spectrum display is a good indicator for the presence of coherent signals in the spill data stream. The determination of a specific frequency´s intensity in the spill spectrum is easy and straightforward. It is calculated from it´s relative value to the DC line in the FFT-spectrum. Once the FFT-transform with appropriate windowing and eventually padding of zeros is obtained, it is a nice tool to have the ability to filter "numerically" in the frequency domain and to compute the inverse FFT (IFFT)-transform of the "manipulated" data. The influence of a single spectral component or a group of harmonics can be shown easily. See Fig. 2 for example.

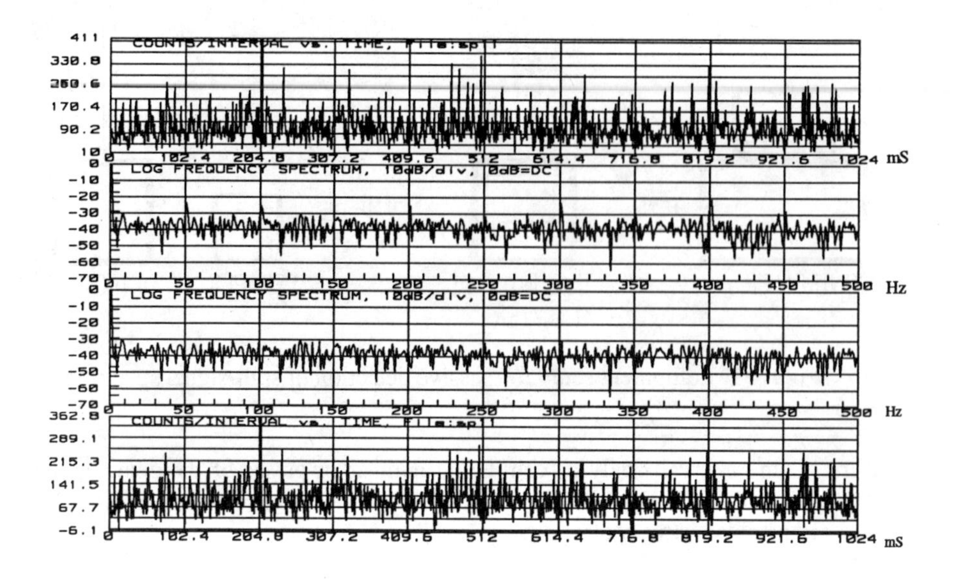

Figure 2. First (upper) trace: Time domain display of a spill, second trace: the calculated FFT-transform (spectrum), third trace: all harmonics of 50 Hz power line frequency 30 dB numerically attenuated, and last trace: IFFT-transform (time domain) showing the simulation of less power ripple in the spill-signal. The spill fluctuations are clearly of non-coherent type.

The time-domain display, the histogram, and the spectrogram can be displayed at the same time. This makes a nice tool for the operators to check for parameter changes, if an irregularity occurs in the sensitive particle extraction process.

RIPPLE EXCITATION

The measurement of spill properties using passive detection is by it´s nature restricted to signals contained in the spill data stream. The determination of beam extraction properties which are essential in the contruction of a spill feedback regulator is by no means easy if only the present detector data available. In the recent years, active excitation of a particle beam became more and more important for beam instrumentation. BTF-measurements using longitudinal and transversal exciters are in operation for many years. A precise beam position measurement using quadrupole modulation is one of the newer examples (2). SIS operation showed the need for a measurement of the effective ripple on the particle spill during slow extraction.

The measurement of the magnet power supply ripple was based on a substitution method. The detector/analyzer setup already described was added by a generator that injected an extra ripple of known frequency in the magnet power supplies of the various magnet groups. The measured extra ripple at the generator frequency detected at the target was then adjusted to be equal to the ripple frequency of interest. Using the known transducer gain between generator amplitude and magnet current, the effective ripple in the spill at all interesting frequencies has been measured. With anti-phase injection, the rejection of a single ripple frequency component was also shown (3).
The equipment setup for the subtitution method is shown in FIG. 3:

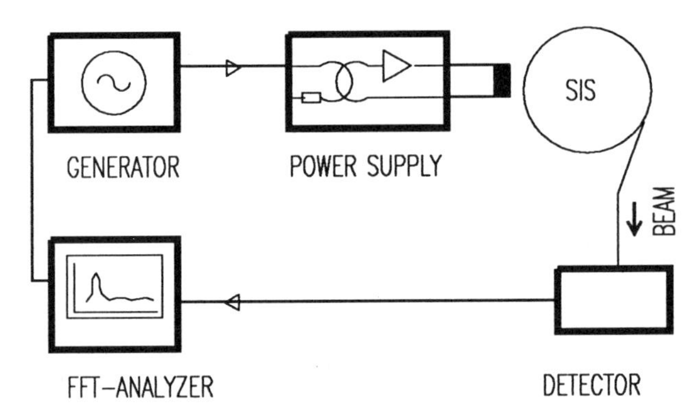

Figure 3. Equipment setup for active ripple excitation

RIPPLE TRANSFER DURING SLOW EXTRACTION

With the application of the FFT-analyzer´s build-in source as the generator, and using a stepped, SIS-cycle-triggered frequency sweep, the forward transfer function of the particles during resonance extraction was obtained in the low frequency range up to 1 kHz (4). The Bode-plot is show in Fig. 4. The measurement of the group delay is the most interesting result of this measurement. The group delay is defined as

$$Tg = - d\varphi/d\omega \qquad \textbf{Eq.2}$$

where
Tg - group delay
$d\varphi$ - phase deviation
$d\omega$ - frequency span $* 2\pi$

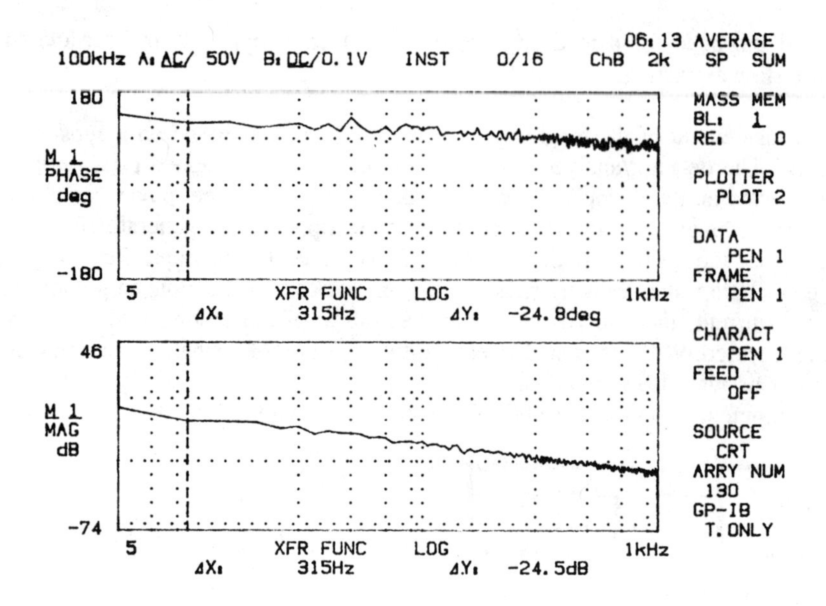

Figure 4. Forward transfer function during resonance extraction

The transfer functions of all "black boxes" of Fig. 3 except the spill forward transfer function during resonance extraction can be measured seperately. Solving the measured closed loop forward-transfer function for the particle transfer function delivers the group delay as well as phase/amplitude frequency response of the particles during resonance extraction.

At SIS, we obtained a delay of roughly 50µS. This value is equivalent to about 50 revolutions of the particles. The delay sets an upper limit for the feedback bandwidth of a spill regulator.

As a hands-on-approach for a spill regulator, a feedback amplifier was used to replace the FFT-analyzer/generator in the closed loop of Fig. 3. Another FFT-analyzer allowed the observation of the feedback effect on the spill signal. It´s output is shown in Fig. 5:

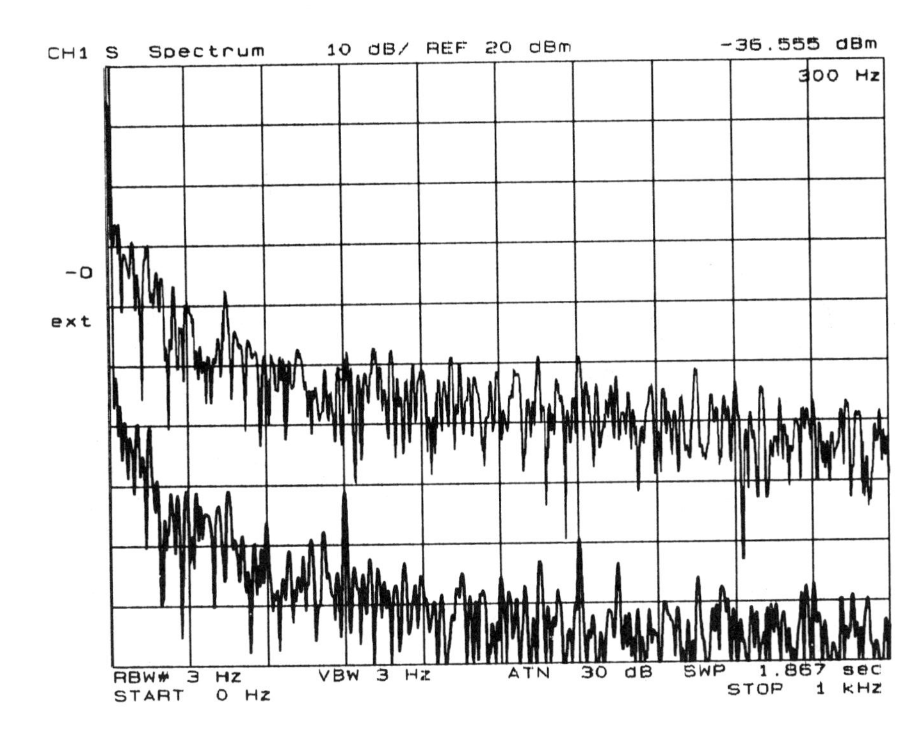

Figure 5. Closed-loop spill regulation using a bandpass feedback-amplifier. Upper trace: feedback on, lower trace: feedback off. Both traces offset for clarity.

The construction of a spill-regulator is not yet completed. The extraction-transfer-function is not linear and the frequency response is very limited. However, a spill-regulator will improve the spill structure in the lower frequency range and is a means to produce more constant average spill-rates at variing synchrotron intensities.

REFERENCES:

1. P. Burla et. al., "Power Supply Study at the SPS", Geneva 1994, CERN SL/94-11 (AP)
2. L. Arnaudon et. al., "A New Technique to center the LEP beam in a quadrupole", Proc. DIPAC 1993, Montreux 1993
3. P. Moritz, "Determination of SIS Magnet Power Supply Ripple using the extracted Beam Signal", GSI Scientific Report 1993, Darmstadt 1994
4. J. Pinkow, P. Moritz, "Spillregelung", GSI-SIS20074.BEX, Darmstadt 1994

Design and Performance of the Beam Loss Monitor System for the Advanced Photon Source*

Donald R. Patterson
Advanced Photon Source, Argonne National Laboratory, Argonne, IL 60439

Abstract

The design of the beam loss monitor system for the Argonne National Laboratory Advanced Photon Source is based on using a number of air dielectric coaxial cables as long ionization chambers. The coaxial cables are multiplexed into a high sensitivity DC current-to-voltage converter, which provides an output proportional to the average loss rate over the length of the multiplexed cable. Losses of sufficient amplitude generate measurable voltage pulses on the coaxial cable at a location near the loss point. Multiplexed pulse timing circuits determine the location of the losses by measuring the time at which these voltage pulses arrive at the beginning of the coaxial cable. The loss monitor system has been tested on the SPEAR accelerator at SSRL and was demonstrated to be as sensitive as the DCCT. Preliminary performance data from the APS injector show that the sensitivities of the current-to-voltage converter circuit are about ten picoamperes of loss monitor signal per picocoulomb per second beam loss rate. The corresponding pulse sensitivity is about 28 μV pulse amplitude in the coaxial cable per picocoulomb of loss. Both these sensitivities are at 300-MeV beam energies. The loss monitor has proven useful in initial commissioning of the injector. Further data will be available as accelerator construction and commissioning continue.

INTRODUCTION

Knowledge of beam loss rates and the approximate location of beam losses has been found to be an important diagnostic tool during commissioning and operation of many accelerator systems. This information is used to determine the cause of catastrophic beam loss events and to improve beam steering and control. The Advanced Photon Source (APS) system designers anticipated that such information would be critical during the initial commissioning and optimization of the APS accelerator(1). Therefore, the APS loss monitor system was designed to provide relative measurements of loss rates and an indication of the location of loss events throughout the accelerator. Large losses can be located more precisely than small losses. The system has no personnel safety function, but was found to be an economical method of providing data on potential radiation hot spots, especially in the rf/extraction building.

*Work supported by the U. S. Department of Energy, Office of Basic Energy Sciences, under contract no. W-31-109-ENG-38.

SYSTEM DESIGN

The APS loss monitor system is similar in many ways to loss monitor systems in use at Brookhaven(2,3,4,5) and SLAC(6,7,8), and has been described in more detail earlier(9). The system is based on a 7/8-inch air dielectric coaxial cable used as an ionization chamber. Five hundred volts DC is applied to the center conductor of the cable. The cable shield is grounded. An ionization gas consisting of a mixture of 95% argon and 5% carbon dioxide at 8 psig is passed through the cable. The average current flowing through the ionization gas is measured. This average current is proportional to the average beam loss rate along the cable. The cable is laid parallel and as close as feasible to the vacuum chamber (subject to mechanical constraints) to maximize the system sensitivity.

When accelerator conditions generate a large, localized beam loss, a correspondingly large, localized ionization is produced in the coaxial cable ionization chamber. The position of the ions in the coaxial cable and, therefore, the location of the beam loss is determined by measuring the time between the arrival of a beam bunch at the beginning of a coaxial cable ionization chamber and the arrival of the resulting voltage pulse at the same end of the cable.

Up to seven coaxial cable ionization chambers are multiplexed into a single

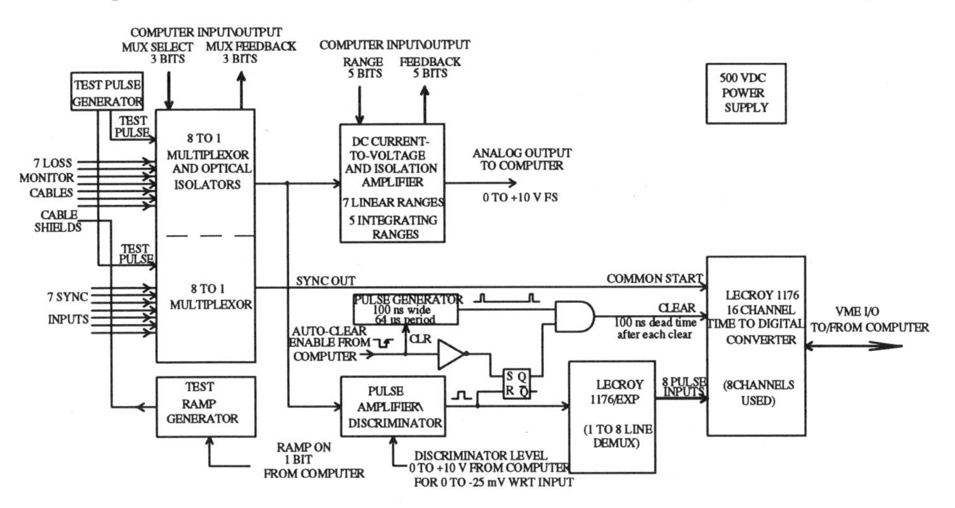

Figure 1: Block Diagram of the APS Loss Monitor System

electronics package. The package is shown in block diagram form in Fig. 1. The DC-coupled current-to-voltage amplifier and input multiplexor are floating on the output of the 500-V power supply. This is necessary so that the high voltage can be applied to the cable center conductor rather than to the outer shield, which could be perceived as a personnel safety hazard. The current-to-voltage amplifier consists of a field-effect transistor input (low leakage) operational amplifier with seven linear feedback ranges and five integrating feedback ranges, plus one integration reset range (see Fig. 2). These ranges are selected by low leakage reed relays. With careful design and layout, including the use of Teflon printed circuit board materials, input currents as low as 10 pA are easily measured. The linear ranges are well suited for operating conditions and locations where particle bunches repeatedly pass the coaxial ionization chamber with a repetition frequency of 60 Hz or higher. Under these conditions the time-averaged ionization current is continuously available at the output of the beam loss monitor in the form of a DC voltage.

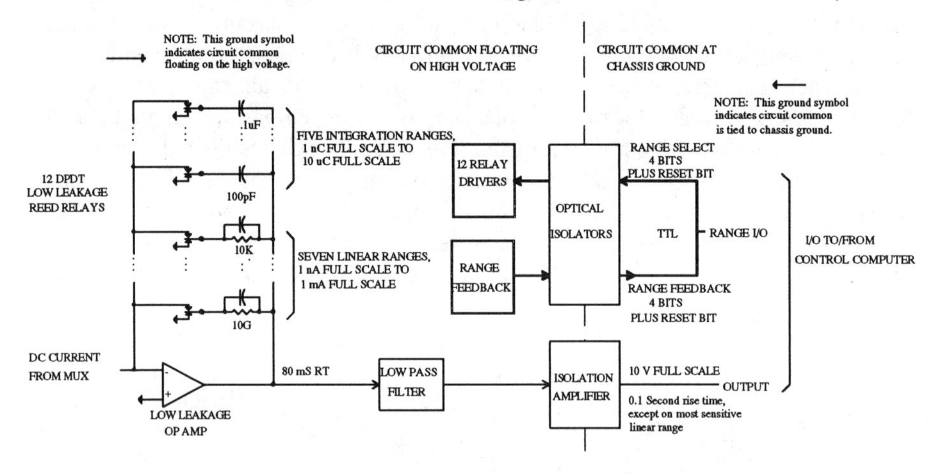

Figure 2: DC Coupled Current-to-Voltage Converter

Other accelerator subsystems, such as the transport line from the positron accumulator ring to the synchrotron, normally experience repetitive particle bunches at a 2-Hz rate. If the linear ranges were used for these subsystems, the resulting output signal would be a pulse that the control system would have to rapidly sample and integrate or peak detect to obtain a signal proportional to the beam loss rate. To simplify the measurement under such conditions, five integrating ranges have been added to the DC-coupled current amplifier. The 1-mA linear range is also used as an integration reset range to discharge the selected integration capacitor.

The amplifier/discriminator circuit contains the high frequency pulse amplifier and discriminator needed to convert the low amplitude pulses from the coaxial ionization chamber into the logic level signals required by the time-to-digital converter. Since only the fast-rising edges of signal pulses are of interest, these circuits include components that implement high-pass filters to reduce response to slower noise signals. The discriminator threshold voltage is set by the control computer.

Pulse timing nominally begins when a particle bunch passes one end of the coaxial ionization chamber, generating a timing synchronization signal. The required reference signal is derived from the nearest convenient beam position monitor. During the timing interval the times of arrival of all pulses on the coaxial ionization chamber are recorded. Each pulse represents a localized beam loss at a position easily calculated from the recorded time interval. Four cables are used in the synchrotron and twenty cables without pulse timing are used in the storage ring to avoid potential pulse timing ambiguities caused by the existence of multiple particle bunches.

TEST RESULTS

Tests of the loss monitor concept were conducted in the SPEAR accelerator at the Stanford Synchrotron Radiation Laboratory (SSRL) in February 1993. Three loss monitor cables were installed parallel to the vacuum chamber beginning at the storage ring injection point and running downstream for 100 feet. One cable was installed outside and above the ring, one outside and below the ring, and one inside and below the ring. Each cable was nominally 25 inches from the vacuum chamber, subject to mechanical constraints that sometimes required significant deviations from the nominal. During these tests, the accelerator operators maneuvered the stored and injected charge with the intent of producing stable losses of known amplitudes in known locations. All measurements were made at 2.28 GeV using 100% argon gas in the cables.

DC signal sensitivities measured for the three cables ranged from 4.2 to 7.7 picoamperes of loss monitor signal per picocoulomb per second beam loss rate. This factor of two difference in sensitivity is considered to be inconsequential. Therefore cable locations in the APS were determined primarily by mechanical considerations. This DC signal sensitivity is about equal to the sensitivity obtainable from the SPEAR direct-coupled current transformer (DCCT). Both can measure loss rates with lifetimes in the 8- to 10-hour range. However, the DCCT requires averaging times of about 20 seconds at low loss rates to remove noise and cannot determine the location of the loss. The loss monitor prototype was able to measure the loss rates with an averaging time of about 1/3 second and could localize the loss to the length of the coaxial cable.

Pulse signal sensitivities in the three SSRL cables with 100% argon gas fill were measured to be 71 to 110 microvolts pulse amplitude in the coaxial cable per picocoulomb of loss. Measurements in the APS linac test stand at 50 MeV showed that pulse amplitudes in 95% argon, 5% carbon dioxide fill gas were six times larger than comparable pulse amplitudes measured in 100% argon. The pulse technique at SSRL with 100% argon did not have enough sensitivity to be useful during stored beam mode, even when lifetimes dropped to less than one minute. It was, however, very useful during injection, showing that the injected charge bunches in SPEAR are frequently lost over several circuits of the storage ring rather than all at once at injection. Figure 3 shows an oscilloscope trace of pulse signals on two of the three loss monitor cables during normal injection. Note that some of the injected charge is lost during initial injection but the majority of the charge makes one pass around the storage ring before being lost near the injection point.

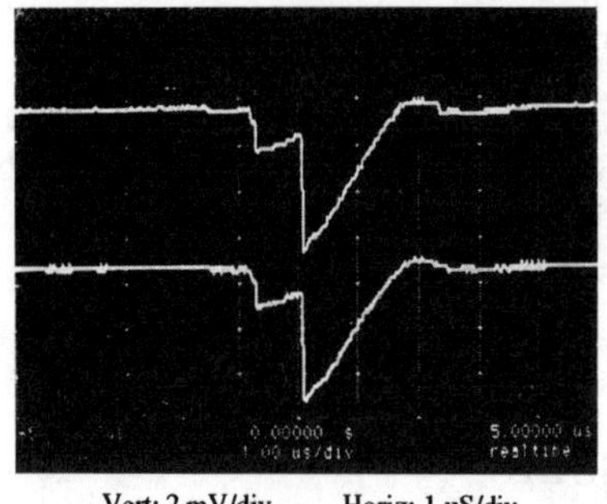

Vert: 2 mV/div Horiz: 1 uS/div
Figure 3

As of this writing, the loss monitor system has been used successfully in support of the commissioning of the APS injector components, including the low energy transport lines, positron accumulator ring, and booster synchrotron. Figure 4 shows early data taken to verify proper operation of the loss monitor. It is a strip chart made of the analog signal obtained with the beam hitting an inserted screen. At the middle of the trace, the screen was removed. The resulting reduction in loss monitor signal verified that the signal was valid and that the loss monitor was functioning as intended.

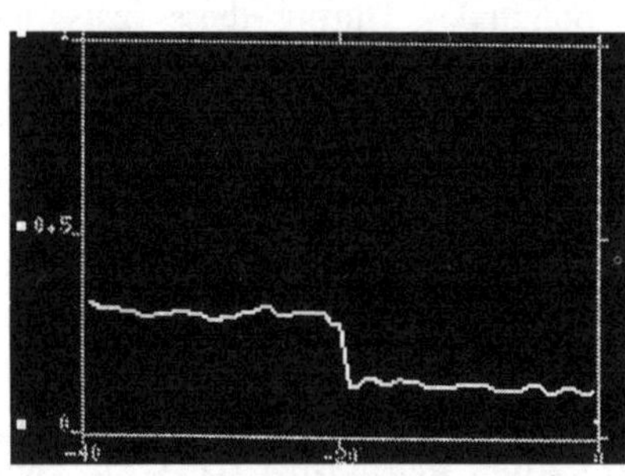

Vert: nanoampers Horiz: seconds
Figure 4

Accelerator operations to date have not allowed much quantitative data to be collected on the performance of the loss monitor. This is because the accelerator pulse repetition rate has been in the 2- to 10-Hz range rather than the designed 60-Hz rate. At these frequencies the DC output from the loss monitor contains a large ripple at the pulse repetition rate. This ripple is aliased by the 6-Hz data sampling rate, making data interpretation nearly impossible. The loss monitor integrating ranges are intended for low pulse repetition rates, but the controlling software for the integrating ranges does not yet exist. However, much useful information has been derived from relative loss rate indications. These relative indications have been a significant tool in early commissioning. More quantitative data will be collected in the future as accelerator operating conditions allow.

One quantitative measurement of the DC current sensitivity was made by intentionally directing a 300-MeV, 10-mA, 30-ns wide, 10-Hz repetition rate electron beam from the linac into the beam dump at the last linac dipole. This

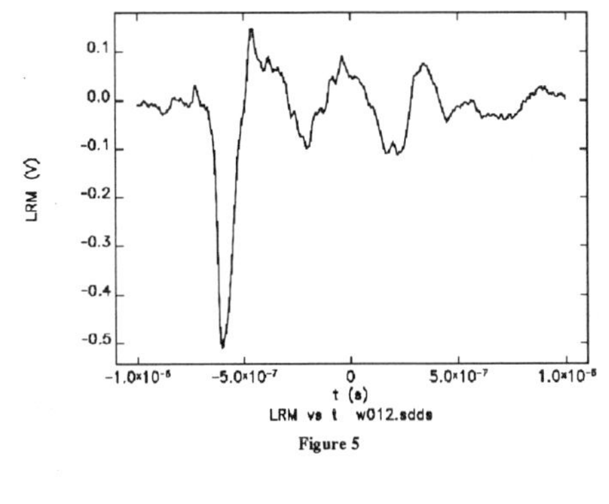

Figure 5

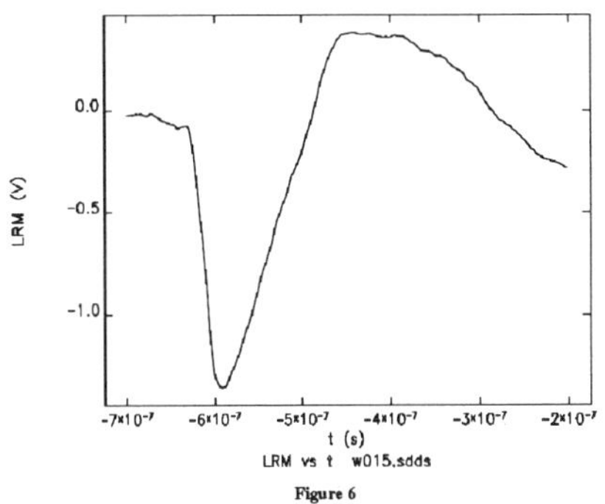

Figure 6

resulted in a charge loss rate from the beam of 3000 pC/s. The measured ionization current in the loss monitor cable was 30 nA, for a loss monitor sensitivity of 10 picoamperes of loss monitor signal per picocoulomb per second beam loss rate. Of course, this sensitivity varies with beam energy, mass between the loss point and the loss monitor cable (shielding and showering effects), and other causes.

Figure 5 shows the voltage output from the pulse amplifier and high pass filter circuits in response to a localized loss signal generated from a 300-MeV, 450-pC electron bunch 30 ns long. The amplifier has a high frequency gain of 40. The fast falling edge of the waveform contains the timing information for determining the location of the loss, and was observed to move relative to beam timing as the loss location was changed. The amplitude of the pulse was used to calculate a sensitivity of 28 μV pulse amplitude at the coaxial cable per picocoulomb of loss at 300 MeV. The lower frequency ringing shown on the trace is because the coaxial cable is not terminated at low frequencies because of the use of DC-blocking capacitors in series with the 50-Ω terminating resistors. This ringing contains no information and can be ignored.

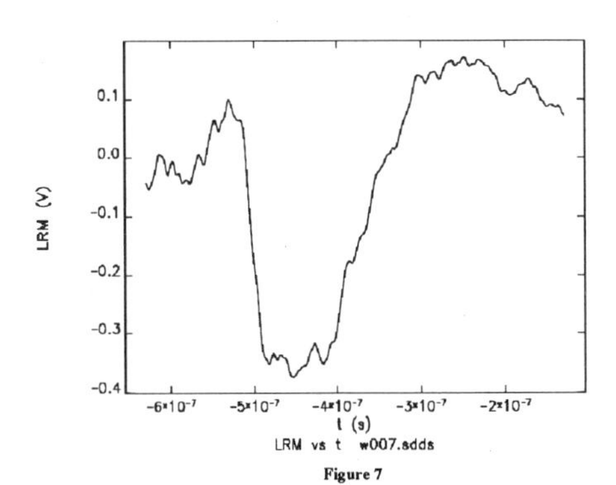

Figure 7

Figures 6 and 7 show two loss events on a faster time base. Figure 6 appears to be from a localized loss event, while Fig. 7 appears to be from a more spatially diffuse loss event. Note that bunch lengths for both Fig. 6 and 7 were 30 ns, limiting the spatial resolution to no better than about 15 feet.

CONCLUSION

Tests of a prototype loss monitor system at SSRL have verified that the concept will work and that signal sensitivities will be within usable ranges. The loss monitor system has been installed and operated in the APS injector. It has proven valuable in initial commissioning of the injector, and some quantitative data has been collected. Further quantitative testing awaits more prototypic operating conditions and the completion of the control and data acquisition software.

---

1. A. H. Lumpkin, et al., "Overview of Charged-Particle Beam Diagnostics for the Advanced Photon Source (APS)," Proceedings of the Fourth Annual Workshop on Accelerator Instrumentation, Berkeley, CA, AIP Conf. Proc. 281, American Institute of Physics, NY, 1992, p.150-157.
2. R. L. Witkover, "Beam Instrumentation in the AGS Booster," Proceedings of the Third Annual Workshop on Accelerator Instrumentation, Newport News, VA, AIP Conf. Proc. 252, American Institute of Physics, NY, 1991, p. 188-202.
3. E. R. Beadle, G. W. Bennett, and R. L. Witkover, "The AGS Booster Beam Loss Monitor System," Proceedings of the 1991 IEEE Particle Accelerator Conference, 91CH3038-7, San Francisco, CA, p. 1231-1233.
4. E. R. Beadle and G. W. Bennett, "The AGS Booster Radiation Loss Monitor System," Proceedings of the Second Annual Workshop on Accelerator Instrumentation, Batavia, IL, AIP Conf. Proc. 229, American Institute of Physics, NY, 1990, p. 35-47.
5. J. Balsamo, N. M. Fewell, J. D. Klein, and R. L. Witkover, "Long Radiation Detector System for Beam Loss Monitoring," IEEE Transactions on Nuclear Science, **Vol. NS-24**, No. 3, June 1977, p. 1807-1809.
6. R. G. Jacobsen and T. Mattison, "Beam-Loss Monitors in the SLC Final Focus," Proceedings of the 1989 IEEE Particle Accelerator Conference, **Vol. III**, p. 1539-1541.
7. J. Rolfe, et al., "Long Ion Chamber Systems for the SLC," Proceedings of the 1989 IEEE Particle Accelerator Conference, **Vol. III**, p. 1531-1533.
8. Max Fishman and Daryle Reagan, "The SLAC Long Ion Chamber System for Machine Protection," IEEE Transactions on Nuclear Science, June 1967, p. 1096-1097.
9. D. R. Patterson, "Preliminary Design of the Beam Loss Monitor System for the Advanced Photon Source," Proceedings of the Fourth Annual Workshop on Accelerator Instrumentation, Berkeley, CA, AIP Conf. Proc. 281, American Institute of Physics, NY, 1992, p.194-203.

A Cryogenic Dose Calorimeter For Pulsed Radiographic Machines

Scott Watson, Karl Mueller, and Todd Kauppila
Los Alamos National Laboratory, Los Alamos, NM 87544

Abstract

Calorimetry is the most direct, absolute technique for absorbed dose measurements. To improve the measurement accuracy for use with quantitative radiography, a calorimeter has been developed for LANL's pulsed radiographic machines which produce bremsstrahlung radiation fields of 50-200 Rad per pulse at 1 meter from the source. This paper descibes the theory of operation, the calorimeter design, and presents results from the PHERMEX accelerator.

INTRODUCTION

Radiation fields are functions of time, space, and energy. Secondary effects are used to measure these fields including: photographic exposure, ionization, heat calorimetry, photo-electric effects, and carrier mobility in diode junctions. We summarize the strengths and weaknesses of each in Table 1.

Technique	Temporal Response	Spatial Resonse	Spectral Response	Sensitivity	Calibration Standard
Electronic	Excellent	Poor	Good	Good	No
Photographic	None	Excellent	Poor	Excellent	No
TLD's	None	Poor	Poor	Excellent	No
Ion Chamber	None	None	Poor	Excellent	Yes
Calorimetry	Poor	Poor	None	Poor	Yes

Table 1. Radiographic Measurement Techniques

Calorimetry is the primary standard for absolute calibration and belongs at the top of the measurement chain.

The absorbed dose is obtained by measuring a temperature change, ΔT, in a target with known specific heat capacity (1), Cp, as in Eq. 1 below:

$$D = C_p \Delta T \qquad (1)$$

One limitation of calorimetry is low sensitivity; an absorbed dose of 100,000 Rad yields a temperature rise of only 0.24 K degrees in water.

Fortunately, for pulsed radiation sources, this lack of sensitivity is partially offset by the near-adiabatic equilibrium of the target material.

In high-energy radiography, the cgs unit for absorbed dose, the Rad, has largely replaced the older exposure dose, measured in Roentgens. The latter is based upon ionization in dry air under charge equilibrium conditions. The extremely long range of secondary electrons in air makes the measurement of exposure dose a practical impossibility for energies above 1MeV. Furthermore, while the Roentgen is defined only for air, the absorbed dose is independant of material. The conversion of absorbed dose in material A to the absorbed dose in material B can be accomplished using the following identity relating the dose and the mass-energy absorption coefficients:

$$\frac{D_A}{D_B} = \frac{(\mu_{en})_A}{(\mu_{en})_B} \tag{2}$$

To compare absorbed dose measurements with exposure dose measurements, we use the following:

$$D_{Rad} = 86.9 \frac{(\mu_{en})_{medium}}{(\mu_{en})_{Air}} X_{Roentgens} \tag{3}$$

DESIGN

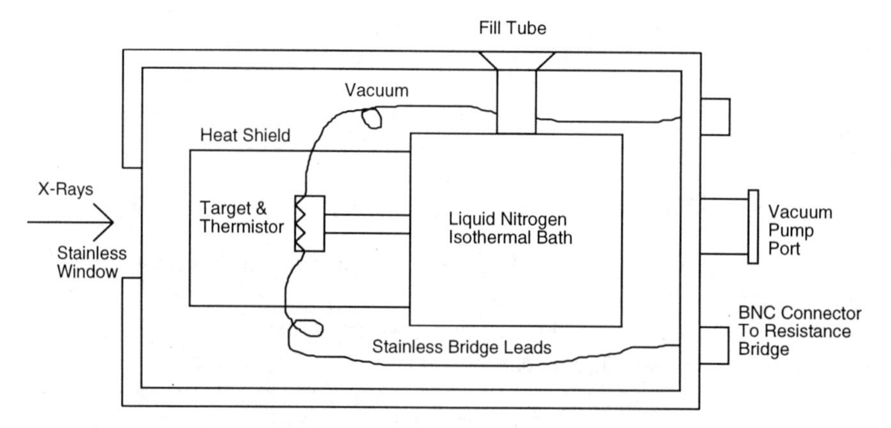

Figure 1. Calorimeter Functional Schematic

A functional drawing of the calorimeter is shown in Fig. 1. Copper and platinum are suitable target materials. Copper has a low specific heat, C_p=189 J/kg-K at 75K, and an extremely high coefficient of thermal conductivity.

Platinum has an even lower specific heat, C_p=84 J/kg-K at 75K, but suffers from poor thermal conductivity. We tested both target materials.

Liquid nitrogen is used as an isothermal bath for two reasons: the specific heat reduced 50%, and the sensitivity of platinum resistance thermometers goes up a factor of 4 relative to room temperature values. Consequently, nearly an order of magnitude increase in sensitivity can be gained at 75K.

To reduce gas conduction losses, a 10^{-3} Torr vacuum chamber, with a 0.13mm stainless X-ray window, is used. To minimize conduction losses, 6-ohm, stainless wire is used for the bridge connections. The thermistor is a 300-ohm platinum RTD. Finally, the evacuated chamber is packed with aluminized-mylar to reduce radiative losses.

The thermodynamics of this calorimeter are analogous to the impulse response of the circuit in Fig 2.:

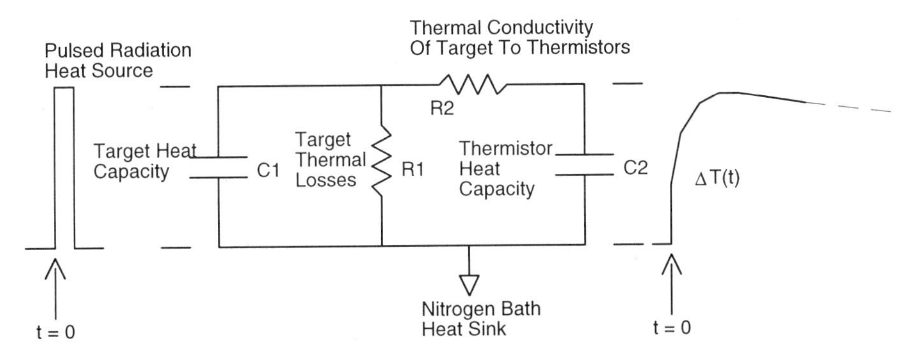

Figure 2. Equivalent Circuit Model

For thermistors with much less mass than the target material, i.e. C2 << C1, and in good thermal contact with the target, i.e. R2 << R1, the recorded temperature as a function of time, T(t), is related to the total absorbed dose, D, in a target with heat capacity C_p by the following:

$$T(t > t_o) \cong 75^\circ + \frac{D}{C_p}\left[1 - \exp(\frac{-t}{R_2C_2})\right] + \Psi \exp\left(\frac{-t}{R_1C_1}\right) \tag{4}$$

$$T(t \leq t_0) \cong 75^\circ + \Psi \exp\left(\frac{-t}{R_1C_1}\right) \tag{5}$$

This temperature characteristic is illustrated below in Fig 3. Notice that R1C1 >> R2C2 indicating that the calorimeter is properly designed.

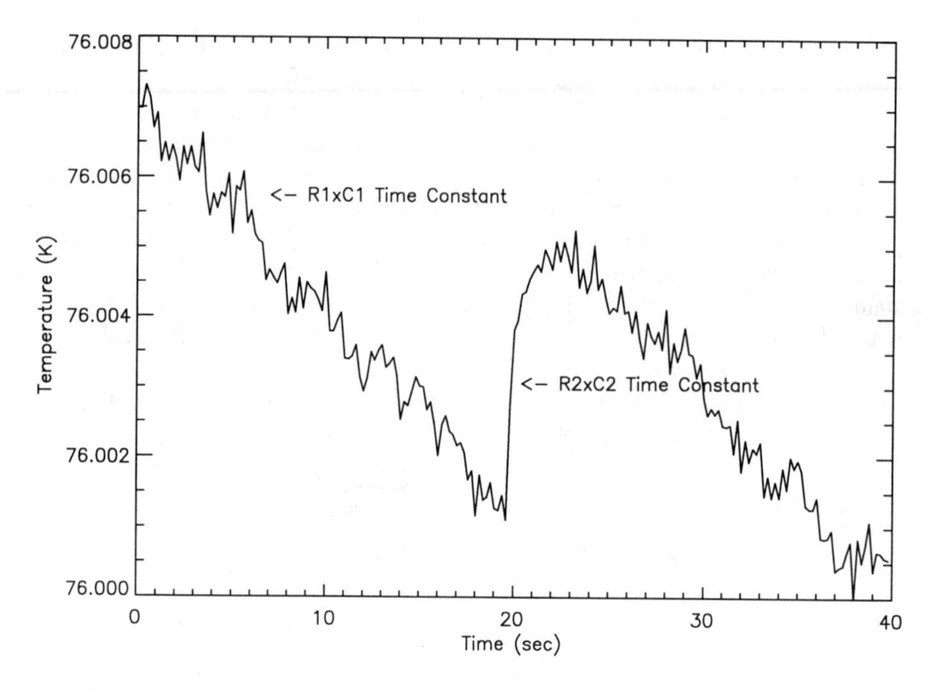

Figure 3. Temperature Characteristic Showing Thermal Time Constants

To obtain the absorbed dose, we can make the following differential temperature measurement:

$$\Delta T = T(t > t_o) - T(t < t_o) \cong \frac{D}{C_p}\left[1 - \exp\left(\frac{-t}{R_2 C_2}\right)\right] \tag{6}$$

for time near the radiation pulse time, t_o, (i.e. R2C2 << t << R1C1), Eq. 6 reduces to:

$$\Delta \mathrm{T} \cong \frac{\mathrm{D}}{\mathrm{C_p}} \tag{7}$$

Equation 7 assumes that the longitudinal energy deposition, dE/dx, is constant in the target material. However, it is necessary to correct for self absorption. The intensity as a function of areal mass, $I(\ell)$, can be approximated by the following:

$$\mathrm{I}(\ell) = \mathrm{I_o} \exp(-\mu_{\mathrm{avg}} \ell) \tag{8}$$

We apply a correction by finding the average intensity in the target material with thickness ℓ, with an average attenuation coefficient μ_{avg} and normalizing against the incident intensity, I_o. The correction term, k, is then given by the following:

$$k = \frac{\mu_{avg}\ell}{1 - \exp(-\mu_{avg}\ell)} \tag{9}$$

For a 3-mm-thick platinum target, this correction factor is k=1.12.

We measured the temperature coefficient of the platinum resistors to be K_R=1.29 ohm/K (for a nominally 300 ohm RTD) which compares favorably with the manufactures published number of 1.31 ohm/K (2).

To determine the absorbed dose from a measured resistance change, ΔR, in thermistors with thermal coeffiecients K_R, in a target with heat capacity Cp, placed d meters from the source we use:

$$D = \frac{k\Delta TC_p}{d^2} = \frac{k\Delta RC_p}{K_R d^2} \tag{10}$$

For our platinum target equation 10 becomes:

$$D_{Rads} = 7.3x10^3 \frac{\Delta R}{d^2} \tag{11}$$

EXPERIMENT

Our present radiation measurement utilizes a Compton diode cross calibrated against type-AA X-ray film with 1mm lead screens. The film is exposed with a Co60 source which is then calibrated against a NIST-traceable ionization chamber. This process is subject to a variety of systematic errors (film chemistry, idealized geometry, mono-energetic vs spectral source information etc) which conspire to make the measurent accurate to approximately +/-20% with a relative error between calibrations of +/-15% (3).

An experiment was conducted at the PHERMEX (4) comparing the present dosimetry technique with the calorimeter. For these tests, the calorimeter target was placed 1m from the radiation source - 25MeV bremstrahlung. A sweep magnet was used to keep the electron beam from impinging on the calorimeter target, and an aluminum filter cone was used. The data was recorded on a remote computer connected to the resistance bridge via HPIB. The remote computer was programmed with the HP-VEE software environment for data acquisition, and the IDL software package was used for data analysis. The experimental setup is shown below in Fig.4.

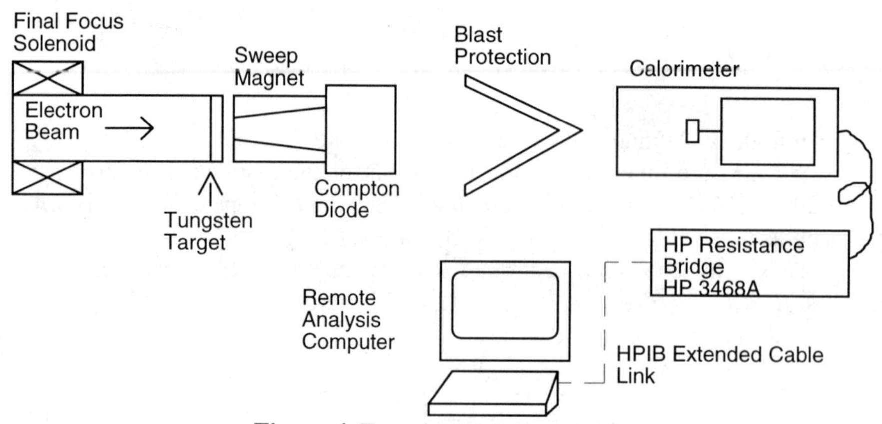

Figure 4. Experimental Arrangment

These results of this experiment are summarized below in Fig 5. A typical calorimeter measurement, corrected for the target thermal time constant, is shown below in Fig 6. The RMS error for a single measurement is +/- 2mK. Time-averaging the temperature data for 10 secconds yields an overall sensitivity of +/- 200uK which corresponds to +/- 2 Rad for the target placed 1m from the source.

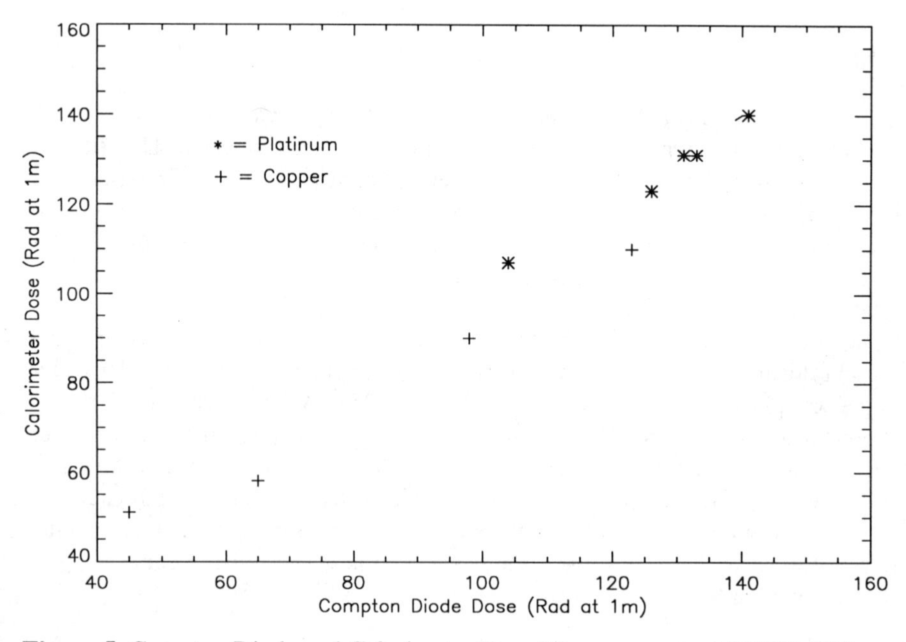

Figure 5. Compton Diode and Calorimeter Dose Measurements at PHERMEX

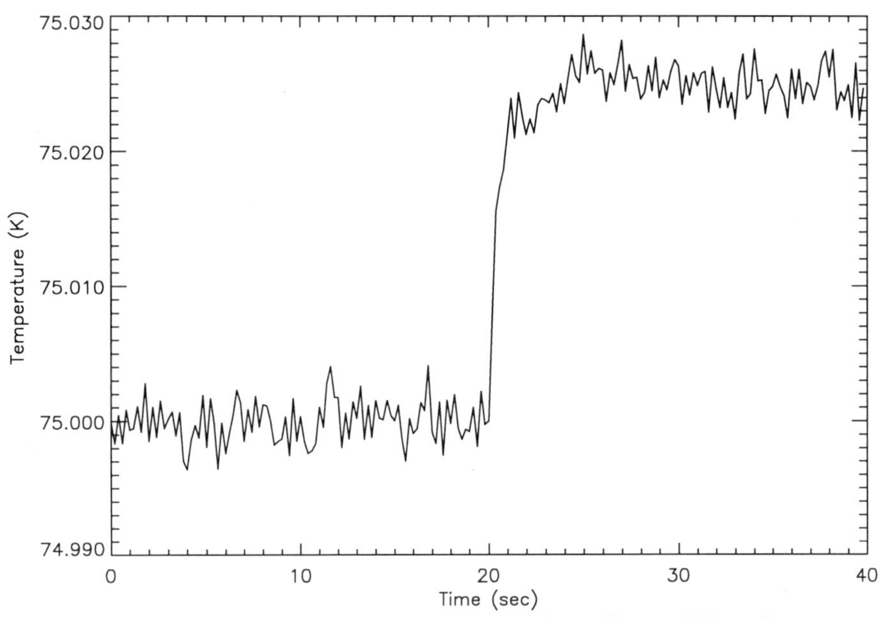

Figure 6. Typical Dose Measurement Corrected For Thermal Time Constant

CONCLUSIONS

Quantitative radiography requires that the radiation field be characterized-in time, space, energy, and amplitude. This information can be obtained by combining calculated energy spectral characteristics (5), electronic temporal information, spatial information from X-ray film, and absolute calibration with a high sensitivity calorimeter. This calorimeter can characterize the radiation fields at PHERMEX with +/-3% absolute accuracy, and with 1 Rad sensitivity.

REFERENCES

1. Shani, G. "Radiation Dosimetry Instrumentation & Methods," CRC Press. (1991).
2. Omega Catalog. "The Temperature Handbook," Omega Technologies. (1994.)
3. Carlson, R. L., "Dose vs. Film Density," LANL Memo no. DX-11:94-156.
4. Venable, D., et. al., "PHERMEX: A Pulsed High-Energy Radiographic Machine Emitting X-rays," Los Alamos National Laboratory report LAUR-3241. (1967).
5. Hawkins, A., et. al. "Absolute Bremsstrahlung Energy Spectral and Dose Distributions," Los Alamos National Laboratory. M-4:GR-93-09. (1993).

Design and Test Results of a Beam Monitor for the CERN Linear Collider Test Facility

Yan Yin, TRIUMF 4004 Wesbrook Mall,Vancouver, BC, V6T 2A3, Canada
Fritz Caspers, Elmar Schulte, CERN, CH-1211 Geneve 23, Switzerland
Tord Ekelöf, Dag Reistad, Uppsala University, P.O.Box 533,S-751 21, Sweden

Abstract

The monitor for the CERN Linear Collider Test Facility (CTF) requires measurement of the position and intensity of each micro-bunch (3-50 ps FWHM) at a spacing of 330 ps. The charge of a single bunch is from 10^9 to 10^{11} electrons. A monitor designed to meet these requirements is presented. Particular emphasis is given to the sensitivity and frequency response appropriate to such short bunches.

Introduction

The ordinary button-type pickup senses the signal across the capacitance between the button and the wall. A resonance may occur if the structure of the button and its housing form a cavity.(1) When the bunch length is much longer than the button size, the beam frequency spectrum does not extend up to the resonance frequency, then the resonance is not excited. However, when the bunch length is comparable to the button size, the resonance may be excited; and if it is within the frequency of interest, it may disturb observation of the beam signal. Therefore the pickup has to be carefully designed to avoid such parasitic resonance.

Beam Parameters

The CTF beam parameters are: (2)

Number of bunches in the train:	1-24
Bunch length (FWHM)	3-50 ps
Bunch repetition frequency in the train:	3 GHz
Train repetition rate:	10 Hz
Minimum clear aperture:	40.5 mm
Beam energy:	4.5 Mev

Normally, the bunch spacing is 333 ps, each bunch has a charge ≥ 9 nC, and a bunch length (FWHM) ≤ 12 ps.

The following calculation derives the beam signal properties.

1). The electron bunches have a Gaussian longitudinal distribution, i.e. the charge density per unit length is: $\rho(x)=\frac{Q}{\sigma\sqrt{2\pi}}e^{\frac{-x^2}{2\sigma^2}}$, and $\int \rho(x)\,dx= Q$. When $x=0,\quad \rho(x)=\rho_{\max}=\frac{Q}{\sqrt{2\pi}}\frac{1}{\sigma}$, which is the peak charge density of the bunch. With a charge per bunch of 9 nC, each bunch contains 5.6×10^{10} electrons. When σ is 25 ps, the peak current is 144 A. When σ is 6 ps, the peak current is 598 A.

2). The frequency spectrum is also a Gaussian distribution with a half width of $1/\sigma$, where σ is half the rms bunch length in units of time. For a bunch with σ=25 ps, equivalent to 40 GHz, the wavelength λ=0.75 cm. The image current broadening effect is $\delta=r/(\sqrt{2}\,\gamma\beta c)$. It will be 5.4 ps. for a pulse of 12 ps in a pipe of radius of r=2 cm, and the total duration of electric fields on the detector will be 22.8 ps.

Pickup Design

The beam is highly relativistic. Since the pickup detects the near field of the beam, only the TEM mode is considered here. Therefore, we can represent the field of the beam by that from an infinitely long line charge; and from this calculate the E field near the pickup and the current induced in the detector. Because the bunch is much shorter than the pipe diameter, the bunch spectrum contains frequencies above the pipe cutoff. These high order modes could be induced and travel in the pipe as wakefields, such wakes have been observed and it is a future project to eliminate them.

The amount of signal the monitor picks up not only depends on the source field, but also on the design of the detector. In order not to form a cavity in the housing of the pickup, which might cause ringing, the pickup is made with a coaxial cone shape and directly welded into a Kaman feedthrough, which is useable up to 40 GHz (50 GHz is specified by the manufacturer Kaman Instrumentation Co). The cone shaped pickup has a smooth transition to the feedthrough. The ratio of the cone radius, a, to the outer conductor radius, b, is kept constant at b/a=2.3 along the cone axis in order to keep the impedance constant at 50 Ω. See fig.1.

The peak of the induced signal on this kind of pickup can be estimated by the following model, which has been developed since the 1970's. The beam is considered to be a constant current source, therefore the pickup with a 50 Ω load can be studied with a short-circuit equivalent circuit.(3) The TEM wave of the beam moves along the center of the beam pipe, see fig.1, and we define the time when the wavefront passes the center line of the pickup as t=0. The induced signal depends on the wave's angle of incidence; which is the angle between the normal to the ground plane and the direction of propagation of the field. In our case, the wave moves parallel to the ground plane, which is the pipe wall, so the incident angle is $\theta=\pi/2$. The current signal, I_{sc} , induced by a very short electric field pulse E_0 can be expressed by the following formula: (4)

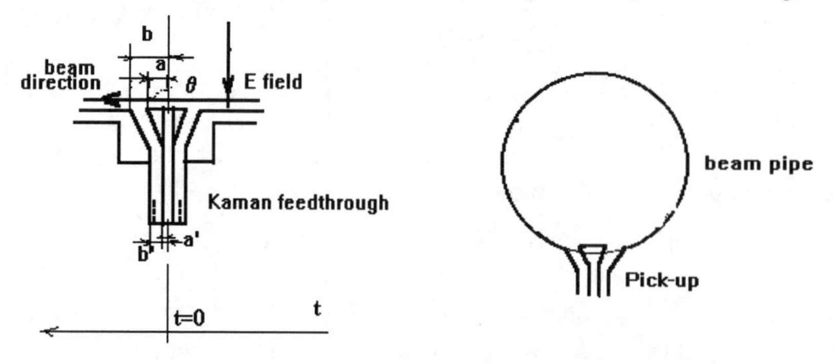

Fig.1. Cone pickup made with a Kaman feedthrough

$$\frac{I_{sc}}{E^i}=\frac{c}{2\zeta_0}\begin{cases}0, & -\infty<t<-\frac{b}{c}\\ \frac{\pi}{2}+\sin^{-1}\frac{ct}{b}, & -\frac{b}{c}<t<-\frac{a}{c}\\ -\sin^{-1}\frac{ct}{a}+\sin^{-1}\frac{ct}{b}, & -\frac{a}{c}<t<\frac{a}{c}\\ -\frac{\pi}{2}+\sin^{-1}\frac{ct}{b}, & \frac{a}{c}<t<\frac{b}{c}\\ 0, & \frac{b}{c}<t<\infty\end{cases}$$

where the impulse $E^i=\int E_0(t)\,dt$, E_0 is the incident field at the pickup aperture and ζ_0 is the free space impedance of 377Ω. Strictly for this formula to apply, the pulse duration must be $\leq b/c$ .For a long pulse, its high frequency component satisfied above requirement should be applied. The physical reason for a field integral appearing in the formula is as follows. The pickup is based on the principle of EM field induction. When a bunch passes the pickup, the pickup sees a TEM wave, changing with time. The varying electrical field, which is perpendicular to the pickup surface, induces a changing magnetic field around the pickup, and this varying magnetic field induces a current in the pickup to form the output voltage. Hence the pickup signal gives an approximation to the derivative of the incident signal. Because in our case the bunch length is comparable to or even smaller than the pickup, the pickup senses the field induced by the whole bunch, and therefore an integral should be applied to the whole bunch.

The above formula shows that the impulse response is a doublet whose time duration is $2b/c$. By integrating the pickup output, the original shape of the pulse can be recovered.

Let us estimate the voltage response of the pickup to a pulse propagating down an antenna placed along the axis of the beam pipe. The peak current occurs at roughly $t=-a/c$ and is given by $\frac{I_{sc}}{E^i}=\frac{c}{2\zeta_0}\left[\frac{\pi}{2}-\sin^{-1}\frac{a}{b}\right]$. I_{sc}/E^i can be transformed to the ratio V_{pu}/V_{tr} where V_{pu} is the voltage at the output of the pickup and V_{tr} is the voltage applied to the wire antenna which simulates the beam. The calculation shows that the ratio of peak to incident signal (V_{pu}/V_{tr}) should be 1.37 percent. The measurement, using a pulse of 25 ps FWHM, shows that the pickup gives out 1 percent of the incident signal, see fig. 2 and 3. The discrepancy is because the above formula is for an impulse with wavelength much shorter than the pickup size whereas in our case the bunch length is the same as the aperture of the pickup.

The frequency response is limited by the size of the pickup and the height of the cone.(4) The minimum cone height h is determined by the maximum allowable difference Δ_x between the path length along the inner and outer conductors. For small cone angles this is given by $\Delta_x=(1/h)\left[(b-b')^2-(a-a')^2\right]$. Ideally Δ_x should be much less than the smallest wavelength present in the spectrum., so that the signal induced in the cone will come in a TEM mode. In our case, b = 3.9 mm, a = 1.8 mm, b'= 1.45 mm a'=0.79 mm, and h is 4 mm. Thus Δ_x is 1.25 mm, which is much less than the wavelengths of interest.

Bench Test Setup and Results

The bench tests were performed with two HP network analyzers, one of 6 GHz and one of 40 GHz bandwidth; both have time domain low pass impulse modes. The 6 GHz network analyzer can synthesize a 150 ps impulse, while the 40 GHz analyzer can synthesize a 25 ps impulse. The monitor consists of a 5 cm long section of beam pipe with the pickup and feedthrough. Two 0.5 m long pipes are connected to either side of the monitor, and the ends terminated in two cones to form smooth transitions to the cable of the analyzer and a 50 Ω termination. A ¼ inch pipe passes along the center to form a 100 Ω coaxial structure, which is tapered to 50 Ω at both ends. Because the pipes are very long, reflections from both ends can be easily distinguished from the parasitic resonance at the pickup. In order to suppress the resonance caused by the setup, some microwave absorbent material was used.

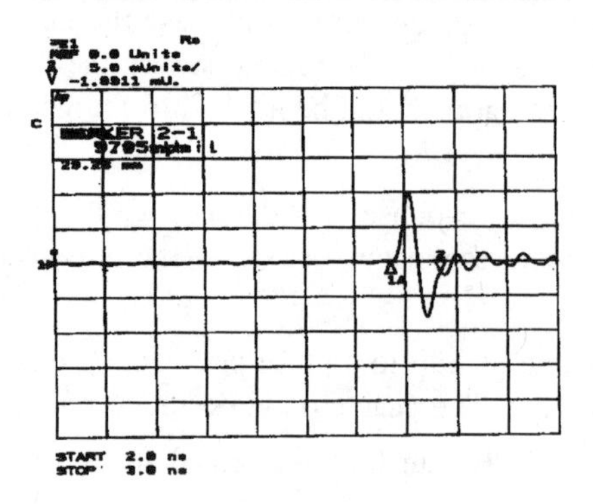

Fig.2 Cone pickup response to synthesized pulse with BW of 40 GHz

Fig.3 Synthesized Incident pulse with BW of 40 GHz

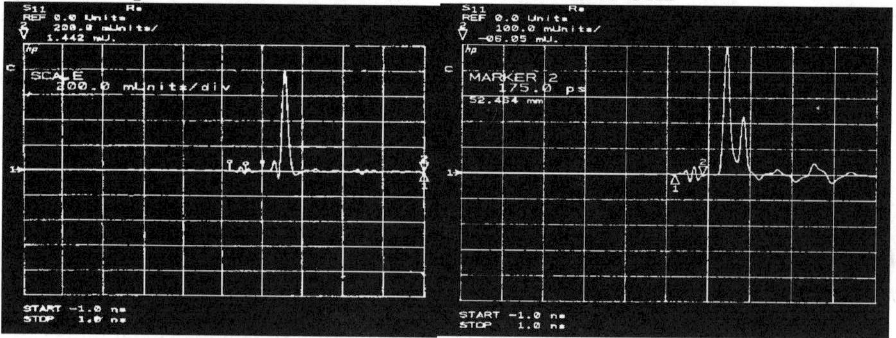

Fig.4 S_{11} for the cone pickup

Fig.5 S_{11} for Kaman feedthrough only

Fig. 4 and 5 are measured with signal directly fed into the pickup and the feedthrough only to see the reflections. The marks in the figures show only a tiny reflection from the Kaman feedthrough vacuum seal, and this shows that the feedthrough is 50 Ω up to our measurement limit of 40 GHz.

When the wavelength is much longer than the pickup size, the shape of the pickup is not important, and so simpler geometries such as the button and short rod can be used in place of the cone. In addition to the cone pickup discussed earlier, 3 prototype pickups (with no vacuum requirement) were manufactured in order to do some preliminary bench tests and make comparisons. The feedthroughs of these prototypes are ordinary SMAs used for printed circuit boards. Among the pickups, one is a 20 mm long stripline of width 4 mm, one is a short rod 5 mm long, and the last is a 4 mm ordinary button protruding 3 mm into the pipe without a smooth transition. These three prototypes were measured only with the 6 GHz analyzer and without the microwave absorbent material. The stripline pickup shows a periodic oscillation behind the doublet response because the geometrical distortion caused by a 4 mm wide stripline in a 40 mm diameter pipe cannot be neglected any more.

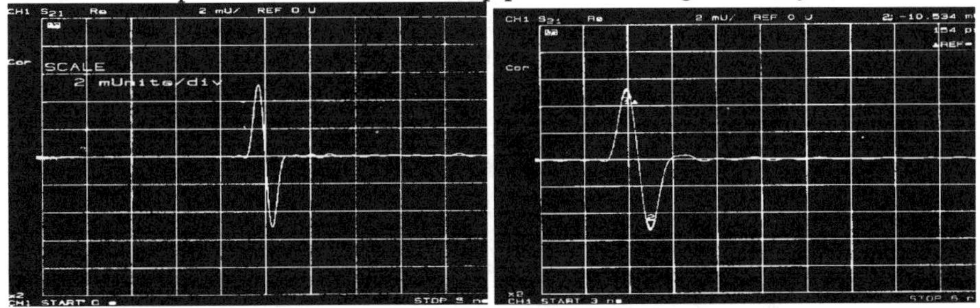

Fig.6 The cone response at 6 GHz, with microwave material.

Fig.7 The small regular button at 6 GHz without microwave material

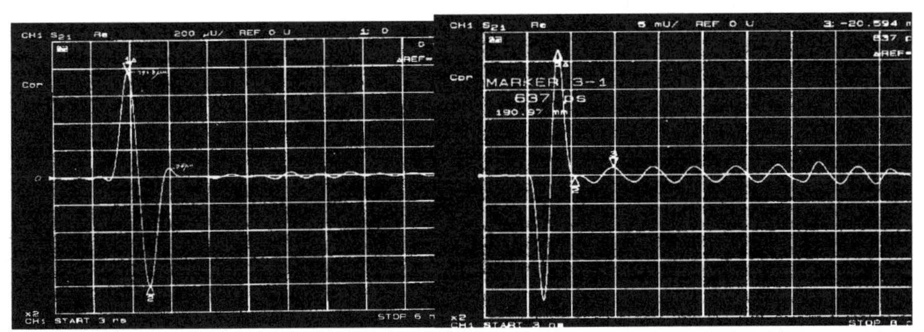

Fig.8 Response of a short rod at 6 GHz.

Fig. 9 Response of stripline at 6 GHz

Beam Tests Results

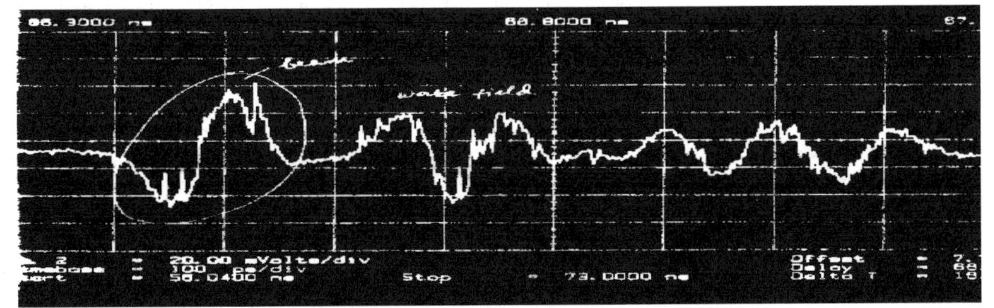

Fig.10 Test results with CTF beam

The cone pickup made with a Kaman feedthrough was installed in the CTF and the beam signals were observed see fig.10. Due to the use of a 15 meter long dispersive cable, the signal was stretched to 150 ps. After the beam signal, there is irregular ringing for 100 ns caused by wakefields. The wakefields can be distinguished from the ringing of the pickup because the waveform is aperiodic and commences almost a half wavelength after the beam induced doublet. The beam tests made with CTF in single bunch mode suggest that strong wakefields may cause problems for beam measurements in the multi-bunch mode. Further study is required to devise a method to extract the multi-bunch beam signal from the strong wakefield signals presently observed in the CTF.

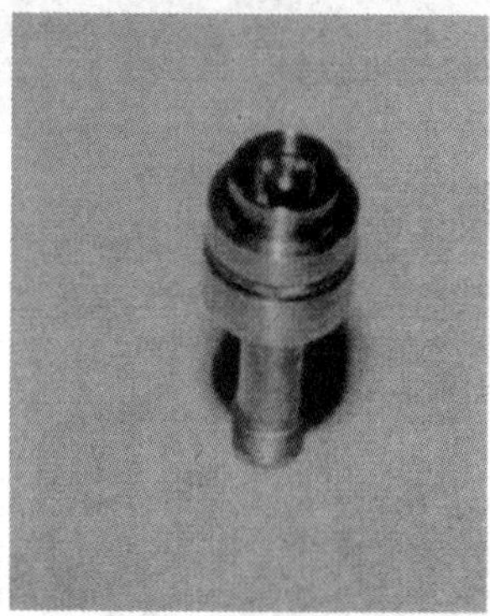

Fig.11. The pickup installed on CTF. **Fig.12**. The cone pickup .

Acknowledgment

This work is a cooperative project between, Uppsala university and CERN. The authors thank J.P.Delahaye(CERN) for supporting the project and TRIUMF for help with manufacturing the prototypes of the pickup. We also thank Jim Hinkson and Bob Hettel for help with some of the measurements performed at LBL and SLAC and Shane Koscielniak for very helpful discussions.

[1]Walter Barry "*Broad-band Characteristics of Circular Button Pickups*" BIW92, Berkeley, P.175-184.

[2] J.P.Delahaye, K.HÜbner, "*Proposal of collaboration between Uppsala University and CERN*", memorandum 10 Jun, 1993.

[3] Carl Baum et al. "*Sensors for electromagnetic pulse measurements both inside and away from nuclear source regions*". IEEE Trans.on Antennas and Propagation Vol. AP-26, No.1, Jan.1978.

[4]D. Lamensdorf, "*The transient Response of the Coaxial Cone Antenna*" IEEE Trans. Antenna Propagation.,AP-18, Nov.1970, P 799-802.

Design, Construction, and Wire Calibration of PAR BPM Striplines*

W. Sellyey, D. Barr, and L. Erwin

Advanced Photon Source, Argonne National Laboratory, Argonne, IL 60439

INTRODUCTION

The Positron Accumulator Ring (PAR) is part of the APS injection system. It receives 24 30-ns FWHM bursts of 450-MeV positrons, and compresses them into 6-nC, 290-ps rms bunches. Striplines were selected as beam position monitors (BPMs) to assure that good position sensitivity is achieved. This paper will describe the design, construction, and wire calibration of the 16 PAR BPMs. It will be demonstrated that all relevant stripline parameters can be determined by solving the two-dimensional LaPlace equation. This was done numerically using the electrostatic part of the PE2D computer program. The construction of the units will be briefly discussed. Wire calibration data on one of the final units will be compared with theory at four frequencies.

THEORY

In the design of the striplines, the beam was taken as traveling at c, in a uniform cross section beam tube. Thus the Fourier decomposition of electromagnetic fields contains only transverse electromagnetic (TEM) waves. For example, the Fourier component of the electric field can be represented by

$$\bar{E}_T\ (x,\ y)\ e^{i\omega(t\ -\ z/c)}, \tag{1}$$

where $\bar{E}_T$ has components only in the transverse x-y plane. It is straightforward to show that the stripline design process reduces to a two-dimensional electrostatic problem. To do this, one breaks the ∇ operator into transverse and longitudinal components:

$$\nabla_T = \hat{i}\,\frac{\partial}{\partial x} + \hat{j}\,\frac{\partial}{\partial y} \quad \text{and} \quad \nabla z = -i\,\hat{k}\,\frac{\omega}{c}\ . \tag{2}$$

Using the general expression of $\bar{B}$ in terms of a vector potential $\bar{B} = \nabla \times \bar{A}$, one obtains $\bar{B}_T = \nabla \times \bar{A}_z$ (with $\bar{A}_T = 0$ because $\bar{B}_z = 0$). Using Maxwell's equation for a Fourier component, one can show that a scaler function exists such that

---

*Work supported by the U.S. Department of Energy, Office of Basic Energy Sciences, under contract no. W-31-109-ENG-38.

$$E_T + i\omega \overline{A}_z = \nabla\psi \tag{3}$$

and thus

$$\nabla_T^2 \psi(x,y) = -\nabla_T \cdot E_T = 0 \quad \text{and} \quad \overline{A}_z = \psi/c\ . \tag{4}$$

We thus need to deal only with the two-dimensional La Place equation.

To accomplish this, it is useful to use Green's reciprocity theorem for electrostatics. Consider a system of n charges q_i at whose location the potential is Φ_i. Now replace these with a new set of charges q_i' which results in potentials Φ_i'. The theorem states that

$$\Sigma q_i \Phi_i' = \Sigma q_i' \Phi_i\ . \tag{5}$$

This can be generalized to sets of conductors with charges q_i and q_i' at potentially Φ_i and Φ_i', or a mixed set of conductors and charges can also be used.

For the beam-stripline system, there are five conductors (see Fig. 1). Besides the four striplines, the fifth conductor is the beam tube. This will always be taken as being at ground potential and thus it will never explicitly appear in a sum of the reciprocity theorem. What we would like to obtain is the signal induced by the beam on the striplines. The first step is to put one volt on each strip and numerically solve for the potential distribution. Call this distribution $v_0(x,y)$. Now we ground the four striplines and put in the charge q at x,y. Charges Q_1, Q_2, Q_3, and Q_4 will be induced on the four strips. Reciprocating gives that

$$Q_1 + Q_2 + Q_3 + Q_4 + qv_0/\text{volts} = 0. \tag{6}$$

We get distribution v_y with strip voltages $V_1 = V_2 = -1$ and $V_3 = V_4 = 1$. Then v_x using $V_1 = V_4 = -1$ and $V_2 = V_3 = +1$, and finally v_q with $V_2 = V_4 = -1$ and $V_1 = V_3 = +1$. The charges can be solved for:

$$Q_1 = -\tfrac{1}{4}[v_0 + v_x + v_y + v_q]\ \text{q/volt}, \qquad Q_3 = -\tfrac{1}{4}[v_0 - v_x - v_y + v_q]\ \text{q/volt}, \tag{7}$$
$$Q_2 = -\tfrac{1}{4}[v_0 - v_x + v_y - v_q]\ \text{q/volt}, \qquad Q_4 = -\tfrac{1}{4}[v_0 + v_x - v_y - v_q]\ \text{q/volt}.$$

For a transmission line in vacuum, the impedance is $Z_0 = 1/cC$ where c is the speed of light and C is the capacitance per unit length. In our case, the capacitance per unit length of a strip is obtained by numerically evaluating $I = \oint \bar{E} \cdot \bar{n} dl$ around an electrode when the potentials v_0, v_x, v_y, and v_q are numerically determined. The impedance is then given by $Z_0 = 376.73\Omega/I$ (SI units). It is important to note that this impedance depends on which of the v's is being evaluated. In particular, for the PAR striplines, the impedance corresponding to

v_0, v_x, v_y, and v_q are R_0 = 50.0Ω, R_x = 45Ω.0, R_y = 45.0Ω, and R_q = 43.0Ω (± 0.2Ω), respectively.

Each strip in a stripline unit is connected to a 50 ohm signal cable through a vacuum feedthrough. The other end of the strip is grounded. Thus the induced current from the beam sees the R = 50-Ω signal cable impedance in parallel with a transmission line shorted to ground at one end. Further, there may be additional capacitances and inductances in the area where the stripline is connected to the 50-Ω cable. These can be taken as part of now a complex impedance R.

Since there are four strip impedances, R_o, R_x, R_y, and R_q, there will be four different impedances seen by the beam-induced current. Call these Z_0, Z_x, Z_y, and Z_q. One gets

$$Z_0 = R_0/(\alpha + R_0/R) \quad \text{and} \quad \alpha = (1 + e^{-2ikl})/(1 - e^{-2ikl}) \tag{8}$$

and to a good approximation

$$Z_x = Z_0(1 + \delta x) \quad \text{and} \quad \delta x = \frac{s_x}{R_0}\left[\frac{1}{\alpha + R_0/R} - 1\right] \tag{9}$$

with similar expressions for Z_y and Z_q. Here $s_x = R_0 - R_x$, 1 is the strip length and k is the wave number. We will allow for the possibility that the four strips are not identical. Corresponding to Z_0 there will now be Z_{01}, Z_{02}, Z_{03}, and Z_{04}. Similarly for Z_x, Z_y, and Z_q. Since $s \ll R_0$, $|\delta| \ll 1$. Variations in Z_x, Z_y, and Z_q among the strips will thus be dominant by variations in Z_0. Thus the δ's will be taken as being the same for all four strips.

The current on strip 1 will be cQ_1. Q_1 is made up of four components associated with v_0, v_x, v_y, and v_q to each of which corresponds a different impedance Z_{01}, Z_{x1}, Z_{y1}, and Z_{q1}. One only measures the magnitude of the voltage and to a good approximation

$$|V_1| = (c/4)\,(q/\text{volts})|Z_{01}|[v_0 + v_x(1 + \text{Re}\delta x) + v_y(1 + \text{Re}\delta y) + v_q(1 + \text{Re}\delta q)] \tag{10}$$

with similar expressions for $|V_2|$, $|V_3|$ and $|V_4|$. Define $\Sigma = |V_1| + |V_2| + |V_3| + |V_4|$, $\Delta V_x = |V_1| + |V_4| - |V_2| - |V_3|$ and $\Delta V_y = |V_1| + |V_2| - |V_3| - |V_4|$. The ratio $r_x = \Delta V_x/\Sigma$ is particularly sensitive to motion in the horizontal direction, while $r_y = \Delta V_y/\Sigma$ is sensitive to vertical displacement. Substituting the voltage magnitudes, one obtains

$$r_x = (v_x/v_0)(1 + \text{Re}\delta x) + F_x(|Z|) \quad \text{and} \quad r_y = (v_y/v_0)(1 + \text{Re}\delta y) + F_y(|Z|). \tag{11}$$

F_x and F_y involve ratios of two impedance terms. The denominator is always $|Z_{01}| + |Z_{02}| + |Z_{03}| + |Z_{04}|$, and the numerator always involves two terms subtracted from two other terms such as $|Z_{01}| + |Z_{02}| - |Z_{03}| - |Z_{04}|$. All but one term is also multiplied by a term like v_y^2/v_0^2. If the striplines were identical, these terms would clearly be zero. At the frequencies being used here, the primary cause of differences would be the strip impedances.

If we write $R_{0i} = R + \varepsilon_i$ where R_{0i} is R_0 for the ith strip in a BPM, one can show that the one term not multiplied by a v factor is

$$r_{x0} = (\varepsilon_1 + \varepsilon_4 - \varepsilon_2 - \varepsilon_3)(1 + \cos 2kl)/8R \tag{12}$$

and a similar term for y. Neglecting the other terms involving $|Z_0|$'s, we write

$$r_x = (v_x/v_0)(1 + \mathrm{Re}\delta_x) + r_{x0}. \tag{13}$$

Near the origin, one can also approximate the x coordinate by

$$x = a_0 + a_1 r_x = (a_0 + a_1 r_{x0}) + a_1 (v_x/v_0)(1 + \mathrm{Re}\delta_x). \tag{14}$$

The term a_0 is a geometric offset resulting from the mechanical displacement of the whole BPM unit relative to the desired origin. $a_1 v_{x0}$ is due to stripline electronic asymmetries. It will be referred to as electronic offset. The electronic offset is frequency dependent, and thus one could, in principle, identify it by making measurements at different frequencies.

There are five terms which are being neglected in the last expression for r_x. The difference over sum factors in these are, in fact, electronic offsets divided by a coefficient like a_1. In principle, analysis of experimental data near the origin could be used to determine these and then one could include these terms in the full analysis.

DESIGN, CONSTRUCTION, AND WIRE CALIBRATION

The beam tube in the PAR is rectangular in shape with rounded corners. The horizontal and vertical dimensions are 122 mm and 38.5 mm, respectively. It is not practical to put electrodes on the vertical sides of the beam tube because of synchrotron radiation and low signal sensitivity. Thus, all four electrodes of the BPM were placed on the horizontal surfaces. In order to not affect the cross-sectional area available for the beam, and to minimize beam impedance, the strips are recessed so they are flush with the beam tube. All striplines are shorted to the beam tube at one end, and at the other are connected to 50-Ω feedthroughs and then to 50-Ω cables.

Figure 2 shows the arrangement that was chosen. It represents a cross section of the beam tube in the region of the striplines. Three criteria were chosen to guide the design process. The impedance, R_0, of the stripline was to be 50 Ω. The second criteria was that the sensitivity in both the horizontal and vertical direction be the same. The third was to make the intercepted charge as large a fraction of the total as possible. There are more geometric variables in the problem than there are constraints. Thus, there is no unique solution to the problem. As a consequence, some parameters were chosen simply to make construction to acceptable tolerances straightforward. One constraint that limited the geometry of the recess in which the strips were placed was the fact that the BPMs needed to fit inside of the PAR quadrupoles; otherwise there would have been no place in the ring to put them. Figure 2 shows the final geometry.

The full BPM units were constructed in sections. First, units containing a pair of strips in their recesses were constructed. Figures 3 and 4 show a unit like this in transverse and longitudinal/vertical cross sections. The housing and strips were first machined out of Inconel. The feedthroughs were welded in first, then tantalum clips were welded to the molybdenum feedthrough center conductors. These clips are separated from the body (ground) of the cavity by one-fourth of their width to approximate a 50-Ω impedance. Finally, the strips were welded in place, and the clips were silver brazed to the strips. All welding and brazing was done using an electron beam. The BPMs were completed by welding two of these housings to the top and bottom of appropriately machined rectangular PAR beam tube sections.

The feedthroughs were 50-Ω SMA, manufactured by Kaman Instruments Inc. Feedthroughs manufactured both in Europe and Japan were compared to the Kaman units using a time-domain reflectometer (TDR). The TDR was a Tectronics SD-24 plug-in unit in a 11802 oscilloscope. The Kaman SMA was exposed to 170 Mrd of Co-60 radiation, and no difference in vacuum or electrical (TDR) characteristics were detected as a result of the irradiation. It was judged that the Kaman unit was superior to the others.

The wire calibration process involves stretching a wire along the longitudinal axis of the stripline. TEM waves are established between the wire and the stripline body to simulate a Fourier component of the beam. To assure that TEM waves were established in the region of the strips, extensions were added to the BPM units. With these extensions, there was at least 8 in of beam tube on either end before the strips were encountered by any signal traveling down the wire.

The drive signal was supplied to one end by a 50-Ω coax cable. The signal on the center conductor went to the wire after passing through a resistive matching network. The shield went to a 6 in x 6 in aluminum plate which was perpendicular to the wire. There was about one inch of space between the plate and the beginning of the extensions. On the other end, there was again a 1-in space, a plate, and a resistive matching network which supplied a signal to a 50-Ω coax. Four heavy braids were connected across the 1-in gap on both ends. They had enough slack so the stripline body could move by several inches.

The wire was fixed in position at both ends during a calibration run. The BPM was attached to a stepper-motor-driven x-y platform. An HP 8753C network analyzer was used as the signal source. A switch matrix was used to select one of the four striplines; the other four were terminated in 50 Ω. The selected signal went to the A input, while the signal from the output and of the wire went to R. The quantity A/R was measured and recorded. The switch matrix, stepper motors, and network analyzer were all under computer control. Data was taken over the range ± 15 mm in x and ± 10 mm in y in 21 x 21 evenly spaced positrons.

RESULTS AND ANALYSIS

A total of 17 striplines were calibrated using the wire method. All were run at 117 MHz because this is the frequency at which the BPM electronics operates. A few were also calibrated at other frequencies. This enabled us to check the theory as a function of frequency. We will discuss the data and results for one particular stripline, serial no. VM-P-S2-08, for which data was taken at four frequencies: 4.89, 23.92, 117, and 352 MHz.

The measured data consisted of a normalized voltage magnitude for each strip for each position of the wire. The normalization was done with respect to the signal coming out of the wire in the wire calibration stand. This normalization assured that drive power variations from the network analyzer and possible impedance variations in the wire/beam-tube system do not affect results. The quantities Σ, ΔV_x, and ΔV_y, and then $r_x = \Delta V_x/\Sigma$ and $r_y = \Delta V_y/\Sigma$ were calculated as a function of x and y position. Two polynomial fits to the data were then generated. Each contained 36 terms starting with a constant and containing r_x and r_y in all combinations up to 7th order. One fit was for x in terms of r_x and r_y and the other fit was for y. The results of this fit were used to generate the graph in Fig. 5 in which the horizontal axis shows the ratio r_x and the vertical shows the position x. The curves are drawn for various values of r_y.

Probably the largest error in the construction of the BPMs was the separation of the top and bottom housing units; this was intended to be 38.2 mm. The average separation of the actual units turned out to be 35.0 mm ± 1.28 mm. For VM-P-S2-08 the separation was 36.62 mm, a 4% difference. In the vertical direction, this means about a 4% difference in the design and actual r_y's. Therefore, the design r's cannot be used to compare with experiment. A new set was calculated for VM-P-S2-08. As was shown earlier, these need to be multiplied by $(1 + \mathrm{Re}\delta)$. In the present case

$$\mathrm{Re}\delta = -.05(1 + \cos kl) \tag{15}$$

for both x and y. Using a strip length of $l = .231$ m, $(1 + \mathrm{Re}\delta)$ is .900, .901, .929, and .998 for the frequencies of 4.89, 23.92, 117, and 352 MHz, respectively.

The resulting r's were also fit to 36-term polynomials to generate x and y coordinates. To compare theory and experiment, a set of differences using the polynomials was generated. Figure 6 shows this difference for 117 MHz in x. Table 1 summarizes the results of the analysis.

Table 1. Results Summary

f(MHz)	Average % Deviation		Offset (mm)	
	x	y	x	y
4.89	1.5	1.5	-.154	.037
23.92	.89	1.4	-.147	.021
117	.89	1.3	-.134	.078
352	.77	1.4	-.122	.204

The listed average deviations exclude offset. The % deviation is calculated by dividing the average deviation by 7.5 in x and 5 in y. These are the average ranges over which x and y are evaluated. The x and y offsets are also listed. The frequency variation of these offsets may explain, at least in part, why the fits to the data are not better than they are. These imply that the electrical offsets are not zero, and thus the neglected five terms in the r's are not zero. Another reason for the disagreement has to do with additional manufacturing errors. For example, the top and bottom half of VM-P-S2-08 are displaced by .25 mm with respect to each other. Thus the theoretical r's are inaccurate resulting in disagreement with the measurements

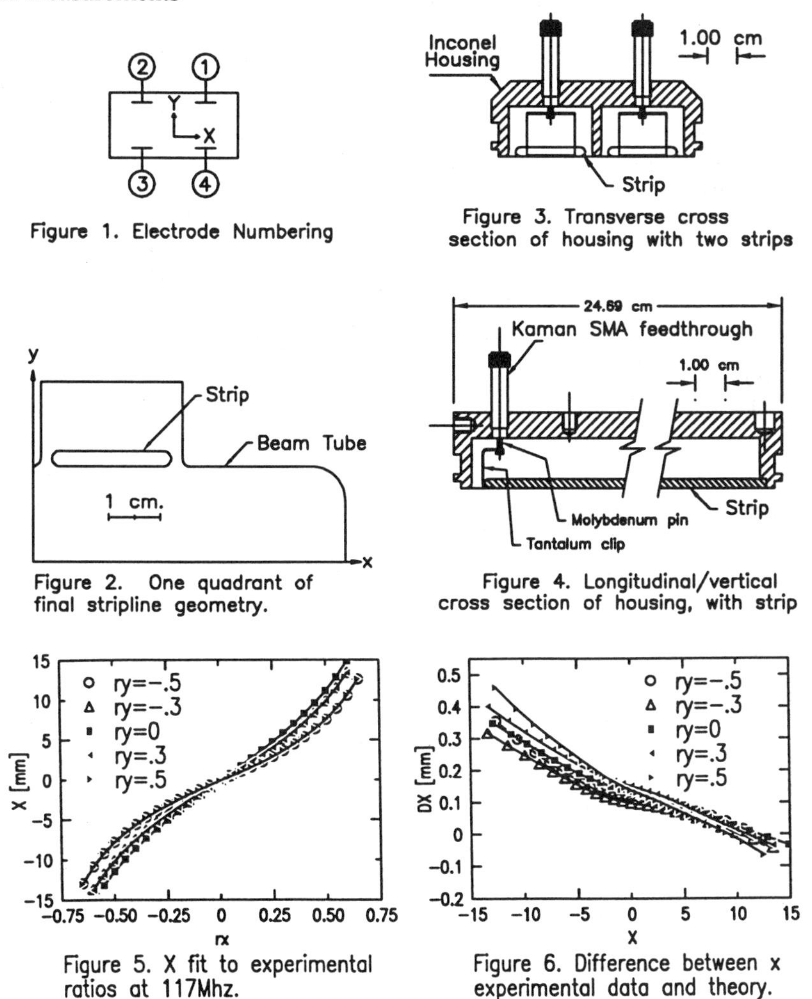

Figure 1. Electrode Numbering

Figure 3. Transverse cross section of housing with two strips

Figure 2. One quadrant of final stripline geometry.

Figure 4. Longitudinal/vertical cross section of housing, with strip

Figure 5. X fit to experimental ratios at 117Mhz.

Figure 6. Difference between x experimental data and theory.

Resolution and Drift Measurements on the Advanced Photon Source Beam Position Monitor*

Y. Chung and E. Kahana
Argonne National Laboratory, Argonne, IL 60439

Abstract

The resolution and long-term drift of the Advanced Photon Source (APS) beam position monitor (BPM) electronics were measured using the charged particle beams in the ESRF storage ring with various beam current and configurations (single bunch, 8 and 16 equally spaced bunches, and 1/3-fill). The energy of the stored electrons was 6 GeV. The integrated BPM electronics system as used for this work is capable of measuring the beam position on a turn-by-turn basis, which can be accumulated for N turns ($N = 2^n$, $n = 1, 2, \ldots, 11$). Estimation of the BPM resolution apart from the low-frequency beam motion was made by measuring the standard deviation in the measured beam position with different Ns. The analysis of the results indicates a BPM resolution of $18/\sqrt{N}$ [μm] for the APS storage ring, which is equivalent to $0.07\,\mu\text{m}/\sqrt{\text{Hz}}$. For the miniature insertion device BPM with 2.8 times higher sensitivity, the resolution will be $0.02\,\mu\text{m}/\sqrt{\text{Hz}}$. The long-term drift of the BPM electronics independent of the actual beam motion was measured at 2 μm/hr, which settled after approximately 1.5 hours. This drift can be attributed mainly to the temperature effect. Comparison of the results with the laboratory measurements shows good agreement. Implication of the BPM resolution limit on the proposed global and local beam position feedback systems for the APS storage ring will also be discussed.

INTRODUCTION

Measurement of the particle beam position in accelerators is an essential part of the beam diagnostics system, and the beam position monitors (BPMs) provide the basic diagnostics tool for commissioning and operation of accelerators. One of the primary applications of the BPMs is the stabilization of the particle and X-ray beam positions through feedback.[1,2] The specification of the required BPM resolution for beam position feedback in the Advanced Photon Source (APS) storage ring is listed in Table. 1.

The BPMs in the storage ring (SR) and injector synchrotron (IS) of the APS have four button-type pickups mounted directly on an elliptically shaped vacuum chamber.[3] The button size and the sensitivity of these BPMs are listed in Table. 2. The theoretical calculation was done analytically with an elliptic vacuum chamber with no photon exit channel,[4] which shows significant difference from the measured value in the case of the insertion device (ID) BPM due to the large aspect ratio.

---

*Work supported by the U.S. Department of Energy, Office of Basic Energy Sciences, under Contract No. W-31-109-ENG-38.

In this paper, we will describe measurements of resolution and long-term drift of the APS BPM electronics using the charged particle beam at ESRF. Several bunch configurations as well as various bunch currents were used. Design, development, and preliminary tests of the BPM electronics are described in Ref. 5, 6, and 7.

Table 1: Specification of the beam position feedback systems.

	Global DC	Global AC	Local
Orbit measurement device	All of the rf BPMs	rf BPMs (1/sector)	rf BPMs X-ray BPMs
Correctors	All correctors (320)	Subset of correctors	Local bump
Specified orbit measurement resolution	25 μm	25 μm	1 μm
Achievable resolution	5 μm	5 μm	1 μm
Required range of correction	± 20 mm	± 500 μm	± 100 μm

Table 2: Parameters for the APS BPMs in the storage ring (SR), injector synchrotron (IS), and insertion devices with 12-mm and 8-mm gap.

	SR	IS	ID (12 mm)	ID (8 mm)
Button radius (cm)	0.5	0.5	0.2	0.2
S_x (cm^{-1})	0.57	0.70	2.08	3.44
S_x^* (cm^{-1})	0.58	0.70	1.63	–
S_y (cm^{-1})	0.53	0.58	1.51	1.47
S_y^* (cm^{-1})	0.55	0.57	1.54	–

(*: measured sensitivity)

EXPERIMENTAL SETUP

The schematic of the experimental setup for BPM resolution measurements is shown in Fig. 1. We used two BPMs in cell 15, separated by a few cm, in the upstream of ID 15. The buttons on these BPMs are of the same type as those on the APS storage ring and injector synchrotron. The filter-comparators (FCs) are located in the tunnel below the BPMs. The raw button signals are filtered through a bandpass filter of 10 MHz with the center frequency of 352 MHz. The output signals from the filter-comparators, Δ_x, Δ_y, and Σ, are sent to the signal conditioning and digitizing units (SCDUs) in the VXI crate in the instrumentation area outside the tunnel. The fourth cable carries the test/trigger signal.

The SCDUs have analog and digital sections. The analog section processes the rf signals from the FC and outputs the normalized signal V_a with the full scale of ±1V. If we let V_d be the output from the 12-bit digitizer, we would have

$$V_a[\text{Volts}] = \frac{V_d[\text{Counts}]}{2{,}048} = \frac{16}{\pi}\tan^{-1}\left(\frac{\Delta}{\Sigma}\right), \tag{1}$$

where Δ is either Δ_x or Δ_y. Using the first-order approximation and putting

$$\frac{\Delta}{\Sigma} \approx Sd \qquad (d = \text{beam position}), \tag{2}$$

we have

$$d[\mu\text{m}] = 0.96\frac{V_d[\text{Counts}]}{S[\text{cm}^{-1}]}. \tag{3}$$

S is the BPM sensitivity as shown in Table 1. With $S \approx 0.55$ cm^{-1} for the APS storage ring, we would have the maximum raw resolution of approximately 1.7 μm. For the ID BPM with 12-mm gap, it would be 0.6 μm.

The SCDU can be remotely set up such that only either Δ_x or Δ_y can be processed, or both of them in an alternate manner. The analog output V_a is then digitized with 12-bit resolution and stored in registers. These registers are read periodically by the memory scanner in the same crate. A maximum of nine SCDUs can be installed in each VXI crate. The VXI crate housing the SCDUs and the memory scanner is connected via GPIB to a computer for data acquisition and control.

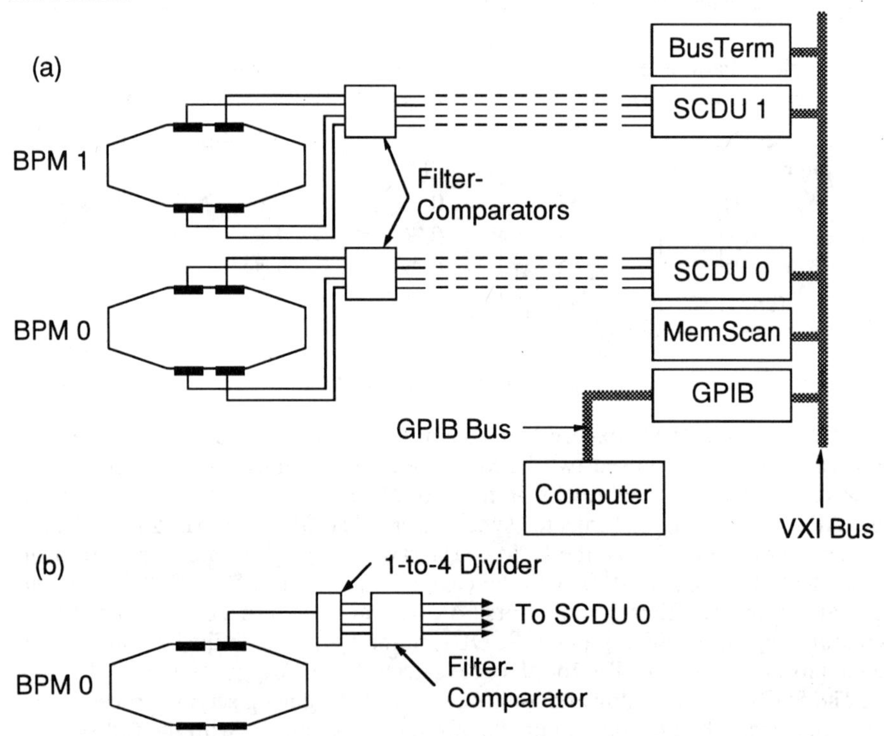

Figure 1. The experimental setup for the BPM resolution and drift measurements: (a) normal connection and (b) 1-to-4 voltage divider connected to a single button was used to reject the actual beam motion for BPM 0.

RESULTS

Figure 2 shows the results of resolution measurement as a function of the beam current with 1/3 of the ring filled. Each data point is the average of N = 2048 samples, which corresponds to 4096 turns in the x-y toggling mode. One milliampere of stored current corresponds to 2.8 nC and one count corresponds to approximately 1.5 μm. The deviation in the x-position is consistently high regardless of the bunch current for both SCDUs, which is due to the actual slow beam motion. The deviation in the y-position shows difference between the two SCDUs. This might be due to the sensitivity in timing to trigger the SCDU 0. With larger current, the effect gets smaller and the resolution is comparable to SCDU 1.

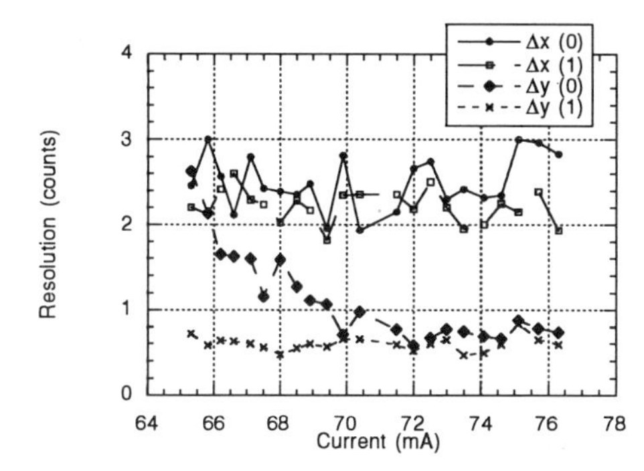

Figure 2. The result of BPM resolution measurement with 1/3-fill and N = 2048 samples. One count corresponds to approximately 1.5 μm.

The result of BPM resolution measurement with a single bunch is shown in Fig. 3. The stored beam current was 5.8 mA, which corresponds to 16.2 nC. Assuming noise reduction scaling proportional to $1/\sqrt{N}$, we can write beam position measurement resolution $\delta_{x,y}$ as

$$\delta_{x,y}[\text{Counts}] \approx \frac{10}{\sqrt{N}}. \tag{4}$$

The deviation of the measured resolution from Eq. (4) is due to the actual beam motion, which is larger in the x-plane. Measurements with different single bunch currents (2 mA $< I <$ 8 mA) indicated little change in resolution, possibly due to the limiting amplifier in the processing electronics.

Let us now convert Eq. (4) to the unit of $\mu\text{m}/\sqrt{\text{Hz}}$. The frequency response of the simple averager on the SCDU can be written as

$$|H(f)| = \frac{1}{N}\frac{\sin(N\pi f/F_s)}{\sin(\pi f/F_s)}, \tag{5}$$

where F_s is the sampling frequency. In the x-y toggling mode with a single bunch, it is equal to 178 kHz for the ESRF. Then, the bandwidth (-3 dB) corresponds to approximately $Nf = 1.39F_s/\pi \approx 79$ kHz. From Eq. (3), the BPM resolution is

$$\delta_{x,y}\left[\frac{\mu\text{m}}{\sqrt{\text{Hz}}}\right] \approx 10\cdot\frac{0.96}{S}\cdot\sqrt{\frac{1}{Nf}} \approx 10\cdot\frac{0.96}{0.7}\cdot\sqrt{\frac{1}{7.9\times10^4}} \approx 0.05. \tag{6}$$

We used the sensitivity $S \approx 0.7$ cm^{-1} for the ESRF. For the APS storage ring with larger circumference and with vacuum chamber dimension slightly larger than the ESRF, we would have $\delta_{x,y} \approx 0.07$ μm/$\sqrt{\text{Hz}}$.

The BPM resolution can be improved by increasing the BPM sensitivity S. For the APS miniature ID BPM, the sensitivity is about 1.6 cm^{-1}, which will give 0.02 μm/$\sqrt{\text{Hz}}$ resolution if the scaling relation in Eq. (4) is assumed. Since the BPM resolution is inversely proportional to $\sqrt{F_s}$, it can also be improved by increasing the sampling frequency F_s. The minimum sampling time by SCDU is 1.2 μsec, and therefore, with the 3.68-μsec revolution time in the APS storage ring, there can be as many as three triggers per turn if the bunches are grouped into three equally spaced clusters.

Measurement of the Σ signal as a function of the single bunch current is shown in Fig. 4. The logarithmic fit to the data gives

$$\Sigma\,[\text{Counts}] \approx 570 + 580\,\log(I[\text{mA}]). \tag{7}$$

As a comparison, we made separate measurements in the laboratory using a CW rf source (-10 dBm with 40 dB amplifier) feeding a 1-to-4 divider connected to a filter-comparator. The result indicated seven counts of resolution with N = 2. Average Σ was 930 counts, which is equivalent to 4.2 mA according to Eq. (7). This is in good agreement with Eq. (4).

Figure 5 shows the BPM resolution measurement with eight equally spaced bunches in the ring as a function of N. The SCDU 0 was balanced with a 1-to-4

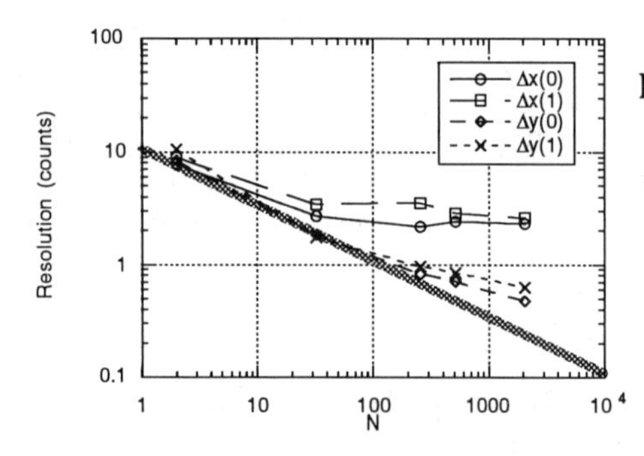

Figure 3. The result of BPM resolution measurement with a single bunch (I = 5.8 mA). The thick line indicates $1/\sqrt{N}$ scaling for noise reduction.

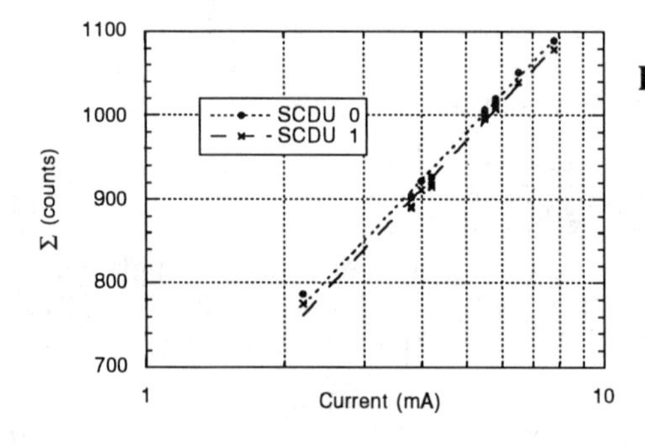

Figure 4. Measurement of the Σ signal as a function of the single bunch current.

divider as shown in Fig. 1(b), and therefore shows worse resolution than SCDU 1. However, the resolution of SCDU 0 keeps improving after $N = 256$, while that of SCDU 1 does not. Resolution of about 0.9 counts at $N = 2048$ corresponds to $0.13\,\mu m/\sqrt{Hz}$, which under normal connection would be about $0.04\,\mu m/\sqrt{Hz}$. Similar measurements with $I = 60$ mA in 16 equally spaced bunches showed roughly the same behavior without improvement in resolution.

The long-term drift of the BPM electronics was measured with the SCDU 0 using the 1-to-4 divider. The stored beam current was $I = 102.5$ mA with 1/3-fill. In order to reduce the short-term fluctuation, the highest number of averages $N = 2048$ was used. The result in Fig. 6 shows the long-term drift is about 1.5 counts/hr, which is equivalent to 2 μm/hr.

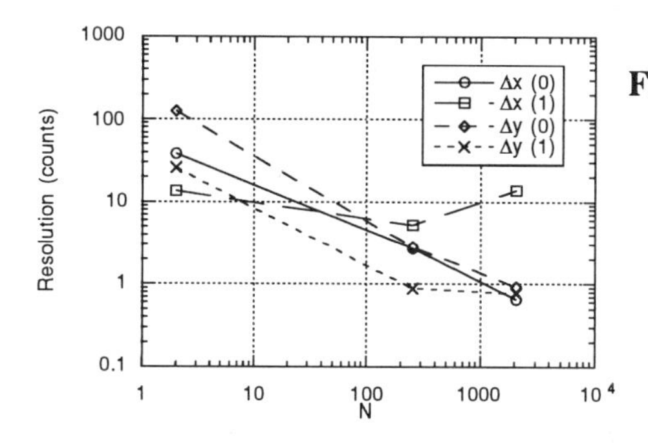

Figure 5. The result of BPM resolution measurement as a function of N with multi-bunch (8) configuration and $I = 30$ mA. The SCDU 0 was balanced with a 1-to-4 divider.

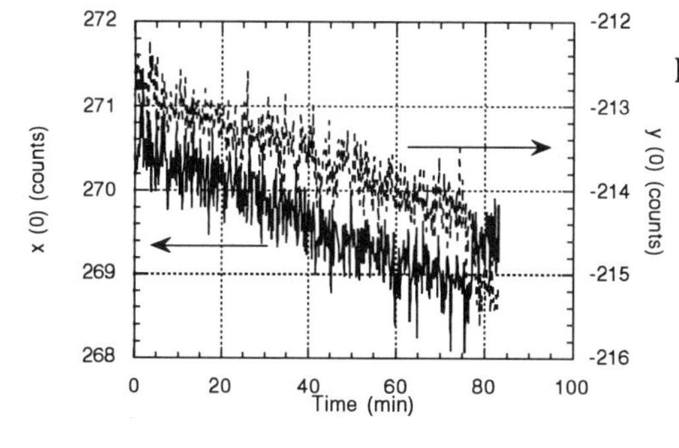

Figure 6. Electronics drift measurement with 1/3-fill, $I = 102.5$ mA, $N = 2048$. The SCDU 0 was balanced with a 1-to-4 divider.

SUMMARY AND DISCUSSION

In this work, we described measurements of resolution and long-term drift of the Advanced Photon Source beam position monitor electronics using the charged particle beam at ESRF. Several bunch configurations, single, 8, 16, and 1/3-fill, as well as various bunch currents were used for the resolution measurements. The long-term drift was measured with the stored beam current of 102.5 mA with 1/3-fill.

With single bunch current of 5.8 mA (Q = 16.2 nC), $0.05\mu m/\sqrt{Hz}$ of resolution was obtained. This is equivalent to $0.07\,\mu m/\sqrt{Hz}$ for the APS storage ring. Changing the bunch current between 2 and 8 mA showed little difference in the resolution, possibly due to the limiting amplifier in the processing electronics. Scaling this result to the miniature ID BPM in the APS storage ring, the resolution would be $0.02\,\mu m/\sqrt{Hz}$ due to higher BPM sensitivity. This result is in good agreement with separate laboratory measurement conducted with a CW rf source.

The long-term drift of the BPM electronics independent of the actual beam motion was measured at 2 μm/hr, which settled after approximately 1.5 hours. This drift can be attributed mainly to the temperature effect.

In the APS storage ring, the rf BPMs will be used for global and local beam position feedback to stabilize the particle and X-ray beams. The correction bandwidth is expected to be approximately 100 Hz with the projected sampling frequency of 4 kHz. Within this bandwidth, the resolution would be better than $0.07\times\sqrt{100} = 0.7$ μm for regular BPMs and 0.2 μm for miniature ID BPMs. Beam position perturbation larger than this resolution will be corrected by feedback.

ACKNOWLEDGMENT

The authors would like to thank the ESRF staff for their hospitality and collaboration during our stay in Grenoble, France, for this work. Thanks also go to J. Galayda, G. Decker, and A. Lumpkin for their continued support and interest.

REFERENCES

1. Y. Chung, "A Unified Approach to Global and Local Beam Position Feedback," *Proceedings of the 4th European Particle Accelerator Conference*, London, 1994.
2. Y. Chung, "Beam Position Feedback System for the Advanced Photon Source," *Proceedings of the 5th Beam Instrumentation Workshop,* Santa Fe, New Mexico, 1993.
3. G. Decker and Y. Chung, "Progress on the Development of APS Beam Position Monitoring System," *Proceedings of IEEE Particle Accelerator Conference*, San Francisco, 1991, p. 2545.
4. Y. Chung, "Theoretical Studies on the Beam Position Measurement with Button-type Pickups in APS," *Proceedings of IEEE Particle Accelerator Conference*, San Francisco, 1991, p. 1121.
5. E. Kahana, "Design of the Beam Position Monitor Electronics for the APS Diagnostics," *Proceedings of the 3rd Beam Instrumentation Workshop,* Newport News, Virginia, 1991, p.235.
6. E. Kahana and Y. Chung, "Test Results of a Monopulse Beam Position Monitor for the Advanced Photon Source," *Proceedings of the 4th Beam Instrumentation Workshop,* Berkeley, California, 1992, p. 271.
7. E. Kahana, Y. Chung, A.J. Votaw and F. Lenkszus, "Configuration and Test of the APS Storage Ring Beam Position Monitor Electronics," *Proceedings of the 5th Beam Instrumentation Workshop,* Santa Fe, New Mexico, 1993.

SSRL Beam Position Monitor Detection Electronics*

J. Sebek, R. Hettel, R. Matheson, R. Ortiz, J. Wachter
Stanford Synchrotron Radiation Laboratory, P.O. Box 4349, Bin 99,
Stanford, CA 94309-0210

Abstract

As part of a program to improve its orbit stability SSRL is redesigning its detection electronics for its beam position monitors (BPMs). The electronics must provide highly reproducible positional information at the low bandwidth required of an orbit feedback system. With available commercial technology, it is now possible to obtain highly resolved turn by turn information so that this electronic module can also be used to measure beam dynamics. The design criteria for this prototype system and performance of the analog section of the processor is discussed.

INTRODUCTION

SPEAR is a 3 GeV electron storage ring used for synchrotron radiation. It was originally built as an $e^- - e^+$ collider for high energy physics, and its BPM detection electronics was designed to differentiate between the signals from the two particles. All of the BPM inputs are multiplexed into one large switching matrix and processed by one set of electronics. We are redesigning the electronics to improve processor speed, dynamic range, and resolution(1)(2). In addition to providing highly resolved positional information under normal operation, the system must be able to detect low current orbits for injection studies, etc. Table 1 lists the relevant parameters used for the design and Fig. 1 gives the flow of the processed signal.

Table 1. SPEAR BPM Parameters

Parameter	Symbol	Value	Unit
Energy	E	3	GeV
Radio Frequency	f_{RF}	358.54	MHz
Harmonic Number	h	280	
Revolution Frequency	f_{rev}	1.2805	MHz
Nominal Beam Current	I_{nom}	10–100	mA
Number of BPMs		40	
Resolution		10	μm
Channel Isolation		> 80	dB
Detector SNR @I_{nom}	SNR	> 126	dB/Hz
Dynamic Range		40	dB

*Work supported by the Department of Energy, Office of Basic Energy Sciences, Division of Material Sciences.

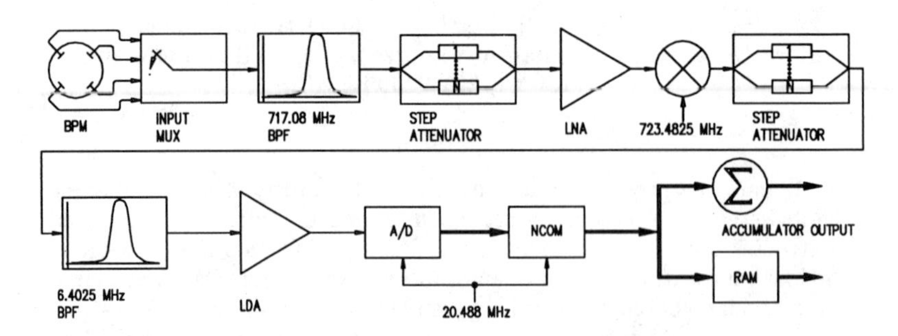

Figure. 1. Processor Block Diagram

BPM SIGNAL SPECTRUM

The periodic nature of the beam in a storage ring means that the signal spectrum on a BPM will be periodic. The spectrum of the 'reference particle' is a sequence of signals at the harmonics of the fundamental frequency, f_{rev}. The amplitudes of the individual harmonics are determined by the frequency response of the pickup electrode. For a bunched beam of many particles in many buckets, the spectrum becomes slightly more complex. Although the locations of the frequencies do not change, their amplitudes now depend on bunch shape and fill pattern. In all cases, the amplitudes of the harmonics incident on the BPM are multiplied by the Fourier transform of the bunch length. For multiple bunch fills, the signals from the various bunches add coherently and modulate the amplitudes of the harmonics with the Fourier transform of the bucket fill pattern. Since the ring can only contain a finite number of bunches (the harmonic number, h), this modulation repeats with f_{RF}. In particular, all harmonics of f_{RF} carry the information of the DC current of the beam. These are the only harmonics of f_{rev} guaranteed to be non-zero for arbitrary fill patterns.

The spectrum at each harmonic is almost, but not quite a pure frequency. Transverse oscillations of the beam give rise to amplitude modulation of the BPM signals and create betatron sidebands around the f_{rev} harmonics. Longitudinal oscillations give rise to phase modulation and synchrotron sidebands.

RF SIGNAL PROCESSING

The periodic nature of a storage ring determined our choice of a harmonic processing system that detects the power in an appropriate frequency bandwidth. Since we determine the beam position by the difference over sum technique(3)(4), we multiplex the signals as early as possible to minimize errors in the signals due to electronic variations. Our RF processing is designed to provide a high quality, narrow bandwidth signal for our f_{IF}.

Processing Frequency

For the reasons given above, we process a harmonic of f_{RF}. The decision as to which harmonic to process was a tradeoff of engineering considerations. SPEAR has several BPMs near the RF cavities, where the evanescent fields from these cavities provide a strong beam-independent signal at the RF, so we rejected processing f_{RF}. Although our BPM buttons are more sensitive to the higher beam frequencies, we chose the second RF harmonic, 717.08 MHz, for two reasons. First, our signal processing electronics will be housed in the control room, typically 100 m from the buttons, and attenuation due to the cable length greatly increases with frequency. Second, the size of our beam pipes gives a typical vacuum chamber cutoff frequency of about 1 GHz. Discontinuities and structures in the vacuum chamber support higher order modes at these frequencies that contaminate the fundamental signal on the BPMs produced by the image charges of the beam.

Signal Multiplexing

The main purpose of the BPM system is to provide highly resolved information about the orbit of the beam. For our vacuum chamber size, resolution of 1 micron beam motion means a variation in the difference signals of about 50 ppm from the 4 buttons of each BPM. The design of the electronics attempts to minimize the potential for systematic errors that could prevent high resolution measurements. Therefore, we have multiplexed as much of the button processing as possible. The current system multiplexes all BPM buttons into a single processor. We will initially commission the new processor with this same arrangement, but then may build one processor per BPM.

The BPM signals are multiplexed at the input. Since this is the only part of the circuitry that is not common to all of the buttons, we desire the technology with the most consistent and repeatable characteristics. We chose GaAs FETs over PIN diodes because of the independence of the FET video impedance with respect to signal level. Standard isolation per switch at our processing frequency is only about 35 dB, so our design cascades three absorptive switches in series for each button. The selected signal passes through its three switches to the processor, while the other buttons see the 50Ω on chip terminator of their first switch. Based on the signal level out of the button, we do not expect to violate any voltage or power breakdown levels of the devices. During initial commissioning of the processor, the multiplexer will be in the control room, but we will study the feasibility of placing it in the ring. If it can be adequately shielded from radiation, the multiplexer will only require one high quality cable from each BPM to the control room. The only signal paths that will vary from button to button will then be a short cable to the multiplexer and the switches themselves.

RF Conditioning

The processor will heterodyne the RF signal down to an f_{IF} of 6.4025 MHz, where its amplitude will be measured. We use a dielectric resonator band pass filter, a 5 section Chebyshev filter with a 1% bandwidth and 7 dB insertion loss at 717.08 MHz, to limit the out of band input power. In order to condition this signal for a 10 dBm image reject mixer, we want to set its power level to a maximum value of -10 dBm. We use a combination of a FET step attenuator and a low noise, fixed gain amplifier to keep this level in range. Measurements show that, for our normal operating range of currents, we will always have a -10 dBm signal at the input of our mixer. We chose a fixed gain amplifier and attenuator arrangement because of its overall lower noise figure than that of a variable gain amplifier. We chose a step attenuator over an voltage controlled variable attenuator because we require the more constant attenuation provided by the digital control rather than the fine adjustment offered by variable control. The operating values of the input power were chosen to be well below the 1 dB compression point of the amplifier and the 3rd order intercept of the mixer in order to maximize linearity of the system.

IF SIGNAL PROCESSING

Our f_{IF} was selected so that, using available commercial technology, we could digitize it directly without sacrificing the resolution of our IF signal. Proper selection of frequency within this range gives us highly resolved, wide bandwidth signals with a minimum of processing overhead.

Analog Considerations

Recent technological advances have produced monolithic 20 MHz, 12 bit A/Ds at reasonable prices. Therefore we tried to select our f_{IF} below 10 MHz, the frequency above which the A/D performance starts to roll off. The lower f_{IF}, however, the harder it is to reject through filtering the mixer image of the desired frequency. Our RF bandpass filter rejects this image at the input by $\sim$40 dB, but by using an image-reject mixer, we reduce the IF image by another 30 dB. We chose f_{IF} to be 6.4025 MHz as a reasonable compromise where very high quality commercial video opamps and digitizers are available while good image rejection is still possible with simple circuitry.

Digital Considerations

For any BPM system which is used to control the beam, one needs to digitize the information at some point and pass it on to other elements of the control system. Available technology now makes it reasonable to digitize the signal at the IF. This allows great flexibility in terms of selecting signal bandwidths to optimize SNR, response time, etc., for various applications. In particular, in addition to providing the information needed for our low-bandwidth orbit feedback, this technique allows us to use this processor to detect, with high accuracy, single turn phenomena for machine physics studies. By digitizing the IF, we also have a system with only one non-linear component, the mixer, thereby improving our system linearity.

Since our digital signal processing starts with the digitization of the IF, we have chosen it, and therefore the mixing frequency, to optimize this processing. When Fourier transforming band-limited data, one assumes that the signal that is transformed is a portion of an infinitely periodic signal. This is not the case for signals with arbitrary frequency content, and this periodic assumption 'contaminates' the transform with non-existent frequency components that are needed to make the sample periodic. To minimize this problem, one applies a 'window' to the data which de-emphasizes the ends of the data sample. This windowing also contaminates the data, but hopefully less than an unwindowed sample. The signals we measure, however, are extremely periodic and we have access to the ultimate system clock, f_{RF}. By making use of this periodicity, we can choose to sample a signal that is periodic with respect to our clock, so that this signal truly is a portion of an infinitely periodic signal. With this method, we get a faithful frequency decomposition of the signal without windowing. Since our signal is coherent while noise is incoherent, N samples per revolution will increase our SNR by $\sqrt{N}$. Therefore, we chose our digitization frequency as $16f_{rev}$, or 20.488 MHz.

When a perfect periodic signal is digitized, the output codes will have a periodic fixed quantization error. To minimize this error, the digitizer should sample the signal at as many values as possible. This means that the periodicity of the sampler should be as relatively prime as possible to the periodicity of the signal. Based on the criteria of image rejection, analog signal fidelity, periodicity, and digital fidelity, we chose $5f_{rev}$, 6.4025 MHz, as our f_{IF}.

IF Analog Conditioning

The remainder of the analog processing optimizes the signal for the digitizer. We use a lumped element band pass filter at 6.4 MHz to pass the output of the mixer. This filter needs only act as an anti-aliasing filter for the digital processing that follows. Its specifications are not very strict since the nearest aliased frequency of f_{IF} is 14.0855 MHz. We set its bandwidth to $\sim f_{rev}$, since we want the ability to observe signals change that quickly. (In fact, we have been very conservative in all of our analog filtering specifications. Since each revolution harmonic carries the same spectral information, we are detecting synchronously with the ring RF, and the button response is essentially constant over the small bandwidths we are considering, the only contamination we would get from aliased signals is a uniform increase or decrease in the detected signals of all buttons. The major danger in this is that two signals may be exactly out of phase and cancel, but contamination on the order of $\sim$40 dB would not affect our detection resolution.) In the IF we again use a combination of a digital step attenuator and fixed gain amplifiers. Although there are variable gain video opamps with the same noise performance as fixed gain opamps, we are more confident in keeping the system gain constant with the step attenuators. We use a low distortion, low noise amplifier to boost the IF signal to the 1V nominal input value desired by the digitizer.

IF Digital Processing

We digitize the data at a high rate to improve the SNR of the system, but it would be very expensive to keep and process the entire Nyquist bandwidth. From a beam dynamics point of view, all desired information is stored within a bandwidth of f_{rev}. Further, it is not clear what information we would ever need to investigate that happens faster than f_{rev}. We therefore use a digital mixer, the Harris HSP45116 numerically controlled oscillator/modulator (NCOM), to beat our f_{IF} down to baseband once per revolution period.

The NCOM takes as input the stream of 12 bit digital words from the A/D, internally multiplies them with the sine and cosine of f_{IF}, accumulates them 16 samples at a time, and then outputs 16 bit words that represent the amplitudes of the quadrature components of f_{IF} during the previous f_{rev} period. (Its rejection of the other harmonics passed by the analog anti-aliasing filter is $\sim$90 dB.) We then use an AMD29200 32-bit microcontroller to accumulate these amplitudes and store them in DRAM. The accumulated sum is the filtered value of f_{IF}, the width of which is determined in software by the number of samples taken. We can look at single turn dynamics of the BPM signal by analyzing the stored NCOM output. One can observe the betatron oscillations by measuring the turn-by-turn amplitude modulation of the data, and can observe the synchrotron oscillations by measuring the phase modulation. The microcontroller will also handle the low level control of the switches and attenuators, and communicate with the rest of the crate via high level commands. It can even implement a phase-locked loop on the NCOM that can keep the Q signal zeroed to reduce the amount of data needed to transfer to the rest of the control system during normal operation.

SYSTEM TIMING

The timing generation of the system is straightforward. To generate the synchronous signals for our clocks and local oscillators, we divide down either f_{RF} or $2f_{RF}$. (For our prototype system, we used direct digital synthesis (DDS) chips to generate the locked frequencies from f_{RF}.) Switching of electrodes will all be done at increments of the revolution period and the processor will sample each electrode for multiples of this fundamental period. These values can, of course, be dynamically changed through software. We plan to package this controller in a format that will interface to a VME environment. Once this decision is finalized, we will use standard interface logic to connect the processor to the control system.

SYSTEM TEST RESULTS

We were able to test a prototype version of the analog portion of the processor during SPEAR's 1994 run by parasitically observing signals from one BPM with 55 mA of current in the machine. At this current we required 29 dB attenuation in the signal path to set our IF signal at the 1V level desired by the A/D, so that our measured analog path SNR will hold down to ~2 mA. For a 10kHz RBW, our signal measured ~70 dB above the noise floor at the IF output. Our signal was clean enough to see the amplitude and phase oscillations on the beam. If we need a greater SNR, we can trade off with the current system dynamic range. We saw no evidence of any problems due to processing at a harmonic of f_{rev}. Direct feedthrough of f_{IF} was ~58 dBc, which we feel can be further reduced by addition of appropriate filters. The other noticeable product, probably a mixer IMD was ~65 dBc. As discussed earlier, neither of these should be a problem.

We are continuing our development of the processor. The one area we feel we can still improve significantly is our multiplexing switch. We achieved ~60 dB isolation between channels, but subsequent bench tests lead us to believe that with improved layout and shielding we may meet our 80 dB design goal. We have yet to characterize the repeatability of the switch, but feel that the best way to do that is with the finished processor. We also will fine tune our IF stage amplifier and filter. Finally, we must still input this signal into the digital processor, but based on the signal purity we observed from the beam tests, we are confident that most of the problems with it can be debugged on the bench.

ACKNOWLEDGEMENT

We would like to thank Max Cornacchia for the encouragement and support he has given this project, H.-D. Nuhn for many helpful discussions, and Theresa Sims, Ramona Theobald, and Kane Zuo for their assistance.

REFERENCES

1. R. Hettel, J. Corbett, D. Keeley, I. Linscott, D. Mostowfi, J. Sebek, and C. Wermelskirchen, "Design and characterization of the SSRL orbit feedback system," in *EPAC Conf. Proc.*, EPAC, 1994.
2. J. Sebek, R. Hettel, H.-D. Nuhn, and B. Scott, "SSRL beam position monitors," in *EPAC Conf. Proc.*, EPAC, 1994.
3. R. Shafer, "Beam position monitoring," in *AIP Conf. Proc.* *249*, pp. 601–636, AIP, 1992.
4. R. Littauer, "Beam instrumentation," in *AIP Conf. Proc.* *105*, p. 430, AIP, 1983.

Experience with Commissioning and Operational Performance of The New Electron-BPM Processing Electronics For The Daresbury SRS

R. J. Smith, J. R. Alexander
DRAL Daresbury Laboratory, Daresbury, Warrington WA4 4AD, UK.

Abstract

A major upgrade of the electron BPM signal processing and readout system has recently been installed and commissioned on the Daresbury Synchrotron Radiation Source (SRS). It provides the improved orbit measurement characteristics required for the global feedback system currently being implemented. The system is based on the amplitude detection of sum (Σ) and difference (Δ) signals for both horizontal and vertical position, generated from the BPM pickup buttons by 180° RF hybrids. Dual channel processing units, for horizontal and vertical position signals, are located at each of the 16 BPM locations together with 16-bit ADCs providing the input to the data acquisition system. New processing electronics has been developed which gives measurement repeatability and noise levels of around 1μm, while covering a 30dB range of electron beam current with only a 5μm change in indicated beam position. The system also has the flexibility to provide other general BPM functions, such as tune measurement and first turn injection signals. Prior to installation, each set of processing electronics was bench commissioned and calibrated using test procedures automated by LabVIEW®. The system has now completed beam performance tests which confirm that it meets its specification, and accelerator physics experiments have begun to implement orbit stabilisation for user beams.

INTRODUCTION

The SRS electron storage ring is equipped with 16 horizontal and 16 vertical electron beam position monitor (BPM) vessels: one of each in each machine straight. The monitors use 2 or 4 capacitive button-type pickups, the outputs of which are combined in adjacent passive 180° RF hybrids, to give sum (Σ) and either horizontal or vertical difference (Δ) signals. These signals are amplitude detected and digitised, and the beam position determined from (K x Δ/Σ) - O where K and O are calibration and offset constants measured for each combination of BPM vessel, processor unit and hybrid. The use of hybrids ensures a wide band linear response and provides beam-current independent measurement of beam position.(1)

As with many other synchrotron radiation sources throughout the world, the SRS is undergoing an upgrade program aimed at providing stabilisation of the closed orbit, via feedback control of beam position.(2) This has necessitated the development of BPM signal processing units with significantly improved performance in terms of position sensitivity, repeatability, dynamic range and acquisition rate. This has been achieved, the units have been installed, and commissioning of the position control systems is at an advanced stage.

Figure 1 shows the new arrangement of signal processing for each pair of BPMs. The RF hybrids are unchanged, but the old arrangement of a coaxial relay RF multiplexer routing the pairs of Σ and Δ signals to a two-plane central detector has been replaced by local signal processors with associated ADC and multiplexer.

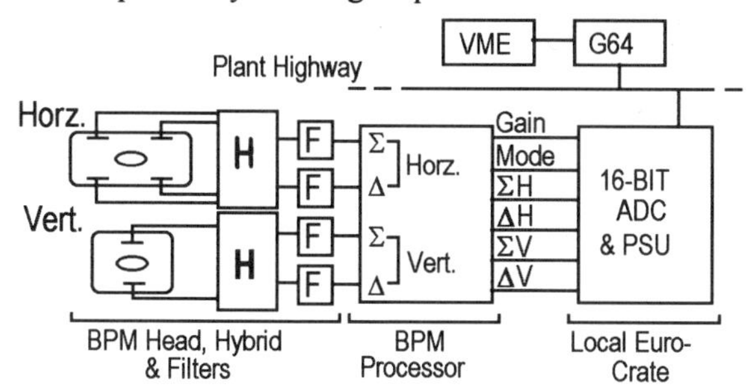

Figure 1. Showing layout of a single processor and its associated ADC Crate with interconnection to the plant highway.

The ADC card, which was designed in-house, multiplexes the 4 detected signals to a 16-bit ADC, provides control of the signal processor's gain and mode (data-acquisition or calibrate), and interfaces to the standard SRS plant control digital highway.(3) Four control highways are used, driven from two G64 bus systems. Microprocessors in these scan the BPMs, calculate average Σ and Δ values if required, calculate Δ/Σ, and store these values in a database which is accessed by the beam-steering control processors (based in VMEbus) over Ethernet.

NEW BPM SYSTEM PROCESSING UNITS

The new detector electronics is housed in a multi-compartment RF case.(4)

Figure 2. New Daresbury Distributed BPM Processor Unit.

Major Specifications for the system are:

Dynamic Range	10-400mA :low gain
	1-40mA :high gain
Position Sensitivity	≤ 0.05mA.mm
Resolution	≤ 10μm
Stability (for fill period)	≤ 20μm
Absolute Cal. Accuracy	≤ 100μm, rms

Each processor unit has two channels, allowing it to cater for one horizontal and one vertical BPM input per straight. The hybrid outputs are filtered before being down-converted to low frequency, where they are low-pass filtered, and detected using a fast comparator and multiplication stage. Two selectable gain ranges provide for multi and single-bunch SRS operation. A single unit (excluding input filters) can be seen below in Fig. 2.

Down Conversion Stage

The down converter stage shown in Fig. 3. is a direct derivative of the phase sensitive amplitude detector system previously installed on the SRS. The hybrid pre-processed Σ and Δ signals are initially narrow band-pass filtered (≈1%) to suppress the 3MHz orbit frequencies generated during SRS single bunch operations. These signals then pass into the inputs of the down converter stage along with a local oscillator (Lo) signal, set at 500.5MHz (i.e. machine frequency + 500kHz). No signal path RF amplification is required but a high output low noise RF amplifier is used for the local oscillator drive to provide sufficient level for the RF mixer.

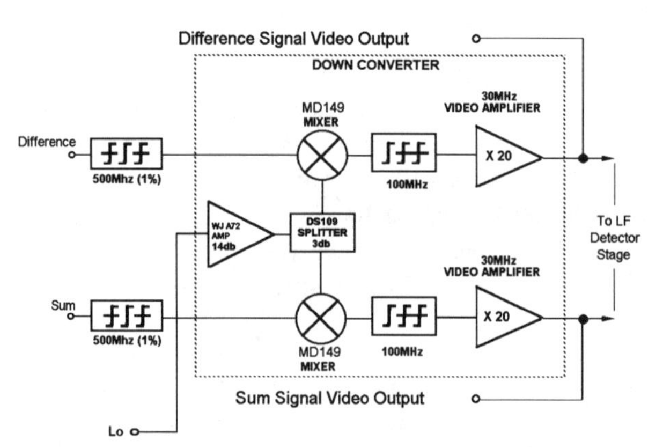

Figure 3. Down Converter Stage showing component layout and signal connections. Video outputs provide for facilities such as direct measurement of tune and wideband studies by switching in of different Lo frequencies.

The use of this basic building block as a two channel frequency converter, shows the following performance, with a noise figure of 3dB (Fig. 4).

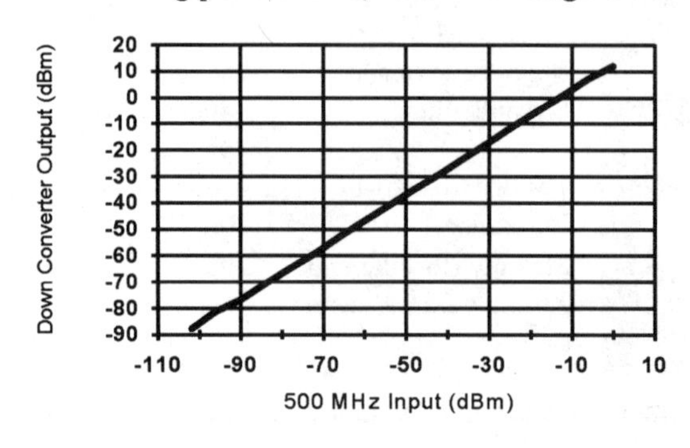

Figure 4. Response of Down Converter Stage (IF = 500kHz) showing 100dB linear range for constant Lo level in heterodyne mode.

Low Frequency (LF) Phase Sensitive Detector Stage

The down converted Σ and Δ signals are fed via video amplifiers to the key component of the new system, shown in Fig. 5. A commercial, wide-band analogue multiplier (AD539) is configured as a dual channel phase sensitive detector, which processes both signals. The larger Σ signal is used to drive a fast (ecl) amplitude comparator to provide a synchronous reference signal for the multiplier. This ensures constant amplitude and phase (<<5° change) over an input range of some 20 to 3000mV (peak) at the LF detector inputs. The overall dynamic range of the units however is restricted to typically 30dB in this application to allow operation from low power ±5.0 Volt linear supplies.

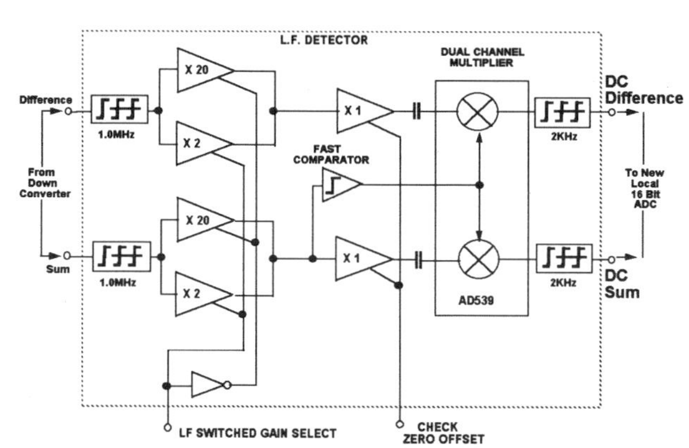

Figure 5. Dual Channel Low Frequency Detector unit showing general system layout for signal paths. Multiplexed gain selection amplifiers allow low gain multi-bunch (x2) and high gain (x20) operation.

The measured dynamic operating range of a typical unit is shown by Fig. 6, for one channel operating in multi-bunch mode. The output is expressed in terms of beam offset. Signal levels (Δ/Σ) corresponding to 1.6mm offset in the new detectors are used to illustrate the response of the detector for real storage ring BPM vessels.

COMMISSIONING

Due to the high precision and resolution required of the new system, each processor unit was commissioned and calibrated on the bench prior to installation. This procedure was doubly necessary, since the local oscillator system, supplying all the processors is distributed through a series coaxial and directional coupler network. This meant that local oscillator power levels vary (±3dB) from the optimum level from location to location. Each processor unit was therefore calibrated for the location into which it was to be installed, and the appropriate power level, measured from the installed Lo supply network, used during testing. In order to complete the calibration of all 32 channels of the new system in the time available, an automated calibration set-up using the graphical data acquisition software LabVIEW® was developed.(5) Virtual Instruments (VIs) were written to control several GPIB instruments, to simulate beam signals, measure the BPM processor responses and present the calibration data graphically. Using simulated input levels with a fixed ratio of 10:1 for Σ:Δ, plots were produced for every unit in high and low gain to assess the operational dynamic range (±5μm). These were scaled to reflect a 'standard' calibration value for the SRS BPM vessels. The channel calibration figure, which varies between processors as a result of component and construction tolerance, was also found from this data and combined with the actual vessel calibration figures for use during the conversion of the measured data to actual position from vessel centre. Typical channel to channel variations observed during calibration produced a spread of some 200μm. Overall variations in operational dynamic range were found to be from 28dB to 36dB, with 30 to 32dBs being more typical results for most channels. A typical calibration curve produced from recorded LabVIEW® data is shown in Fig. 6.

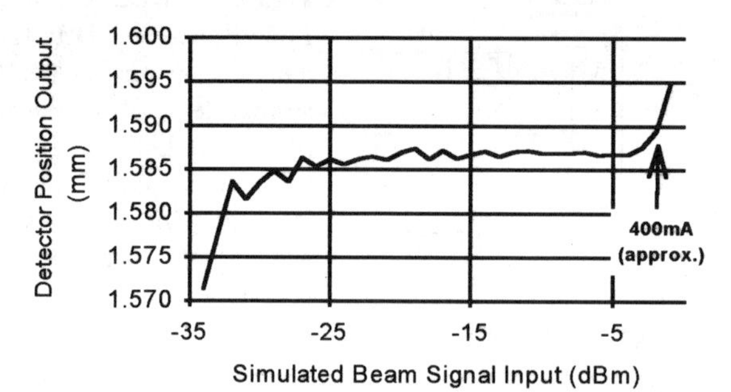

Figure 6. Typical Processor response calibration curve showing detector operation over 30dB range (±5μm) for simulated beam Σ and Δ signals (automated data taken using LabVIEW® virtual instrumentation).

Calibration in this way allows the Σ channel gain to be adjusted to set the optimum operational range for beam current, without compromising signal levels for the Δ channel. For the SRS, the gain was selected for a maximum beam current of around 400mA, at which point the processor unit Σ output becomes saturated due to the low voltage power supplies used. At the lower end, of beam current, performance falls off due to insufficient Σ level to drive the LF. detector fast comparator stage. All channels within the installed system are set to have the same gain. It is possible that a modification of the Σ channel gain may be necessary for the vertical BPMs, as signal levels here are approximately 6dB higher than those from the horizontal. This is largely due to the fact that the vertical BPM pickups require only single element hybrids, whereas those in the horizontal plane use a multi-element unit. Typically the changeover point from high to low gain for single to multibunch operation is approximately 35mA.

OPERATIONAL PERFORMANCE

The installation of the new electron BPM processor system was accompanied by a new suite of software on the SRS controls computer in order to take and present the data now available. The BPM system, now operating continuously with data update rates of 2Hz is able to present a view of beam movements over an entire beam lifetime, and demonstrate the beam position movements and drift over that period at every BPM location. An example of this is shown by Fig. 7 below.

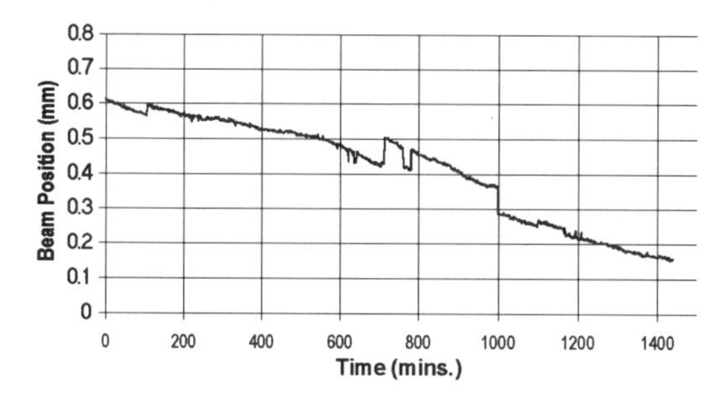

Figure 7. Plot of data from a horizontal BPM, showing Beam Movement over a 23 hour user beam lifetime.

With commitments to implement global horizontal feedback in the near future, the systems noise performance is of prime importance. This is demonstrated below by Fig. 8 for the horizontal plane with 16 averages taken for each reading, with 600 readings taken at a rate of 1 per second.

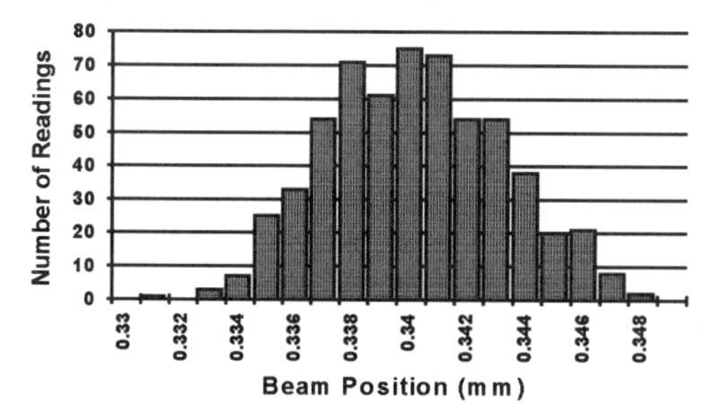

Figure 8. Histogram plot for 600 data points over 10 minutes showing typical horizontal BPM system noise. The RMS noise is only 3μm, which should be compared to the horizontal electron beam size of ≈1.0mm (sigma).

Similarly the noise performance for the vertical plane is shown below in Fig. 9. This shows a total spread in readings of only 4μm.

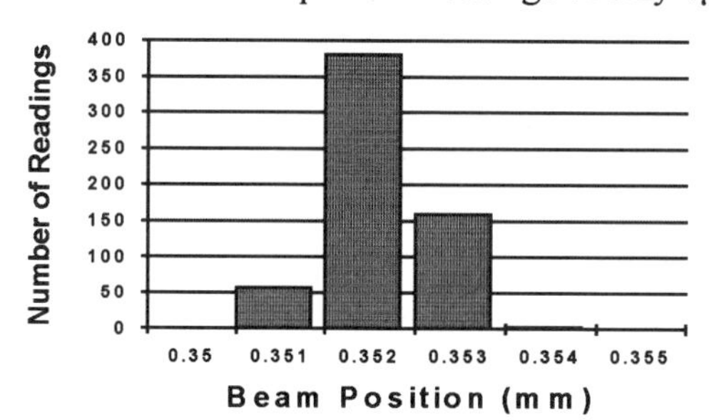

Figure 9. Histogram plot for 600 data points over 10 minutes showing typical vertical BPM system noise. The RMS noise is <1μm, which should be compared to the vertical electron beam size of ≈100μm (sigma).

Several experimental runs of global horizontal (and vertical) feedback have taken place during SRS beam study periods. The ability of the system to correct artificial orbit distortions has been proved. The major operational mode is to maintain the beam orbit to within a few microns of its initial 'set' position. Figure 10. shows data plotted for all horizontal BPMs during the first demonstration of feedback.

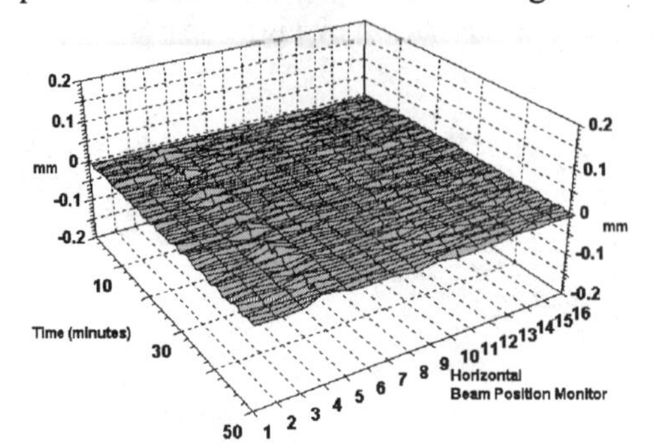

Figure 10. Horizontal orbit drift over 50 minutes with feedback in operation. Maximum error is < ±15μm with an RMS error between 2.5 and 5μm. This should be compared with orbit drifts of approximately ±100μm over a normal user beam lifetime.

It is of interest to note that a small amount of noise can be seen on BPM 4, which was found to have been caused by pick-up of external, electrical noise which has subsequently been resolved, and is not a true beam effect.

CONCLUSIONS

Operational performance targets for the new processing electronics and ADC system have been met or exceeded, and the system is now in everyday use for beam position correction. Experimental results for global feedback have proved very encouraging, and the system has demonstrated its ability to reduce beam drift from >100μm to <15μm (peak) over a user beam period. The 34dB dynamic range and RMS noise levels of 3μm and 1μm for the horizontal and vertical planes respectively, verify a high level of system performance and justifies the approach taken. Additionally the simple, flexible design, and down converting mode of operation of the BPM units will allow its implementation on future storage rings.

REFERENCES

1. Ring, T and Smith, R.J., "Orbit Measurement Techniques at Daresbury", ABI proceedings, KEK, Japan, April 1991.
2. Martlew, B.G. et al, "Development of Global Feedback for Beam Position Control in The Daresbury SRS Storage Ring", EPAC proceedings, London, UK, 1994.
3. Ford, L.M. et al, "A High Precision Digitiser for The New Orbit Processing Electronics at Daresbury", EPAC proceedings, London, UK, 1994.
4. Smith, R.J. et al., "The Implementation of a Down Conversion Orbit Measurement Technique on The Daresbury SRS", EPAC proceedings, London, UK, 1994.
5. Clarke, J.A. and McIntosh, P.A., "The Application of LabVIEW® for Data Acquisition at an Accelerator Laboratory", EPAC proceedings, London, UK, 1994.

Position Monitoring of Low Intensity Beams Using A Digital Frequency Down Converter

Hengjie Ma, Craig Drennan
Fermi National Accelerator Laboratory, PO Box 500, Batavia, IL

Abstract

In monitoring the position of very low intensity beams a signal processing scheme similar to that used in an amplitude-comparison monopulse radar may be employed. In this scheme, an I-Q demodulator for both the sum and difference channels and a phase detector are needed to detect the beam position. It is complex and costly to implement the signal processing with discrete analog components. However, a newly available HSP50016 Digital Down Converter (DDC) chip has provided an attractive alternative. This DSP chip processes the digitized output of the IF section by first converting the signal to baseband using an in-phase/quadrature mixer and then filtering the result with a combination of a programmable high decimating filter and a fixed FIR shaping filter. The accuracy of the quadrature demodulation, nearly ideal filter shape factor and filter reject-band attenuation make the DDC a favored choice over a discrete analog design in an application dealing with very weak beam signals.

INTRODUCTION

It is desired that Beam Position Monitor (BPM) detectors, such as those used in the Switchyard area at Fermilab, be used in the Fixed Target Areas. Upcoming experiments such as KTEV and E-815 will be seeing higher intensity beam making this type of device feasible. The advantage of using a BPM is that the detector is non-intrusive. This means that material is not placed in the path of the beam causing secondary emissions which increase the uncertainty of the experiments' data.

Even though the beam intensities in the Fixed Target area will be higher they are still in or just below the lowest range of the current Switchyard BPM system. There is a need to take position and intensity measurements for three cases; a fast spill of beam, 10^{12} protons over a 1.5 millisecond interval, a slow spill of $2x10^{12}$ protons over a 20 second interval, and a slow spill of beam, 10^{11} to 10^{12} protons over a 20 second interval

Proposed here is a digital signal processing scheme to directly compute the difference-to-sum ratio (Δ/Σ). The digital scheme is aimed at meeting the required sensitivity, noise suppression and flexibility for the lower beam intensity case.

DESIGN MOTIVATION

The functionality of a BPM is similar to that of a simple monopulse tracking radar which detects the azimuth of targets using signal amplitude comparison. Therefore, some of its signal processing schemes and circuit designs can be directly applied to our beam position monitor (1), (2), (3).

The current signal processing scheme has evolved from investigation of the following issues.

Measurement Error Due to Channel Mismatch (4)

The resonant beam detector currently used in Fermilab Switchyard has a receptive sensitivity of typically -120dBm for a slow spill of 10^{12} protons over 20 seconds. Therefore, the total system gain will have to be at least 120dB, and the bandwidth less than 10Hz in order to get usable measurement data. For the conventional AM/PM BPM, increased gain and reduced bandwidth often result in the instability of instrument zero. To eliminate zero-drift due to channel mismatch the conventional signal processing scheme of one-step AM-to-PM conversion is modified into a two-step conversion of AM-to-Δ/Σ followed by Δ/Σ-to-PM with the signal amplification, filtering and frequency down-converting done between the two conversions.

Bias of Estimate Due to Noise (5)

In the Fixed Target area applications, the lower beam intensities will not provide good signal-to-noise ratios out of the detector. Noise, assumed to be a zero-mean random process, has a modulation effect on both the amplitude and phase of the BPM signals,(6),(7). Furthermore, noise in the two channels are statistically dependent to some degree due to imperfect channel isolation. With the phase detection method used to perform Δ-over-Σ normalization in the AM/PM BPM, the obtained estimate contains a certain bias. This estimation bias is dependent on both beam position and intensity. Statistical analysis suggests that this dependence increases with the poorer signal-to-noise ratios associated with lower beam intensities.

This particular estimation bias problem is avoided by handling the Δ-and-Σ directly using a quadrature mixer for down conversion and computing the Δ-over-Σ ratio with a digital processor.

Advantages of Using Digital Signal Processing Methods

While the cost of digital signal processors, dedicated function DSP devices, and analog-to-digital converters have dropped, their performance has increased to the point where there is a clear advantage in using these digital devices in earlier stages of the electronics. Advantages of the digital I-Q detection method proposed here include the following.

a) The overall performance of the digital detection beyond the ADC is highly repeatable and is not subject to drift over time and temperature.
b) Beyond the ADC the signal is not subject to reduction of the signal-to-noise ratio before the final filter due to the low noise floor and high spurious free dynamic range of the demodulator.
c) The digital filter bandwidth can be manipulated remotely through a digital communications port. This aids in optimizing the tradeoffs between noise suppression requiring narrow bandwidths and observation of changes in beam position or beams of short duration requiring wide bandwidths.
d) Filters can be realized with excellent specifications allowing in many cases relaxation of the specifications on the analog filter stages.
e) Converting the signal information to base-band and decimating the stream of signal samples allows relaxation of the data processor speed and memory requirements.

CIRCUITRY DESCRIPTION AND IMPLEMENTATION

Tunnel Unit

Figure 1 is the block diagram of the analog RF front end for the direct Δ-Σ BPM system currently being developed. The two plates of the resonant beam detector couple the time-harmonic fields generated by the passing beam bunches. The ratio of the two output signal amplitudes, A and B, is related to the displacement of the beam x by approximately $0.67x = 20\log(A/B)$. There is a significant capacitive coupling between the two plates of the beam detector. To increase the detection sensitivity to beam displacement and increase the output, a tuning network is used (8). C1, L1, and C2, L2 form the two pararell resonant tanks for plate A and plate B respectively. Their center frequency is 53.104 MHz. The unloaded Q is about 200, and the pararell resonant impedance is about 10 K ohm. This yields a bandwidth of 270 KHz. Trimer capacitor C3 decouples plates A and B. The impedance match between the resonant tanks and the 50 ohm input imdedance of the hybrid junction is accomplished through the taps on the coils. A 180° hybrid junction converts the two output signals from the detector into the difference and sum signals.

The distance between the beam detector and the instrument rack is several hundred feet. There is certain amount of signal loss on the cables and noise pickup at the interconnections. To avoid the further SNR deterioration due to the long distance signal transmission, it may be necessary to place the two RF preamplifiers near the detector in the tunnel. MITEQ low-noise/60dB RF amplifier AU-4A-0110 was chosen for the pre amplification. Its survival under the radiation is still under the study.

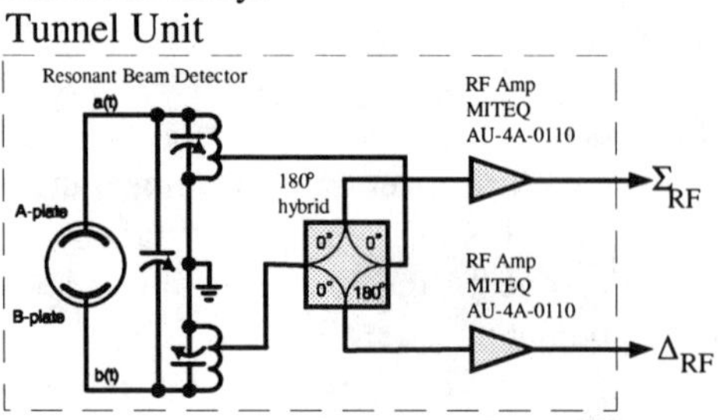

Figure 1. Block Diagram of the Tunnel Unit.

RF-to-IF down-converter

The RF-to-IF conversion and IF amplification are done with two NE615 VHF receiver chips. An input tuning network provides a preliminary noise/interference suppression and the impedance match between the 50-ohm cables and the mixer inputs. The conversion gain is about 13dB. The IF amplifier has a gain of 39dB. The insertion loss of the two stage ceramic IF filters is 17dB. Therefore, the net gain of the RF/IF stage is 35dB, and the bandwidth is about 5KHz. U5 and U6 boost the IF signal amplitude to a proper level for the digitization.

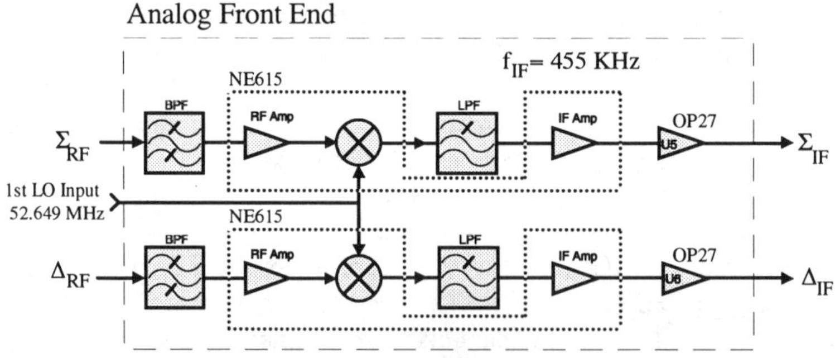

Figure 2. Analog Front End Block Diagram.

PLL LO signal generator

During the flat top period of the beam spill, the Tevatron RF can vary slightly. In order to prevent the IF from drifting out of the IF filter pass band, the LO frequency tracks the Tevatron RF. This is accomplished with a phase-locked loop based LO signal generator.

The phase-locked loop based LO signal generator unit provides not only the LO frequency for the analog down converter but also the clock for the digitizer and the I-Q demodulator/filter in the DDC's. The IF reference frequency is derived from a crystal oscillator of 1.82 MHz (IF x 4). A crystal-controlled VCO generates the LO frequency, which is 53.104MHz – 455KHz =52.649MHz and varies within 50 ppm around the center frequency. The LO is mixed with Tevatron RF to generate the actual IF. The actual IF is compared with the IF reference frequency. The resultant error provides the base for adjusting the frequency of the VCO. The phase detection in the PLL is implemented with digital circuits. It has a frequency-tracking capability, which ensures an adequate frequency-capture range. To minimize the phase noise in the generated LO, the bandwidth of the loop filter is set to be less than 1KHz.

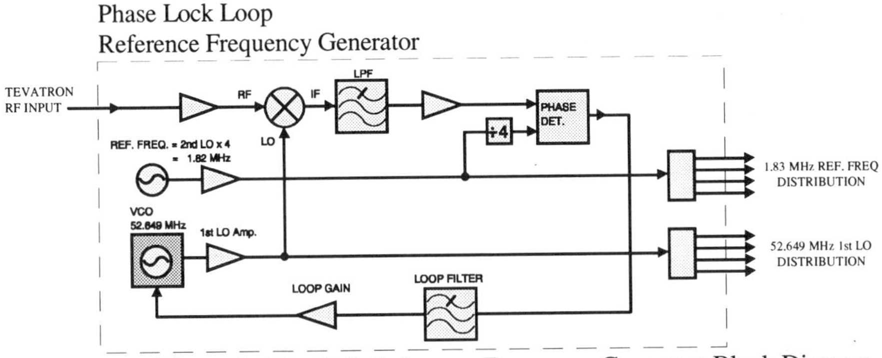

Figure 3. The PLL Reference Frequency Generator Block Diagram.

Digital IF Stage and IQ Detection

The sum and difference signals out of the analog IF stages are scaled for input to the +/- 2.5V input of the Analog to Digital Converters (ADC's) using the dual-channel variable-gain amplifier, U7. The gain for each channel is variable from 0 to +41dB and is controlled by the output of the data processor.. The proposed circuit will use an HI5800 12 bit, 3 MSPS sampling ADC. The device will be operated at 1.82 MSPS,(9).

The Digital Down Converter (DDC) is a single chip synthesizer, quadrature mixer and high decimation lowpass filter. The 12 bit sampled data stream from the ADC is set into the 12 most significant bits of the DDC's 16 bit input. The quadrature mixer down converts the amplitude information at the IF frequency to baseband (DC). The complex result is lowpass filtered and decimated with identical real filters in the in-phase (I) and quadrature (Q) processing chains. Lowpass filtering is accomplished via a five stage high decimation filter (HDF) followed by a fixed finite impulse response (FIR) filter. The combined response of the HDF and FIR results in a -3dB to -102dB shape factor of better than 1.5. The stop band attenuation is greater than 106dB. The composite pass band ripple is less than 0.04dB,(10).

The end-to-end noise floor of the HSP50016 is greater than 100 dB below full scale and the spurious free dynamic range (SFDR) of the internal modulation process is greater than 102 dB. The frequency selectivity of the DDC is less than 0.00032 Hz at 2.73 MSPS.

The combination of down conversion and decimation is an efficient way to handle a bandpass signal centered at the IF frequency,(11). Decimation is the process of lowpass filtering the signal that was translated to DC and re-sampling the signal at a lower sampling rate. By using multiple stages of filtering and decimation, filters with sharper frequency cutoffs can be realized with fewer multiplication's. Also, decimation serves to reduce the data rate that a microprocessor or microcontroller must handle.

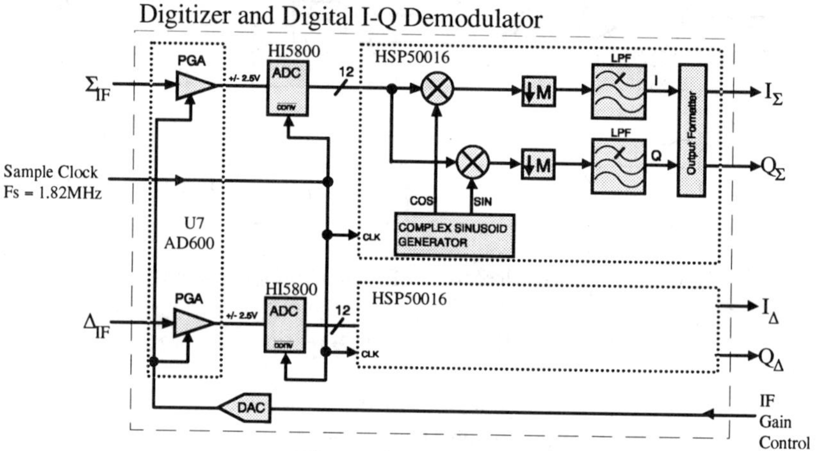

Figure 4. Digital IF and IQ Detection Block Diagram.

There is a direct relationship between the decimation rate for which the HSP50016 is set and the resulting bandwidth of the lowpass filter response. This relationship is -3dB BW = 0.13957 Fs / R and -102dB BW = 0.19903 Fs / R, where Fs is the input sampling rate and R is the HDF stage decimation factor.

The Data Processing Section

The I data and Q data are output from the HSP50016 with a synchronous serial connection directly to the data processor or to a FIFO buffer. The processor would also need to setup the Control Words of the DDC. These seven 40 bit control words determine the operation of the down converter, the decimation rate and filter bandwidth, and the input and output formats.

CONCLUSIONS

Experimenters in the Fixed Target Area wish to take advantage of the non-intrusive beam position and intensity measurements provided by the BPM's like those currently used in Switchyard. However, the current AM/PM method in use now is not expected to operate at the lower beam intensities and poorer signal-to-noise ratios in these areas. This is because of the instability of the bias of the signal estimates. The direct Δ/Σ method involves only a linear conversion and does not have this estimate bias problem. Therefore, averaging more signals together or decreasing filter bandwidth should produce better estimates of the signal mean.

The availability of less expensive analog-to-digital converters with higher accuracies and higher sample rates allow digitization of signals at the standard IF frequencies. The HSP50016 allows accurate and efficient handling of the stream of digital values representing the signal. The magnitude and relative phase of the Δ and Σ signals is preserved in the I-Q demodulation process and the ratio Δ/Σ can be computed in the data processor or at a higher level of the data processing system to determine beam position.

REFERENCES

1. Kahana, E. and Chung, Y., "Test Results of a Monopulse Beam Position Monitor for the Advanced Photon Source," Proceeding of Accelerator Instrumentation Fourth Annual Workshop, Berkeley, CA 1992, pp. 271- 278.
2. Ma, H. and Moore, C., " A Beam Position Monitor for Low Intensity Beams," Proceeding of Accelerator Instrumentation Workshop, Santa Fe, NM, 1993.
3. Skolnik, M.I. , Introduction to Radar Systems, 2nd Ed., McGraw-Hill, 1980.
4. Ma, H., "Report III on Switchyard Beam Position Monitor," Fermilab-TM-1904, January 1993.
5. Ma, H., "The Effect of Beam Intensity on the Estimation Bias of Beam Position," Fermilab-TM-1905, December 1991.
6. Shanmugan, K.S., Breipohl, A.M., Random Signals: Detection, Estimation and Data Analysis, John Wiley & Sons, 1988, New York.
7. Shanmugan, K.S., Digital and analog communication systems, John Wiley & Sons, 1979, New York.
8. Kerns, Q., et. al., " Tuned Beam Position Detector for the Fermilab Switchyard," PAC 1987, Vol. 1, pp. 661-663.
9. HI5800 Data Sheet, Harris Semiconductor, FN 2938.4, December 1993.
10. HSP50016 Data Sheet, Harris Semiconductor, FN 3288.3, June 1994.
11. Crochiere, R.E. and Rabiner, L.R., "Multirate Digital Signal Processing", Prentice Hall, 1983.

An Overview of High Input Impedance Buffer Amplifiers for Wide Bandwidth Signals

David W. Peterson
Fermi National Accelerator Laboratory, P.O. Box 500, Batavia, IL 60510

Abstract

Recent evaluations of wide bandwidth (<100 kHz to >150 MHz), high input impedance (>10 kΩ) amplifiers are shown. Included are discussions of noise performance, dynamic range, phase linearity, cost and other considerations. Accelerator applications for such amplifiers include diagnostic pickups, filters, and signal distribution systems. Operational experiences with pickup preamplifiers in the Fermilab Antiproton Source Damper systems are described.

INTRODUCTION

One of the improvements to the Fermilab Antiproton Source Accumulator Damper System involves upgrading the performance of the pickup preamplifiers. The damper pickups are electrically short stripline pairs, and therefore high impedance preamps are needed to provide flat response at low frequencies. The input impedance of the amplifier should be greater than 10 kΩ and preferably 100 kΩ or more. The amplifier must have flat phase and frequency response from 200 kHz to 150 MHz and be capable of driving 50 Ω cable. Input signals as high as 8 volts peak to peak are possible.[1]

The preamplifier currently in use consists of a differential transistor input pair driving a common emitter output. Frequency response is flat from 100 kHz to over 200 MHz. The input resistance is approximately 12 kΩ. The requirements for high input dynamic range and large output swing into 50 Ω dictate that the output transistors be heavily biased and the entire unit operate from ±24 Volt supplies. The large amount of power dissipation in the transistors and bias network requires the use of forced air cooling.

Three issues regarding the operation of the present amplifiers are to be addressed in the new design: the high power dissipation has raised some concern about reliability, the low frequency response of the damper system would improve if the preamp input impedance were higher, and since only the difference output is available, a new preamp configuration is required to utilize active cancellation of the common mode signal.[2] The cancellation scheme involves adding part of the sum mode signal (with the appropriate phase shift) to the difference signal to null out the common mode component in the difference signal. An examination of commercially available buffer amplifiers has been enlightening and revealed some devices which are suitable for this application.

Work supported by the U.S. Department of Energy under contract No. DE-AC02-76CH03000.

GENERAL CONSIDERATIONS

Noise Performance

The primary contributions to the overall noise in the circuit are from the thermal noise of the resistors and the internal current and voltage noise sources of the amplifier.

The root mean squared (rms) value of the thermal noise voltage due to a resistance is given by

$$e_n = (4kTBR)^{0.5}, \tag{1}$$

where : k is Boltzman's constant (1.38 x 10 $^{-23}$ joules/K),
T is the absolute temperature, K, of the resistance,
B is the noise bandwidth in Hz,
R is the resistance in Ω.[3]

It should be noted that the noise bandwidth is the integral of the system frequency response from zero to infinity. For a single pole circuit this reduces to

$$B = \frac{\pi}{2} \cdot f_{3dB}, \tag{2}$$

where f_{3dB} is the -3dB bandwidth of the circuit. [4]

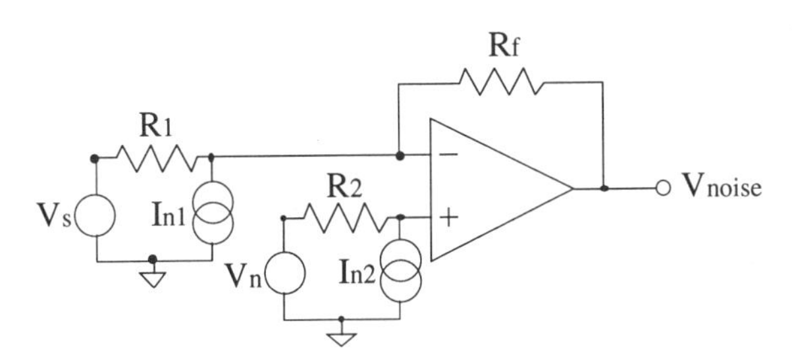

Figure 1. A General Op Amp Noise Model.

A general model of op amp noise is shown in Fig. 1.[5] The three resistors, R_1, R_2 and R_f contribute to the total output noise as

$$V_{OR_1} = (4kTBR_1)^{0.5} \cdot G, \tag{3}$$

$$V_{OR_2} = (4kTBR_2)^{0.5} \cdot (G+1), \tag{4}$$

$$V_{OR_f} = (4kTBR_f)^{0.5}, \tag{5}$$

where

$$G = R_f / R_1 . \tag{6}$$

The current noise sources of the amplifier, I_{n1} and I_{n2}, and the voltage noise source, V_n, contribute to the output noise as

$$V_{OI_{n1}} = I_{n1} \cdot R_1 \cdot G = I_{n1} \cdot R_f, \tag{7}$$

$$V_{OI_{n2}} = (I_{n2} \cdot R_2) \cdot (G+1), \tag{8}$$

$$V_{OV_n} = V_n \cdot (G+1). \tag{9}$$

The total rms output noise due to all these contributions is

$$V_{noise} = \left(V_{OR_1}{}^2 + V_{OR_2}{}^2 + V_{OR_f}{}^2 + V_{OI_{n1}}{}^2 + V_{OI_{n2}}{}^2 + V_{OV_n}{}^2\right)^{0.5}. \tag{10}$$

The desire for high input impedance can negate any benefit of a low noise amplifier. For example, a 100 kΩ resistor at room temperature has a thermal noise voltage density of about 40 nV/√Hz. If the amplifier has a current noise density of 4 pA/√Hz then the resistor and amplifier combination will produce 400 nV/√Hz of noise at the input! However, this simple model does not tell the whole story; in many high impedance applications the capacitance of the input circuit and the amplifier will cause the total noise to roll off rapidly at higher frequencies.

A parameter commonly known as the noise resistance is expressed as

$$R_n = V_n / I_n, \tag{11}$$

where V_n and I_n are the amplifier noise densities seen earlier.[6] This is the optimal matching resistance for lowest amplifier noise figure.[7] Note that this is not the source resistance for minimum absolute noise but the best match to minimize the amplifier's noise contribution in proportion to the total noise of the system. It can be seen from Eq. 1-10 that the lowest absolute noise occurs as the resistor values go to zero. The noise resistance also gives a measure of the input resistance below which one should be concerned primarily with voltage noise specifications and above which one should be concerned with current noise specifications.

Current feedback operational amplifiers have many good qualities but low current noise is generally not one of them. Current noise densities are approximately 6 times greater in current feedback devices than in comparable voltage feedback amplifiers.

Dynamic Range

The noise generated in the input circuit and amplifier sets the lower limit of the usable signal range. In many types of accelerator applications the signal will vary over an extremely wide range due to differences in beam intensity, position or structure. RF bunching can induce extremely large signals on beam pickups.

In the case of the P-Bar Damper system, the preamps must be able to handle the full bunched beam signal at each of the pickup plates. This signal can be as high as 8 Volts peak to peak. The preamp noise must also be as low as possible to minimize the power requirements of the final amplifiers for proper damping of coherent transverse modes.

Phase Linearity

The requirement for flat phase at low frequencies is usually not a problem with most amplifiers. However, one must be careful to select appropriate components for power supply bypassing and DC blocking. For example, a 0.1 μF DC blocking capacitor on the output of the amplifier may introduce undesired phase shift at low frequencies in a 50 Ω system.

High frequency phase performance is dictated by the internal structure of the amplifier and the general circuit layout. Some of the buffer amplifiers have large amounts of input capacitance and in combination with the input impedances can cause gain peaking at high frequencies. The addition of a series input resistor of the proper value can help reduce the amount of peaking.

Cost

In certain cases the performance requirements can only be met by a custom built buffer amplifier. It is estimated that the cost of the original damper preamps is over US $1000 each due to specially matched transistor pairs, high frequency printed circuit boards, custom enclosures, assembly time and nitrogen gas cooling. Some of the plastic dual in-line package (DIP) buffer amplifiers cost as little as $2.59 each.

OPERATIONAL EXPERIENCE

Other Differential Amplifiers

Before the decision was made to try active cancellation of the common mode signal, a variety of commercially available differential amplifiers were investigated. Even though they were eventually found to be unsuitable for the damper preamplifier, two devices are worth mentioning for their unique characteristics. The first is the μA733, a differential video amplifier from Signetics and Texas Instruments similar to the NE/SE592.[8,9] Various filter configurations can be realized by adding reactive components at the gain select pins. The other is the NE/SA5209 differential wideband variable gain amplifier from Signetics.[10] It has an input impedance of 1.2 kΩ and a bandwidth of 850 MHz. It is used at Fermilab to buffer the differential output of the Analog Devices AD640 120 MHz logarithmic amplifier.

Low Voltage Buffer Amplifiers

Buffer amplifiers operating from ±5 volt supplies are very common and exhibit many desirable characteristics. The widest bandwidth units will typically operate in this voltage range since the small geometry of high speed semiconductor devices will not usually tolerate higher supply voltages. The power dissipation is lower and surface mount (SO) packages are more common for these devices. The Analog Devices AD9630, similar to the Comlinear CLC110, is an example of one of these devices that met almost all the requirements for the damper preamps.[11] Figure 2 shows the measured frequency response of this device. The -3 dB point is at 904 MHz.

Higher Voltage Buffer Amplifiers

It may be necessary to operate at supply voltages greater than ±5 volts to handle the input voltage range or the output drive requirements. Amplifiers operating at higher voltages will often have package formats capable of greater heat dissipation. These may involve extra pins in a surface mount or DIP package to conduct the heat away from the device or a flat pack or metal can package which can be attached to a heat sink. The BUF634 from Buff Brown comes in either an 8 pin SO or DIP packages or a 5 pin TO-220 package.[12] Figure 3 shows the frequency response of the Burr-Brown OPA633 and the BUF634. The measured -3 dB bandwidth for the OPA633 is 180 MHz and for the BUF634 is 169 MHz. Note that both Fig. 2 and Fig. 3 use the same scales.

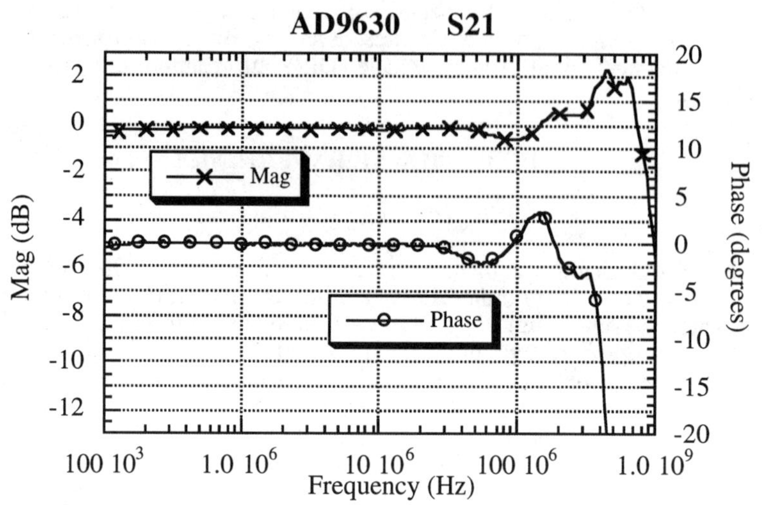

Figure 2. Analog Devices AD9630 Frequency Response.

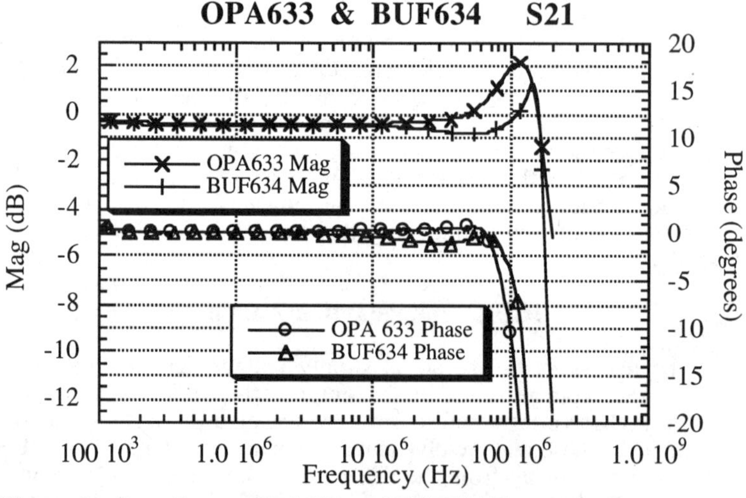

Figure 3. Burr-Brown OPA633 and BUF634 Frequency Response.

In certain cases the measured bandwidth may not agree with the manufacturer's specification. This was observed with some of the older open loop hybrid buffers.

While AC performance is important, DC parameters should not be ignored. Input bias currents can cause large offset voltages, particularly in circuits with high impedance AC coupled inputs. Typical bias currents range anywhere from 1.5 to 30 μA.

Other Circuits

The use of high impedance buffer amplifiers is certainly not limited to pickup preamplifiers. Some of the lower frequency devices are intended to be used as current followers in op amp circuits. A high speed differentiator has been constructed using two AD9630s, a capacitor and a resistor. Performance is very uniform to 200 MHz.[13] A signal multiplexer system is currently being implemented using buffer amplifiers to tap off transmission lines to create lossless signal splitters. The signals are then distributed to a switch matrix that routes them to a variety of test equipment. The buffer amplifiers create a very compact and well isolated splitter with low loss and bandwidth from DC to over 300 MHz.

SUMMARY

A variety of high input impedance buffer amplifiers have been evaluated. Some of the top contenders for use in the P-Bar Damper preamplifiers are shown in table 1.

The selection of an appropriate amplifier requires careful consideration of many device parameters, including current and voltage noise, input impedance, input bias, bandwidth, output drive capability, power supply voltages, package formats and cost.

Table 1. Characteristics of Selected Buffer Amplifiers.

Device	Mfr.	3dB BW (MHz)	Input R (Ω)	Input C (pF)	Noise V (nV/√Hz)	Noise I (pA/√Hz)	Output I (mA)	Supply (±V)
AD9630	Analog Devices	750	150 k	1	2.4	3.6	50	7
BUF601	Burr-Brown	900	2.5 M	1	4.8	*	20	6
BUF634	Burr-Brown	160 @ 100 Ω	8 M	8	4	*	250	18
OPA633	Burr-Brown	260	1.5 M	1.6	20	*	100	20
CLM4102	Calogic [14]	400	150 k	6	15	*	200	17
CLM4121	Calogic	180	500 k	3.5	15	*	140	15
CLM4133	Calogic	350	10 G	3	10	*	200	15
CLC115 (Quad)	Com-linear [15]	270	750 k	1.6	3.0	2	60	7

* information was unavailable at this time.

ACKNOWLEDGMENTS

The author wishes to thank Ken Fullett for his many helpful suggestions, Jim Budlong and Pat Sheahan for their work on the P-Bar Damper system and Bob Vargo for his help with the amplifier measurements.

REFERENCES

1. J. Petter, J. Marriner, and J. McCarthy, "Transverse Beam Dampers for the FNAL Antiproton Rings," in Proceedings of the 1987 IEEE Particle *Accelerator Conference,* (IEEE, New York, 1987), pp. 791-793.
2. D. McGinnis, Private communication at Fermilab, Summer 1994.
3. *1992 Amplifier Reference Manual,* (Analog Devices, Inc., Norwood, MA, 1992) Ch. 2, pp. 25-26.
4. P. Gray and R. Meyer, *Analysis and Design of Analog Integrated Circuits*, (John Wiley & Sons, New York, 1977), Ch. 11, pp. 659-660.
5. W. Kester, *System Application Guide,* (Analog Devices, Inc., Norwood, MA, 1993) Ch. 10, p. 26.
6. *ibid.*, Ch 8, p. 5.
7. P. Horowitz, and W. Hill , *The Art of Electronics*, Second Edition, (Cambridge University Press, 1989), Ch. 7, pp. 428-452.
8. *1987 Linear Data Manual Volume 2: Industrial*, (Signetics Corporation, Sunnyvale, CA, 1987), Ch. 4, pp. 245-250.
9. *Linear Circuits Data Book*, (Texas Instruments Inc., Dallas, TX, 1984), Ch. 5, pp. 101-108.
10. *RF Communications Handbook* , (Signetics/Philips Corporation, Sunnyvale, CA, 1991), pp. 35-49.
11. *1992 Amplifier Reference Manual,* (Analog Devices, Inc., Norwood, MA, 1992) Ch. 2, pp. 461-466.
12. *Burr-Brown Integrated Circuits Data Book, Linear Products 1994*, (Burr-Brown Corp., Tucson, AZ, 1994), Ch. 3.
13. K. Fullett, J. Budlong, Private communication at Fermilab, Summer 1994.
14. *Summer/Fall 1994 Shortform*, (Calogic Corp., Fremont, CA, 1994)
15. *1993-94 Databook*, (Comlinear Corp., Fort Collins, CO, 1993), Ch. 6, pp. 15-18.

Front-End Electronics for the Bunch Feedback Systems for KEKB

E. Kikutani, T. Obina, T. Kasuga, Y. Minagawa, M. Tobiyama
National Laboratory for High Energy Physics,
Oho 1-1, Tsukuba-shi, Ibaraki, 305 Japan
L. Ma
Institute of High Energy Physics, Academia Sinica,
Beijing 1000039, China

Abstract

At KEK, a double ring collider for the B-meson physics, KEKB, will be constructed. One of the main features of the accelerator system of KEK B-factory (KEKB) is that the stored current is very high. Owing to this, strong coupled bunch instabilities will occur. We will install bunch feedback systems to suppress these instabilities. In this paper, we will present the characteristics of the front-end of the feedback systems.

INTRODUCTION

In the last spring, a B-Factory project at KEK, named KEKB, was officially approved. KEKB consists of two storage rings, Low Energy Ring (LER) of 3.5 GeV and High Energy Ring (HER) of 8.0 GeV with one interaction point and linear accelerators as injectors. In order to achieve a very high luminosity, so many bunches will be stored in the two rings. Owing to this it is very likely that strong coupled bunch instabilities will occur. We will install bunch feedback systems(1) to suppress these instabilities.

The main machine parameters concerning the feedback systems are listed in the following table.

ring name	LER	HER	
beam energy	3.5	8.0	GeV
RF frequency	508.887		MHz
harmonic #	5120		
circumference	3016		m
# of bunches	~5000		
bunch spacing	2		nano sec.
# of particles/bunch	3.3×10^{10}	1.4×10^{10}	
rad. damp. time (long.)	40	20	milli sec.
synchrotron tune	$0.015 \sim 0.02$		

Note that the bunch spacing is only 2 ns and that the number of bunches is about 5000. The design of the feedback systems should be very tight from a technical point of view, due to this short bunch-spacing and the large number of bunches.

DESIGN OF THE FRONT-END ELECTRONICS

Concepts

In this paper, the term, "front-end electronics", means the circuitry between pickup electrodes and an analog-to-digital converter (ADC) which outputs the digital signal proportional to the position (horizontal, vertical, longitudinal) of a bunch. The circuitry mainly consists of passive components but often also of some auxiliary active components.

For both the longitudinal and transverse directions, the technique used in the electronics is to detect the phase difference between two quasi-sinusoidal signals which are made from pickup-signals. The frequency of these sinusoidal signal is called the detection frequency. The phase-difference measurement is done by a usual technique, namely by using a set of a double balanced mixer (DBM) and a low pass filter.

For the longitudinal case, the principle is quite simple. The longitudinal position (i.e. a timing lag of a bunch relative to the synchronous particle) is detected as the phase difference between the beam signal and a timing-reference signal. On the other hand, the principle in the transverse case is slightly complicated: we use the so-called AM/PM method(2).

FIR filter with cables and power combiners

The quasi-sinusoidal signals, which were mentioned above, can be made by several methods. Among these methods, we adopt a very simple one: the signal from a pickup is divided into a few branches by a power combiner and summed up again by another power combiner. If the lengths of the cables which connect the two combiners are chosen to be $a + n\lambda$, this system become effectively an FIR filter which is favorable to the detection frequency. Here a is an offset of the lengths, λ the wavelength of the detection frequency and n is the cable number. The principle of the technique will easily be understood by the right block diagram of Fig. 1. Also the multi-electrode method (the left diagram in Fig. 1) is possible. In this case the lengths of the cables from the electrodes to the combiner must be rightly chosen. Of course, we can draw higher power out of the beam with the latter method.

Description of the circuits

In Fig. 2, a block diagram of the circuit for the longitudinal direction is shown. The circuit compares the phases of the pickup-signal and a reference signal, which is made by frequency-multiplying the rf-signal for the acceleration. Strictly speaking, the output depends not only on the longitudinal position but also on the bunch current. But we do not care this dependence, because the role of the front-end electronics is to provide the kicker of the feedback system with the input information: position detection itself is not a required function to the front-end electronics. Naturally the feedback gain will

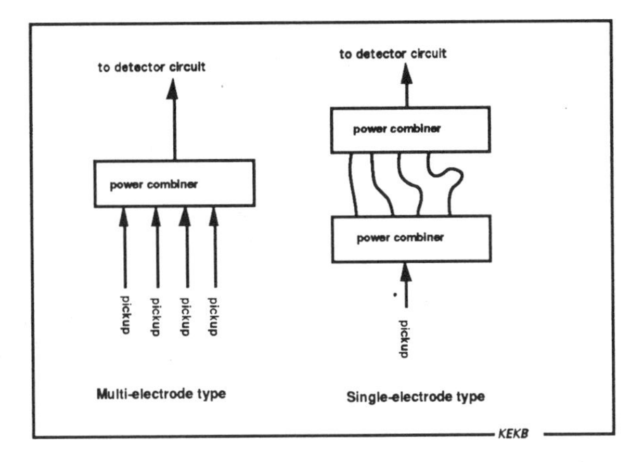

Figure. 1. Two types of the cable-combiner FIR filters.

depend linearly on the bunch current but the growth rate of the instabilities will also depend on the bunch current.

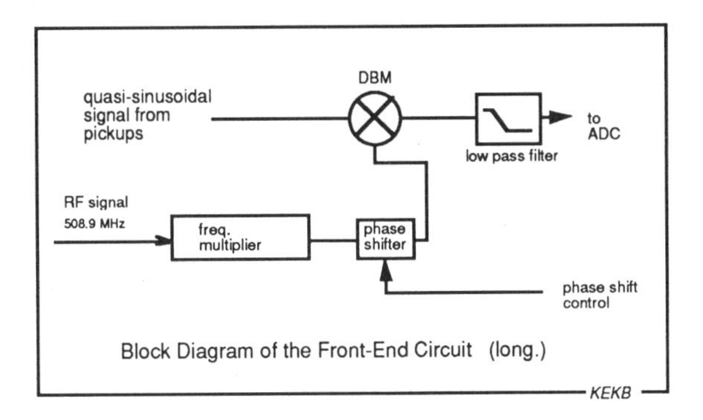

Figure. 2. Front-end circuit for the longitudinal position detection.

For the transverse case, the circuit is the AM/PM as mentioned above. In Fig. 3, a block diagram of the circuit is shown. Two quasi-sinusoidal signals to be compared are made by the AM/PM hybrid out of two pulse-trains from pickups. The output of the AM/PM hybrid will be proportional to the square of the bunch current without any regulation. In order to avoid this steep dependence, the pulse heights of two inputs of the DBM are regulated by limiting amplifiers. In high frequencies such as 1 GHz or higher, limiting amplifiers which preserve the phases are not easy to make. We must carefully select the model of these amplifiers.

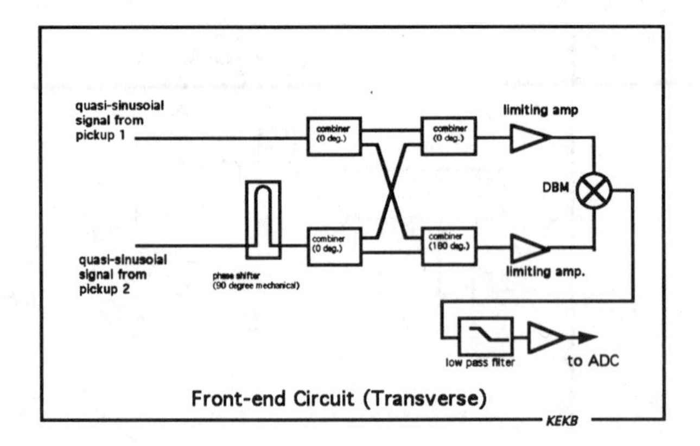

Figure. 3. Front-end circuit for the longitudinal position detection.

Choice of the detection frequency

So far, we have described the principles of the position-detection technique used in the front-end electronics. Here we explain how to choose the detection frequency. There are a few reasons to prefer higher frequencies, that is, shorter wavelengths, as the detection frequency. The first reason is that a given number of the sinusoidal cycles in a pulse train lasts for a shorter time with a higher detection frequency. This means the better isolation between two signals from successive bunches. Secondly, we can obtain a higher output for the same timing lag with a higher detection frequency, because the phase difference become larger under a higher frequency. This is advantageous to the longitudinal position measurement.

But we must keep our mind on the cut-off frequency of the beampipe as a waveguide. If we use the frequency higher than the cut-off frequency, the signal will be covered with noise and it becomes impossible to measure the position in the bunch by bunch manner. Also we must be aware of the fact that higher frequencies are, in general, less easy to treat considering the characteristics of circuit components and cables. As a result, frequencies, 2$\sim$3 GHz, are the candidates for the detection frequency. The decision of the frequency will depend on how we can make the beampipe small.

Candidates of the ADC for our purpose

Because the bunch spacing is only 2 ns, we must choose the ADC with the conversion rate higher than 500 Mega samples/s. Our candidates of the ADC, that are commercially available, are SONY 1276K and Tektronix TKAD10C. Both of them have 8-bit resolution that is high enough for our usage. The analog bandwidth is 1.2 GHz and 300 MHz for the Tektronix and the SONY, respectively. The Tektronix ADC may be preferable for our purpose because of its wide bandwidth. But stable delivery is not hopeful for the Tektronix. Hence, our temporary conclusion is to use the SONY.

Candidates of the type of pickup electrodes

The stripline and the button are candidates for the type of the pickup electrodes. The output of a stripline can have a desired frequency spectrum by choosing its length suitably. It is an advantage of the striplines. In addition, the relatively higher output power is available from striplines. But from the view point of delicate design of the electrode, the button is preferable. Precise shaping of an electrode is essential to make a good pulse shape where the term good pulse shape means the shape free from ringing. We now study the characteristics the pulses from the button-type electrodes by comparing various button shapes and various feedthrough connectors by the analytical and experimental methods(3).

BEAM STUDIES

Following the concept which we have explained, we made prototypes of the front-end electronics for both the longitudinal and transverse directions. The performance of these circuits were tested using the beams of the TRISTAN Main Ring and PF Ring at KEK.

Experimental Conditions

The detection frequency was chosen to be 1.5 GHz, which is 3 times of the rf frequency. The reason why we adopted this frequency is that the typical size of the beampipe is 10 cm in diameter both in TRSITAN Main Ring and in PF ring. The corresponding cut-off frequency is about 1.6 GHz.

The types of the pickups used in the experiments were a train of striplines for the experiments in TRISTAN Main Ring and a train of button electrodes in PF Ring. The lengths of the striplines are adjusted so as to a peak gain is obtained at 1.5 GHz.

The circuit components are as follows. The model of power combiners for the FIR filters is Mini Circuits ZB8PD-4. It is a 1:8 combiner/divider with the frequency range of from 1 kHz to 2 GHz. The DBM's were R&K M21L and R&K M21. The recommended power level for the input, "RF", of the DBM is 7dBm and 17dBm for M21L and M21, respectively. The low-pass filter at the end of the circuit was TAMAGAWA ULF-174. The cut-off frequency was chosen to be 1 GHz. The limiting amplifier for the transverse circuit was Watkins-Johnson CLA-45-1, which works in the frequency range of 0.8$\sim$ 4.2 GHz.

In the experiments only one bunch was stored in the ring and the number of particles in a bunch was almost the same as that of the KEKB rings. The longitudinal and transverse oscillation were artificially excited by shaking the rf phase (for the longitudinal case) or by kicking a bunch by the transverse kicker electrodes.

The data acquisition was done by the ADC, SONY 1276K, with 4kbyte memory, which is packaged as a CAMAC module. Because the output of the front-end circuit is rather low (particularly in the transverse measurement) comparing with the input range of the ADC, we put a 15 dB amplifier to fit the signal level. Even though the ADC has a capability of digitizing in 500MHz, the trigger for the ADC was the revolution frequency of the accelerators, 100 kHz for TRISTAN Main Ring, and 1.2 MHz for PF ring.

Results

One of the checking points of our technique is if it is possible to make a clean pulse-train by the cable-combiner system. The typical output of the cable-combiner filter is shown in Fig. 4. Note that the main part of the oscillation lasts for only 3 ns. This means that we can detect the position bunch by bunch, if successive bunches are separated by 4 ns at least.

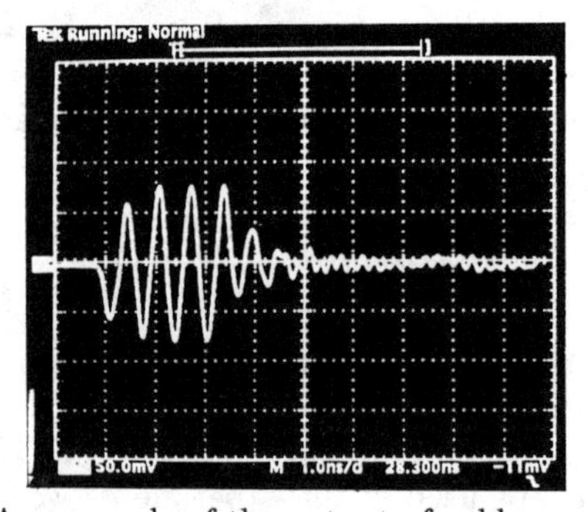

Figure. 4. An example of the output of cable-combiner system.

The power spectra of the longitudinal and transverse oscillations are shown in Fig. 5. The left graph is for the longitudinal case. The data was taken when a bunch oscillates with the amplitude of 10 ps. The peak appears in the graph corresponds to the synchrotron tune of TRISTAN MR. From this graph we know that the noise level is lower than the signal by about -40dB. The right graph is for the transverse oscillation observed in PF Ring. The amplitude of the oscillation was about 100μm. Again the noise level is lower than the signal by -40dB.

Through these experiments we found that the figure, 3, is acceptable for the number of cycles in the quasi-sinusoidal signal. For the longitudinal case, the figure, 2, can be hopeful but not established.

DISCUSSIONS

From the results of the experiments with beams, we conclude that measurement of the position of a bunch is possible when the bunch spacing is longer

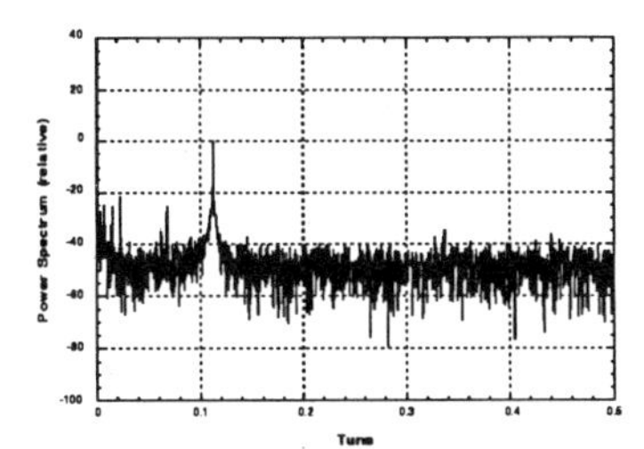

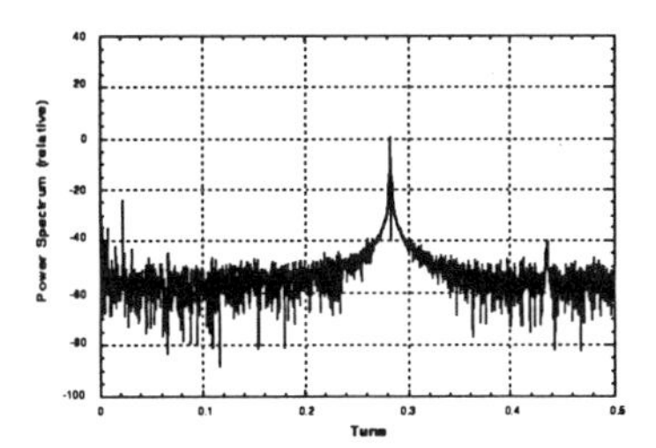

Figure. 5. Power spectrum in the longitudinal (left) and transverse (right) oscillation. The level of the highest peak is normalized to 0 dB.

than 4ns. However, the spacing will be 2 ns in KEKB. Then our next step should be to establish the technique for the 2 ns spacing. A possible way is increasing the detection frequency from 1.5 GHz to 2 or 2.5 GHz. In order to make it feasible the smaller beampipe must be used around the pickups. Within a few months we will make experiments with new smaller beampipe to examine the technique.

SUMMARY

We designed the front-end electronics for the longitudinal and the transverse systems. For the longitudinal, it uses the simple phase-comparing method and in the transverse so-called the AM/PM methods is adopted. In order to survey the performance of these circuits, we made experiments in TRISTAN Main Ring and in PF Ring, both at KEK. Through these experiments we found that our circuits can catch the bunch oscillation up to the bunch frequency of 250 MHz. At the next stage, we will establish the technique to detect the oscillation under the bunch frequency of 500 MHz.

REFERENCES

1. E. Kikutani et al., “Development of the Bunch Feedback Systems for KEKB” to be appeared in *Proceedings of the European Particle Accelerator Conference*, (London, 1994), also available as KEK Preprint 94-49.
2. The AM/PM technique is explained, for example, in a paper by R.E. Shafer, “Beam Position Monitoring”, in “AIP Conference Proceedings 212”.
3. T. Obina et al., ‘Oscillation Detection Part of the KEKB Bunch Feedback Systems”, in preparation.

A Time-Gated Integrator Array for Beam Profile Measurement

Hengjie Ma, John Marriner
Fermi National Accelerator Laboratory*
P.O.Box 500, Batavia, Illinois 60510, U.S.A.

Abstract

A 96-channel time-gated integrator array has been developed for taking the beam profile data from a segmented wire ionization chamber. The parallel readout scheme has the advantage of very-low input impedance following the long cables between the detector and integrator. The cross-talk in therefore reduced. The design of the integrator array has emphasized speed and accuracy as well as the flexibility of use. The integrator array also includes the features of auto-stop and calibration current source for the convenience in its use. Some special issues as its application in a particle accelerator have been considered in the design.

INTRODUCTION

The segmented wire ionized chamber (SWIC) has been proved to be an important instrument for measuring the beam profile, specially in the low beam intensity areas such as the Switchyard. A microprocessor-based SWIC scanner was developed as the data readout device in 1979 . (1) Recently, it has been proposed to develop a new readout device. The device should have very low input impedance so that some economic multi-conductor cables can be used without sacrificing channel isolation. It is also desired that the device be an independent unit which can be easily interfaced with the existing control devices such as CAMAC cards. The new readout device should be able to detect very-low intensity beams, and it should be fast enough that the time-window of the measurement can be set very narrow. With these ideas, a ninety-six channel integrator array has been design, and the prototype for the proof-test has been constructed.

---

* Operated by University Research Association, Inc.
under contract with U.S. Department of Energy

CIRCUIT DESCRIPTION

The integrator array consists of a 96-channel integrator bank, control interface and threshold detector for the auto-stop function. Figure 1 shows the block diagram of the device. The integrators for the ninety-six channels are identical. The core component of the circuit is a switched integrator chip ACF2101 from Bur-Brown. It is a dual-channel integrator, and has three on-chip switches for each channel; sample/hold switch, reset switch, and select switch at the output end for multiplexing.(2) A two-channel group is comprised of an ACF2101 and a quad-switch chip MAX 327. Its detailed schematic is shown in Figure 2. In the rest part of this section, some thoughts in the design are discussed.

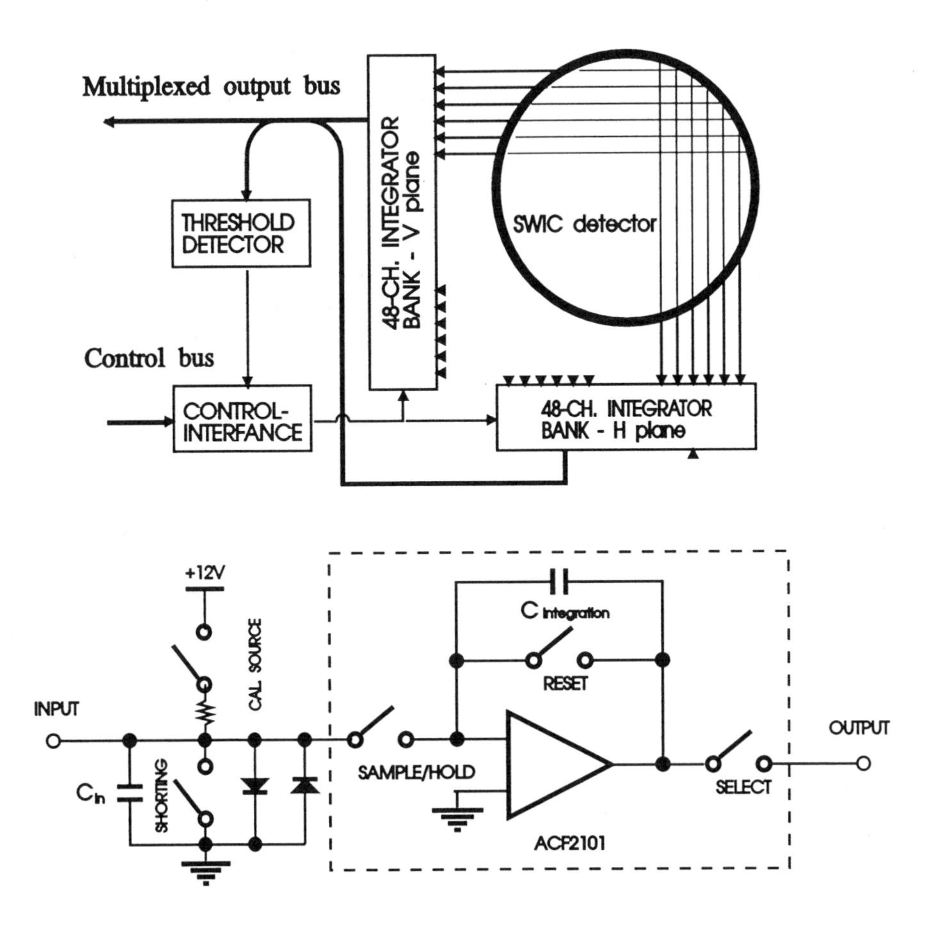

Figure 1. (Top) Block diagram of integrator array as the SWIC data readout device (Bottom) A single integrator circuit.

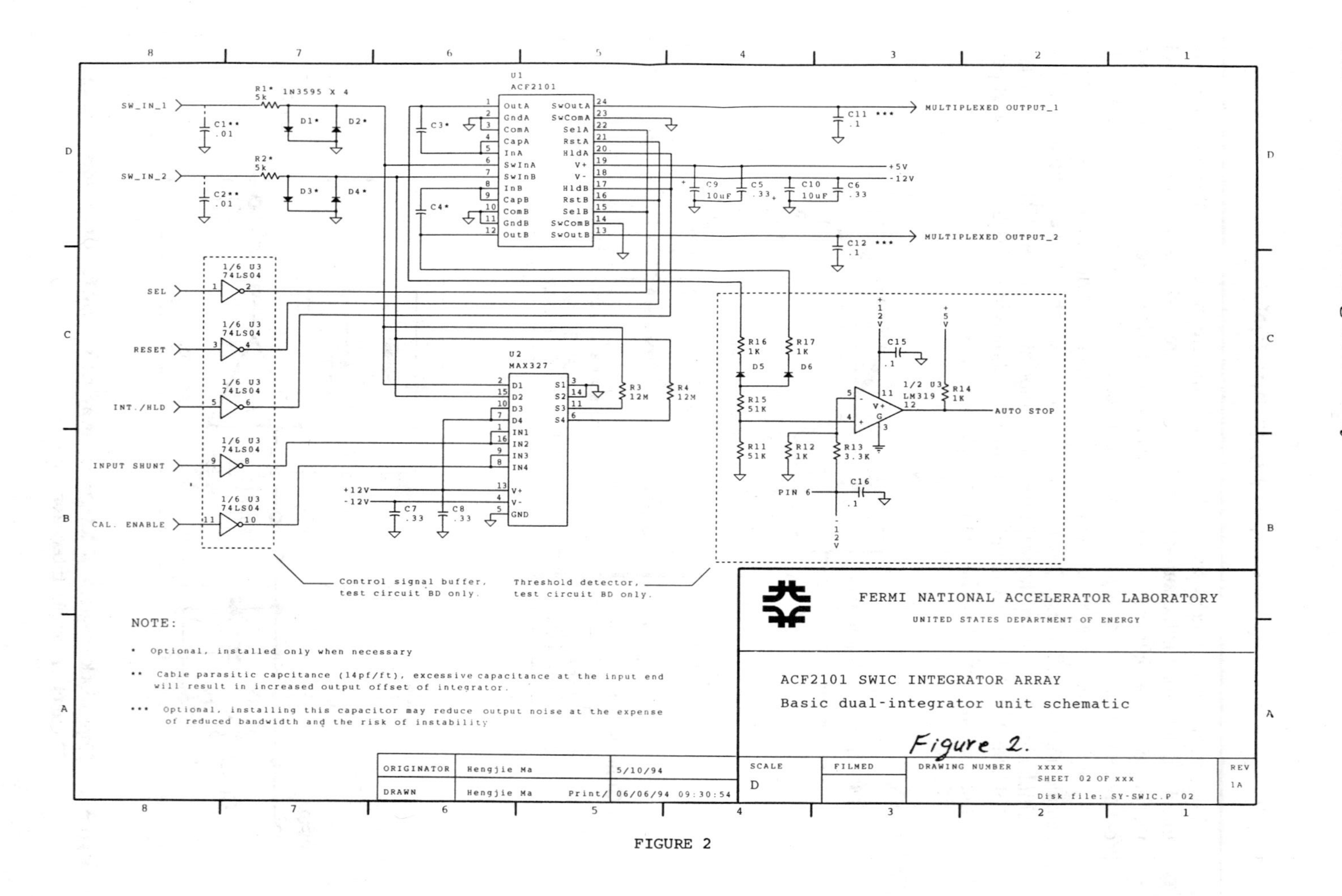
U1 ACF2101
MULTIPLEXED OUTPUT_1
MULTIPLEXED OUTPUT_2
SW_IN_1
SW_IN_2
1N3595 X 4
SEL
RESET
INT./HLD
INPUT SHUNT
CAL. ENABLE
1/6 U3 74LS04
U2 MAX327
1/2 U3 LM319
AUTO STOP
Control signal buffer, test circuit BD only.
Threshold detector, test circuit BD only.
NOTE:
* Optional, installed only when necessary
** Cable parasitic capcitance (14pf/ft), excessive capacitance at the input end will result in increased output offset of integrator.
*** Optional, installing this capacitor may reduce output noise at the expense of reduced bandwidth and the risk of instability
FERMI NATIONAL ACCELERATOR LABORATORY
UNITED STATES DEPARTMENT OF ENERGY
ACF2101 SWIC INTEGRATOR ARRAY
Basic dual-integrator unit schematic
Figure 2.
ORIGINATOR Hengjie Ma 5/10/94
DRAWN Hengjie Ma Print/ 06/06/94 09:30:54
SCALE
D
FILMED
DRAWING NUMBER xxxx
SHEET 02 OF xxx
Disk file: SY-SWIC.P 02
REV
1A

FIGURE 2

Input Fault-Protection

The SWIC detector is a high-impedance current-source device. When the integrator following it is in the hold mode , the output of the SWIC detector is open-circuited, and the output voltage will rise almost to the voltage of its B-power supply which is from 400 to 2000 volts. That will destroy the circuit immediately. There are several methods of protecting the integrator circuit. One of them is to parallel a big capacitor to the input as suggested by the manufacturer of ACF2101.(3) The capacitor will absorb the charges from the detector during the hold period. The value of the capacitor is determined by the maximum hold period. Obviously, this scheme will not work if the integrator does not operate periodically, or the hold-period is too long as the case of the Switchyard application. A solution for the case of the long hold-period is adding a short-circuit switch at the input end. The shorting switch closes down when the sample/hold switch of the integrator opens up, and vise versa. An ultra-low leakage switch chip MAX 327 from Maxim is chosen for this purpose. MAX 327 is a quad-switch chip. Two of the four switches on one chip are used for shorting the inputs of the two integrator channels provided by one ACF2101 chip. Beside the risk of the input overvoltage presenting during the normal operation, there is also a very slim chance that a broken wire of the signal plan in the detector contacts the high-voltage plan. Adding two shunt diodes D_1 and D_2 as shown in Figure 2 may be a solution for this case. However, the leakage of the diodes is unacceptable when in the beam intensity is very low, which is the case for the Switchyard. Therefore, the protection diodes as D_1 and D_2 will not be installed on the units for the Switchyard application.

Input Cable Stray Capacitance

The cables connecting the SWIC and the integrator chassis have a typical stray capacitance of about 14pf/ft. The lengths of the cables in the Switchyard are several hundred feet. Hence, the loading capacitance presenting at the input end of the integrator would be as much as ten thousand pico-Farads. In contrast to that, the integrating capacitor could be as small as 100pf for handling high-speed sampling or very-low intensity beams. The charges stored in the cables are accumulated over the long hold-period prior to the current integrating cycle. Therefore, without clearing the accumulated charges in the cable, the measured the beam profile would have a certain histogram content, instead of the desired instant picture of the beam profile for the moment of the measurement. Furthermore, the measured profile may also be distorted by the inconsistent input dc offset of the integrator chips. The shorting switch at the input end can discharge the cables during the hold-period. This is another reason that the short-circuit switch should be added.

Output Multiplexing & Noise Reduction

The output multiplexing can be achieved by the series switch at the output end provided by ACF2101. This feature is useful when the data acquisition system does not have enough input channels. The intrinsic noise in the output of ACF2101 is determined by the ratio of the capacitance at the input end to that of the integration capacitor. If the total capacitance at the input end, including the cable stray capacitance, is 10,000pf and only the 100pf internal integration capacitor is used, the minimum output noise for the hold mode will be about 1mV. The noise level can be reduced by adding a parallel capacitor C_{11} and C_{12} at the output end. However, the trade-off will be the reduced bandwidth and therefore the reduced speed, as expected.(4), (5)

Threshold Detection for Automatic Stop

In some special circumstances, the precise moments of stopping the integrator is difficult to determine in advance. It is therefore desired to have the integrator be able to stop automatically at the moment when the built-up voltage on any of the ninety-six channels reaches the preset threshold value (for example, 75% of full range). In the integrator array, this function is realized with a logic OR-ring formed by diode D_5 and D_6 connected to the integrator outputs before the multiplex switches. The level of the threshold is set by comparator U_3.

Current Source for calibration

At the input end of each integrator channel, there is a 12-Megohm resistor (R_3 or R_4) in series with a ultra-low leakage switch provided by U_2 (MAX327) connecting the input end cross the +12V power supply. It forms an 1-μA current source for the calibration. The current source is needed to provide a reference for correcting the errors in the uniformity of the integration capacitors and the dc offsets of the integrator chips. It can be turned on and off by one of the four switches provided by U_2.

INTEGRATOR OPERATION MODES

Operating the switches of an ACF2101 produces little charge transfer. (2) However, a certain output dc offset was still observed during the test on the prototype. It has been found that the dc offset is caused by the offset voltage of the integrator input. The offset voltage charges the input capacitance including the stray capacitance of the input cables. The resultant output dc offset is inversely proportional to the capacitance of the integrator capacitor. When the integration capacitor is greater than 1000pf, the offset becomes very small. In our test, it is less than 2 mV. However, if an application requires a smaller integration capacitor such as 100pf, the increased output offset would become a problem.

Fortunately, the SWIC application allows using a special operating sequence (special mode) to reduce the output offset significantly. For the integrator array, a standard operating sequence (standard mode) would start with the sample/hold and reset switch opened, and the shorting switch closed. The reset switch is first closed for a moment to clear the integrator capacitor. The shorting switch is then opened up and the sample/hold switch is closed to start an integration process. After a certain time, the sample/hold switch is opened up to end the integration and hold the output voltage.

The special operating sequence for the SWIC application starts with the rest, sample/hold, and shorting switch closed. Therefore, the integration capacitor is already cleared, and the integrator is in the mode of an unit-gain amplifier. The shorting switch is first opened up, and 100 μ-second later the reset switch is opened up to start the integration. After a certain time, the sample/hold switch is opened up and the shorting switch is closed down. The integrator is in the hold mode. After the data is taken, the sample/hold, shorting, and rest switch are re-closed, and the integrator is ready for the next cycle.

SPECIFICATIONS AND PRELIMINARY TEST

A prototype containing only two channels was built for the proof-test. The resut is summarized in follows:

1. Input dynamic range	10 pA $\sim$ 100 μA
2. Input current polarity	Unipolar, positive
3. Output voltage swing	0 ~ -10V
4. Integration time range	10 μsec. ~ 1 sec.
5. Integration capacitor	100pf (internal), 0.001 or 0.01 μF (external)
6. Output offset	15 mV (internal cap. only/special mode) 2 mV (w/1000pf ext. cap./special mode) 150mV (internal cap. only/standard mode) -15 mV (w/1000pf ext. cap./ stanard mode)

7. Output noise level — <1mV (during hold period)
$10 + C_{in}/C_{integrator}$ * external noise level at input port (mV, during integration period)

8. Total output droop 1 sec. after reset — <2mV (int. cap. only)*
<1mV (w/1000pf ext. cap.)*

*Test condition: integration capacitor was charged to 8V, standard operating mode.

REFERENCES

1. Higgins, B., " User's Manual for microprocessor-based SWIC scanner," Fermilab Technical Memo TM-868, 1900.000, April 2, 1979.

2. ACF2101: Low noise, dual switched integrator, Data Sheet, Bur-Brown Co., 1993.

3. DEM-ACF2101BP evaluation fixture, Data sheet LI-415, Bur-Brown Co., May, 1992.

4. Baker, B., " Comparison of noise performance between a FET transimpedance amplifier and a switched integrator," Bur-Brown Co. Application Bulletin AB-057A, Jan., 1994.

5. Baker, B., " Improved noise performance of the ACF2101 switched integrator," Bur-Brown Co. Application Bulletin AB-053, May, 1993.

Ion Probe for Beam Position and Profile Measurement

John A. Pasour and Mai T. Ngo
Mission Research Corporation, 8560 Cinderbed Rd., Newington, VA 22122

Abstract

We describe a non-perturbing diagnostic to measure the position and profile of tightly bunched (μm-sized) beams. The probe consists of a finely-focused, low-energy (≤30 keV) ion beam that is injected across the path of the beam to be measured. The deflections of the ions are measured with a gated microchannel plate (MCP) detector and recorded with a video camera and frame grabber. By appropriately selecting the ion species and energy, the probe can be used with beams having a wide range of charge density. We will describe two operating regimes. The first holds for very short duration bunches, with the deflection scaling simply as the bunch charge divided by the bunch radius. More generally, the deflection depends on bunch length and must be solved numerically. We are now building a probe, using a specially-designed ion gun, that will be tested at SLAC. We will present the theory of operation of the device, discuss various possible configurations, and describe the components we are incorporating in the initial version of the probe.

INTRODUCTION

The general form of the ion probe diagnostic that we are developing is shown in Fig. 1. This diagnostic uses a tightly focused ion stream that is injected across the beam tube in the path of the high-energy beam. The probe ions are deflected by the high-energy beam, and the direction and magnitude of the deflection are directly related to the spatial and temporal charge distribution of the accelerated beam. Easily-resolved deflections can be produced by microbunches with total charge on the order of a nCoul and pulse durations of a few psec. The geometry of Fig. 1 will be used throughout. The high-energy bunch to be diagnosed travels along the z axis. The probe ions are injected in the y direction, but displaced from the axis in the x direction by an amount x_0, the impact parameter. The bunch has a radius r_b and a duration τ_b, corresponding to a length $L_b = c\tau_b$. The ions are deflected through an angle θ by the electromagnetic fields of the bunch. For highly relativistic bunches, the deflection does not depend on bunch energy.

We have previously developed an electron probe (1) which has many of the desirable features required for this application. However, the electron probe is restricted to use with beams which are not too tightly focused, because the space-charge potential of a strongly-bunched beam reflects the probe electrons unless their energy is very high. A probe of positive ions promises to have important advantages over an electron probe, including 1) ability to penetrate dense electron

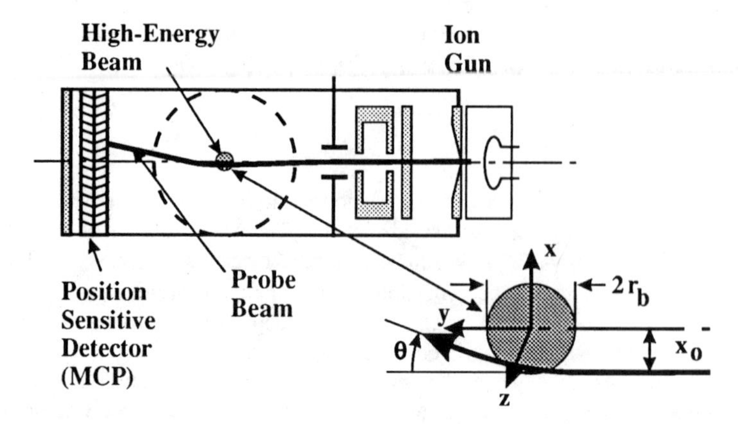

Figure 1. Geometry of the ion probe.

bunches, 2) less sensitivity to stray magnetic fields, 3) greater signal-to-noise ratio (via transit time isolation), 4) broader operating range (ability to adjust ion mass as well as energy), and 5) increased resolution. The trade-off is that a suitable ion gun is more complex than the electron gun.

We have analyzed this probe analytically and numerically. In either case, the fields of the bunch, as derived in Ref. 1, are used to calculate the ion deflections. We will present selected results of these calculations to illustrate the sensitivity and predicted operational characteristics of the probe.

IMPULSE MODEL

We begin by considering the probe behavior in the limit of a very short duration bunch, in which case the ion deflection can be considered to be an impulse. The situation is as illustrated in Fig. 2. On the timescale of the bunch, the ions of velocity v appear stationary, and the deflected ions are those that are within the interaction length L shown when the bunch passes. These ions arrive at the detector over a period of time given approximately by L/v. Thus, even for a psec bunch, the deflected ion pulse duration is typically a few nsec. The maximum deflection is acquired by the ions in the peak field, i.e., those at the edge of the bunch for a uniform bunch or those at r = 1.585 σ for a gaussian beam.

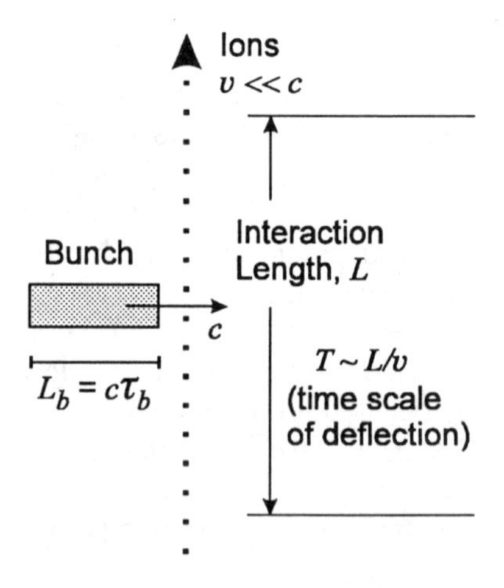

Figure 2. Diagram of interaction.

From the Lorentz force equation, we have for a uniform beam that the maximum deflection velocity is given by

$$\Delta v_{x_{\max}} = \frac{Ze}{Am_p} E_{x\max} \Delta t \approx \frac{Ze}{Am_p} \frac{I_b}{2\pi r_b \varepsilon_0 c} \tau_b \; , \tag{1}$$

where Ze and Am_p are the charge and mass of the ion, and I_b is the current of the electron bunch. In Eq. 1 we have assumed that the impulse duration is simply the bunch duration. Expressing the ion velocity in terms of the ion energy $W = eV$, where V is the ion gun voltage, and using the bunch charge Q_b, we obtain the following theoretical relationship for the maximum deflection angle:

$$\theta_{\max} = 748^{\circ} \sqrt{\frac{|Z|}{AV_{[\mathrm{kV}]}} \frac{Q_{b[\mathrm{nCoul}]}}{r_{b[\mu\mathrm{m}]}}} \, . \tag{2}$$

Similarly, for a gaussian beam, the maximum deflection angle is

$$\theta_{\max} = 337^{\circ} \sqrt{\frac{|Z|}{AV_{[\mathrm{kV}]}} \frac{Q_{b[\mathrm{nCoul}]}}{\sigma_{[\mu\mathrm{m}]}}} \, . \tag{3}$$

Comparing these simple expressions with numerical calculations of the actual ion deflection, as shown in Fig. 3, shows that this simple model is valid over a wide range of parameters. In general, the model is valid so long as an ion does not move appreciably for the duration of the bunch; i.e., when $r_b/v >> \tau_b \Rightarrow r_b/L_b >> v/c$.

If this condition is not satisfied, the deflection depends on the bunch duration as well as the charge and radius. Figure 4 shows the numerically-calculated maximum deflection as a function of pulse duration for several sets of parameters. Clearly, as the ion velocity is increased (either by increasing the voltage or decreasing the mass), the impulse model begins to lose its validity. By properly choosing the operating regime, one can in effect select whether or not the probe is sensitive to bunch length.

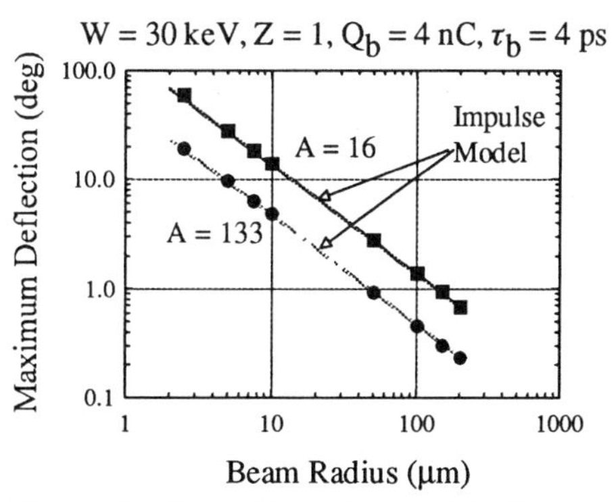

Figure 3. Test of impulse model. Points are from numerical calculations.

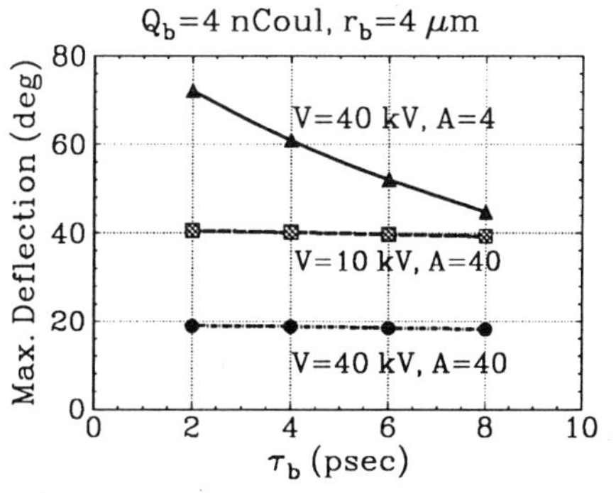

Figure 4. Dependence on bunch length.

MODES OF OPERATION

In normal operation, a small-diameter ion beam ($r_{ion} << r_b$) is scanned across the bunch path (assumed to be repetitive and reproducible), and the maximum deflection as a function of impact parameter x_0 is determined. For ions charged oppositely to the bunch, the deflection peaks at the location of peak field (not always true for identically-charged ions). Thus, the value of x_0 for maximum deflection determines the beam size, the value for no deflection corresponds to the bunch centroid, and the deflection amplitude provides a measure of the bunch charge (and possibly bunch length) and/or a more precise measure of bunch radius if Q_b is already known.

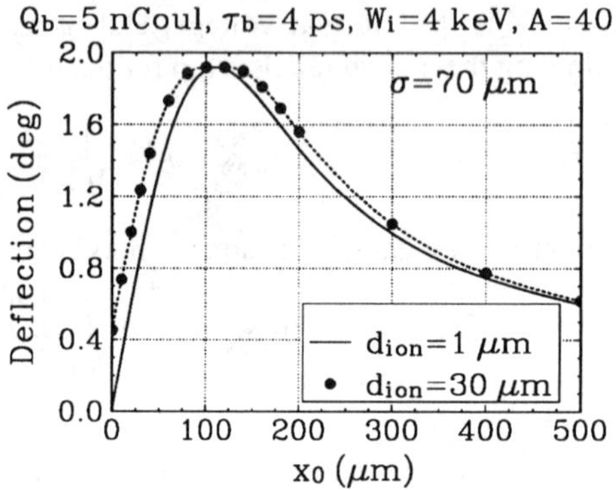

Figure 5. Deflection vs. impact parameter with two ion beam diameters.

A sample curve representative of the parameters of our planned SLAC experiment is shown in Fig. 5. Here, the effect of a finite ion beam size is also shown. The 30-μm beam results in a slight broadening of the deflection curve, but the location and magnitude of the peak deflection are unchanged and well defined.

It is possible to use an ion beam somewhat larger in diameter than the bunch, in which case the ions striking the detector can provide a measure of the maximum deflection angle from a single bunch. This operating mode is particularly important for μm or sub-μm bunches, because it is difficult to generate ion beams this small. An example of this mode is shown in Fig. 6, where we have assumed a 50-μm-diameter ion beam. The calculations use 500 ions equally distributed from $x = 0$ to $x = 25$ μm, and we show a histogram plot of the ion positions 1 cm beyond the intersection of the two beams. The width of the bins is 20 μm, which is a good approximation for

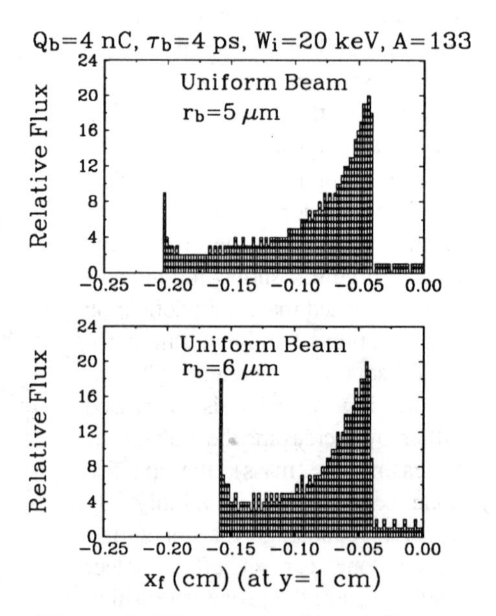

Figure 6. Histograms of ion deflection with a large-diameter ion beam.

the resolution of an MCP detector. As with the small-diameter beam, the largest deflection angle gives a good measure of bunch radius, but now the ions do not have to be scanned. (However, the ion density must be sufficiently large to ensure that a reasonable number of ions are in the peak field of the bunch). The large step that occurs in both plots at $x_f \sim -0.04$ cm is simply due to the finite size of the ion beam. It represents the deflection that occurs at $x_0 = 25$ μm, which is about the same in both cases because the enclosed charge at that value of x_0 is the same. (If the calculation were performed over a much longer time interval, the points near the origin would also be much more numerous, because there would be many ions that do not interact strongly with the bunch.) The relative number of ions deflected through the various angles also gives an indication of bunch profile. For example, a gaussian bunch produces a flatter deflection vs. impact parameter curve, resulting in a weighting of the histogram toward larger deflections than in the case of the uniform beams shown here. However, this information may be difficult to extract from laboratory measurements.

It is also possible to monitor the ion energy rather than just the deflection angle to determine bunch parameters, because the electric field that causes the deflection also changes the ion velocity. Figure 7 illustrates the effect. The left graph shows the ion energy after the interaction with the bunch as a function of time for two different bunch radii ($t = 0$ is arbitrarily chosen here). The first ions to interact with the bunch have their energy decreased, because they are beyond the accelerator axis when the bunch arrives and are therefore pulled back toward the bunch. Ions injected later have not yet reached the accelerator axis when the bunch arrives, so they are accelerated. The right graph shows the maximum and minimum ion energy as a function of bunch radius. Typically, the net change in ion energy varies with bunch parameters in the same way as the maximum deflection angle does. Thus, a larger diameter bunch or lower bunch charge provides a smaller change in ion energy. Again, the ion beam can be larger than the bunch, but a sufficient number of ions must pass through the center of the bunch trajectory to ensure an accurate measurement (so a minimum ion density is required).

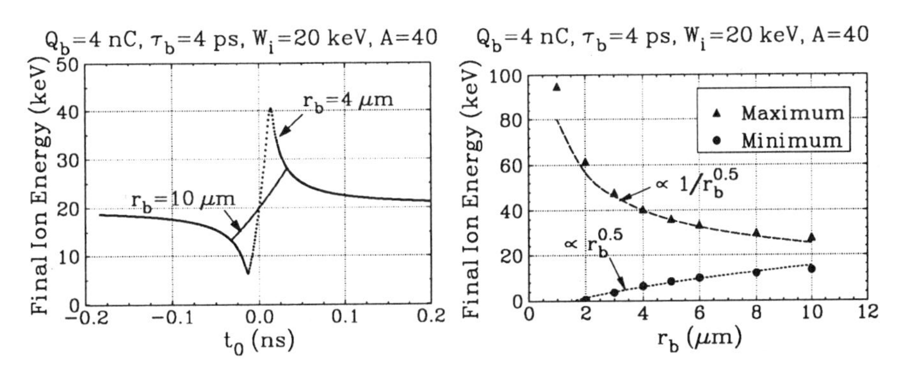

Figure 7. Ion energy variation during interaction with a bunch.

Note that for the parameters in the right hand graph, some ions are reflected when the bunch radius is decreased below 2 μm. The probe could be configured to monitor these reflected ions, which might actually be preferred for sub-μm bunches. By injecting the ions with a velocity component parallel to the accelerator axis and displacing the ion gun and the detector in z, the z coordinate of the reflected ions at the detector becomes a direct measure of reflected ion energy.

PROTOTYPE PROBE DESIGN

We are developing an ion probe to be tested at SLAC at the end of the two-mile accelerator. At this position, the bunch has a typical cross section of 50 $\times$ 100 μm, a length of 1 mm (3 psec), and up to 5 nCoul of charge. From the above results, an ion probe utilizing Ar^+ ions at $\leq$10 keV should produce easily observable deflections of tens of mrad. Such deflections allow us to place the detector 15 to 30 cm from the beam tube axis and have the deflected ions remain within the field of view of a 40-mm-diameter MCP. This spacing is sufficient to allow good transit time isolation via a gated detector (the ions arrive at the detector long after the bunch has passed). Gating not only minimizes the background signal, it decreases the spot brightness produced by undeflected ions. Ideally, the detector should be on only during the time that deflected ions are arriving, which is typically a few nsec. We are incorporating a strip-line gated MCP, provided by Lawrence Livermore National Laboratory (P. Bell), that can easily be gated on for 5 to 10 nsec.(2)

To test the performance of the detector in the radiation environment, we have installed a package consisting of the gated detector, a video camera, and an ion pump and ionization gauge next to the beam line at SLAC. Background noise does not appear to be a problem based on preliminary measurements with this apparatus. The assembly is being left in place and monitored periodically to determine the lifetime of the components.

The ion beam must satisfy a number of stringent requirements. Obviously, the ion beam divergence must be much less than the deflection angle to be measured. There must also be enough ions in the vicinity of the bunch as it passes to provide good statistics. This imposes a minimum current density requirement. Furthermore, the number of deflected ions reaching the detector must be large enough to overcome whatever noise is present. The current density condition is straightforward to analyze, resulting in a value of several mA/cm^2 for our parameters. To demonstrate this, we have performed a simple calculation using a 35-nA, 30-μm-diameter ion beam (5 mA/cm^2), with the ions initially equally spaced in an array. An expanded view of the ions that have undergone the largest deflection is shown in Fig. 8. Here, each point is an individual ion, and we are looking in the plane of the detector (xz plane, deflection direction is downward). Clearly, this current would be sufficient assuming a detector that can respond to individual ions (as the MCP can), provided that the background noise is small. By gating the detector and using transit time isolation as described above, we believe that this will be the case.

To generate an ion beam for this application, we are building an ion gun that uses a multicusp plasma source (from K.N. Leung at Lawrence Berkeley Lab) together with an optical system that we are developing from a design by R. Keller, also of LBL. The gun has a long focal length (~18 cm), so that the ion beam waist will be at the accelerator beam tube axis. This design will also allow the detector to be placed up to 30 cm from the accelerator axis. A scaled diagram of the entire configuration is shown in Fig. 9. The gun is predicted to provide a beam having a current density >50 mA/cm^2 and a divergence of several mrad (depending on the size of the aperture).

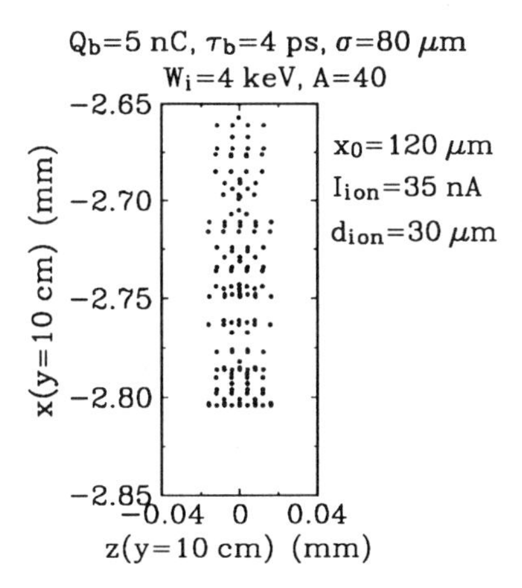

Figure 8. Deflected ions as seen by a detector 10 cm from the accelerator axis.

ACKNOWLEDGMENTS

This work is supported by the U.S. Department of Energy under its Small Business Innovative Research program. We thank Marc Ross and Douglas McCormick of SLAC, Ka Ngo Leung and Roderich Keller of LBL, and Perry Bell of LLNL for their assistance in developing and testing the prototype diagnostic.

REFERENCES

1. Pasour, J.A. and Ngo, M.T., "Nonperturbing electron beam probe to diagnose charged particle beams," Rev. Sci. Instrum. **63**, 3027 (1992).
2. Bell, P.M., et al., "Measurements with a 35-psec gate time microchannel plate camera," SPIE **Vol. 1346**, Ultrahigh- and High-Speed Photography, Videography, Photonics, and Velocimetry '90, p.456 (1990).

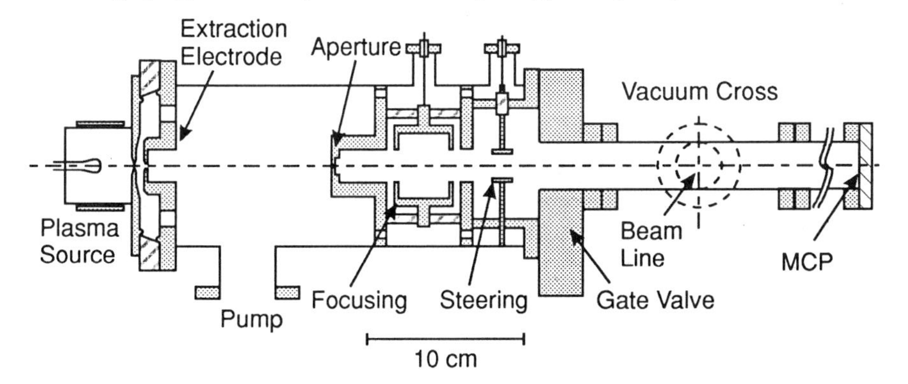

Figure 9. Schematic diagram of prototype ion probe (to scale).

Fermilab Booster Ion Profile Monitor System Using LabView

J. R. Zagel, D. Chen, and J. Crisp
*Fermi National Accelerator Laboratory, P.O. Box 500, Batavia, IL 60510**

Abstract

The new Booster Ion Profile Monitor has been implemented to simultaneously capture both horizontal and vertical profiles at a once-per-turn sample rate, throughout a Booster cycle. The system uses LabVIEW software running on a MacIntosh Quadra 650 talking to both VME and CAMAC hardware. Microchannel plate voltage is turned on just prior to making a measurement and automatically turned off when the measurement is complete. This action allows using a high gain while preserving microchannel plate lifetime. The data captured may be archived for later analysis. Current analysis available include position, emittance/sigma, 2D color intensity plot of raw data, and single turn profiles for any turn during the cycle.

INTRODUCTION

The Booster Ion Profile Monitor is the result of much previous work. Originally installed in the Fermilab Anti-Proton source,(1) the hardware has been modified and reinstalled in the Booster.(2) It has been used to make measurements of transverse profiles and emittances.(3) The current system is now operational in both horizontal and vertical planes and is interfaced to the Fermilab accelerator control system (ACNET). The system is now comprised of a cost reduced set of preamplifiers in the tunnel and new data acquisition hardware. We have used commercial hardware and software where possible to minimize development time and cost.

An amplifier is required to convert the 140 nanoamp (namp) signal to the 0 to 5 volt 50 Ω input of the A/D cards. In the Booster Ionization Profile Monitor, up to 48 anode strips with 1.5 mm spacing are used to measure the transverse profile of the proton beam. Because the profile can be measured each Booster turn, or about 1.6 μsec at extraction, each strip requires a dedicated amplifier with sufficient bandwidth.

Data is acquired using 4 channel, 1 MHz OmniByte Comet digitizers that are provided a synchronous gate once per Booster revolution. They contain enough on-board memory to capture a complete Booster cycle, on the order of 20,000 turns in 33 msec. The digitizers reside in a VME chassis along with timing cards that provide gates derived from the accelerator clock system (TCLK). This chassis is controlled by a MacIntosh Quadra 650 computer running National Instruments' LabView. Programming of the system was accomplished in this graphical language to take advantage of its vast library of routines and rapid prototyping capabilities. A side benefit of this is that it is a natural way for an engineer to accomplish a measurement system without having to resort to C or assembly code for the majority of the project. Enhancements were produced within the Instrumentation Department,(4) to allow the presentation of results and control of the measurements through ACNET.

---

* Operated by the Universities Research Association under contract with the U. S. Dept. of Energy

SYSTEM DESIGN

Microchannel Plates

The current gain of a microchannel plate, (MCP), can be estimated from the following equation, (5)

$$gain = \frac{I_{out}}{I_{in}} = .85\left(182 + \frac{V_{mcp}}{440}\right)^{\frac{1}{2}}\left(\frac{V_{mcp}}{440}\right)^{9.5}$$

For the Booster Ionization Profile Monitor, (IPM), a comparison of calculated and measured gain as a function of MCP voltage agrees up to about 700 volts where saturation effects become significant, fig 1. With the beam intensity at 3e12 protons, the incident current on the MCP was 0.75 namps for each 1.5x100 mm anode strip. At 720 volts, the initial gain of 1550 provided 1100 namps at the MCP output, significantly greater than the bias current of 675 namps per strip. The MCP will provide this current until the capacitance across the plate begins to discharge. This will cause a non linear gradient across the channel resulting in reduced gain and a smaller signal current.

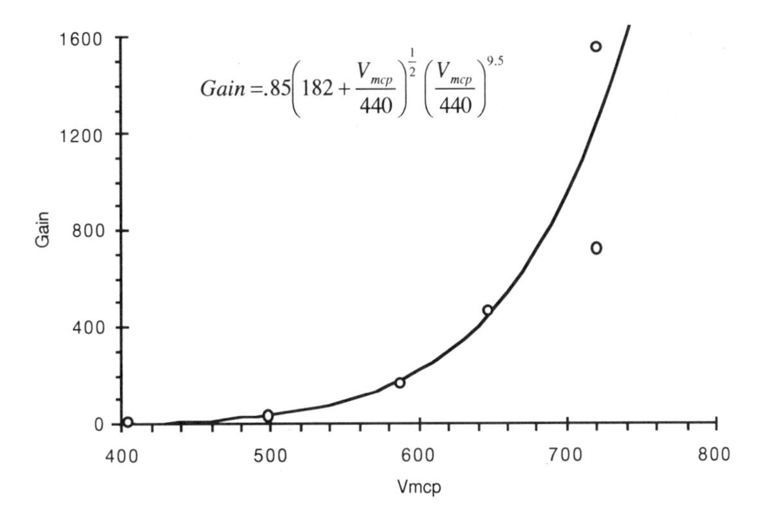

Figure 1. Measured and calculated gain of Booster MCP. The signal was beginning to saturate at the 720 volt bias, both the initial (upper point), and steady state (lower point) gain is plotted at this voltage.

Assuming a relative permitivity of 8.3 for the glass and a thickness of 1 mm, the 80 by 100 mm MCP will have 333 pf of capacitance, accounting for the size and spacing of the channels, or about 6.2 pf for each strip. At the 720 volt bias, the current in excess of the bias current, 1100-675 namps, would discharge 6.2 pf by 100 volts in 1.5 msec, consistent with the observed behavior. The 20 MΩ impedance across the MCP forms a 7 msec charging time constant.

To minimize the effects of saturation it is recommended the signal current be less than 25% of the bias current. For 3e12 protons in the Booster, a 590 volt bias provides 553 namps of current per anode strip, a current gain of 187, and a peak signal of 140 namps.

The signal to noise ratio for the MCP output can be estimated as,(5)

$$\left.\frac{S}{N}\right|_{MCP} = \sqrt{\frac{I_{in}}{e\,1.8\sqrt{\pi}\,B}}$$

$$I_{in} = ion\ current\ striking\ MCP$$

$$B = 3db\ bandwidth\ in\ Hz$$

For the Booster, the incident ion current of 0.75 namps and bandwidth of 150 kHz provides a signal to noise ratio of 100.

Detector

The time required for the signal to cascade through the MCP is only 270 psec with a 30 picosec spread.(5) Thus the MCP itself is fast enough to resolve the longitudinal profile of the 3 to 5 nsec long bunches. The Booster IPM uses an 8 kV clearing field across a 12 cm gap. The time required for an electron, initially at rest, to drift from the center of the gap to the surface of the MCP would be only 2.7 nsec, 114 nsec for a proton, and 484 nsec for a singly ionized water molecule. The polarity of the clearing field is chosen to drive positive ions into the MCP and it is believed, but not known, that the majority are ionized water molecules. With a bunch spacing of 26.4 to 18.9 nsec the spread in drift times and initial velocities should reduce the amount of energy in the anode strip signal at the bunch rate, and its harmonics. In practice, a significant amount remains but it is not clear how much is caused by ions amplified through the MCP and how much is coupling between electromagnetic fields in the detector and the strips or conductors carrying their signal.

The anode, or output, side of the MCP is shielded from the beam side with a double sided circuit card whose top and bottom are coupled through capacitors spaced around the square MCP mounting hole. The top of the board is grounded to the enclosure and the top of the MCP is connected to the bottom of the board. This was done to reduce beam coupling into the signal wires between the anode strips and the vacuum feed through connector. Typical MCP bias voltages are, 590 volts across the plate itself and 230 volts between the MCP and the anode strips.

For a total impedance of 20 MΩ across the 1.02 mm thick, 80 by 100 mm wide MCP, the volume resistivity is 160 MΩ-m making the skin depth 2e6 meters for frequencies much greater than 20 Hz. The top and bottom surfaces of the MCP are coated with nichrome or inconel and have an impedance of 200 Ω/sq. This would require a uniform thickness of 5e-9 meters of nichrome, which has a skin depth of 1.6e-5 meters at 1 Ghz. Thus, shielding the strips from the beam fields or from beam excited modes in the detector is difficult.

Preamplifiers

The signal to noise ratio of the amplifier will be proportional to the square root of the input impedance, provided the dominant noise source is the thermal noise of the impedance itself.

$$V_{signal} = R I_{out} \qquad (I_{out} = MCP\ output)$$

$$V_{noise} = \sqrt{RKTB} = 64 \times 10^{-12} \frac{V}{\sqrt{\Omega\ Hz}} \sqrt{RB} \qquad (at\ 300° K)$$

$$\left.\frac{S}{N}\right|_{Amp} = \sqrt{\frac{R}{KTB}}\, I_{out}$$

For a signal of 140 namps, bandwidth of 150 kHz, and impedance of 5 kΩ, the theoretical signal to noise ratio becomes 400. A good low noise op-amp will have an equivalent noise at the input about equal to that of a 1 kΩ resistor. The resulting signal to noise ratio of an op-amp and two 5 kΩ input resistors is about 200. (Note: S/N is 10 times smaller for 50 Ω than for 5 kΩ.)

The signal must be transported from the anode strips in the detector to the amplifier. Transmission lines having 50 Ω impedance and velocity of 67% of the speed of light will have about 30 pf per foot of capacitance, $C \approx 1/Z_0\ \upsilon$. This capacitance, in combination with the amplifier input impedance, can be used to limit the bandwidth of the signal. For cables much shorter than a wavelength, a good approximation to frequency response is simply equating the 3 dB bandwidth to $1/2\pi RC$, fig. 2. With 4 feet of RG-174 cable and 5 kΩ impedance the bandwidth becomes 260 kHz for the IPM. The cable is 1/4 wavelength long at 41 MHz.

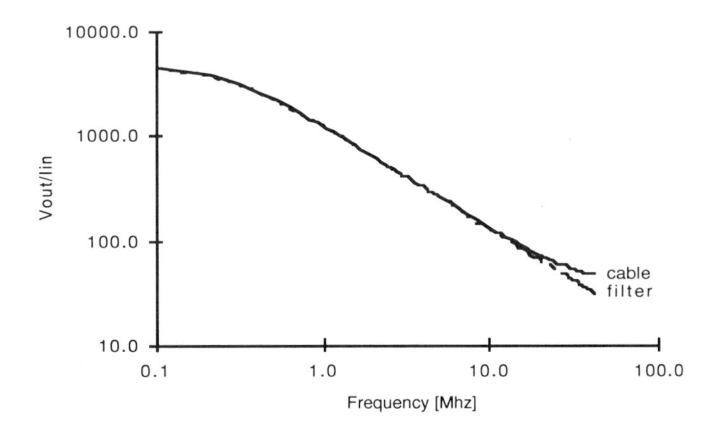

Figure 2. Comparison of output voltage from 5 kΩ resistor terminating 4 ft of cable driven with a current source to first order low pass filter, $f_0 = 260$kHz.

Placing the amplifiers near the detector in the beam enclosure minimizes the possible potential difference between grounds. Nearly one volt at the power line frequency and it's harmonics has been measured between the beam enclosure ground and that in the equipment galleries. The open anode strip makes the noise source impedance large compared to the amplifier input impedance at these frequencies, reducing the effect.

Several hundred millivolts of high frequency components induced by the beam have been measured on the anode strip signals. Although their amplitude was reduced with shielding inside the detector, they could not be completely removed. Because the slew rate of the op-amp was exceeded and faster in the negative direction, the presence of the high frequency components induced a substantial negative offset in the preamp output. To reduce the effect, better shielding was installed in the detector, a 1 MHz low pass filter was placed at the amplifier input, and an op-amp with very large slew rate was selected. Using all three steps significantly reduced, but did not completely eliminate, the problem. A small offset is still experienced near transition where bunches become shortest and higher harmonics are most intense.

Current feedback style amplifiers have the highest slew rate and therefore the best rejection to rf at their input. The AD844 amplifier was selected by virtue of it's low noise. The disadvantage of current feedback amplifiers is that they have significantly higher input current noise, the equivalent of 6.1 namps rms with a 150 kHz bandwidth. Combined with the 140 namp signal level, the current noise results in a signal to noise ratio of only 23. A more typical op-amp, such as the OP-37, has 30 times less current noise but has a slew rate 120 times slower, 17 V/μsec compared to 2000 V/μsec. Current feedback amplifiers also have larger temperature dependent voltage and current drifts. In the IPM these offsets are corrected in software by measuring them each cycle prior to beam injection.

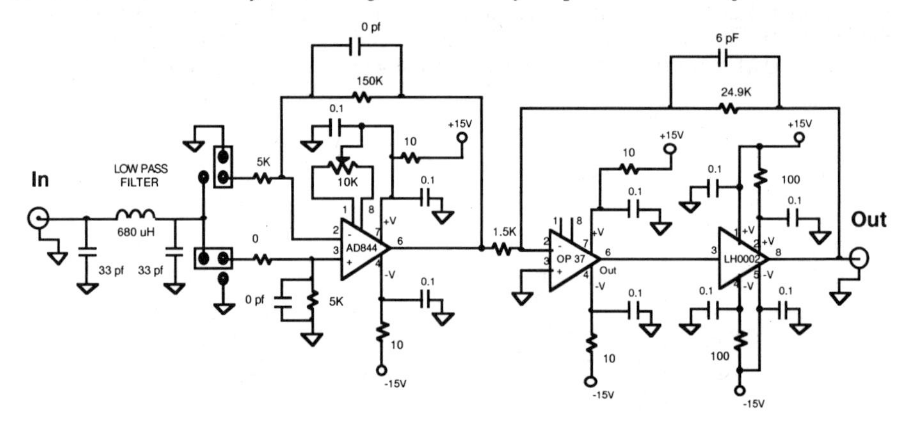

Figure 3. Booster Ionization Profile Monitor Preamp.

The amplifier is inverting so electrons collected on the anode strips produce a positive output voltage. The 150 kΩ feedback resistor on the first stage forms a 240 kHz low pass filter with the AD844 transcapacitance of 4.5 pf. Transimpedance is 2.5 MΩ . The voltage gain of 500 will affect the amplifiers response to noise which may be present at the input. The LH0002 amplifier at the output is used to drive the 50 Ω load.

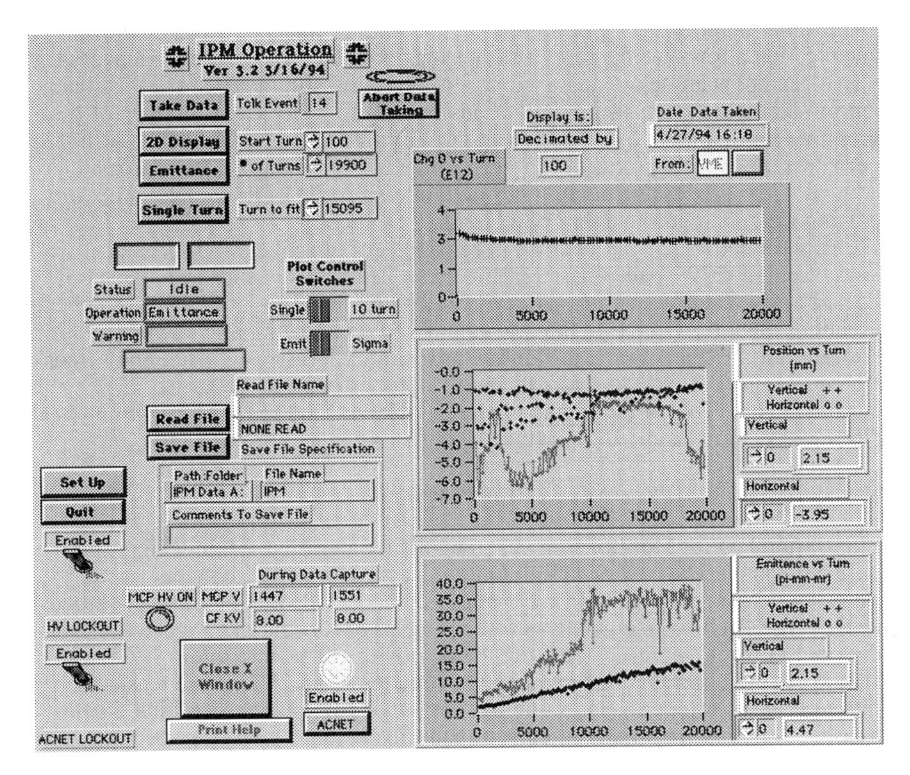

User Interface

The user interface for the IPM is accomplished in 2 ways. A full graphical screen is available using commercial MacIntosh or X-Window software packages. The parameters measured, calculations performed, and settings control are also available through ACNET consoles. Measurements are initiated by determining a TCLK event on which to start data acquisition. MCP bias voltage is applied and the digitizers capture data once per revolution, 2.77-1.59 μsec. 2.5 MBytes of data are transferred from the digitizers into the MacIntosh. Single turn profiles, or emittance/sigma/position vs. turn analysis, is available for the complete, or subset of, the Booster cycle. Data may be archived on an optical disk for later study.

Data Analysis

Data is presented as an array of 20000 samples for each 30 channel plane, both horizontal and vertical. Channels are separated by 1.5 mm. Each 30 channel slice provides a raw, single turn, profile from which a centroid and rms is calculated. Then these values can be used as inputs to a full Gaussian plus linear background fit. The full fit is displayed by default in "Single Turn Profile" mode. It can be turned off to speed up calculation time for "Emittance" and "Position" plotting. At the present time calculated emittance is based on measured sigma and the beta function without dispersion, which does not yield a true emittance.

Raw data includes background and offsets due to bias and drift. The timing of data collection allows the capture of samples prior to beam injection. An average of the first 8 samples are taken and used as a background subtraction for each channel. To minimize calculation time, data is decimated to a practical number of points appropriate for display, on the order of 200 points maximum. If the data of interest covers less than 200 points, then turn by turn information (un-decimated) is displayed. The process of decimating data allows for selecting each "nth" turn or averaging n turns to produce the "Nth" term.

Ions produced when the beam hits residual gas are under the influence of the clearing field and the space charge of the beam itself. This leads to a spread of the collected ions and subsequently the measured sigma is larger than that of the beam. Corrections must be applied to the raw data based on intensity and beam size during the acceleration cycle. We use Monte Carlo simulation to determine the effect of the beam induced field on drifting ions. This effect can be up to 25% to the measured sigma. The correction algorithm is a linear function of the beam intensity and beam size and is applied to data prior to plotting sigma or emittance.

Improvements

Ionization Profile Monitors are designed for the Fermilab Main Ring. The most significant change is to use the High Output Technology, microchannel plate. The impedance across these is 100 kΩ compared to the 20 MΩ of those installed in Booster. The MCP bias current increases linearly with bias voltage but the gain, and signal current, increase with the 10 th power. The maximum MCP gain is limited by the requirement that signal current be less than 25% of the bias current to avoid saturation effects. By reducing the MCP impedance the bias current is increased, allowing larger signal current and thus better signal to noise ratio at the output of the amplifiers.

A planned improvement for the Booster IPM is calculation of a true emittance which includes the dispersion. Using the charge, rf voltage, and bunch length, we can determine momentum spread which leads to the dispersion. Charge and rf voltage signals are now captured with the raw data. The calculation will be added when a true bunch length signal becomes available.

REFERENCES

1. Krider, J., "Residual Gas Beam Profile Monitor", Nuclear Instruments and Methods in Physics Research A278 (1989) pp. 660-663.
2. Rosenzweig, J. B., "Progress Towards a Turn-By-Turn Beam Profile Monitor for the Fermilab Booster", Proceedings Second Annual Accelerator Instrumentation Workshop 1990, Particles and Fields Series 44, pp.328-335.
3. Graves,W.S., "Measurement of Transverse Emittance in the Fermilab Booster", Ph.D. Thesis, University of Wisconsin. May 1994.
4. Blockland, W., "A LabView-based Accelerator Instrumentation Platform, Presented at European Particle Accelerator Conference, June 1994, London.
5. Eberhardt, E.H., "Parameters Pertaining to Microchannel Plates and Microchannel Plate Devices", ITT Electro-Optical Products Division, Technical Note No. 127, August, 1980.

Measuring Micron Size Beams in the SLC Final Focus*

D. McCormick, M. Ross, S. DeBarger, S. Horton-Smith, C. Hunt
T. Gromme, L. Hendrickson, L. A. Yasukawa, G. Sherwin
Stanford Linear Accelerator Center Stanford University.Stanford, California 94309

C. Fritsche, K. McGinnis
*AlliedSignal Inc. Kansas City Division **, Kansas City, Mo. 64141-6159*

Abstract

A pair of high resolution wire scanners have been built and installed in the SLC final focus. The final focus optics uses a set of de-magnifying telescopes, and an ideal location for a beam size monitor is at one of the magnified image points of the interaction point. The image point chosen for these scanners is in the middle of a large bend magnet. The design beam spots here are about 2 microns in the vertical and 20 microns in the horizontal plane. The scanners presented a number of design challenges. In this paper we discuss the mechanical design of the scanner, and fabrication techniques of its ceramic wire support card which holds many 4 and 7 um carbon wires. Accurate motion of the wire during a scan is critical. In this paper we describe tests of stepper motors, gear combinations, and radiation hardened encoders needed to produce the required motion with a step resolution of 80 nanometers. Also presented here are the results of scattered radiation detector placement studies carried out to optimize the signal from the 4 micron wires. Finally, we present measurements from the scanner.

* This work supported by DOE contract DE-AC03-76SF00515
** Operated for the U.S. Department of Energy under Contract No. DE-ACO4-76-DP00613

INTRODUCTION

Wire scanner beam size monitors are used throughout the Stanford Linear Collider (SLC) to make beam profile and emittance measurements.[1,2]. After the conclusion of the 1993 SLC run, the final focus optics were upgraded to provide improved chromatic correction. As part of this upgrade, five new scanners were added to both the north and south final focus.

One of these scanners is located at an image of the interaction point (IP), and it provides a direct estimate of the vertical beam size at the IP. Because of its location, this scanner plays a particularly important role in measuring the properties of the incoming beam and the performance of the final focus optics. At this wire scanner, the design ßx and ßy are small, and the horizontal dispersion is a few centimeters. The aspect ratio of the beam at this scanner is ten to one. The primary goal of this scanner is to measure the vertical beam size and the x-y coupling to within 5% at the full SLC repetition rate and current.

MECHANICAL DESIGN

The location of the IP image is at the center of a large bend magnet. Two types of scanner installations were considered. One design required cutting the steel of the magnet in half, moving the sections apart and building separate coils for each of the resulting halves. A standard SLC wire scanner design, [1] which moves at 45° with respect to the horizontal, would have been used. Another option required boring a vertical hole through the middle of the magnet. A wire scanner with a vertical motion was needed for this design. The impact of the magnet modifications, the performance required of the wire scanner, and the costs of each design were considered. Given that a scanner with vertical motion can make all the required measurements, and that calculations of the effects from the hole in the magnet indicated only minor field distortion, the scanner with the vertical motion was chosen. Measuring the vertical beam size is done using a horizontal wire. The x-y coupling and the x beam size measurements are obtained by using two sets of wires; one at ±5 degrees and another at ±15 degrees with respect to the horizontal wire. The angles of theses wires with respect to the beam must be known to ±3.0 milli radians in order to maintain the required measurement accuracy. The scanner is built so that the angle of the wires with respect to the beam could be measured and adjusted after the scanner installation. The wires are accurately fixed to a ceramic fork which was indexed to surveyor tooling balls on the top of the scanner. Adjusting the angle of the wires is accomplished by having the slide and shaft assembly mounted on a circular saddle. The scanner can be rolled until the wires are at the correct angle with respect to the beam.

WIRE CARD FABRICATION

The wire scanner card is fabricated from a 4.50 x 3.75 x 0.27 inch thick 99.5% alumina substrate (a herman). Machining the herman is done using a CO_2 laser. Parts of the herman are cut away to reveal the outline of the wire card, a two tine fork. The wire card is left attached to the herman in several locations. This provides a means of handling the card without risk of contamination and aids in the installation of the carbon wires. Tooling holes in the card are ground to exact tolerances using a diamond jig grinding machine. These holes are used to mate the wire card precisely to the shaft of the wire scanner. After the wires are attached, the card is scored along the attachment points and broken out of the herman.

V-shaped grooves shown in Fig. 1 were machined into the surface of the card to precisely locate the carbon wires. The grooves are approximately 14 microns wide and 20 microns deep. The machining was performed using a excimer laser with a high precision X-Y table. Since the location of the grooves are referenced to the card's tooling holes, the position of each wire on the wire scanner is known.

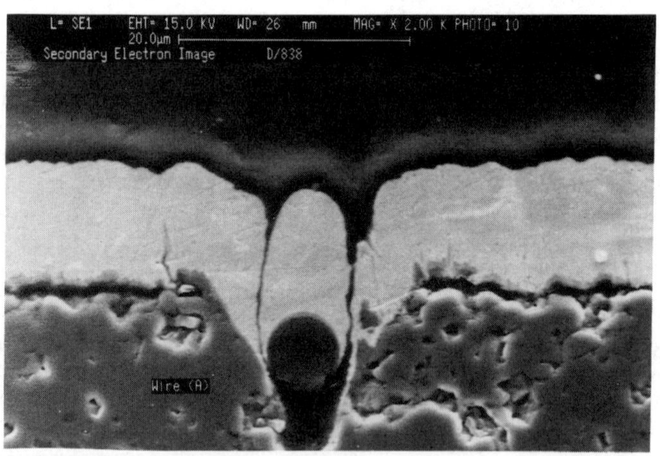

Figure 1. A seven micron carbon wire is positioned in a V groove cut in an alumina substrate.

Attaching the wires to the fork proved to be difficult. The initial approach was to place a mask over the card leaving only the tines exposed. Titanium and then palladium was sputtered over the wire filled grooves. It was believed that this would seal the wires in place with a "blanket" of metalization. Two problems prevented this technique from working. The metalization did not adhere to the tines in the areas where the excimer laser machining was performed. This was found to be caused by a fine non-adhering alumina particulate[3] left behind after the excimer laser machining. This problem was solved by refiring the card at 1400 °C before the wires were placed in the grooves. The particulates were sintered to the card, providing a solid substrate for the metalization process. The metallization adhered to the wires and the tines, but cracks developed between them, and the wires still pulled easily out of the grooves. To solve this, the wires were soldered into the grooves using a 48.5% Sn / 51.5% In solder preform. The solder was reflowed in a nitrogen belt furnace using no flux. A finished wire card is shown in Fig. 2.

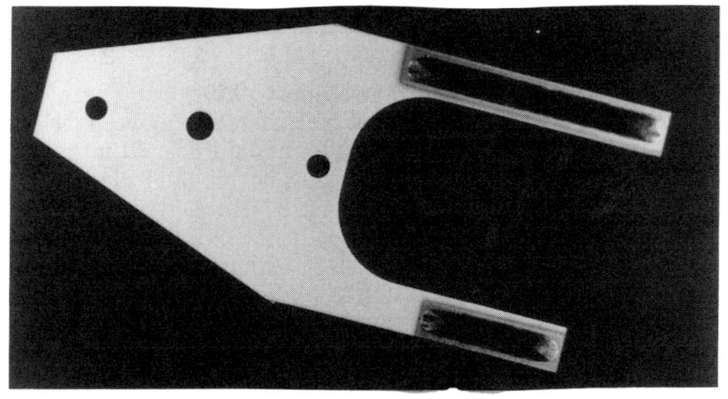

Figure 2. A finished wire card is shown above. The ±15, ±5, and 0 degree wires, and the diamond ground holes which precisely position the card on the shaft the wire scanner are shown.

WIRE SCANNER MOTION

Standard SLC wire scanners use a stepper motor with 200 full steps per revolution. The motor drives a ball screw with a lead of 2mm per revolution, producing a linear motion of 10 microns per full step. The 2 micron vertical spots at the scanner require a much finer step resolution. Microstepping divides full steps of a stepping motor into 125 smaller microsteps. For this scanner, microstepping improved the step resolution to 80 nanometers. A test facility was constructed to measure the uniformity of the motion. A linear optical encoder with a resolution of 50 nanometers was used to measure slide motion.[4]. A least squares linear fit to the encoder data was performed and the residual from the fit to the data is shown in Fig.3. Depending on which part of a normal step a two micron beam sampled, the measurement error could be as large as 30%. These errors were too large to meet the measurement goals for these scanners.

Three attempts were made to reduce this error. The microstepping module used for the tests has a feature that resets the motor to the nearest full step. This forces the scan to begin from the start of a full step. It was hoped that the error would be similar enough from one scan to the next to allow a correction. Unfortunately, the motion varied at different motor speeds and different slide positions and the corrected data had an error equal or larger to the original. Another attempt at

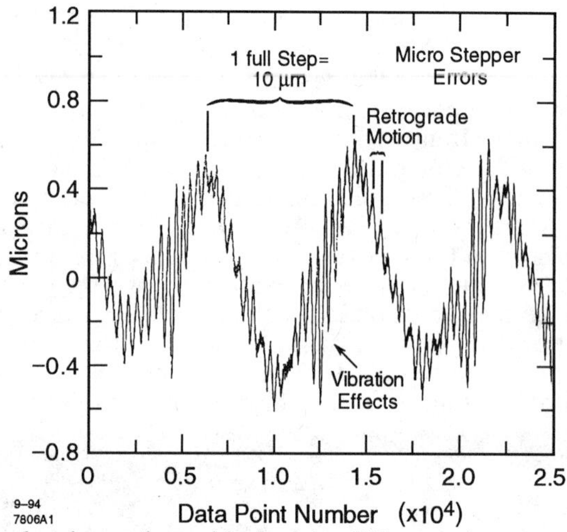

Figure 3. This plot shows the residual of the difference between the position of the scanner recorded by the optical encoder and a linear fit to the data. The stepper motor was using microstepping. The residual clearly shows the microstepper error over a single full step.

reducing the measurement error consisted of using reduction gears instead of microstepping to achieve the required step resolution. Two different types of gear systems were tried. One was a nested set of gears that produced an 80:1 reduction and the other was a rotary index table with a reduction of 180:1. The residuals from both these systems were unacceptable.

These results indicated that it would be unsatisfactory to build a system that relied on the motor step count to determine the position since the motion is uneven. Therefore we decided to measure the actual position of the scanner during the scan. A magnetic linear encoder and interpolating unit was tested [5]. The magnetic encoder was appealing because of its resolution (100nm) and radiation hardness. It has no electronics in the sampling head. It was compared to the optical linear encoder. The residual from the magnetic encoder was about ± 5microns. In spite of the 100nm resolution, the accuracy of the magnetic encoder was insufficient and we decided to install the optical linear encoder on the wire scanner. Radiation dosimetry was installed on the gauge. At present the north final focus encoder has an integrated dose of over 100 kilorads with no observable decrease in performance. It is expected that the semiconductor components in the optical encoder will fail and the unit will have to be replaced after about a year of operation.

At certain motor speeds there is a considerable amount of vibration imparted to the scanner structure. Some tests were done with a laser doppler vibrometer [6] to study the actual motion of the wire during a wire scan. Data from the vibrometer was compared with the data from the optical encoder. The two sets of data showed similar structure but weren't exactly the same. This result prompted an investigation to quantify the amplitude of the vibration and develop techniques to reduce it. In these tests, the scanner was moved at constant speed while the optical encoder sampled the position of the slide. The RMS of the residual to the linear fit was tabulated over all the motor speeds. Various damping mechanisms were fitted to the motor and the scanner and another set of data was taken. Figure 4 shows a plot of the dramatic reduction in the vibration of a damped and undamped wire scanner.

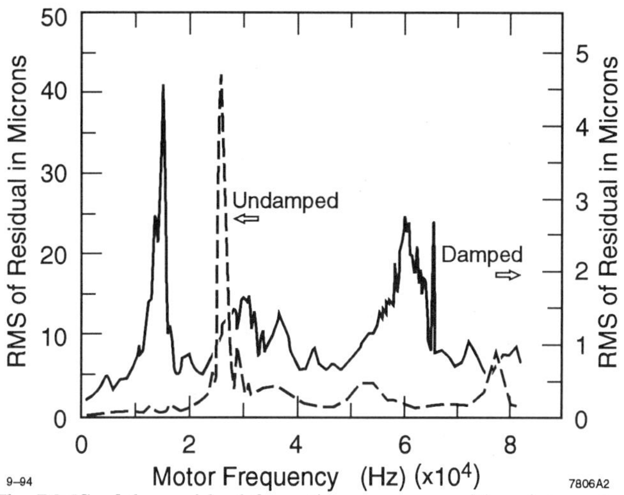

Figure 4. The RMS of the residual from the scanner position data and a linear fit to the data is plotted against the frequency of the stepper motor microsteps. The vertical axis on the left reflects the vibration of the scanner before a damper is added. The axis on the right reflects the reduction of vibration after a rotary damper [7] is attached to the shaft of the motor.

DETECTORS AND SIGNALS

Outside of the IP, these are the first SLC wire scanners to use carbon wires instead of tungsten. In general scanners using tungsten wires have an excellent signal to noise ratio. It was expected that the signal from carbon wires would be about two orders of magnitude less than the signal from the tungsten wires. It was clear that detectors with a gain of at least a factor of 100 would be required. Proportional tube detectors were chosen for their low cost, compact size and rugged construction. The proportional tubes are constructed from a tube 15 cm long with an inside diameter of 1.5 cm and a 40μm sense wire. A mixture of 89% Ar, 10% CO_2 and 1% methane is passed through the tubes at a rate of 0.05 ft^3/hour.

The detectors are placed to intercept bremsstrahlung or lower energy particles scattered from the wire. Since the scanner is in a bend magnet, an excellent location is downstream where the bremsstrahlung from the wire exits the beam pipe. Other detectors were placed on the opposite side of the vacuum chamber from these tubes to detect degraded electrons scattered off the wire. Placing these detectors in the ideal locations has been difficult because of the many beam line components already installed. Background levels in these detectors can be problematic. Proportional tubes used to detect bremsstrahlung from the wires are also sensitive to synchrotron radiation produced by the bend magnet. The tubes placed to intercept the scattered electrons from the wire are also in the path of particles scattered from upstream collimators. Figure 5 shows an example of the signal to noise for good beam conditions. If the beam is unstable or has large emittances, the background in the proportional tubes can increase to a point where accurate beam size measurements are difficult. Figure 6 shows a plot of a skew scan made by scanning all the wires of the scanner. The display also includes: 1) The beam size (WNAMS) and error, for each wire. 2) The beam intensity (TMIT) for each scan. 3) An asymmetry (ASYM) term and error, that indicates the presence of a tail on the beam.

Carbon wires do have an advantage over tungsten wires. Because the signal from the carbon wires is small, wire scans can be done with out turning off the central drift chamber high voltage of the physics detector.

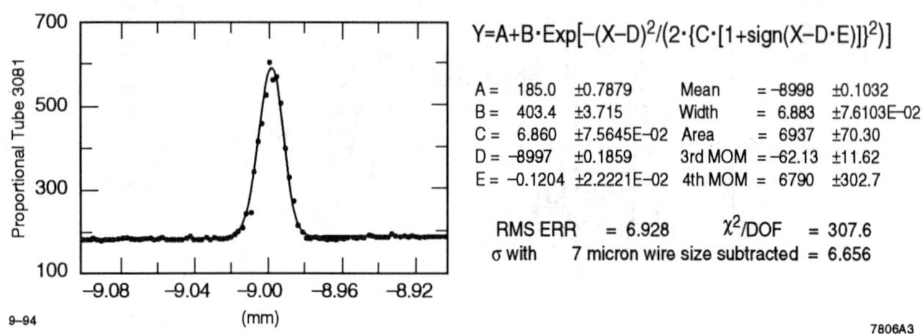

Figure 5. Wire scan using a 7um carbon wire. The elevated background is believed to be due to synchrotron radiation.

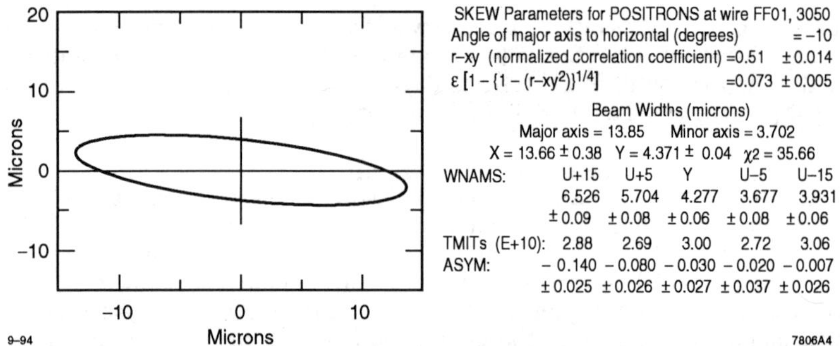

Figure 6. Plot of a skew scan. Results from five separate scans using one wire at each of the five angles are combined to produce measurements of the vertical and horizontal beam size, and the x-y coupling.

PROBLEMS

After operating the scanners for about two months, a number of the 4 micron wires were found to be broken. The wire card was removed and placed in a scanning electron microscope and the broken ends were observed in detail. Figure 7 shows the end of one of the broken wires. The hemispherical shape observed on the broken ends was surprising. It has been proposed that a similar effect to radiation enhanced sublimation [8] of graphite in fusion reactors is responsible for the destruction of the carbon wires.

CONCLUSIONS

This scanner has successfully met its goals and measured spot sizes below three microns and the x-y coupling of the beam. Readout of scanner position during scans is working well and the life time of the electronics in the encoder has been better than expected. The unexpected failure of the carbon wires has caused concern and possible solutions are being examined.

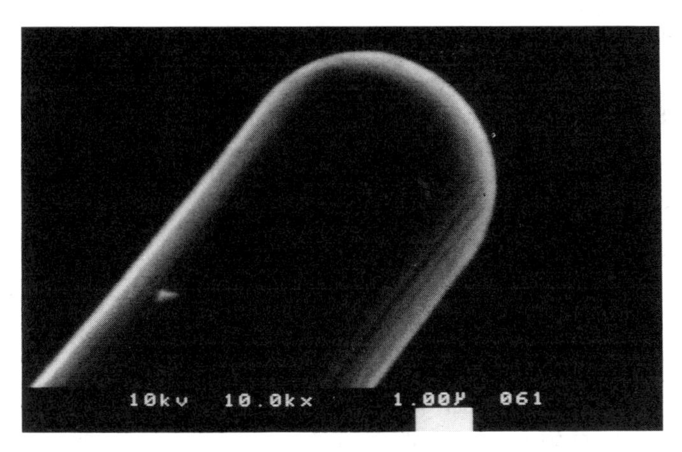

Figure 7. The end of a broken carbon wire is shown under an electron microscope. The wire failed after approximately 2500 scans. For scale, note the 1 micron wide rectangle along the bottom of the picture.

ACKNOWLEDGMENTS

The success of this project is due to the efforts of many accelerator physicists engineers, designers, and technicians. Special mention should be made of the following individuals: P. Emma, C. Field, E. Bong, D. Arnett, Y.Y. Sung, G, Sausa, Janet King, and A. Tilghman.

REFERENCES

[1] M.C. Ross et al,"*Wire Scanners for Beam Size Measurements at the SLC* " IEEE Particle Accelerator Conference. Vol. 2 1201-1203 (1991)

[2] M.C. Ross et al, "*Experience with Wire Scanners at SLC*", SLAC-PUB-6014 December (1992) Invited talk at Beam Instrument Workshop 92 Berkeley, CA October 27-30, (1992)

[3] John E. Smoot et al, "*Excimer Laser Surface Processing of Alumina Substrates*", American Ceramics Society 96th Annual Meeting, April 1994

[4] Heidenhain Inc., 15 commerce Drive Schaumburg, Il 60173 "MT60k optical linear encoder "

[5] Sony Magnascale America Inc. 137 Bristol Lane Orange, CA 92665 "DG50N linear encoder"

[6] Polytec Optronics Inc. 3001 Rehill Ave,, Bld. 5-102 Costa Mesa, CA 92626 "Laser Doppler Vibrometer"

[7] Superior Electric 383 Middle Street Bristol, Conn. 08010 " Lanchester rotary damper"

[8] R. E. Nygren et al, "*Radiation-enhanced Sublimation of Graphite in PISCES Experiments*", J.Vac. Sci. Technol. A8 (3), May/Jun 1990

Diagnostics for High Intensity Beams

L. Rezzonico, S. Adam, M. Humbel
PSI, Paul Scherrer Institute, CH-5234 Villigen-PSI, Switzerland

Abstract

At PSI, the operation of the accelerator system at a proton beam current of 1 mA has become routine. Commissioning as well as operation of the high intensity beam is based on a set of reliable beam diagnostic elements. This paper describes different kinds of wire probes, developed at PSI for high current proton beams. Besides this, a compact probe driving and data taking electronic module is presented and some examples of an advanced presentation of evaluated probe measurement data are given.

INTRODUCTION

The main accelerator in the PSI meson factory is the 590 MeV ring cyclotron, a separate sector cyclotron well suited for high intensity proton beams. In the years 1974 to 1984 the 72 MeV protons injected into the 590 MeV ring cyclotron were produced with injector 1, a conventional AVF cyclotron allowing for beam intensities of 100-200 μA. Since 1984 the injector 2 is used to produce the high intensity beams of 72 MeV protons. The injector 2 is itself a ring cyclotron which uses a 0.87 MeV DC beam coming from a Cockcroft-Walton pre-accelerator. With injector 2 the beam current extracted from the 590 MeV ring cyclotron could be raised to 400-500 μA.(1)

An improvement program aimed at a further raise of the maximum intensity extracted from the 590 MeV ring cyclotron has started in 1990.(2) An important part of this improvement is the enhancement of the RF power fed to the accelerating cavities in order to provide enough power to go into the beam and to allow for an increase of the accelerating voltage. Having 3 out of 4 cavities on high power the beam intensity could be raised to 1 mA in a highly stable operation mode in 1994. With all 4 cavities fully equipped and after some further tuning we expect to achieve the goal of a beam intensity of 1.5 mA until 1996.

These steps to increase the beam intensity could only be successful because they have been accompanied with a careful redesign of the beam diagnostic equipment in the accelerators and in the beam transfer lines.(3)

Experience with setting up high intensity beams has revealed the importance of beam probes that can measure the very beam one wants to produce. All substantial reductions of beam intensity imply changes to many other parameters in the accelerator, resulting in the fact that the beam measured at reduced intensity becomes difficult to compare to the full beam. For most of the new probes a way has been found to fulfill this important wish.

The other essential component, accompanying the diagnostic hardware, is the software for gathering and analyzing the probe data, including elaborate methods for the interpretation of the data to give hints for beam optimisation.

THE PROBE HEAD

General Aspects

The traditional design of a beam probe with a moving block that stops the beam is not feasible for high intensities. For the diagnostics of high intensity beams the two functions of measuring a beam profile and of stopping the beam must be separated and actually should be realized with completely independent devices. In order to stop a beam of more than 100 kW beam power a big metal block with good cooling is needed and the beam spot must be made quite large using a very small incident angle of the beam onto the metal surface. Therefore we have decided to avoid moving beam stops at all and to only provide beam stopping elements at some few fixed locations.

The solutions remaining feasible for differential probes are thin metal blades and elongated double-T-profiles for medium intensity beams and metal ribbons and thin wires for high intensities. The blades and the double-T-profiles have the great advantage that they only need a mounting mechanism holding them from one side, whereas the real high intensity probes all need some kind of a fork.

The thin wires that are most frequently used are from carbon with a diameter of 40 μm and from molybdenum with a diameter of 50 μm. Glueing them to the soft springs that apply the required little bit of tension to the wires is a very delicate procedure.

A method of measuring the profile of a high intensity beam without using a probe head at all is the detection of the light emitted by the residual gas. This scheme is used successfully in the beamline for the 870 KeV DC-beams from the Cockcroft-Walton pre-accelerator.(4)

The Three Wire Probes

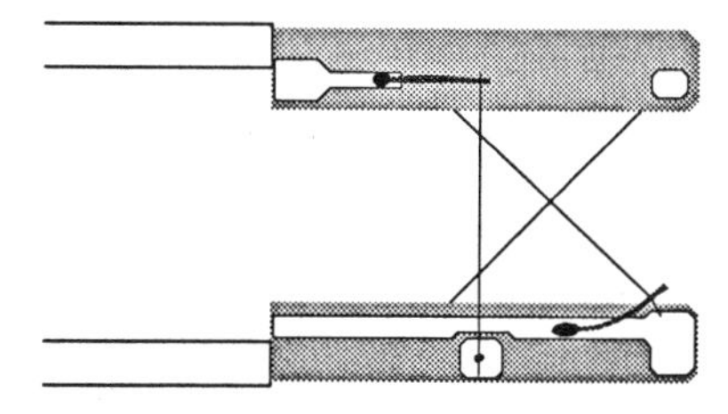

Figure 1. Sketch of the head of a three wire probe. The thin wires and the soft springs are mounted on double sided electronic prints. Each wire has one end fixed and the other end mounted on a soft spring. The wires do not touch each other as they are on different levels wrt. to the drawing plane.

Conventional radial probes in cyclotrons provide some information on the vertical behaviour of the beam. This is not possible with just one thin wire spanned vertically through the beam median plane. When the signals from three wires crossing the median plane at different angles are compared, the vertical position and width of the beam can be reconstructed through calculation. A schematic view in Fig. 2 explains the principle this reconstruction is based on.

The reconstruction of the vertical beam behaviour relies on the fact that the individual turns on the three signals can be separated, identified and then correlated. In cases where successive turns overlap we did not yet find a method appropriate for the reconstruction. This is the reason why in the 590 MeV ring cyclotron the extraction probes and the long radial probe have only one wire. As an exception, one extraction probe has three wires and therefore it can provide measurements that allow to make experiments on algorithms for the reconstruction of the vertical beam behaviour when turns are overlapping.

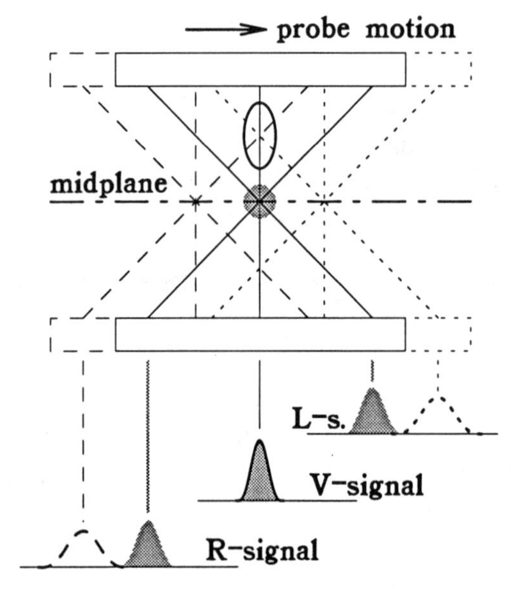

Figure 2. Demonstration of the principle, how a three wire probe can reconstruct the position and height of the beam in the vertical direction. A beam above the midplane (outlined ellipse) has its peak on the L-signal shifted to a higher radius (dotted lines) and its peak on the R-signal to a lower radius (dashed lines) relative to the synchronism of the three peaks for a beam in the midplane (shaded). The vertical width of the beam can be detected from the relative width of the L- and R-peaks compared to the V-peak.

THE PROBE-MOTION MECHANICS

For the new probes all sliding vacuum seals were avoided; the long radial probes therefore need a long vacuum box to allow their parking outside the beam region. Particular care should be taken for a proper guiding of the cables that join the moving head to the fixed vacuum feedthrough.

At injection and at extraction the construction of a probe as a moving fork with a length below 40 cm is feasible. To build a moving fork of a length of almost 3 meters would clearly be impossible. Only an extremely heavy construction of such a fork would provide sufficient stiffness. A cute solution to this problem has been found with the two long branches of the fork being guided by rails above and below the midplane. The rails are fixed to the sectormagnets and must be well adjusted wrt. to each other (see Fig. 3).

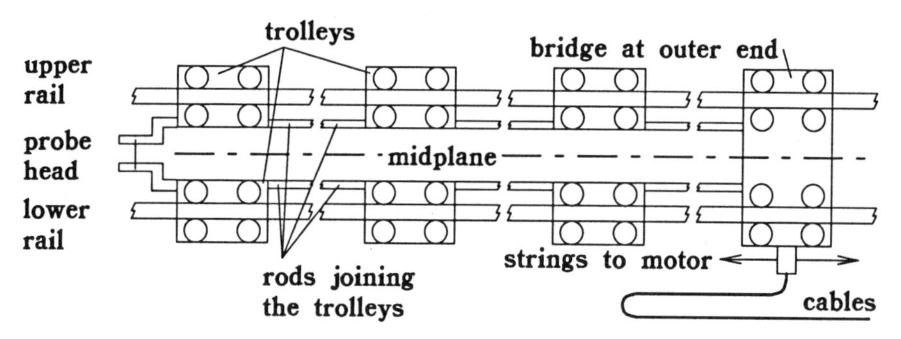

Figure 3. The principle of a long fork guided by a pair of rails: Instead of providing sufficient stiffness in the fork itself, the two rods of the long fork each joins a series of 4 little trolleys. Because of the trolleys, the vertical and azimuthal position of each point on the fork is defined by the pair of rails, while the bridge between the upper and lower part and the rods joining the trolleys have to assure the radial position.

Another highly specialized probe motion mechanics had to be constructed for the profile monitors located in the high energy beamline near the main pion production target (see Fig. 4). Because of the high activation to be expected for the beamline elements in this region, all parts of these devices that should be accessible had to be positioned 2.5 m above the level of the beamline and the assembled devices were all put in position from above with the crane. For all beam probes, but in particular for those who are buried in the shielding of the target region, a high reliability is extremely important. The probe motion mechanics in operation at PSI therefore look as if they would have to move much heavier things than just a probe head, and the experience over the years has proved that this design policy was right.

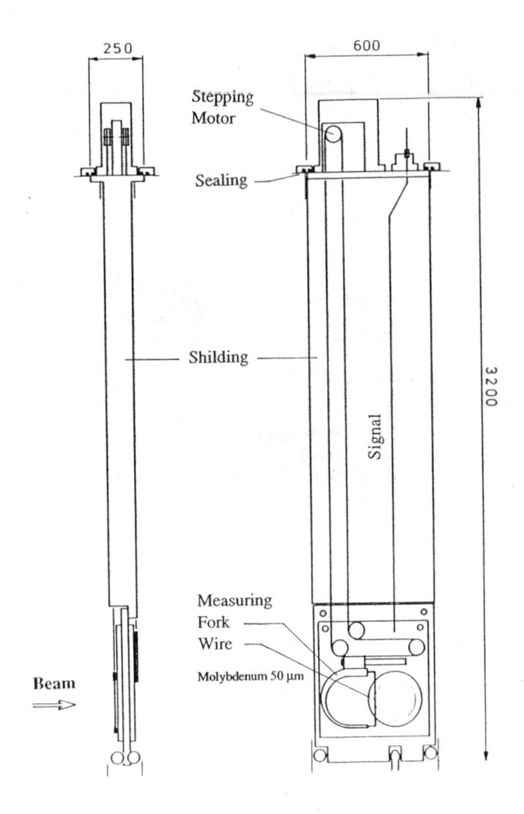

Figure 4. A highly specialized profile monitor to be used in the beamline near the target station. The motor and the vacuum seals are located 2.5 m above the level of the beamline. The fork of the probe head with the spanned wire must therefore be moved with a system of pulling wires. The volume above the beam level until the upper level of the motor and the vacuum seals contains the local shielding for this part of the beamline.

THE ELECTRONICS FOR PROBE DRIVING AND SIGNAL PROCESSING

In the new generation of beam probes each device is driven by the same type of universal CAMAC module which also handles the filtering, the current to voltage conversion and the digitalisation of the probe signals into a local memory (see Fig. 5). The local processor (HPC46003 from National Semiconductors) controls the probe motion (acceleration, constant speed, deceleration, and also calibration), it provides the signals that inform the main machine interlock about a probe intersecting the beam which means that an enhanced beam spill is temporarily legal, it organizes the transfer of probe trace data from the ADC to the dual port RAM and it has to handle several exception conditions. Due to this module the program on the control computer can require a probe trace to be measured with a simple command, wait for its completion and then read back all the data from the dual port memory. With this integrated module, and due to the high speed of 25 cm/s, taking a full trace of the long radial probe takes less than 20 seconds.

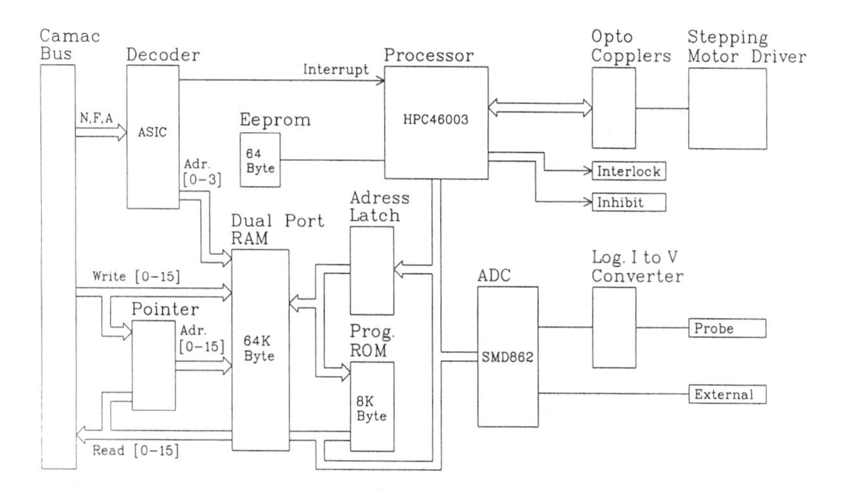

Figure 5. Block diagram of the CAMAC module for probe driving and signal processing. Directly connected to the CAMAC bus are the command decoder ASIC, the pointer memory and the dual port RAM. The main processor directs the probe motion, issues the notify-signals to the main machine interlock and steers the transfer of probe readings from the ADC to the dual port RAM. The stepping motors are galvanically insulated from the CAMAC module with optocopplers. Due to the high dynamic range required, a logarithmic conversion of current into voltage is applied.

THE INTERPRETATION OF THE PROBE SIGNALS

An extremely important component, accompanying the beam diagnostic hardware, is the software for gathering and analyzing the probe data. The aim of this software is to transform the peaks (or series of peaks) from differential beam probes into hints about the status of the beam that are easy to understand.

Several steps are included in the processing of the probe data. First the noisy raw data have to be smoothed using digital filtering. Without smoothing, and without the various peak recognition criteria that have evolved from many experiments, the next processing step which decomposes the probe trace into a series of individual peaks would fail quite often. A Least Squares fit is then made to obtain some simple characteristic numbers describing the beam properties, such as the average beam width, the average radial gain per turn, and the centering parameters A and B (the cosine and sine amplitudes of the betatron oscillation). For three wire probes the reconstruction of the vertical beam behaviour is included. A well designed graphical presentation of the probe tracing results further helps the operator to get an overview on the beam status (see Fig. 6).

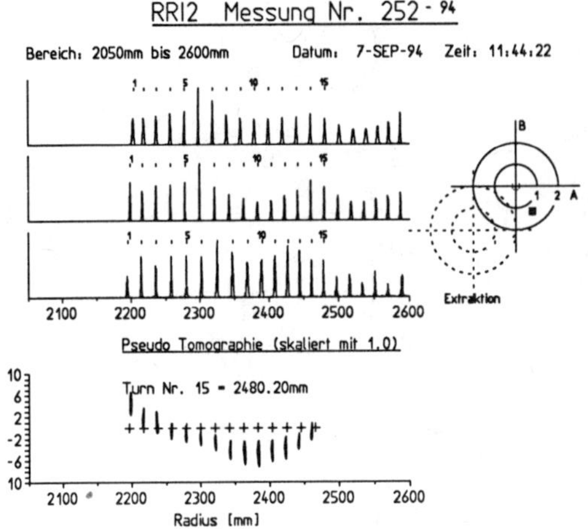

Figure 6. The display of a radial probe trace at injection into the 590 MeV ring cyclotron including a display of derived beam properties. The reconstructed behaviour of the beam in the vertical direction is shown below the radial traces from the three wires. The little square in the display on the right hand side of the picture, resembling a shooting target, indicates the radial centering error in the variables A/B (cosine and sine amplitudes), well known by the operators.

CONCLUSIONS

The new design of the beam probes for the PSI accelerator facility has been highly successful, the probes are fast, reliable and most of them can work with the full beam. Together with the programs for the interpretation of the probe data, the new probes are an important tool for the operator. The high quality of the beam diagnostics hardware and software has therefore formed a basis for the reliable operation of the PSI accelerators at high intensities.

ACKNOWLEDGEMENTS

The authors want to thank R. Erne and M. Graf for their excellent work in the mechanical construction of the beam probes, U. Frei and U. Müller for the highly reliable electronics and W. Joho and T. Stammbach for many stimulating discussions.

REFERENCES

1. U. Schryber, "Upgrading the SIN Facilities to Higher Intensities", 10th Int. Cyclotron Conf., 1984, East Lansing, USA, pp195-202.
2. T. Stammbach, "Experience with the High Current Operation of the PSI Cyclotron Facility", 13th Int. Cyclotron Conf., 1992, Vancouver, Canada, pp28-35.
3. L. Rezzonico, "Beam Diagnostics at SIN", 11th Int. Cyclotron Conf., 1987, Tokyo, Japan, pp457-460.
4. L. Rezzonico, "A Profile Monitor Using Residual Gas", 12th Int. Cyclotron Conf., 1989, Berlin, Germany, pp313-316.

MECHANICAL DESIGN CONTROL AND IMPLEMENTATION OF A NEW MOVABLE INTENSITY PROFILE BEAMLINE MONITOR FOR THE TRIUMF PARITY EXPERIMENT 497

Thomas C. Ries
TRIUMF, 4004 Wesbrook Mall, Vancouver, B.C., Canada, V6T 2A3

Abstract

Two new movable beam intensity profile monitors have been installed into the TRIUMF Parity Experiment 497 Beamlines. Each unit serves two functions. Firstly, the beam median position, in a plane normal to the beam, is detected by split plate Secondary Emission Monitors. This information is used to lock the beam into the position of the movable monitor to within a few µm's via high band width ferrite core steering magnets operating in tandem in a closed loop servo feedback control system. Secondly, the beam profile and intensity is detected via a multi-wire secondary emission non-movable monitor, where the data provides high precision values regarding centroidal positions and profiles. The centroid position of the beam is statistically determined to an accuracy of +/- 10 µm from a data record length of 1 second. The design of each device adheres to strict standards of mechanically rigid construction. The split plate SEM accuracy and repeatability is better than 15 µm with an absolute resolution limit of 0.4 µm. Maximum travel is 2 inches in the vertical plane. Since the device is mechanically modular and both degrees of freedom are combined into a single mechanical unit, fast and easy handling is possible for maintenance in radioactive areas. The actuators are dc servo motors with tachometers driven by linear servo power amplifiers. These amplifiers are used in lieu of pulse width modulated amps. to eliminate electrical noise produced by the switching circuits. Position sensing is done by variable reluctance type absolute rotary encoders providing 16 bit resolution over the full range of travel. Positioning is done manually using a self centring potentiometer on the control panel that provides a +/- velocity command signal to the power amplifiers. This configuration ensures good controllability over a very large range of positioning speeds hence making 0.4 µm incremental positioning possible, as well as, fast relocations over large relative distances. The precision movement and jitter was measured in the laboratory. Examples will be given of the monitor use with beam.

INTRODUCTION

The design of two new intensity profile beamline secondary emission monitors (SEM) with precision adjustment capability in the plane orthogonal to the beam axis was required to replace an existing makeshift unit with poor mechanical stability, control, and maintainability as well as low manufacturing precision of the profile harp monitor. It serves two basic functions. It must facilitate determination of beam intensity, profile and position via multi-wire (or thin foil strip "harp monitor") SEM's in both the horizontal and vertical directions to an accuracy of about ±15 µm relative to some fixed reference frame, as well as, provide beam centroid position signals via movable split plate SEM's for positioning the beam relative to the harp monitor. This is accomplished by using the interpreted data from the split plate SEM's to lock the beam onto the position of the movable split plate monitor to within a few microns via high band width ferrite core steering magnets operating in tandem in a closed loop servo feedback control system. Figure 1 shows how the monitors fit into the beam steering configuration in the beamline.

The overall requirements for the new monitor are that it must serve the same functions as the existing monitor, but include the following considerations;

- a more stable vibration free platform,
- better and more reliable remote position control,
- improved serviceability,
- better quality and more precise harp monitor foils,

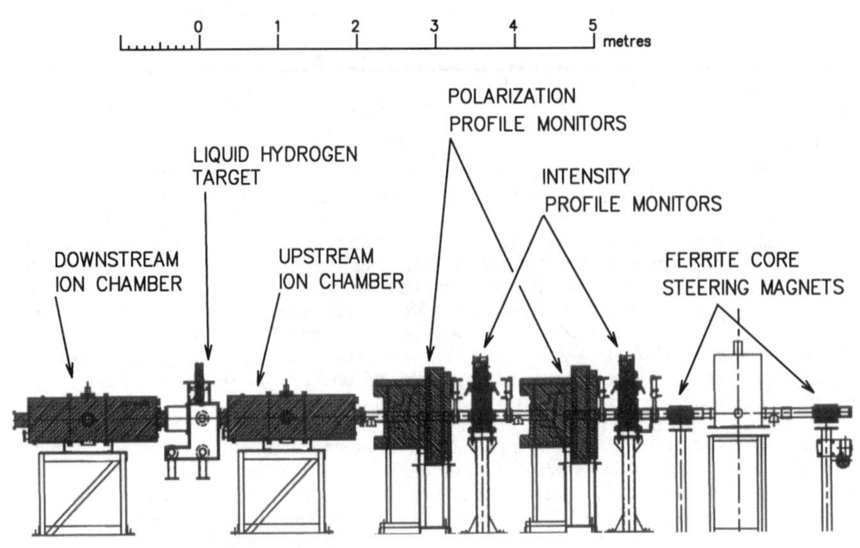

Figure 1 Parity Experiment 497 Beamline Equipment.

- improved cabling system to reduce electrical noise,
- increased mechanical travel of movable monitor pack,
- increased structural stiffness and mass to absorb externally induced vibrations,
- minimize mechanical backlash to facilitate reproducibility.

The following performance specifications are therefore designed to complement these requirements (Note: x = horizontal, y = vertical);

- maximum travel (ie. x, y directions) : 2.0 inches,
- servo positioning speed : 0.4 μm/s - 0.02 in/s,
- travel time for maximum travel : 2.0 minutes,
- position resolution limit : 0.4 μm,
- velocity transducer : analog tachometer (2.0 vdc/1000rpm),
- motor controller : velocity servoed bi-directional linear power amplifier,
- input motion command : bi-directional manual velocity control or computer generated velocity command input via DAC interface,
- position transducer : variable multi-turn absolute encoder with 16 bit resolution binary output from translator module,
- position output : decimal readout calibrated (scale and offset) to monitor position with 16 bit binary output connector to CAMAC input.

The monitor system was engineered to meet these specifications and is described in the following sections.

MECHANICAL DESIGN

Two Intensity Profile Monitor (IPM) box units are needed to steer the beam according to the requirements of the Parity Experiment #497. Figure 2 shows a general assembly of the monitor, vacuum box and stand in the beamline.

Note the structurally rigid construction to minimize externally induced vibrations. As well, the vacuum pumps in the vicinity are mounted on vibration

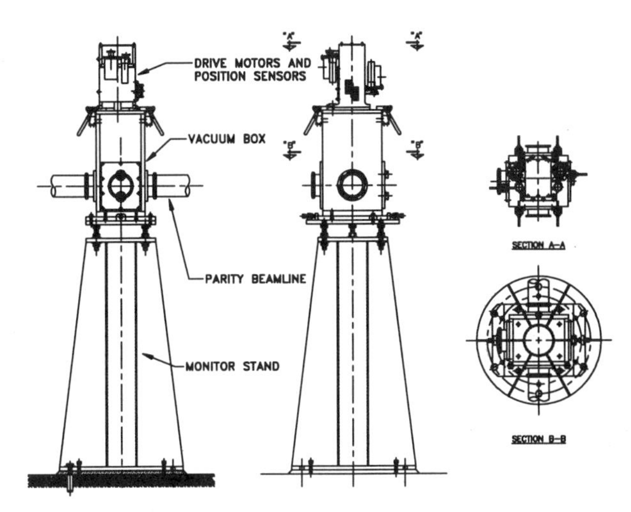

Figure 2 Side, Front, Plan and Section Views, of Intensity Profile Monitor Installation For The TRIUMF Parity Experiment #497.

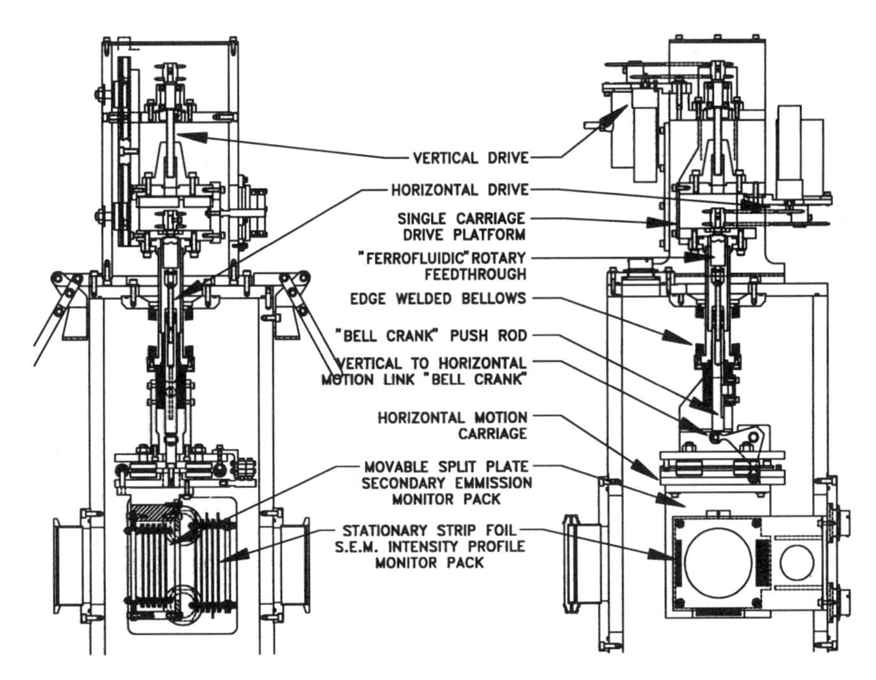

Figure 3 Side and Front Views of the Mechanical Drive Detail Assembly of the Intensity Profile Monitor Unit.

isolators. Ultra quiet turbo pumps using magnetic bearings are mounted directly to the monitor box to maintain the very high vacuum required near the SEM's.

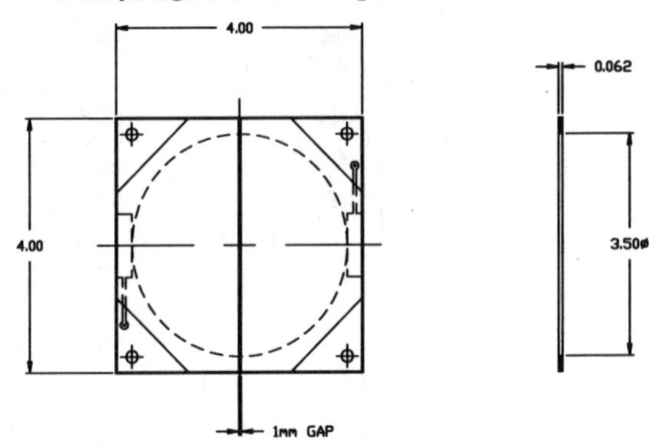

Figure 4 Parity IPM Split Plate Monitor Board.

Figure 3 shows the mechanical drive detail assembly of a single monitor unit. First note that all components of the moving split plate SEM are arranged on a single base plate which also acts as the vacuum seal cover plate for the vacuum monitor box and are oriented toward "heavier gage" construction in order to isolate outside vibrations and to increase overall stiffness. The vertical position encoder and drive servo-motor are fixed to the vertical H-frame and drive the vertical motion carriage, which is guided by preloaded linear bearings, through a 0.375"ϕ-1/32" lead screw. The horizontal position encoder and drive servo-motor are fixed to the vertical motion carriage. Rotary motion going through the "ferrofluidic" vacuum feedthrough is converted to vertical linear motion by a nut and threaded pushrod which is then converted to horizontal motion of the split plate SEM support carriage (which is also guided by preloaded linear bearings) by a "bell crank" arrangement. All electrical connections are made through the 42 pin feedthrough connector in the vacuum base cover plate. The vacuum plate indexes on spring loaded quick register alignment fixtures on top of the vacuum box and are held fixed by standard quick release toggle clamps.

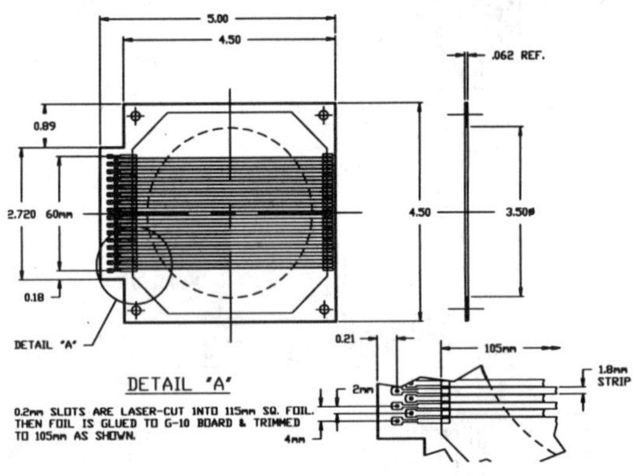

Figure 5 Intensity Profile Strip-Foil SEM Board.

The movable split plate SEM's are attached to the base of the horizontal motion carriage which in turn is attached to the vertical motion carriage and can be positioned to better than ± 15 μm in both vertical and horizontal directions about the beam axis. These monitor boards consist of a thin aluminum foil, between 5 and 25 microns thick with a 1 mm gap cut down the middle, pre-tensioned and glued to one side of a 0.062" thick fiberglas PC board, as shown in Figure 4. The split plate SEM monitor pack assembly with its HV boards running at +300VDC are similarly constructed in cross-section, to the strip-foil harps as shown in Figure 6.

The harp monitors are also SEM's but provide detailed spacial information of the intensity and the profile of the beam. The construction is similar to the split plates except that the foil is precision cut into narrow strips as shown in Figure 5. This is a very difficult procedure that was initially performed with laser beam cutters which yielded a poor quality cut. A method was developed at the Probes Lab at TRIUMF by which the strips are cut very cleanly using a straight knife edge device, which was also developed by the Group, and using conducting adhesives to fasten the ends of the strips.

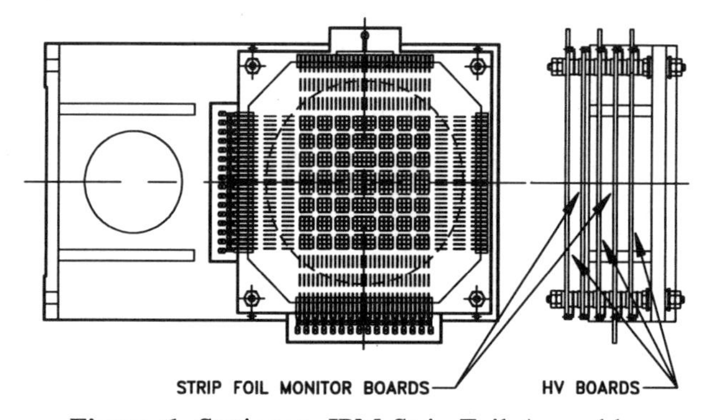

Figure 6 Stationary IPM Strip Foil Assembly.

When assembled the profile monitor pack resembles the split plate unit closely, as shown in Figure 6, and is firmly fixed to a side vacuum seal plate. The unit is independent physically and electrically from the movable split plates and can be easily removed through the side of the vacuum box for servicing. The current signals are removed through two hermetic multi-pin connectors on the side plate.

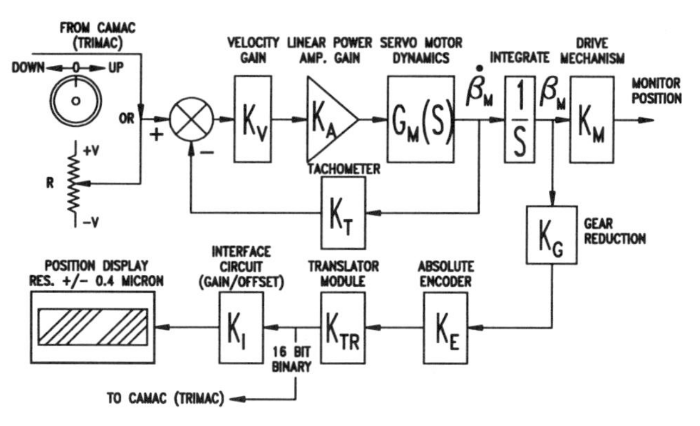

Figure 7 Parity IPM Velocity Control Block Diagram.

CONTROLS

A very large dynamic range is required of the controller to meet two conflicting requirements from very coarse and fast relative positioning (0.02 in/s), to very fine and slow positioning (0.4 µm/s) on a single command button. Velocity control mode, as shown in Figure 7, is therefore used with tachometer feedback to provide a single pole system that is always asymptotically stable and can provide the high resolution control required. The parameters V and K_T are chosen so as to give the desired range on the "spring return to zero" control knob when it is displaced either CW or CCW from its equilibrium zero velocity position as shown. The resulting command voltage, summed with the negative tachometer feedback voltage, will drive the motor and hence the monitor according to the commanded rate. The position is sensed by the variable reluctance type absolute encoder through the gear/screw drive mechanisms and after some interface electronics displayed on an LCD display to an accuracy of ± 0.4 μm. The loop shown in Figure 7 can be connected, if necessary, to a TRIMAC motor controller through the CAMAC data buss system. This will form a closed position feedback loop which can be controlled at a higher level by an external supervisory computer.

TEST RESULTS

Two assembled IPM units were installed into the operational Parity Experiment beamline setup shown in Figure 1 and were used to provide beam intensity, profile, and position information. Some examples are presented below.

As mentioned above the signal from the movable split plate SEM's are used to lock the beam centroid onto them, by appropriately controlling the steering magnets in a closed position servo-loop control. Moving the beam is therefore done to an accuracy that is limited by the mechanical components. Some preliminary calibration data is presented indicating the reproducibility of the positioning in the X and Y directions of one of the split-plate monitors plotted against the beam position as calculated from the strip-foil current signals. Figure 8 shows a hysteresis plot in the X-direction indicating a maximum possible error of ±12.5 μm. This was due to the low preload forces on the mechanism which has since been improved considerably by installing heavier springs. In the Y-directions where the preload is greater, due to the air pressure loading through the bellows, the reproducibility is considerably better, as shown in Figure 9, where the error is only ±3 μm.

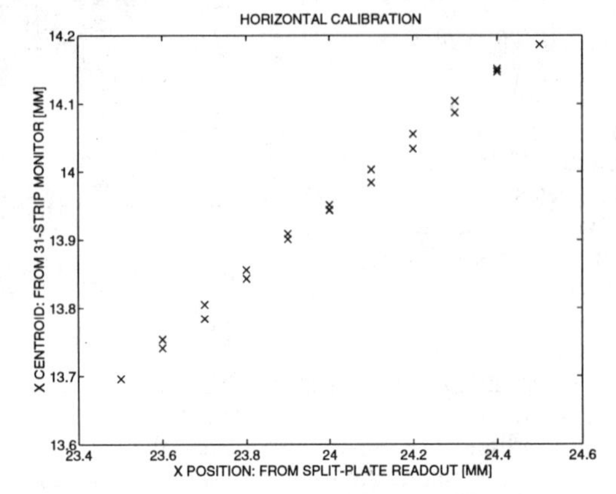

Figure 8 Horizontal Split Plate Calibration Showing a Maximum Deviation From the Mean of About ±12.5 μm including backlash.

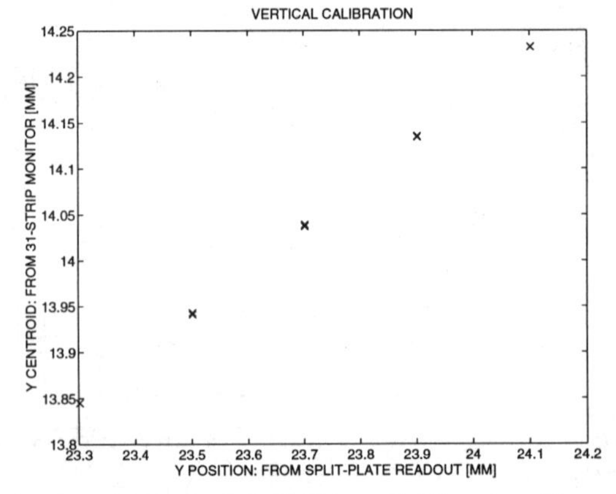

Figure 9 Vertical Split Plate Calibration Showing a Maximum Deviation From the Mean of About ±3.0 μm Including Backlash.

The following example shows the effects of beam profile and position modulation on the shape of the detected profile from one of the X-direction harps. Figure 10 shows the effects of coherent position modulation demonstrating the ability to measure beam position from a distribution of 600 one-second measurements. Coherent width modulation is demonstrated in Figure 11 showing the distribution of 600 one second measurements. The RMS spread is 13μm indicating that a one second measurement can detect coherent beam size changes with a 1σ precision of 13μm.

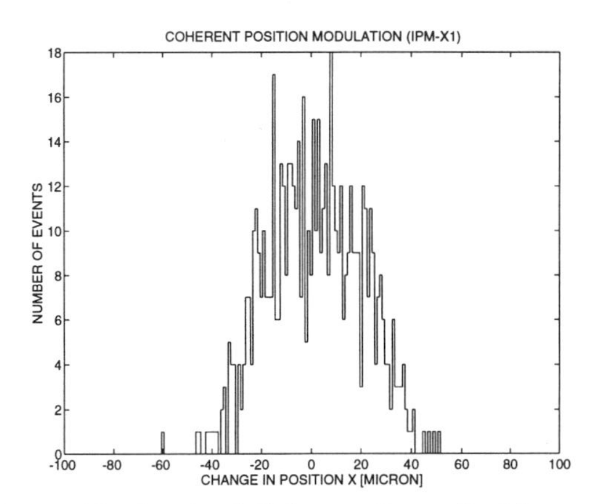

Figure 10 Graph of Coherent Beam Position Modulation vs Number of Event Counts.

CONCLUSIONS

The probe was tested and found to give satisfactory indications of beam position and intensity for the Parity Experiment. Positioning of the split-plate SEM, regarding resolution and repeatability, are well within the required tolerances. Beam test results compare favourably with the existing beamline diagnostics at TRIUMF regarding signal quality and required positional accuracy.

ACKNOWLEDGEMENTS

We would like to thank Des Ramsay for inspiring the design and development of this monitor, and Bill Rawnsley for the electrical design, and the Probes Group for invaluable suggestions, as well as improvements to the strip-foil board.

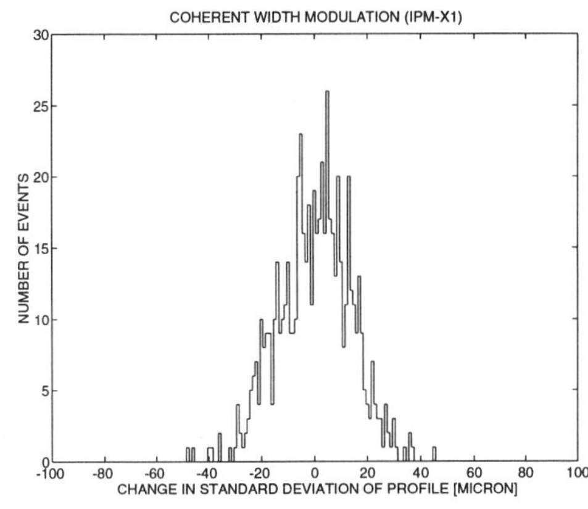

Figure 11 Graph of Coherent Beam Width Modulation vs Number of Event Counts.

REFERENCES

1. Ramsay, W. D.; "Engineering Design Specifications and Requirements", internal communications.
2. Van Oers, W.; "Beam Line Monitor Service Request - Engineering Design Specifications and Requirements", internal memo, July 17, 1991.
3. Ries, T.; "Intensity Profile Secondary Emission Monitor for Parity Experiment #497", Assy.dwg. #D-33605, #D-33606, Design Office drawing list #DL-3564, February 1992.
4. Rawnsley, W.; "Electrical design", internal communications.

Performance of Beam Monitors Used at a Beam Transport System of HIMAC

M.Torikoshi, H.Ogawa, S.Yamada, M.Kanazawa, T.Kohno, K.Noda, Y.Sato, E.Takada, N.Araki, K.Kawachi and Y.Hirao
National Institute of Radiological Sciences, 9-1, Anagawa-4-chome, Inage-ku, Chiba-shi, Chiba 263, Japan

M.Sudo, H.Takagi and K.Narita
Accelerator Engineering Corporation, 9-1, Anagawa-4-chome, Inage-ku, Chiba-shi, Chiba 263, Japan

M.Mizobata(1) and K.Ueda(2)
Mitsubishi Electric Corporation,
(1)Wadasaki-cho, Hyougo-ku, Kobe-shi, Hyougo 653, Japan
(2)Marunouchi 2-2-3, Chiyoda-ku, Tokyo 100, Japan

Abstract

Total 35 beam monitors have been installed along high energy beam transport lines of a heavy ion medical facility, HIMAC, of NIRS. They are able to monitor beam profiles and intensities at a wide range of beam intensity from 1×10^{11}pps to a few thousands pps. A computer specially designed for real time operation of industrial processes is in use for controlling the monitors. Totally the beam monitor system is easy to handle, and has a good performance.

INTRODUCTION

Heavy ion beams have been accelerated at the Heavy Ion Medical Accelerator in Chiba (HIMAC) of National Institute of Radiological Sciences, since November 1993(1). The clinical trial started on 21, June 1994, after several basic biological and physics experiments lasting about 2 months(2). At the trial 290MeV/u ^{12}C beam was irradiated, which has about 15cm range in water.

The HIMAC is a complex of an injector linac cascade(3), two synchrotron rings, a high energy beam transport system and a beam irradiation system. The heavy ion beams with charge to mass ratio of 1/2 can be accelerated up to maximum energy of 800MeV/u by the synchrotron rings. The high energy beam transport system (HEBT) consists of horizontal beam lines and vertical beam lines with total length about 240m(4). The configuration is shown in Fig.1.

In the HEBT system, the beams have to be switched from one therapy room to another within 5 minutes by changing only current of switching magnets(4). The beams are to be tuned spatially with better than 1mm accuracy so as to deliver the beam without beam loss. At present 35 beam profile monitors and 10 beam intensity monitors have been installed. Since the heavy ion beam intensity is relatively low compared with usual proton beam intensity,

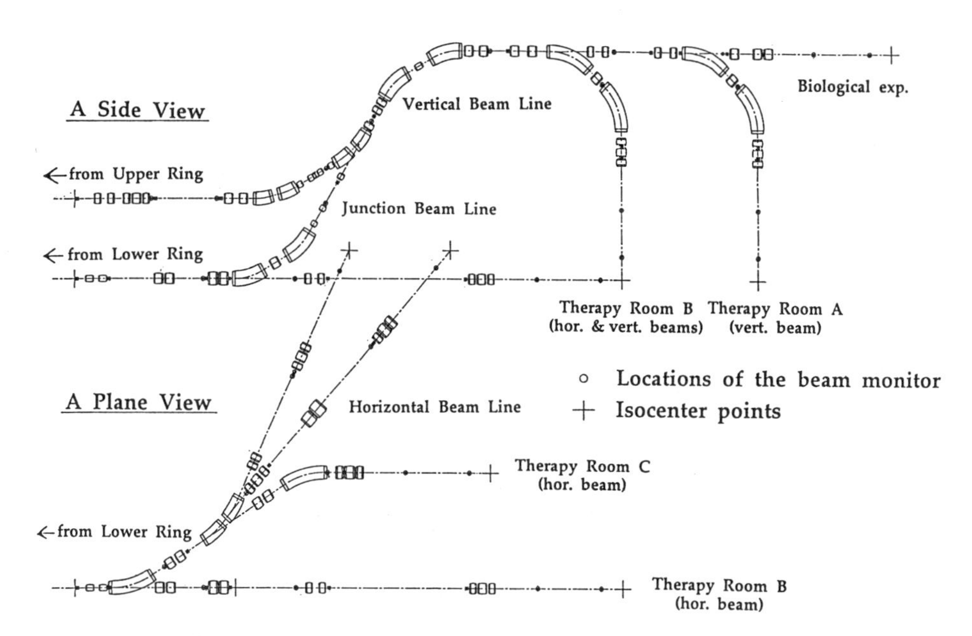

Figure 1. Configuration of the HEBT. The horizontal beam lines transport beams from lower synchrotron ring to two therapy rooms (B and C) and a physics experiment room. The beams from upper ring are delivered vertically to two therapy rooms (A and B), and horizontally to a biological experiment room. A junction line connects the horizontal line to the vertical lines. The therapy room (B) is equipped with a horizontal beam port and a vertical one, so the beams can be irradiated in both directions simultaneously.

the beam monitors are required to cover a wide intensity range. To satisfy this requirement, a multiwire proportional counter (MWPC) is adopted as the profile monitor because of its gas amplification, and a plane parallel ionization chamber as the intensity monitor because of its larger output current.

A prototype of the monitors was built in 1990 to prove fundamental performances(5). Furthermore all monitors were tested by using X-ray beam to check their integrated performances in advance. The performances of the monitors are described including the results of the prototype and of the X-ray beam.

BASIC DESIGN OF THE BEAM MONITOR SYSTEM

Beam Profile Monitor

The design goal of the MWPC was to cover a wide beam intensity range from 1×10^6 to 1×10^{11} pps in order to monitor profiles of various kinds of beam. An approximate formula of Diethorn(6,7) predicts a gas amplification by using a few parameters, such as wire spacing, diameter of wires and anode cathode gap of the MWPC, and characteristic constants of gasses. The parameters were determined to obtain the gas amplification of about 5,000 at less than applied voltage of 2.7kV to cover the above range.

The MWPC has two orthogonal signal wire planes. Each plane has 32-wire with spacing 2mm apart, the effective area of the plane is 64mm $\times$ 64mm. The wire is 20μm diameter of tungsten. Cathode planes also consist of 50μm diameter wires of Be-Cu. The anode cathode gap is 4mm. The outline of the MWPC is shown in Fig.2. The mixed gas of Ar(80%) and CO_2(20%) is used. The gas flows continuously through the vessel at the flow rate of 10cc/min.

Figure 2. MWPC as a sensor head of the beam profile monitor.

Beam intensity monitor

The plane parallel ionization chamber gives relatively large output currents in comparison with the Faraday Cup or a secondary electron emission monitor. However, an intensity range of the sensitivity is limited by ion-recombination in the chamber. Thus the monitor was designed to cover with a beam intensity range of 1×10^7 to 1×10^{10} pps.

The ionization chamber has an effective area of 100mm diameter. An anode plane is made of copper-laminated polyimide sheet. Cathodes consist of wires to reduce multiple scattering of the beams. The anode-cathode gap is 7.5mm each side. Electric current signals induced on the both sides of the anode are read together into electronics. The ionization chamber was installed on back of the MWPC in a common vessel. The typical applying voltage is about 1.5kV at the intensity of 1×10^8 pps.

Electronics

Induced current on a wire of the MWPC is integrated in an RC circuit in an interval of 10msec $\sim$ 1.5sec. Generated current in the ionization chamber is converted to voltage by an Ope-amp with gain of $\times$1,$\times$10,$\times$100 and $\times$1000. The signal of 10V of the full scale is converted to 11-bit signal. Additional three bits, a start trigger, EOC and a flag bit indicating kind of data, are put together with the 11-bit signal. Parallel 14-bit signal is sent in a sequence of

64 profile data and 1 intensity datum (or dummy datum) to buffer memories through RS422A transmission lines.

The electric elements of R and C were selected to keep the RC constants within 0.2%. Furthermore the electronic elements are also selected to make leak currents low.

Operation

The monitor is controlled remotely and locally. In the remote control system, a UNIX computer designed for industrial processes is used as a main computer of the HEBT. The operating system of UNIX with a real time feature enhances the real time processes of the monitors. Local control panels were set on the floors of the beam line to operate the monitors locally. They also interface with the computer. The system is shown schematically in Fig. 3.

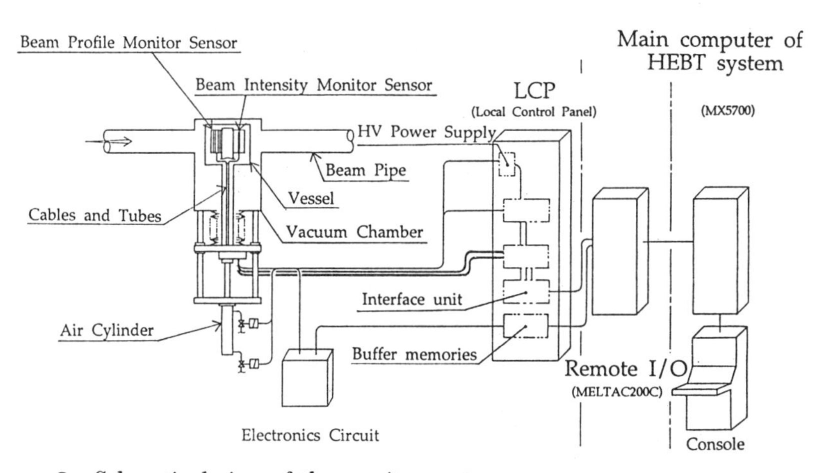

Figure 3. Schmatical view of the monitor system

In the daily beam tuning, the monitor is operated by using CRTs with touch panels. The status of the monitor is also displayed on the CRTs. The monitor is plunged into the beam line by an actuator within 2sec. After plunging the monitor into the beam line, preset voltages are applied automatically or controlled manually. Applying about 200V gives good beam profiles of 290MeV/u C beam with intensity of 1.8×10^8 pps used for the therapy.

PERFORMANCE OF THE BEAM MONITORS

The profile monitor

In Figure 4, the gas amplification factor is shown as a function of applied voltage to the cathode. In the case of X-ray beam, the gas amplification factor was derived from comparison with the integrated charge and the initially produced charges measured at plateau region. In the case of C beam, the initially produced charge at the first ionization was calculated with aid of Bethe-Bloch's formula, while the C beam intensity was measured by the beam intensity monitor installed with the profile monitor mentioned above. The figure also shows gas amplification factors of the prototype profile monitors with 70MeV proton

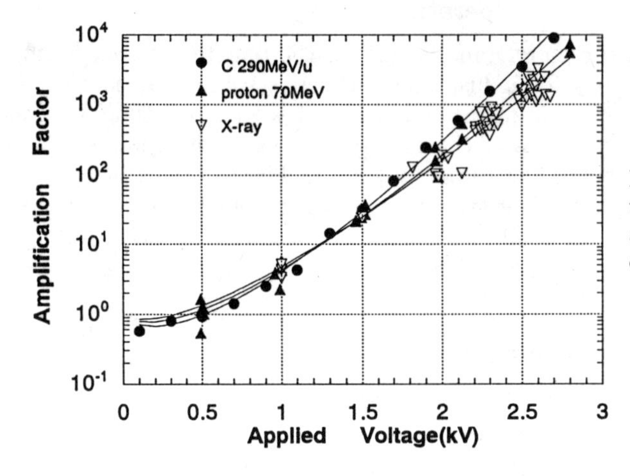

Figure 4. Gas amplification factor as a function of applied voltage to the cathode.

beam from a cyclotron of NIRS. The amplification factor reaches about 9,000 at a maximum voltage 2.7kV at the C beam. In the figure, the data are fitted by a solid-line curves derived from the Diethorn's formula. All fitted curves close each other. It means these monitors have identical operational characteristics.

Typical noise level is 1 digit of full 2048 digits. The noise level is independent of the time width of integration. It means that leak currents in electronic elements does not contribute to the noise, or are very small. According to simple estimations, the leak currents are supposed to be less than 0.1nA each channel. Cross talk between adjacent channels is less than -60dB. To keep the noise level low, noisy electronic elements were selected out. And insulation between the anodes and the cathodes were kept more than $5 \times 10^{12}\ \Omega$ to reduce leak currents between them. Because of the low noise level and the small cross talk, the profiles can be observed clearly even though the peak height of the profile is less than 0.1V.

The monitor could observe the profiles of the lower intensity beam than about 5,000 pps required for LET measurements, because of the gas amplification factor of about 10^4 and amplifying display scale about 10^3 times by software.

The intensity monitor

Correlation of output current and beam intensity (linearity) was measured by using He, C and 70MeV proton beams as shown in Fig.5. The 70MeV proton data were taken by using the prototype monitor. The dashed line in the figure shows linear correlation. Including all data the standard deviation from the linear correlation is about $\pm 6.6\%$. If the data of the intensity region of higher than 1×10^8pps is selected, however, the standard deviation is about $\pm 3.1\%$. It is supposed that the large deviation at the lower intensity region is caused by leak currents of the Faraday Cup.

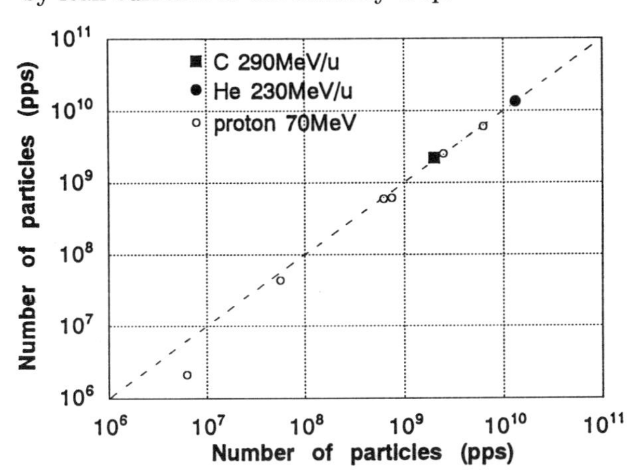

Figure 5. Correlation of output current and beam intensity. The abscissa indicates the beam intensity measured by the Faraday Cup, and the ordinate indicates the intensity measured by the intensity monitors.

MWPC EXTENSION FOR CONTINUOUS MONITORING

Non-destructive profile monitors are proposed to use at right before the therapy rooms. The non-destructive monitors can observe the beam profiles continuously without affecting the therapy during the irradiation. The MWPC was in use in the atmosphere as the prototype with 290MeV/u C beam. The gas amplification factor was measured to be about 40 at maximum. The test proved the monitor to be useful to observe the profiles at more intense beams than 1×10^6 pps. Thus it could be a good candidate for the non-destructive monitors.

SUMMARY

Performance of the beam monitors is summarized as follows,

1. The profile monitor covers a wide intensity range of $1\times 10^{11} \sim 5\times 10^3$ pps. This coverage is obtained by the high gas amplification factor, the low noise level and the small cross talk.
2. The output current of the intensity monitor is linear with the beam current within about $\pm 3\%$ at the more intense region than 1×10^8 pps.

This linearity is good enough to tune the beams.

3. The beam monitor system is controlled easily, so the HEBT system gives a good tunability and reproducibility of the beams.

ACKNOWLEDGMENTS

We would like to express our gratitude to the members of Research Center of Charged Particle Therapy of NIRS for helpful discussion. We also wish to thank the members of Accelerator Engineering Corporation for their warm support.

References

[1] K.Sato et al., "Status report on HIMAC", Proc. 4th European Particle Accelerator Conf., London, 1994, to be published.

[2] T.Kanai, "Heavy-ion radiation therapy", Phys. in Med. Bio. **39a**(1994)480

[3] A.Kitagawa et al.,"Commissioning of injector linac of the HIMAC", Proc. the Fifth Japan-China Joint Symp. on Accel. for Nucl. Sci. and Their Appl.,(1993)55

[4] K.Noda et al.,"Beam transport system for the HIMAC", Proc. the 13th Int. Conf. on Cyclotrons and their Applications (1992)625

[5] M.Torikoshi et al.,"Development of beam profile monitor for HIMAC", Proc. the 8th Symp. on Accelerator Science and Technology, (1973)317

[6] T.Tomitani, "Analysis of potential distribution in a gaseous counter of rectangular cross-section", Nucl. Inst. and Meth. **100** (1973)179

[7] W.Diethorn, NYO-6628 (1956)

Beam Diagnostics in the Ion Accelerator

E.Serga, V.Novikov
IHEP, Protvino, Moscow Region, Russia

I.Churin
JINR, Dubna, Moscow Region, Russia

P.Cantoni
INFN, Italy

P.L.Frabetti
Universita di Bologna, Italy

R.Tonini
Universita di Modena, Italy

Abstract

For ion beam adjustment and monitoring the diagnostic system based on Faraday cups was built. This system permits to measure the mass spectra, the size, the current of the ion beam and size the scanned area. The result of measurements are presented.

INTRODUCTION

The ion accelerator at the Modena University is used for the ion implantation into solid and has a wide range of beam characteristics(1). The ions are from H^+ to Se^{+++} with energies from 15 to 600 KeV and ion current from 4 to 400 μA. The RMS. of the beam is about 3-5 mm. The wafers to be implanted are of 150 mm in diameter and must be irradiated with the nonuniformity of 1 % by two-coordinate electrostatic scanning system. For a fine implantation of the wafers, it is very important to control the ion beam characteristics from the ion accelerator. The control system of te ion implanter is built around the IBM PC/AT-486 running under UNIX and microcomputer system for different accelerator subsystem (vacuum, ion source, scanning system etc.). For beam adjustment and monitoring the diagnostic system based Faraday cups and CAMAC was constructed and installed. During the implantation the diagnostic system monitors and displays in on-line the mean beam current, its size, position and the size of the scanned area. Detailed analysis of the ion beam is made using the stored data on the personal computer. This paper describes the system configuration and its performances. The preliminary results of measurements are also presented.

SYSTEM CONFIGURATION AND ITS PERFORMANCES

System configuration

A schematic diagram of the diagnostic system is illustrated in Fig. 1. One of the goal of this work was to create low cost monitoring system. It is made up of six Faraday cups, 12 current to voltage convertors, LeCroy model 5249B 12 channel ADC(11 bit resolution, 108 μs conversion time, 0.1-5 μs gate for each channel), TGC trigger-generator to synchronize the process of the measurement and the crate CAMAC with the CAEN C111 controller for connecting to the personal computer. Faraday cups is mounted in vacuum chamber, the first one, FD1, is located just after 20° analyzing magnet. By this one the mass specter of the ion beam is measured. FD2-FD5 is located 0.7 m downstream from the scanning system. The arrangement of FD2-FD5 is shown in Fig. 2b. As the scanning system deflects the ion beam to irradiate the wafer with fine uniformity we can measure all characteristics of the ion beam without stoping of the irradiation. By FD6 the mean beam current is measured. FD6 is the insulated vacuum chamber to keep the wafer.

Operation of the monitoring system

The most interest is the measurement of the ion beam characteristics in during time the irradiation of the wafers. In order for this happen, the two-coordinate electrostatic scanning system is used. As the generator feeds the 5 KHz operating voltage to the vertical and horizontal plates (see Fx and Fy in Fig. 2a), Fx and Fy have little difference of $\Delta F = F_x - F_y = 25\ Hz$ the trajectory will be as it is shown in Fig. 2b. The period of total scanning (will say "frame") is 40 ms=1/25 Hz. The period of one scanning (will say "line") is 200 μs=1/5 KHz, respectively. Beginning of the frame is determinated from the condition Fx and Fy is in phase. To get profiles in the vertical directional we must measure the current from FD3 and FD5 at the section A-A (see Fig. 2b). This is reached in the following way: the synchronization of measurements is made from Fx and additionally the output pulse is delayed by 50 μs from the input pulse. As the line period is about 200 μs and we have individual ADCs for each channel this interval is sufficient that the currents from FD3 and FD5 to digitize and transmit to the personal computer. For every line we will have one measured point. Total number of measured points for the frame will be about 200. For the horizontal direction it is made the same way.

Calculation of the ion beam characteristics

The Fig.3. shows the current from FD3 and FD5 as a function of the point(line) number. From these results all characteristics of the ion beam may be calculated by following formulas: position of the peaks is

$$Y_m = \frac{\sum_k (k + 50m) y_{k+50m}^2}{\sum_k y_{k+50m}} \tag{1}$$

were k=1..50 is the point number,
m=0..3 is the peak number.
line spacing is

$$\Delta y = \frac{L}{(Y_2 - Y_1)}, mm \tag{2}$$

were D is the spacing between FD3 and FD5 in mm
size of the scanned area is

$$YY = \Delta y \cdot (Y_1 - Y_0 + Y_3 - Y_2)/2 + L, mm \tag{3}$$

beam deflection from the axis is

$$YY0 = (Y_3 - Y_2 - Y_1 + Y_0) \cdot \Delta y/2, mm \tag{4}$$

irradiation efficiency is about

$$\rho = (L^2/YY \cdot XX) \cdot 100\% \tag{5}$$

RMS. of the peaks are

$$\sigma_m^2 = \frac{(\sum_k (k + 50m + Y_m)^2 \cdot y_{k+50m}^2) \cdot \Delta y^2}{\sum_k y_{k+50m}^2}, mm^2 \tag{6}$$

were k and m is the same.

MEASUREMENTS

Fig. 4 shows an example for scanning Ar^+ beam. This measurement was done with low currents (60 μA). The figure (a) shows that the profiles have Gaussian-like distribution with σ_x of 3.2 mm and σ_y of 7.0 mm. The beam and scanned area characteristics is presented in the Table 1.

Table 1

σ_x	3.223 , mm
σ_y	7.041 , mm
XX	124.978 , mm
YY	148.500 , mm
XX0	2.746 , mm
YY0	-0.037 , mm
ρ	34.484 , %

Fig. 5 shows an example of the mass specter for the 40 keV ion beam. The stored date is processing to get the positions of the peaks in mass/charge units.

CONCLUSION

We have developed and installed the beam monitor system using Faraday caps and CAMAC for diagnosis of the ion beam at the Modena University ion implanter. The system has worked very well in measuring the size, position and scanned area of the ion beam. Also it has given the very simple means for controlling the ion beam as a routine work. As a next step, we are going to investigate the ion beam by 2-dimensional beam monitor that will be installed in future and to find the way to get more efficiency operation of the ion implanter.

ACKNOWLEDGEMENTS

The authors would like to express their thanks to prof. G.Ottaviani for his useful discussions and help.

REFERENCES

1. A.F. Vyatkin, at al., Development of ion implantation equipment in USSR, NIM B55(1991)386-392.

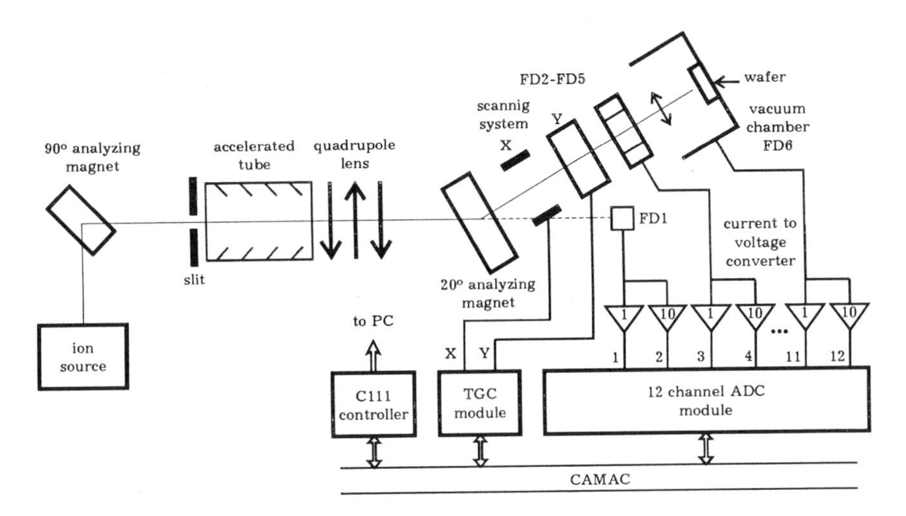

Fig.1. Schematic diagram of the diagnostic system

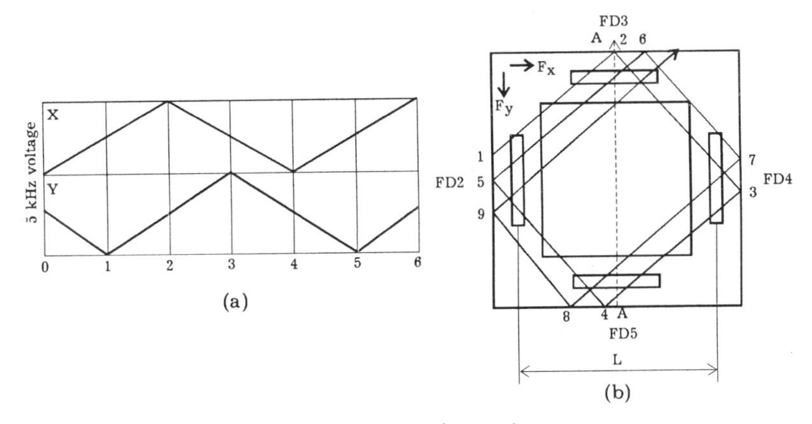

Fig.2. Operation of the scanning system
(a) amplitude of the scanning voltage and
(b) trajectory of the ion beeam.

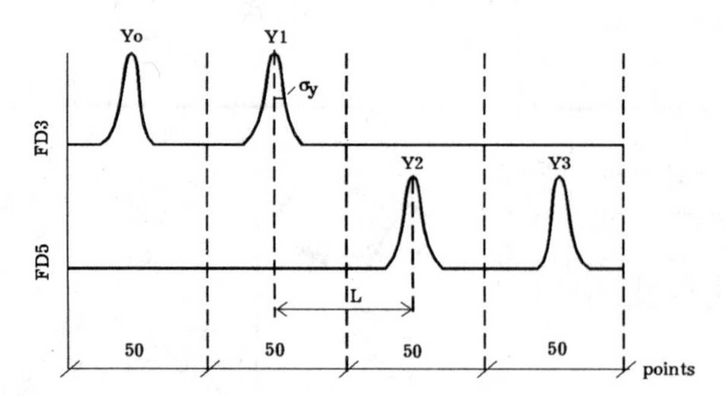

Fig. 3. FD3, FD5 currents vs. the point number.

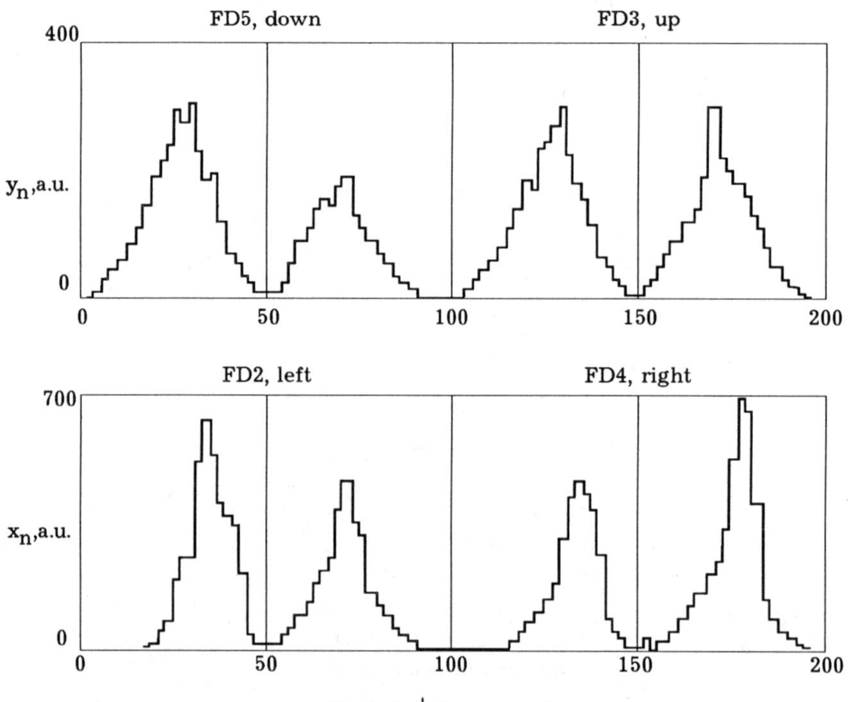

Fig.4. Ar^{+} beam profiles

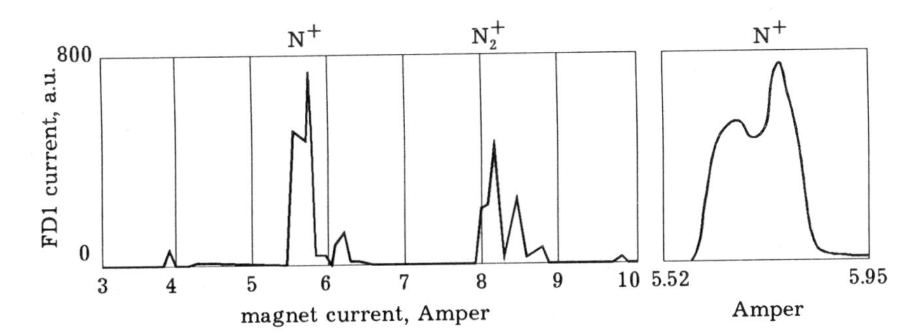

Fig.5. Mass specter of the 40 keV ion beam.

The Monitor System of the 7MeV Proton Linac at ICR

Y. Iwashita, M. Kando, M. Ikegami, H. Dewa, T. Shirai,
S. Kakigi, H. Fujita, A. Noda, and M. Inoue
Accelerator Laboratory Nuclear Science Research Facility,
Institute for Chemical Research, Kyoto University, JAPAN

Abstract

The beam monitor system at ICR 7MeV proton linac is described. Because of the pulsed beam (<1% duty) and the simplicity, fluorescence on a screen plate is observed for the beam profile monitoring. To clear the beam line, the screen can be flipped away by an electromagnetic actuator installed in the vacuum vessel. With this view screen, an emittance monitor is constructed. The beam current can be measured non-destructively by a core monitor. Because the pulse length is rather long (~ 50 μs), a special pre-amplifier using Negative Impedance Converter is designed. Very thin steering magnet with eight poles is also developed to compensate the deflection of the output beam.

INTRODUCTION

The 7 MeV proton Linac at Institute for Chemical Research, Kyoto University consists of a 50 keV proton ion source, 2 MeV RFQ, and 7MeV Alvarez DTL.(1) (See Fig. 1) The linac system is operated in the pulse mode ($\geq$ 180 Hz, $\geq$ 60 μs) at the frequency of 433MHz. The accelerator tanks are in the area surrounded by the radiation shielding wall of 1 m thickness, while the ion source and the Low Energy Beam Transport (LEBT) is located outside of the shielded area. Because of the thickness of the wall and the necessity of the beam measurement after the ion source, the length of the LEBT is rather long considering its beam current (~10 mA or more) and the energy. The LEBT has 45° bending magnet called Mixing Magnet(2), which separates the H_2^+, H_3^+, and so on. Mixing Magnet has the edge angle of 23° at the entrance, which reduces excessive radial focusing. With further installation of another ion source of H^- in the other arm of the Mixing Magnet, the simultaneous acceleration of H^+ and H^- can be performed. This Mixing Magnet also makes the LEBT long.

The proton beam from the ion source is shaped by an Einzel lens and goes

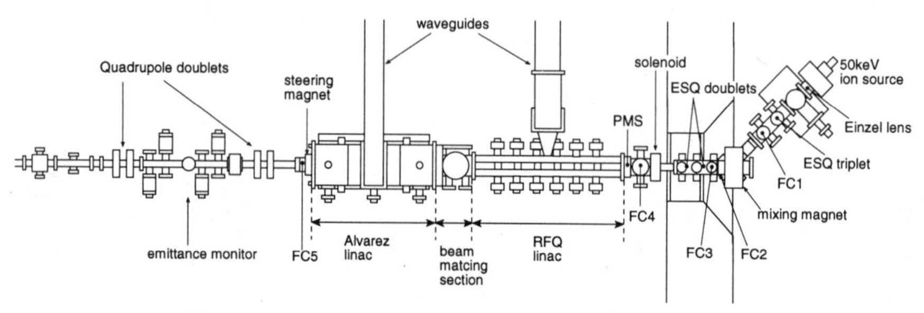

Figure 1. The layout of the 7 MeV proton linac.

through the electro-static quadrupole (ESQ) triplet lens. After the Mixing Magnet, there are two ESQ doublet, one solenoid magnet lens, and a Permanent Magnet Symmetric (PMS) lens.(3)

The RFQ about 2 meter long accelerates the 50 keV proton to 2 MeV, and the Alvarez DTL accelerates it to 7 MeV. Between the RFQ and the DTL, Permanent Magnet Quadrupole (PMQ) lenses and a rebuncher are installed. A thin steering dipole magnet, which has variable strength and steering direction, is installed right after the DTL to adjust its output beam axis . After the steering magnet, a current monitoring beam stopper with an insulated head is installed. For less neutron production, a Tantalum plate is welded on the header surface of this 7 MeV proton Beam stopper.

Because the vanes of the RFQ have an angle of 45 ° from the horizontal and vertical plane " X-Y ", the coordinate system " X'-Y' " of the RFQ and DTL is also 45 ° rotated, and then the output beam from the DTL has the rotated coordinate system. With four 45° skewed quadrupole magnets (two doublets), the twiss parameters of both X' and Y' can be equalized, where the beam is axisymmetric. Between the two doublets, an emittance monitor is installed.

BEAM MONITORS IN THE LEBT

Before the Mixing Magnet, a beam stopper (FC1 in Fig. 1), which is inserted/extracted pneumatically, is installed. Because the stopper axis is insulated from the vacuum flange, the beam current can be measured. On the surface of the stopper, a fluorescent screen plate is fixed, where fluorescence on the plate gives the beam profile. A fine stainless steel mesh covers the plate to reduce the charge up on the screen plate. Without this mesh, the current from the stopper axis does not reflect the beam current because of the secondary electron emission from the plate. FC3 is the same as the FC1 system. FC2 is just an insulated flange on the straight line from the ion source.

Flip Screen Monitor

FC4 is also a current monitoring beam stopper with an insulated copper head. The side of this head is shielded by a copper pipe, which can be biased. Because of the poor vacuum conductance in the RFQ, the cavity is evacuated from both entrance and exit endplates of the RFQ, and then the space before the RFQ is limited. A flip screen monitor is added on the current monitoring beam stopper FC4 (See Fig. 4). The screen sits on the beam line when the stopper is extracted. To clear

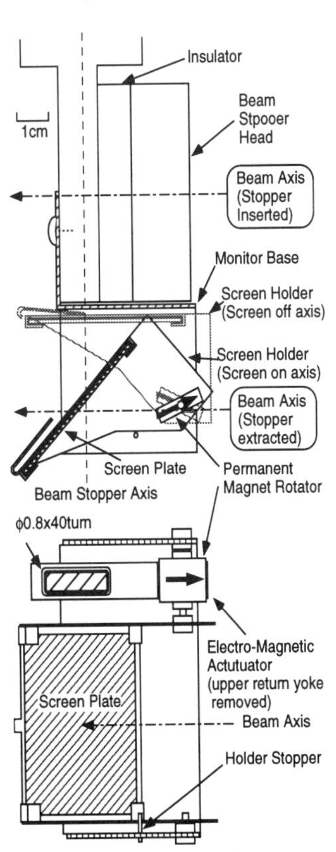

Figure 4. Flip monitor on the beam stopper (FC4).

the beam line, the screen can be flipped away by an electro-magnetic actuator (See Fig. 5). A permanent magnet (Sm Co) rotator sits in a C-shaped electro-magnet. Because of the permanent magnet, the rotator holds its position without any current through the magnet coil. With short pulse current, the rotator can be flipped. It means that no electric power is consumed except for the flipping time. Only electric feedthrough is needed on the vacuum flange to operate this device. This actuator can flip the light screen quickly compared with the conventional pneumatically operated system whose vacuum is shielded by bellows. Because the surface of the fluorescent screen is damaged by high current beam, the dose should be minimized. This quick operation is also preferable in this context. With a capacitor of 0.1 [F] charged up to 20V and a series register of 0.5 [Ω], the electro-magnetic actuator can be flipped. Then the discharging current is estimated at about 40[A] and the duration is 50[ms]. The assembled flip screen monitor is shown in Fig. 6.

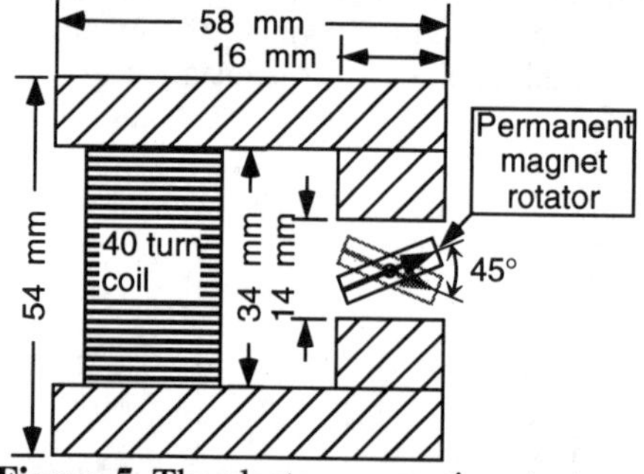

Figure 5. The electro-magnetic actuator.

Figure 6. The flip screen monitor

Fluorescence Screen Plate (4)

The screen plate is AF995R by Desmarquest, which is made of alumina ceramic doped with the chromium oxide. It has high resistance against radiation damages and high temperature resulted by the beam power. The maximum working temperature of the screen is 1800 °C. The spectra of the fluorescence were measured by 50 keV, 2 MeV and 7 MeV proton beam. A monochrometer and a photo multiplier tube were used in the measurements. The result at 50 keV is shown in Fig. 7. The major peak is at 696 nm and the wavelength of this peak is independent of the beam energy. There is a broad peak around

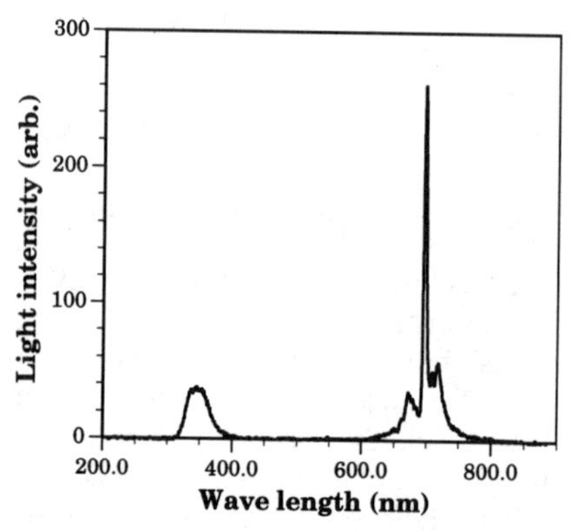

Figure 7. Spectrum of the fluorescence. The proton beam is 50 keV and 0.2 mA.

the 350 nm. The 696 nm peak is very bright and used as the light source of the profile and the emittance monitor in slow measurements. The decay time constants of the 696 nm fluorescence afterglow at 50 keV and 2 MeV are measured as 1 ms and 2.5 ms, respectively. We have tested the 350 nm fluorescence as the light source. The fluorescence of 350 nm has short afterglow in the case of electron beam.(5) The response of the fluorescence is found to be fast enough even for the proton beam.

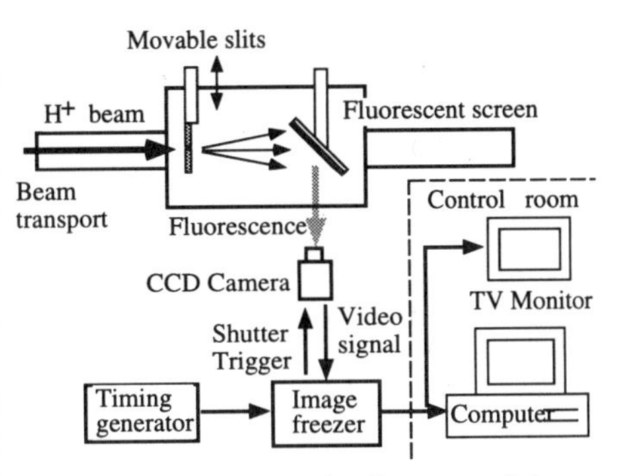

Figure 8. Schematic diagram of the beam emittance monitor system.

Emittance Monitor with Fluorescent Screen and Slits

With this profile monitor (FC4) and a movable slit, the emittance of the beam can be measured. A schematic diagram of the beam emittance monitor is shown in Fig. 8. The slits define the transverse position and spread of the beam. The transverse momentum can be measured by the screen that is located at the downstream of the slits. The fluorescence is observed by a CCD camera and digitized by an image freezer. The image data is analyzed by a personal computer. The interline transfer type CCD sensor is used in the camera. The effective number of pixels is 768 x 494. The resolution corresponds to 0.2 mm/element on the screen. The g value of the CCD camera is 1.0 and the output voltage is proportional to the light intensity. The shutter speed is adjustable from 8.3 ms to 32 μs and the timing is synchronized with the linac operation. The CCD output signals are digitized into 256 steps by the image freezer, and transferred to the computer. The calibrated results of the emittance are displayed on the monitor of the computer.

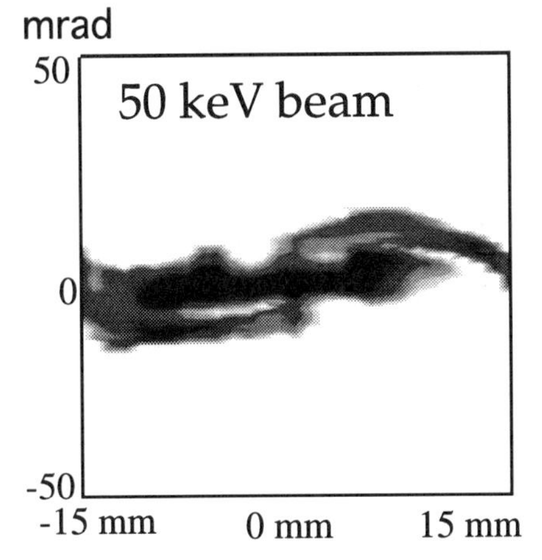

Figure 9. Unnormalized emittance of the 50 keV beam at the entrance of the RFQ linac. The beam current is 5 mA.

The slit plate is made of copper and the width of the slits is 0.2 mm. The distance between the slit and the screen is 500 mm, and then the resolution of the angle is 0.4 mrad. The measurement time is 180 seconds per one axis when the slits are moved with 0.5 mm steps within ±20mm range. Figure 9 shows the measured 100 % emittance of the 50 keV beam at the entrance of the RFQ linac. The beam current is 5 mA. The aberration of the ESQ lenses in the LEBT should be the main cause of the complicated phase space. The 100 % unnormalized emittance is 170 π·mm·mrad.

TOROIDAL COIL PULSE BEAM CURRENT MONITOR (6)

This current monitor has three features; the small core size fitting in the beam matching section (BMS), glass container sealing the vacuum, and the low input impedance I-V converter reducing the droop in the output pulse. Because we want to install the toroidal coil in the PMQ holder in the BMS, the coil must be small, and vacuum tight . The coil was sealed in a glass container as shown in Fig. 10. The core is made of Mn-Zn ferrite (TDK H5C2 T28x13x16). The inner diameter, outer diameter and the length of the glass container are 12 mm, 40 mm, and 70 mm, respectively. The number of turns of the coil is 30, and the inductance L of the coil is 14 mH. The input impedance Z of the I-V converter should be low, otherwise the droop of the output signal, which is expressed as L/Z becomes large. The negative impedance converter (NIC) is applied to achieve the low input impedance. The circuit diagram is shown in Fig. 11. The ratio I_2/I_3 is equal to the ratio R_3/R_2. The input impedance Z of this converter is given by

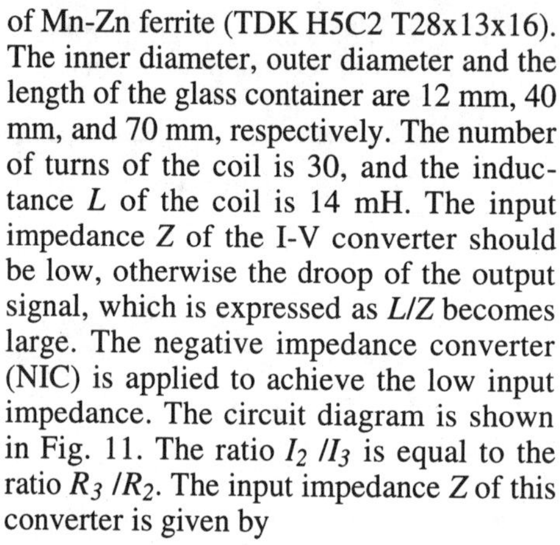

$$Z = \frac{R_1 R_3 - R_2 R_4}{R_3} \quad (1)$$

The R_1 is composed of a fixed resister of 995 Ω, and a variable resister of 10 Ω. This variable resister is adjusted so that Z becomes zero. Then the time constant of the droop become infinity.

The system was tested with a function generator. The test pulse signal fed through the core and the output signal of the I-V converter was measured as shown in the Fig. 12. After adjustment, the time constant of the droop is measured to be 35 ms, which corresponds to 0.4 Ω of the input impedance. Our beam pulse length is 50 μs and the droop was evaluated at 0.2 %. The output of the core monitor and that of the faraday cup right after it was measured simultaneously with real beam current. The result is shown in Fig. 13.

Figure 10. The toroidal coil sealed in the glass container.

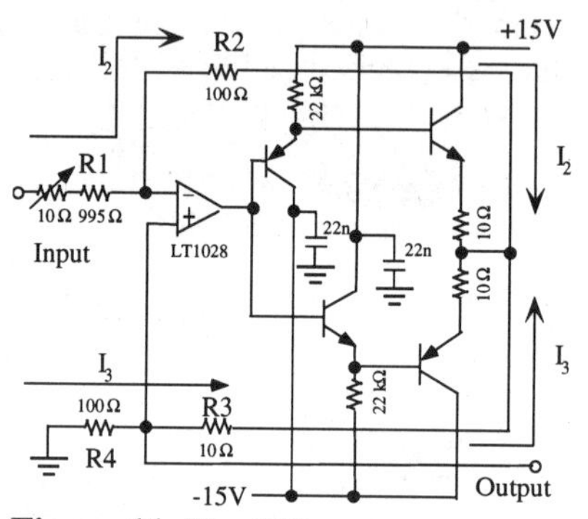

Figure 11. The I-V converter.

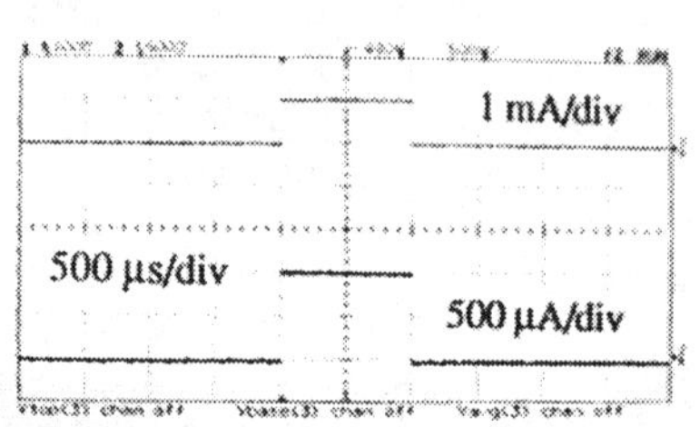

Figure 12. Top: test input current. Bottom: output of the core monitor.

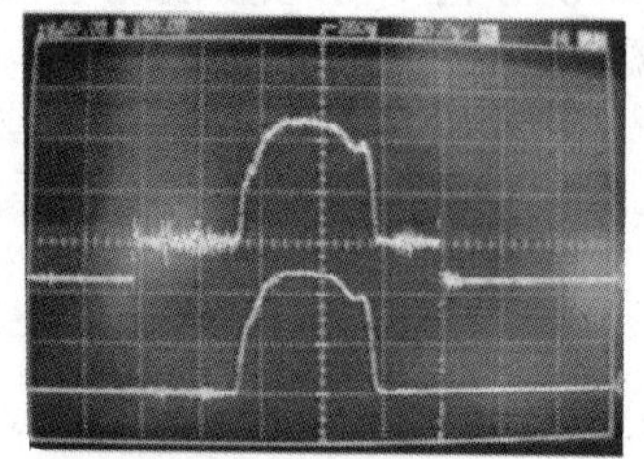

Figure 13. Result of the beam calibration test. Top signal from the core monitor, bottom signal from the faraday cup.

These signals agree very well. This pedestal offset comes from the offset canceling circuit, which removes the fluctuation of the zero level. The peak current measured by the core monitor is 110 μA, and that of the faraday cup was 120 μA. This difference comes form the 5 % gain loss in the offset canceling circuit.

7 MeV EMITTANCE MONITOR

Compact steering magnet

Because the output beam from the DTL diverges rapidly, it should be adjusted by Q lenses. Unfortunately, the output beam from the DTL has a slight angle against the axis, and then the beams are deflected away by the Q lenses. The angle changes with the operating condition. In order to measure the beam emittance, we had to adjust the beam axis, otherwise the beams were out of range for the emittance monitor. Because of the space limitation, single thin steering dipole magnet, which has variable strength and steering direction was needed.

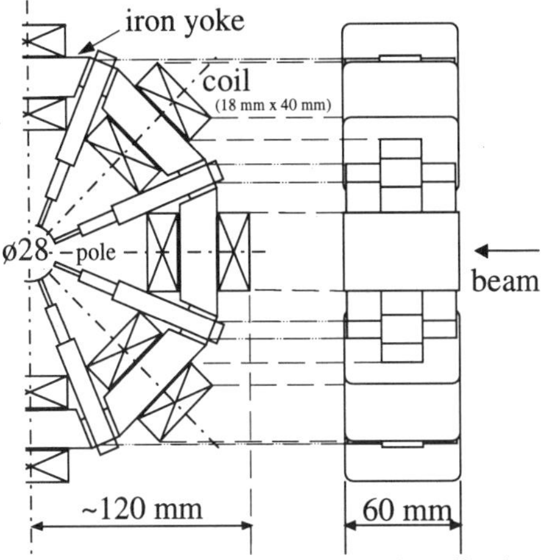

Figure 14. Eight-pole steering dipole magnet.

It can be shown that a circular distribution of radial magnetic field, whose strength of the magnetic field is proportional to the cosine of the azimuthal angle, can produce a dipole field in the circle. For simple fabrication, we adopt "eight pole structure" in which the continuous cosine field distribution is approximated by eight poles as shown in Fig. 14. The strength of the magnetic field at the surface of each pole tip is chosen to be roughly proportional to the cosine of the azimuthal angle. The magnetic field produced by the magnet with the eight pole structure is calculated by PANDIRA as shown in Fig. 15. With this structure, we can change the strength and the direction of the dipole field only by changing the excitation currents. So we can steer the beam for both x and y

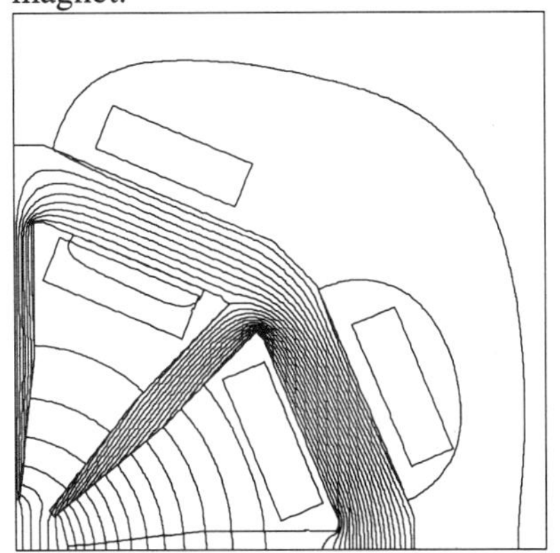

Figure 15. Flux plot of eight pole magnet.

direction by the single element. It is very advantageous to use under severe space limitation.

The steering magnet with the eight pole structure has been fabricated and installed in the beam line. The effective length of the steering magnet is measured to be 97 mm and the magnetic field of 0.4 kG is achieved.

Emittance monitor (7)

The emittance monitor after the Alvarez linac is based on the conventional two-slit method. Distance between the front slit and the rear slit is 1 m and the width of the slit is 0.2 mm. These slits can be actuated in steps of 0.2 mm. Scanning a region of ± 10 mm x ± 10 mrad with 1 mm x 1 mrad intervals, takes about five minutes in measuring the beam distribution in the phase spaces for both x-x' and y-y' plane. Typical example of the measured beam distribution in x-x' plane is shown in Fig. 16. The accuracy of the measured emittance is mainly limited by the accuracy of the current detector, which consists of a beam collecting plate, a current amplifier and 12 bit A/D converter. At the beam current of 0.6 mA, we can measure the emittance with an accuracy of ± 1 π·mm·mrad.. Typical rms emittance for x-x' plane is measured to be 4.8 π·mm·mrad..

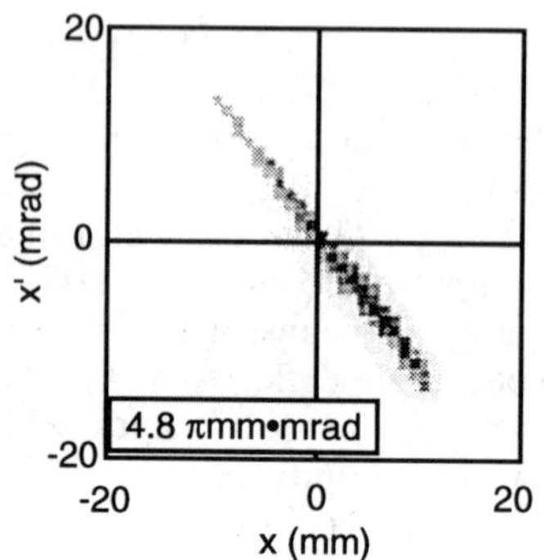

Figure 16. Typical measured emittance at 7 MeV.

REFERENCES

(1) Y.Iwashita, et.al., "Operating Characteristics of the ICR Proton Linac", in 1992 Linear Accelerator Conf. Proceedings, (Ottawa, 1992) AECL-10728 pp. 746-748.

(2) Kando, et. al., "Improvement of the Low Energy Transport System at the ICR 7 MeV Proton Linac", in 1994 Linear Accelerator Conf. Proceedings, Tsukuba, Ibaraki, JAPAN August 21-26,1994.

(3) Y.Iwashita, "Axial Magnetic Field Produced by Radially Magnetized Permanent Magnet Ring", ibid.

(4) T. Shirai, et. al., "Emittance monitor with view screen and slits", in 1994 Linear Accelerator Conf. Proceedings, Tsukuba, Ibaraki, JAPAN August 21-26,1994.

(5) Y. Hashimoto, et al., "Beam Profile Monitor using Alumina Screen and CCD Camera", Proc. of 7th Symp. on Acc. Sci. and Tech., Japan, 314, 1991

(6) H. Dewa, et. al., "Pulsed Beam Current Monitor with a Toroidal Coil", in 1994 Linear Accelerator Conf. Proceedings, Tsukuba, Ibaraki, JAPAN August 21-26,1994.

(7) M. Ikegami, et. al., "Emittance Measurement at the ICR 7 MeV Proton Linac", ibid.

EMITTANCE ESTIMATION FOR ELECTRON ACCELERATORS BY MEANS OF CHANNELING RADIATION

C.Yu. Amosov, B.N. Kalinin, D.V. Kustov,
G.A. Naumenko, A.P. Potylitsin,
V.A.Verzilov and I.E.Vnukov
(Institute for Nuclear Physics,
Tomsk Polytechnical University
Tomsk, Russia)

INTRODUCTION

The existing methods of measurement of linear dimensiosns and divergence of electron beams using the radiation during the crossing of targets by electrons are restricted by angular factor γ^{-1}, where γ is a Lorentz factor of electrons. This fact puts a principle restriction on the low limits of sensitivity of these methods.

The characteristic angle of planar channeling radiation of electrons in single crystals is the Lindhard angle. For a light crystals, such a diamond or silicon, the Lindhard angle for electron energy up to 3 GeV is less then γ^{-1}. This fact we used for measurement of transverse distribution of electron beam.

Until recently for determination of beam profiles and divergence use has been made of EM radiation generated by the interaction of particles with the amorphous target. For instance, in Stanford linear accelerator bremsstrahlung was used to measure e-beam profiles, while in SPS proton accelerator (CERN) it was the optical transition radiation.

We carried out an experiment on determining the divergence of electron beam of 900 MeV using the orientation dependence of PXR yield. An electron beam of variable and controlled divergence is easily obtained by means of multiple scattering of particles in amorphous substancies. This method is simple and convenient, and multiple scattering is easy to calculate.

For this aim, in the experiment in front of the crystalline target we placed an amorphous aluminium plate as a simulator of divergence. The electron beam passing through the amorphous aluminium was broadened due to multiple scattering.

MEASUREMENT OF LINEAR DIMENSION OF ELECTRON BEAM

The experiment for measurement of linear dimension and transverse distribution of electron beam, crossed a crystal target, was carried out on Tomsk Synchrotron. The energy of accelerated electrons was 900 MeV. During the slow throwing of electrons on the target (20 ms) owing to betatron oscillation of electrons they cross a target at a some distance from its borden. It forms therefore a transverse distribution of electrons, crossing the target.

The aim of our experiment was the measurement of this distribution. The scheme of set up of experiment is given in fig. 1.

The electron beam, which cross the diamond crystal at 0.5 mm thickness is blocked by lead scatterer of 4 mm thickness. Step of blocking is 0.25 mm. The root mean square angle of multiple scattering of electrons in the scatterer is 12.6 mrad. The Lindhard angle Θ_L for the plane (100) of diamond crystal and electron energy of 900 MeV is 0.2 mrad.

For each position of scatterer the orientation dependence of planar channeling radiation was measured. The difference between maximum and minimum of radiation intensity in orietation dependence is the intensity of channeling radiation only. For illustration the orientation dependence of planar channeling radiation without scatterer is given in fig. 2. Owing small angle of capture of electrons into channeling state ($\Theta_L \cdot \gamma^{-1}$) the contribution of electrons, scattered by lead, in orientation dependence may be ignored. In such a way maximum intensity of channeling radiation in orientation dependence corresponds with number of nonscattered electrons, crossed the crystal. In fig. 3 is given measured dependence of this value on position of scatterer. Derivative of this dependence, that is given in fig. 4, corresponds with transverse distribution of electrons, crossed the crystal.It is seen in fig. 4 a full width at half maximum of transverse distribution is equal to 0.8 mm.

Pay attention to impossibility to carry out such measurement using another methods. This method used at a linacs make it possible to measure the transverse and angular distribution of electron beams.

PXR-BASED TECHNIQUE FOR DIAGNOZE OF BEAMS OF CHARGE PARTICLES

The measured PXR characteristics have been used to define the divergence of an electron beam. A possibility considered for beam diagnostics used the dependence of PXR characteristics on the angular distribution of incidence of particles and their energy.

As it follows from PXR theory /1,2/ the intrinsic angular distribution of PXR spot is determined by the angle

$$\Theta^2 = \gamma^{-2} + \Theta_s^2 - Re(\chi_0) + \Theta_R^2. \qquad (1)$$

Each term in the right part has a clear physical sens. The value $Re(\chi_0) = -\omega_0^2/\omega^2$ characterises the divergence of pseudophoton beam for charge particle in the medium. Here ω_0 is the plasma frequency, ω is the energy of PXR quanta, Θ_R is the divergence of electron beam and Θ_s is the root mean-square angle of multiple scattering (MS) of particles in the target.

Consequently, measuring the angular distribution of PXR spot, we can obtain information on the value of Θ_R. The same information could be obtained while measuring the orientation dependence of PXR photon yield using a detector of small angular aperture /3/. Among the advantages of PXR over other radiation types we can list its large radiation angles and ,hence,an unambiguous determination of the effect from among the other accompanying processes. Moreover, PXR could serve useful for the diagnostics of the beams of relativistic adrons because of the possibility to use a thin (0.1 mm) crystal target adding negligibly small distortions to the beam characteristics. The angular distribution at the simulator output was count according to multiple scattering theory.

In table 1 the results of experiment are given.

Table 1

	Calculation	Experiment		
Target	Al	Si	Si + Al	Al
Θ		4 ± 0.21	4.25 ± 0.21	
Θ_s	1.35			1.44 ± 0.3

In this table we see a good agreement of calculated and measured divergence of electron beam crossing the crystal.

Using the value of the PXR yield we may assume that the application of PXR-based techniques is expedient for the beams with the flux $> 10^7$ particles per second.

It is worth reporting in conclusion that a recent experiment /4/ has registered PXR of 70 GeV protons.

This work is partly supported by International Scientific Foundation (grant RI 4000).

REFERENCES

1. I.D.Feranchuk and A.V.Ivashin, J.Physique 46(1985)1981.
2. V.G.Baryshevsky and I.D.Feranchuk, Nucl. Instr. and Meth. 228(1985)490.
3. Yu.N.Adishchev, V.A.Verzilov, A.P. Potylitsyn et al., Nucl. Instr. and Meth. B44(1989)130.
4. V.P.Afanasenko, V.G.Baryshevsky, R.F.Zuevsky et al., JETF Pis'ma (in Russian) 54 (1991) 493.

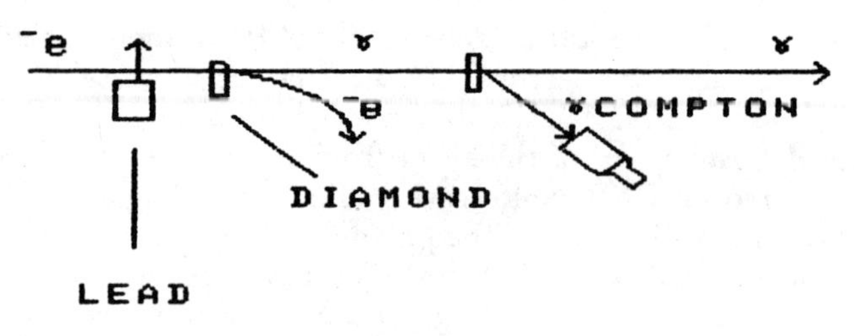
e
γ
γ
e
γ COMPTON
DIAMOND
LEAD

FIG 1.

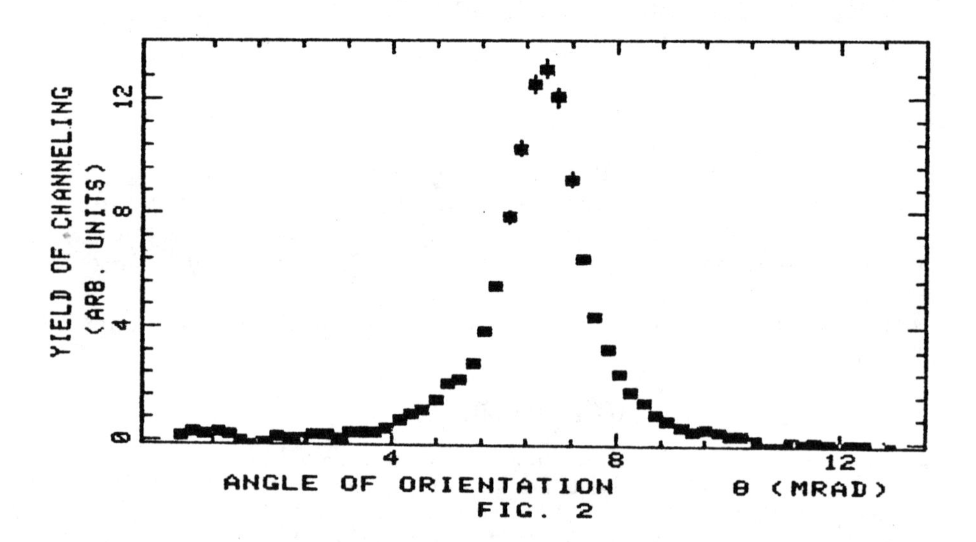
YIELD OF CHANNELING
(ARB. UNITS)
0
4
8
12
4
8
12
ANGLE OF ORIENTATION
θ (MRAD)

FIG. 2

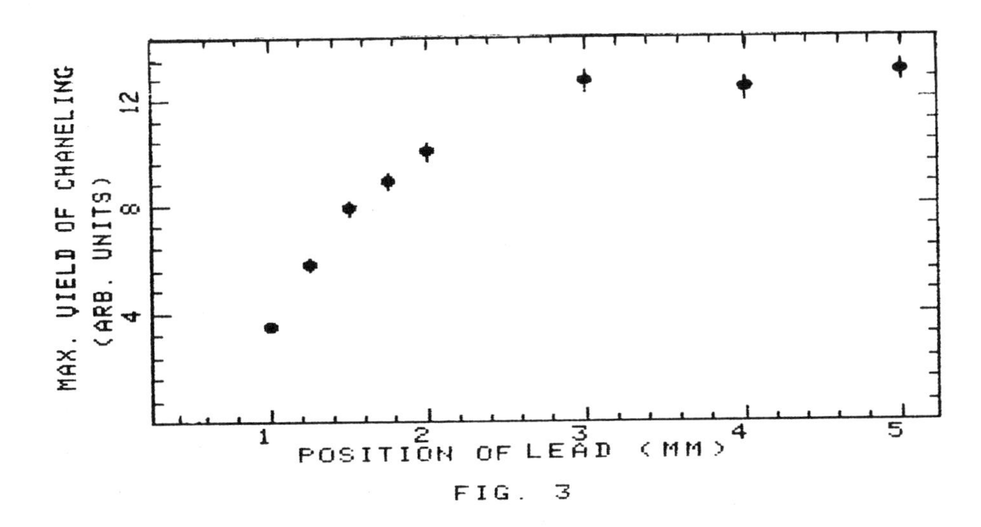
MAX. VIELD OF CHANELING
(ARB. UNITS)
4
8
12
1
2
3
4
5
POSITION OF LEAD (MM)

FIG. 3

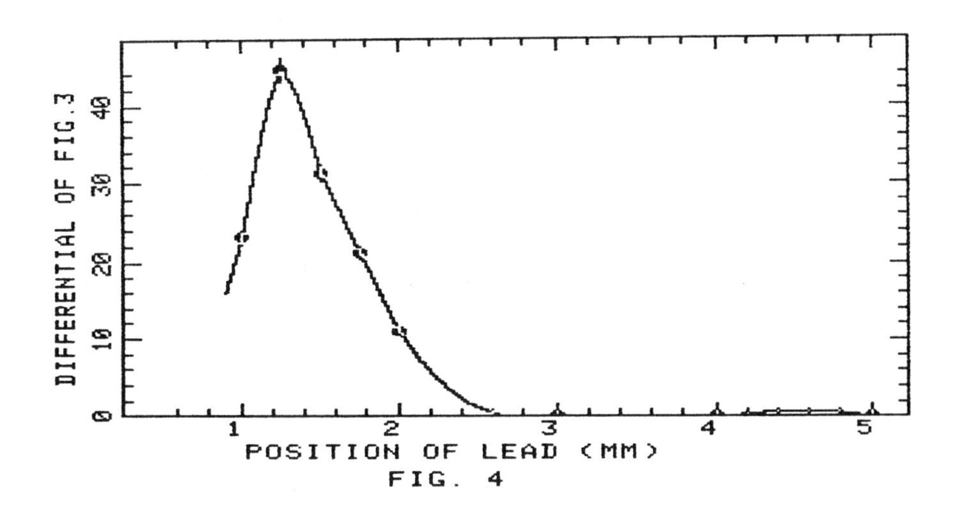
DIFFERENTIAL OF FIG.3
0
10
20
30
40
1
2
3
4
5
POSITION OF LEAD (MM)

FIG. 4

TWO DIMENSIONAL BEAM CURRENT DISTRIBUTION MONITORS

A.Tron
MEPhI, Kashirskoe sh. 31, 115409 Moscow, Russia

Abstract

To control beam shaping in front of a target, to prevent the thermomechanical damage of first neutron source's wall and also to control beam propagation in transport system including particles transverse coupling motion two types of monitors for two-dimensional beam current distribution measurements are offered. The principle operation of one of them consists in recording the density distribution of secondary electrons along a wire emitter scanning the beam. The principle of operation of another one consists in transformation of two-dimensional primary beam current distribution into corresponding electron one which is transferred, registered in discrete points and then approximated. The description of these monitoring systems and their testing results are presented.

INTRODUCTION

For understanding and minimizing beam loss along high power ion linac-neutron sources driver [1, 2] the description of beam behavior in terms of linear uncoupled theory of the transverse motion of charged particles is insufficient. Particles transverse coupling motion is caused by longitudinal magnetic fields or fields which do not have a midplane symmetry with respect to the beam. It occurs, in fact, in every real machine too as the result of distortions and imperfections. The coupling motion makes the behavior of the beam more difficult to understand and the operation of the machine becomes more complicated [3].

In connection with this monitoring of two-dimensional beam current distribution acquires particular importance. Moreover, the measurement of two-dimensional beam current distribution is necessary in front of any target [4], it is necessary for the control of bringing together several beams or measurements the position of each beam in the presence of others [5], for determination of excess of the beam current density above limiting magnitude to prevent, for example, the thermomechanical damage of the first wall of intense neutron sources [6], for operation control of ion sources.

Two types of the monitors for two-dimensional beam current distribution measurement which are realized new principles of operation are presented in this paper. One of them has been successfully tested in the transfer line of the INR linac and another one has been successfully bench tested for the INR intense neutron source.

The former must satisfy the next main requirements. The spatial resolution of the beam current components within the entire ion pipe aperture of 54 mm diameter must be not worse than 1 mm and the monitor's size along the beam must be 0.15 m.

In the latter the rms error of maximum proton beam current density(j_m) measurement must be not more than ±5% and its measurement time - not more than 10 ms. Diameter of measurement area must be not less than 130 mm at the ion pipe diameter of 160 mm.

In both cases the devices must ensure their calibration without disassembling and must disturb the beam negligibly. The base parameters of the monitors must be independent of the beam energy.

Our studies have shown that the known methods and devices for j(x,y) measurements [8, 9] do not satisfy these requirements totally.

BEAM CROSS SECTION WIRE SCANNER

To control more precisely the ion beam propagation in transfer line (with beam energy of 0.75 MeV) of the INR linac containing bending magnets, quadrupoles, solenoids, steering correction magnets a new type of secondary electron monitor for two-dimensional beam current distribution measurements has been proposed [4, 10] and successfully tested. Figure 1 shows layout of the monitor and Fig.2 - photo of the monitor mounted in the transfer line of the INR linac.

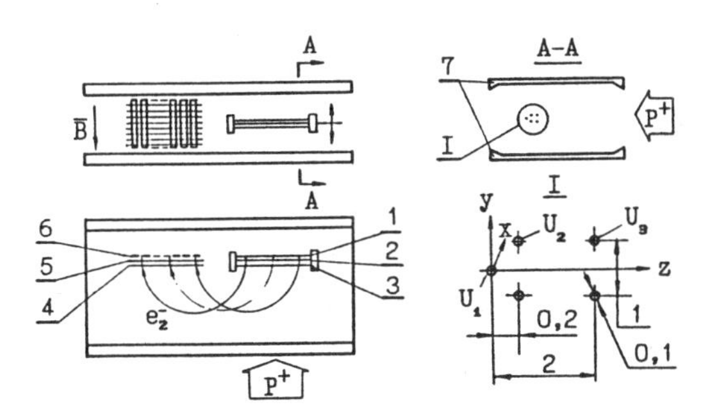

Figure 1. Layout of the beam cross section wire scanner.

Figure 2. Photo of the monitor.

The principle of the monitor operation consists in a recording of the density distribution of the low energy secondary electrons (emitted with energy less than 50 eV) along a wire-emitter (1) scanning the beam perpendicularly to its axis by step motor.

By applying high negative voltage (U_1 is about 1 kV) to the wire (1) the secondary electrons produced as a result of the primary beam-wire interaction are accelerated till the electrodes (3) under ground potential (U_3) and transferred in uniform magnetic field to the plane of multichannel collector (6) with screen (4) and lock (5) grids in front of it. The highly uniform magnetic field in a region of the electrons motion (not worse than 0.3%) is produced by two specially shaped poles (7). To limit the electrons motion in *y* direction a focusing electrodes (2) were installed. Their mutual position (in mm) is shown in Fig.1. For every position of the wire-emitter relative to the beam axis the secondary electron distribution is measured by the sequential registration of multichannel collector currents.

The geometry and potentials of the electrodes were initially determined by computer simulation. Then they were corrected in experiments with the thermoelectrons beam emitted from the emitter (1). Figures 3 and 4 show the results of the calculations which explain action of the focusing electrodes (2) in (y,z) plane: Fig.3 shows the distribution of electrons potential energy in the electrodes electric field, Fig.4 shows the electrons trajectories and equipotential lines of the same electric field.

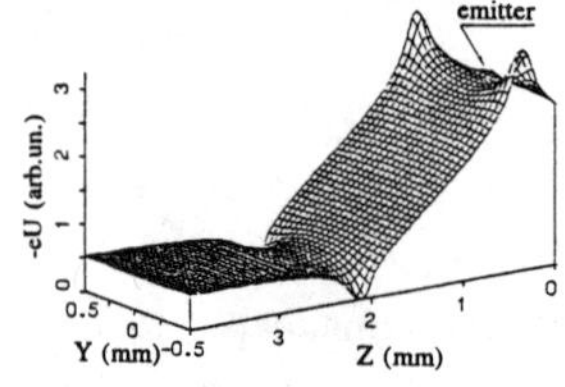

Figure 3. Distribution of electrons potential energy in electric field of primary converter.

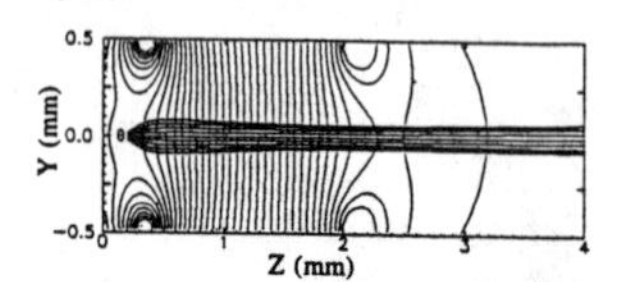

Figure 4. Electron trajectories and equipotential lines of primary converter electric field.

It should note that in the vertical plane (y, z) spatial resolution of the beam current is determined by the emitter diameter which was equal to 0.1 mm. For presented geometry of the monitor's primary converter the secondary electrons flying out from the emitter within the angle of ±30 degrees form the electron beam shown in Fig.4. Electrons with another starting angles (along the wire surface) produce background the current density of which is two orders less than corresponding magnitude in the maximum of main electron beam flux.

The monitor resolution along x axis was determined by the numerical integration of electron motion equation with account of field being resultant of the electrodes electric field, uniform magnetic field and the ion beam one. Initial angle-velocity distribution of secondary electrons on the emitter was taken according to experimental data from [11].

FWHM of secondary electrons distribution along x axis in the collector plane when their initial distribution along the wire-emitter was delta-function is about 0.5 mm.

The distribution have been obtained for electron trajectories radius equaled to 40 mm and for the following ion beam parameters: proton energy 0.75 MeV, the rms beam radius 7.5 mm, pulse current 100 mA.

In the case when the beam space charge effect may be neglected the monitor resolution Δx is mainly defined by initial electrons velocity distribution and its magnitude may be simply estimated [4] as

$$\Delta x = 4z\sqrt{\frac{\Delta W_0}{W}},$$

where z is distance between the emitter and the grid under ground voltage; ΔW_0 is FWHM of initial electron energy distribution (for estimation ΔW_0 may be taken as 4 eV for carbon, tungsten); $W = -eU_1$ is electron energy after acceleration.

However, there is construction factor which limits monitor resolution. This is finite width of strips of the multichannel collector. In our case the strip pitch was 1 mm.

Some results

The monitor is remotely controlled by a personal computer IBM PC and CAMAC. The hardware also contains unit for the monitor calibration. Using the flux of thermoelectrons from the wire-emitter one can check operation of measurement system.

To represent the results of two-dimensional ion beam current distribution measurements graphically both as isometric picture and lines of equal current density in the transverse plane the corresponding software, code "Beamvis", has been designed. In Figure 5 one can see one of the beam image obtained with this monitoring system within the pipe diameter of 54 mm.

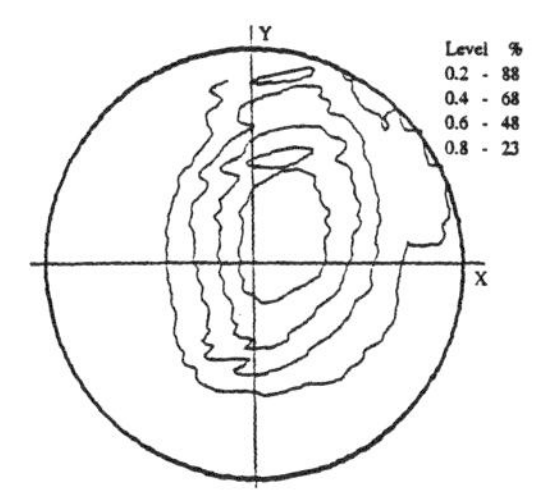

Figure 5. Lines of equal beam current density and beam percent within of them.

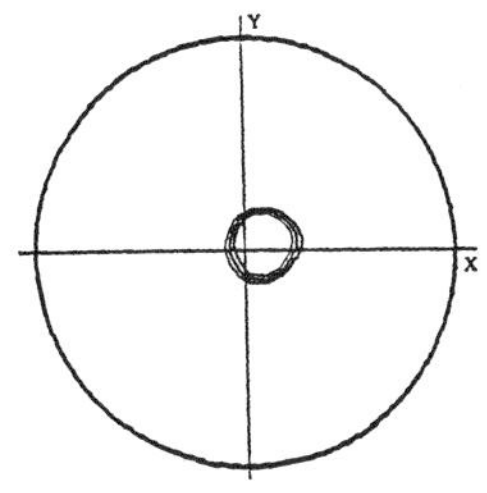

Figure 6. Beam image after providing a diaphragm with hole diameter of 10 mm.

For comparison in Fig.6 is shown image of beam which was obtained after providing a diaphragm with hole diameter of 10 mm in front of the monitor.

There is one more important effect which limits the monitor operation. When the temperature of the wire-target is beyond 2000 K the thermocurrent density can excess the magnitudes compared with the secondary electron one. It ought to note too that for the tension of the wire it is always necessary to know a possible highest wire-target temperature because the limit of the elasticity strongly depends on the target temperature.

In Figure 7 and 8 are shown some results of our studies. Maximum temperature dependencies of tungsten and carbon emitter on time at the 160 MeV proton beam heating are shown in Fig.7, where the beam parameters were the following: the rms radius 2.5 mm, pulse current and its duration - 0.05 A, 0.1 ms. The emitter length is 60 mm, its radius is 0.05 mm. The curves are given for three values of pulse repetition frequency. Thermophysical coefficients for calculations have been taken from [12].

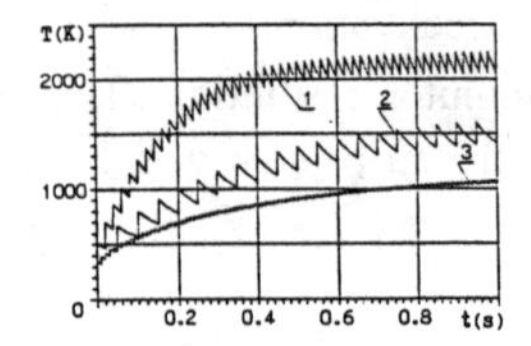

Figure 7. Temperature vs time dependence of targets from tungsten (curves 1,2) and carbon (3) for three pulse current repetition periods: 1- 20 ms; 2- 50 ms; 3- 10 ms.

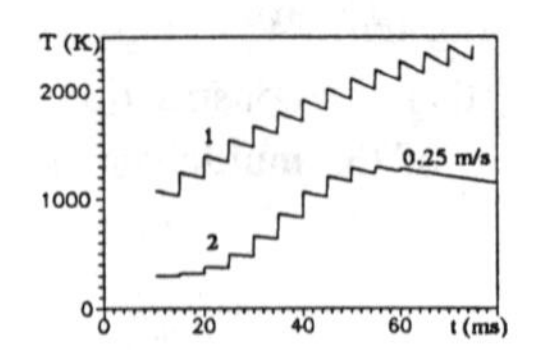

Figure 8. Maximum temperature time dependencies of carbon wire-emitter under the proton beam for two target velocities: 1- 0 m/s and 2- 0.25 m/s.

As it follows from Fig.7 the carbon emitter can operate at more high average beam current than the tungsten one. In Fig.8 is shown temperature dependence of the same carbon wire-emitter for the proton beam with energy 100 MeV, pulse current 250 mA, pulse duration 0.1 ms and its repetition period of 5 ms. The rms beam radius is 2.5 mm.

Thus, the carbon wire-emitter can operate in the high energy part of future ion linac at average beam current of 5 mA. As to our case (0.75 MeV) the tungsten wire-target can stay under the above mentioned INR's beam at the beam pulse repetition frequency up to 50 Hz.

RAPID MONITORING OF BEAM CURRENT DISTRIBUTION

For rapid monitoring of maximum proton beam current density j_m and measurement of two-dimensional beam current density distribution j(x,y) at the entrance of the INR intense neutron source the rapid monitor [6, 7] has been proposed and created the layout of which is shown in Fig.9. All sizes are presented in mm.

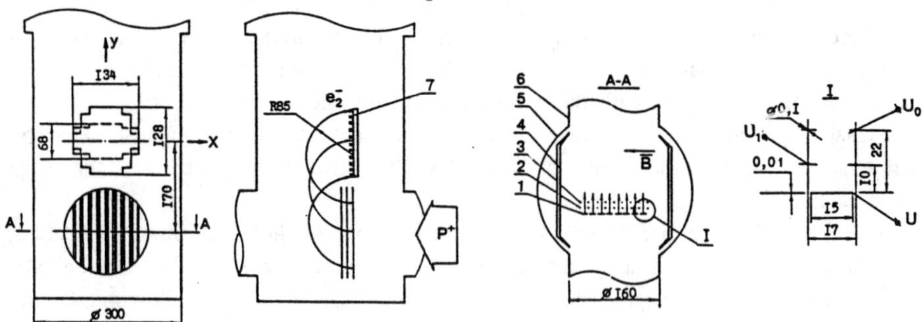

Figure 9. Layout of rapid j(x,y) monitor.

The principle of the monitor operation is the following. Electrons that have been produced as a result of interaction between the primary beam and thin striplike emitters (1) made from 0.01 mm tantalum foil are accelerated from the emitters (1) with negative potential (U) equaled to -4 kV till the electrodes (3) under ground potential (U_0). The focusing of the electron flux in (x,z) plane was realized by installation of additional electrodes (2) with potential (U_1) close to the emitter one.

Using semicircular focusing in uniform magnetic field the electrons are transferred from the ion beam space to the plane of 64-channel collector. The uniform magnetic field is produced by specially shaped poles (4), which in detail are shown in Fig.10.

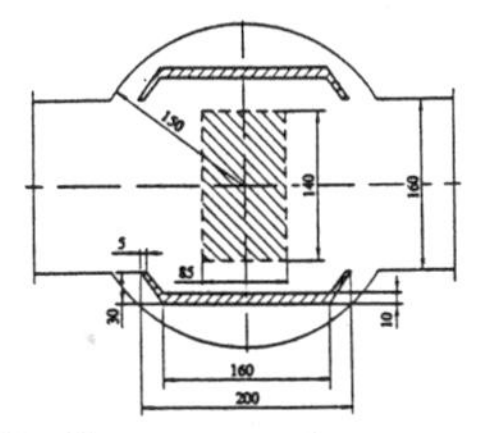

Figure 10. Geometry of magnet poles of the rapid monitor.

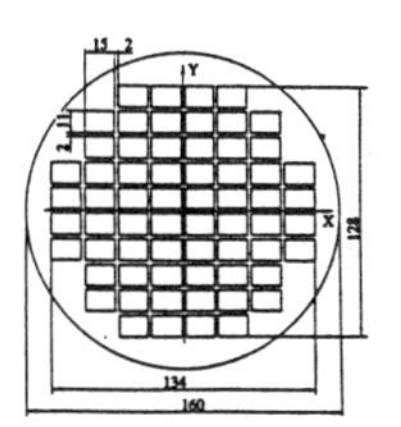

Figure 11. Geometry of 64-channel collector.

In Fig.9 the current collector maximum sizes of 134×128 mm are shown, its geometry is clear from Fig.11. The lock and screen grids are placed in front of the collector (7).

The monitor resolution accounting real initial energy-angle secondary electrons distribution have been defined by numerical simulation. The electrons initial distribution along y was assumed to be delta-function and along x as uniform within emitter strip width.

HWHM of secondary electrons distribution along y coordinate in the collector plane are o.5 mm and along x - 4 mm that are smaller by a factor of 5 as the rms sizes of the ion beam. The monitor resolution is inversely proportional to electrons velocity i.e. by raising emitter potential it may be extremely improved.

The corresponding secondary electron two-dimensional current distribution is registered in discrete points and then approximated by two-dimensional Kotelnikov's series. Monitoring system records the beam current distribution with a high precision for a time less than 10 ms and determines excess of the beam current density above limiting magnitude for less than 1 ms. The device size along the beam is about 200 mm at the diameter of the beam pipe of 160 mm.

Studies have shown that the monitor discussed can be successfully used with slight modification for monitoring of the same transverse sizes proton beam but with pulsed beam current up to 15 A.

ACKNOWLEDGMENTS

Author thanks his colleagues from MEPhI for help in creating of the above mentioned monitors.

Author expresses his acknowledgment to Yu.Ya.Stavissky and S.K.Esin for support of these works which were fulfilled under contracts 91-3-859, 93-3-859, 92-3-956.

REFERENCES

1. Jameson R.A., Lawrence G.P., Schriber S.O., "Accelerator-Driven Transmutation Technology for Energy Production and Nuclear Waste Treatment", in Proceedings of the EPAC92 Conference (Berlin, March 24-28, 1992), pp.230-234.

2. Carminati F., et.al., "An Energy Amplifier for Cleaner and Inexhaustible Nuclear Energy Production Driven by a Particle Beam Accelerator", CERN report, CERN/AT/93-47(ET).

3. Willeke F., Ripken G., "Methods of Beam Optics", DESY report, DESY 88-114, Aug., 1988.

4. Tron A., "Secondary Electron Monitor of Two-Dimensional Beam Current Distribution", in book "Calculation Methods and Experimental Investigation of Linac's systems", Moscow, Energoatomizdat, 1987, pp.41-45 (in Russian).

5. Rossmanith R., "CEBAF Beam Instrumentation", presented at the Workshop on Advanced Beam Instrumentation", Tsukuba, Japan, April 22-24, 1991.

6. Tron A., "Rapid Measurements of Two-Dimensional Ion Beam Current Distribution for Pulsed Neutron Source", presented at the Particle Accelerator Conference, Washington, D.C., USA, May 17-20, 1993.

7. Tron A., "Secondary Electron Monitors in Linacs for Intense Neutron Source", presented at the LINAC94 Conference, Tsukuba, Japan, August 21-26, 1994.

8. Alonso J.R., Tobias C.A., Chu W.T., "Computed Tomographic Reconstruction of Beam Profiles with a Multiwire Chamber", IEEE Trans. Nucl. Sci., Vol.NS-26, 3077 (1979).

9. Komissarov U.P., et.al., "Beam Cross Section Monitor", presented at the 11-th All-Union Particle Accelerator Conference, Dubna, USSR, October 25-27, 1988.

10. Tron A., Vasilev P., "Secondary Electron Monitor of Beam's Two-Dimensional Transverse Current Density Distribution", in Proceedings of the EPAC92 Conference (Berlin, March 24-28, 1992), pp. 1124-1126.

11. Schultz A.A., Pomerantz M.A., "Secondary Electron Emission Produced by Relativistic Primary Electrons", Phys. Rev., Vol. 130, 2135 (1963).

12. Physical-chemical properties of materials, Kiev, "Naukova Dumka", 1965 (in Russian).

Noninterceptive Transverse Emittance Measurement Diagnostic For An 800 Mev H⁻ Transport Line*

D. P. Sandoval
Los Alamos National Laboratory
Los Alamos, NM 87545

Abstract

A nonintrusive diagnostic device that will measure the transverse-phase-space parameters of an 800 MeV H^- beam is under development. The diagnostic device will make the measurements under normal operating conditions and will not perturb the particle beam or produce unwanted radiation. The diagnostic will use the phenomenon of laser-induced photodissociation to sample the H^- beam. This paper discusses the preliminary design of the diagnostic device.

INTRODUCTION

Theoretical models and sparse experimental results show that the Los Alamos Meson Physics Facility (LAMPF) beam distribution is nonlinear in transverse phase space. It is important for transverse phase-space matching of high current beams that the Twiss parameters be understood. For example, matching into a storage ring is very critical; if the beam phase space (Twiss parameters) is understood, correction by nonlinear elements is possible. Similar considerations apply to other high intensity systems that must be understood, such as transport lines or targets.

A diagnostic device in Line D of the LAMPF for measuring the x, x' and y, y' phase space distributions would provide information which may allow improved matching into the Proton Storage Ring (PSR) as well as providing critical information for advanced projects. Ideally, one would like to have a non intrusive diagnostic which provides the users an on-line method for determining the transverse emittance of the particle beam during normal operation (at full current) without perturbing the particle beam. H^- laser neutralization techniques demonstrated on the Ground Test Accelerator (1,2) can provide such a diagnostic device. This device would obviously have the ability to provide beam profiles as well, without producing radiation (unlike the charge-collection wires currently being used elsewhere at the LAMPF).

Advances in laser technology have in the production of highly reliable high-power lasers which require minimal maintenance. This allows one to place a laser near the beamline, eliminating the need for long laser beam transport lines. The result is the ability to use a laser beam to reliably and precisely sample the H^- particle beam with little human intervention.

SYSTEM DESCRIPTION

A laser beam would be focused into a narrow cylinder with a diameter on the order of 0.2 to 0.5 mm. The laser beam will traverse the particle beam

*Work supported and funded by the US Department of Energy, Office of Defense Programs.

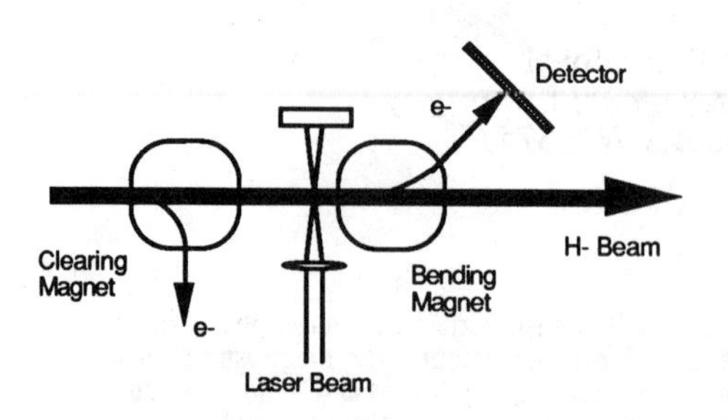

Figure 1. Schematic diagram of the transverse emittance measurement system.

neutralizing a narrow slice (x or y) of the beam. The number of photodetached electrons (and neutral particles) produced is proportional to the particle beam density and the laser energy density. One has the choice of collecting the photodetached electrons or the neutral particles to determine the x' (y') distribution. The laser beam would be scanned across the particle beam to obtain a full x, x' (y, y') distribution (into the page in Fig. 1).

Using the neutrals to measure the x' (y') distribution has the advantage that magnetic or electric fields will not effect the measurement. However, a bend and an offset beam line are required. Electron collection techniques do not require a dedicated offset beam line and can be made more compact longitudinally. Unwanted background electrons can be removed using a clearing magnet. The disadvantage is that stray magnetic and electric fields effect the trajectory of the electrons. This paper will focus on the electron collection method.

EXPECTED SIGNAL AND NOISE LEVELS

The electron signal level is calculated using the H^- beam and laser parameters given in Table 1. A baseline detector configuration is also described and evaluated.

Neutralization Fractions

The neutralization fraction f_{neut} which is defined (2) as the average probability of neutralization for beam particles that have been illuminated by the laser (Eq. 1).

$$f_{neut} = 1 - e^{-\sigma(\lambda)\phi t_i} \quad (1)$$

Where $\sigma(\lambda)$ is the photodetachment cross section for the laser wavelength λ, ϕ is the photon flux and t_i is the illumination time. The neutralization fraction assuming the following conditions was calculated using Eq. 1 with a result of f_{neut} = .99, i.e. 99% of the illuminated beam is neutralized.

Table 1. H^- beam and laser parameters.

Laser energy - 420 mJ per pulse	Neutralization cross section - 3.8 x 10^{-17} cm^2
Laser pulsewidth - 20 ns (t_i)	x rms beam width - 1.8 mm
Laser beam diameter - .3 mm	y rms beam width - 2.1 mm
Laser wavelength - 1.06 μm (Nd:YAG)	x' rms beam divergence - 0.207 mrad
Ion beam energy - 800 MeV	y' rms beam divergence - 0.189 mrad
Ion beam current - 10 mA peak current (I_p)	

Electron Signal Level Calculations

The expected signal levels (electrons at the detector) were calculated assuming the laser and ion beam parameters listed above. The available charge Q_x at the detector can be estimated by using Eq. 2. This is the signal available if this diagnostic is used as a profile monitor. The limits of integration are determined by the laser beam width and location.

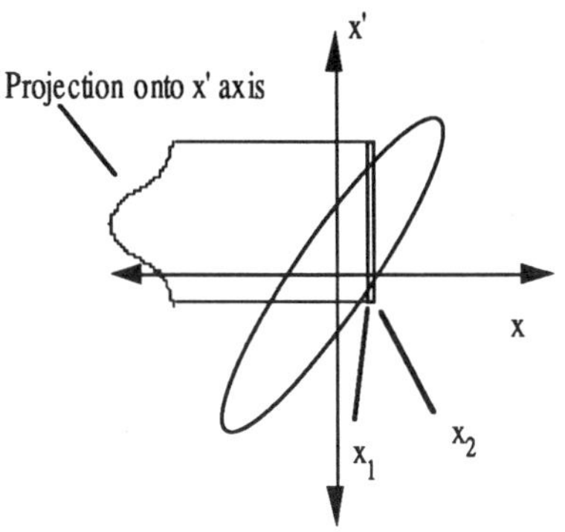

Figure 2. x, x' phase space ellipse.

$$Q_x = I_p t_i f_{neut} \int_{x_1}^{x_2} \frac{1}{\sigma_x \sqrt{2\pi}} e^{\frac{-(x)^2}{2\sigma_x^2}} dx \qquad (2)$$

Where σ_x and is the rms beam width in x, I_p is the peak current, t_i is the laser temporal pulse width and f_{neut} is the neutralization fraction. The limits of integration, x_1 and x_2, are defined by the laser beam diameter in the interaction region. Assuming a normal distribution in x' one can estimate the signal level $S_{x'1x'2}$ per x, x' pixel by using Eq. (3). See Figure 2. If one uses a width in x' of .2 $\sigma_{x'}$ then the limits of integration are x'-.1 $\sigma_{x'}$ to x'+.1 $\sigma_{x'}$ where x' is the position of interest. This would give 20 pixels in x'. The expected signal levels are given in Table 2 for the above mentioned parameters.

$$S_{x'_1 x'_2} = C_x \int_{x'_1}^{x'_2} \frac{1}{\sigma_{x'} \sqrt{2\pi}} e^{\frac{-(x')^2}{2\sigma_x^2}} dx' \qquad (3)$$

Table 2. Expected signal levels in number of electrons for various locations in x x' space. These signal levels are detectable provided they are above the noise level.

	x' location (center)	x' location (1 $\sigma_{x'}$)	x' location (2 $\sigma_{x'}$)	x' location (3 $\sigma_{x'}$)
x location (center)	8.1 x 10^6	5.0 x 10^6	1.1 x 10^6	9.2 x 10^4
x location (1 σ_x)	5.0 x 10^6	3.0 x 10^6	6.7 x 10^5	5.6 x 10^4
x location (2 σ_x)	1.1 x 10^6	6.9 x 10^5	1.5 x 10^5	1.2 x 10^4
x location (3 σ_x)	9.4 x 10^4	5.6 x 10^4	1.3 x 10^4	1.0 x 10^3

Detector Configuration and Signal Level

The baseline detector consists of an intensified camera viewing a sodium iodide scintillator (Figure 3). This detector configuration was selected as the baseline system because it uses components that we have on hand and are familiar with, not because it is the best choice.

The range (depth of penetration) of 435 keV electrons in NaI(Tl) is about 0.58 mm (3). The total path length that the electrons travel is 1.2 to 4 times the range, the ratio being largest for slow electrons in materials of high Z (3). Given the

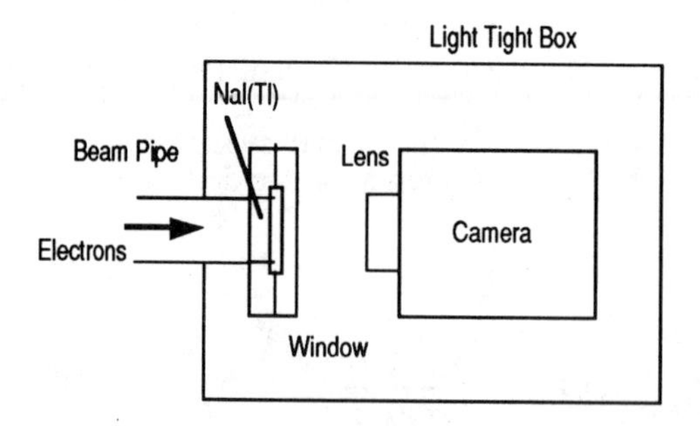

Figure 3. Schematic diagram of the baseline detector system.

above parameters, each pixel (x, x') at the detector (scintillator) will be 40 μm wide by approximately 1 cm in height. The narrow dimension is the dimension of interest (see Figure 4). In order to keep the effects caused by electron scatter in the scintillator to a minimum, the scintillator thickness will be one-half of the required resolution (approximately 20 μm).

NaI(TI) produces about 52000 photons with a mean wavelength of 415 μm per MeV per particle if the particle is completely stopped. The energy loss across the range of the electrons is fairly linear (4). Therefore, the ratio of photons produced is equivalent to the ratio of the scintillator thickness divided by the range of the electrons. Each electron produces about 1800 photons in a 20 μm thick scintillator.

The electrons contained in the phase space pixel located at x = 3σ and x' = 3σ produce 780 x 10^3 photons for each 20 ns laser pulse (at the scintillator). These photons are radiated into 4π steradians and it is assumed that 10% of the photons produced will be transmitted through the lens system to detector (Camera Micro Channel Plate - MCP). A lens system magnification of one-half is also assumed. This yields an energy density of 18.7 x 10^{-12} Joules per cm^2 at the camera MCP for each laser pulse.

Background Signal Level Calculations

It is assumed that a clearing magnet can be placed upstream of the laser-ion beam interaction region to remove unwanted electrons. The rate that electrons are produced by gas stripping Γ_{bk} can be estimated (5) by using Eq. 4.

$$\Gamma_{bk} = \sigma_s \rho x \frac{I}{e} \tag{4}$$

Where σ_s (1.6 x 10^{-18} cm^2) is the stripping cross section ρ is the density of the residual gas (assumed to be 1 x 10^{-7} torr of N_2), x is the path length (assumed to be 10 cm), I is the beam current and e is the unit charge. The resulting signal to background ratio is about 1.8 x 10^8:1 or 165 dB. This assumes that the detector is gated for 250 ns.

Background due to gas stripping should not be a problem if electrons are used as the signal source. If H^0 s are used as the signal source the background due to gas stripping will be much higher due to the longer ion beam path length (x).

Noise Level Calculations

The sources of noise are background electrons, background radiation due to beam spill, photon noise and noise associated with the camera. The intensifier, the CCD array and the camera electronics will be considered as a unit. The camera being considered is an ITT intensified camera (Model F4577) which was used on GTA. Each noise source was considered separately. The results are given below.

Photon Noise

The rms noise level δ, due to the signal radiation is defined to be the square root of the variance of a Poisson distribution $\delta = \sqrt{\overline{n}}$ where $\overline{n}$ is the average number of photons incident on the detector. The rms error E_{rms} at pixel location x = 3σ and x' = 3σ is

$$E_{\text{rms}} = \frac{\overline{n}}{\delta} = \sqrt{\overline{n}} = 279. \tag{5}$$

Background Noise - Extraneous Light

Noise due to background radiation (optical) in many cases can be subtracted from the data (assuming a constant background). Great efforts will be made to eliminate any possible reflections and sources of extraneous light from the viewing area of the cameras. It is assumed that the detector will be built such that all unwanted background light is eliminated.

Background Noise - Beam Induced Radiation

The maximum expected beam spill Q_{spill} over a 1 meter length is 1 nanocouloumb per macropulse (6). Equation 6 is used to calculate the resulting radiation in Rads (6).

$$Rads = \kappa \frac{Q_{spill}}{y^2} \sin^2\theta \; e^{\frac{-\theta}{13^\circ}} \tag{6}$$

The empirical constant k = 0.15 [Rad-m^2/nC], y is the radial distance from the straight beam line to the detector and θ is the viewing angle between the spill and the detector. The optimum viewing angle is 24° with half-max. at about 10° and 50°. The resulting radiation is 15.7 x 10^{-3} Rads per macropulse which is equivalent to 470 x 10^3 mips/cm^2 per macropulse[1] (mip = minimum ionizing particle) or 4.6 mips/cm^2 per μbunch compared to 2.5 x 10^5 electrons/cm^2 per μbunch at the scintillator (at x = 3σ and x' = 3σ). This was assumed to be a worst case scenario (6). It appears that the background caused by beam spill can be neglected.

Camera Noise

The minimum detectable signal level for the intensified camera is 1 x 10^{-12} W/cm^2 at a wavelength of 400 nm. The integration time for a single frame is 16 ms. This gives a minimum detectable energy density per frame of 625 x 10^{-15} J/cm^2 compared to a signal of 18.7 x 10^{-12} J/cm^2. These numbers are for faceplate illumination.

Digitizer Noise

It is assumed that the digitizer resolution will be 2 units out of 256 or better, depending on the digitizer that is selected for the project.

---

[1] A mip is a minimum ionizing particle with 3 x 10^7 mip/cm^2 = 1 Rad.

Signal Integration

The image seen by the camera is depicted in Fig. 4. The image will be a long vertical line with its height corresponding to the particle beam width (at the point of intersection with the laser) and any spread caused by variations in particle velocity in the bend plane. The horizontal dimension contains the angular information (x') of the beam emittance for each x position which is defined by the laser).

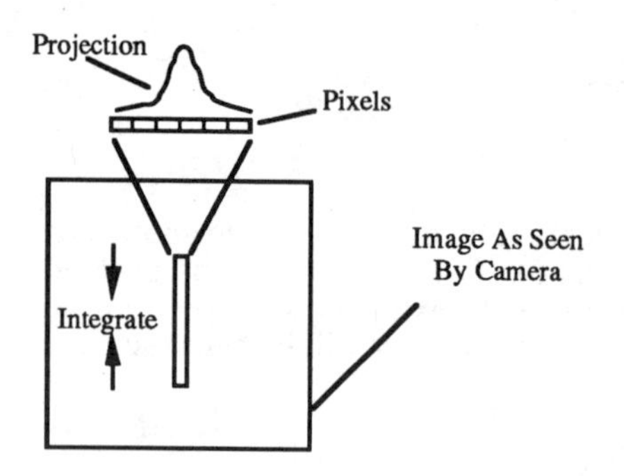

Figure 4. Illustration of a beam image as seen by the camera.

The signal will be integrated in the vertical dimension in order to remove any effects caused by variations in particle velocity. The final data set will be a 1 D array of numbers per image (per laser position).

Assuming random noise, the noise of the overall system will be decreased by a factor of $\sqrt{N}$ (N is the number of rows) when the signal is integrated. By integrating 50 lines the signal-to-noise ratio (SNR) is increased by a factor of 7.

Signal-to-Noise Ratio

The minimum detectable beam radiance is limited by the camera sensitivity and the digitizer resolution. The digitizer reduces the sensitivity of the camera by a factor of two. The sensitivity of the system becomes 1.25 x 10^{-12} J/cm^2. The total system SNR is approximately 15:1 at the $x = 3\sigma$, $x' = 3\sigma$ point in phase space. Integrating the signal (adding 50 video lines in the vertical dimension) increases the SNR to 106:1.

CONCLUSIONS

A viable, laser based diagnostic system can be developed that measures the transverse emittance of the LAMPF beam in Line D. Calculations show that this diagnostic system has a SNR of 106:1 when viewing a single pixel located at $x = 3\sigma$ and $x' = 3\sigma$ in transverse phase space. This implies that the emittance of better than 99.7% of the beam can be measured. The measurement phase space resolutions are x = .3 mm by x' = 41 μrad.

This paper describes a practical approach for making emittance measurements. Improvements to the detector system will increase the dynamic range and SNR. For example focusing the electron beam to a line on the scintillator will increase the irradiance (W cm^{-2}) on the camera MCP, which in turn increases the SNR. Also, using a diode array as the detector would increase the dynamic range. Several options are available and should be considered.

Effects of stray magnetic and electric fields on the electron transport must still be evaluated. The stray fields are not expected to cause complications in making the measurement. The effect of the laser beam on the electron trajectory must also be considered.

This appears to be a viable diagnostic device for measuring the transverse emittance of the LAMPF beam.

REFERENCES

1. R. C. Connolly, et. al "A Transverse Phase-Space Measurement Technique for High Brightness, H^- Beams", Nuclear Instruments and Methods in Physics Research, A312 (1992) 415-419.

2. Y. W. Yuan, et. al. "Measurement of Longitudinal Phase-Space in and Accelerated H- Beam Using a Laser-Induced Neutralization Method", Nuclear Instruments and Methods in Physics Research, A329 (1993) 381-392.

3. J. B. Birks, "Theory and Practice of Scintillation Counting", Pergamon Press, Oxford, 1964.

4. Glenn F. Knoll, "Radiation Detection and Measurement, John Wiley and Sons, New York, 1989.

5. D. R. Swenson, et. al. "Non-Invasive Diagnostics for H^- Ion Beams Using Photodetachment by a Focused Laser Beam" AIP Conference Proceedings 319. Beam Instrumentation Workshop, Santa Fe, New Mexico 1993.

6. Private communication with Mike Plum. Los Alamos National Laboratory, Accelerator Operations and Technology Division.

Device for Electron Bunch Measurement in the Picosecond Region

A.V.Aleksandrov, M.S.Avilov, N.S.Dikansky, P.V.Logatchev, A.V.Novokhatski.
Institute of Nuclear Physics, 630090 Novosibirsk, Russia.

R.Calabrese and V.Guidi.
Dipartimento di Fisica dell'Universita and INFN,I-44100 Ferrara, Italy.

G. Lamanna, G.Guillo and B.Yang.
Laboratori Nazionali di Legnaro,I-35020 Legnaro, Italy.

L.Tecchio.
Dipartimento di Fisica Sperimentale dell'Universita and INFN, Torino, Italy.

Abstract

The method of bunch length measurement in picosecond region using circular scanning in RF cavity was developed. Also the same cavity provides bunch length measurement based on analysis of wake fields, excited in the cavity by travelling bunch. Both methods doesn't need precise synchronization with the beam and were used for investigation of the time response of GaAs photocathode in NEA condition. In this report the description of experimental setup and obtained results are presented.

Introduction

It is planned to use a laser-driven RF gun at the injector complex for Novosibirsk Phi-factory project (1). Attractive opportunity gives GaAs photocathode, which is able to generate a polarized electron beam. However it could be used in RF gun if the time response of GaAs photocathode is short enough and comparable with RF period. An experimental facility has been fabricated to measure the length of electron bunch extracted from GaAs photocathode illuminated by short laser pulse. The method of bunch length measurement using circular scanning by RF cavity was developed. Also fast non destructive method of bunch duration measurement using time-domain analysis of wakefield excited by travelling bunch in the same cavity was tested. Both methods are proved to be reliable and can be used in other experiments.

Circular Scanning in RF Cavity.

The main idea of the method is circular scanning of electron beam travelling in rotating magnetic field of RF cavity. The electrons passing through the cavity along it's axis experience transverse deflection, which direction depends on the longitudinal position of electron in the bunch. As a result longitudinal position of electron transformed to angular position in the plain orthogonal to the axis.

The circular deflection of electron beam in the cylindrical cavity can be performed by transverse magnetic field of TM110 mode with circular polarization. Circular polarization is provided by exciting of two orthogonal modes shifted on $\frac{\pi}{2}$ in phase. Let us consider the movement of electron in the magnetic field with step-like distribution along the z-axis on the assumption that the deflection angle α is small so we can assume longitudinal momentum p_z to be constant.

$$\left.\begin{aligned} H_x &= H_0 \cdot cos(\omega t) \\ H_y &= H_0 \cdot sin(\omega t) \end{aligned}\right\} \quad 0 \leq z \leq d$$

$$H_x = H_y = 0 \quad z < 0, \quad z > d$$

Here, d is the length of cavity gap, ω is RF frequency. In this case, equations of motion can be written as:

$$\dot{P}_x = eH_0\beta_z \cdot sin(\omega t)$$

$$\dot{P}_y = eH_0\beta_z \cdot cos(\omega t)$$

Upon integrating over time from τ_0 to $\tau = \frac{d}{\beta_z \cdot c}$ we obtain the deflection angle:

$$\alpha_x = \frac{eH_0 \cdot \lambda}{\pi\gamma mc^2} \cdot sin\mu \cdot sin(\omega\tau_0 + \mu)$$

$$\alpha_y = \frac{eH_0 \cdot \lambda}{\pi\gamma mc^2} \cdot sin\mu \cdot cos(\omega\tau_0 + \mu)$$

where $\mu = \frac{\pi d}{\lambda\beta_z}$ is transit angle, λ is RF wavelength.

The transverse position of the particle after drift L is described by radius R, which doesn't depend on τ_0

$$R = \frac{eH_0\lambda \cdot L}{\pi\gamma mc^2} \cdot sin\mu$$

and azimuthal angle Θ, which is proportional to τ_0

$$\Theta = w\tau_0 + \mu - \frac{\pi}{2}$$

The maximum deflection is reached when transit angle is equal to $\frac{\pi}{2}$. In this case beam with duration $\Delta\tau$ sweeps in the plane orthogonal to the axis an arc of circumference with sizes:

$$R = \frac{eH_0\lambda \cdot L}{\pi\gamma mc^2}$$

$$\Delta\Theta = w \cdot \Delta\tau$$

When measure the angular size $\Delta\Theta$ we can determine the beam duration $\Delta\tau$.

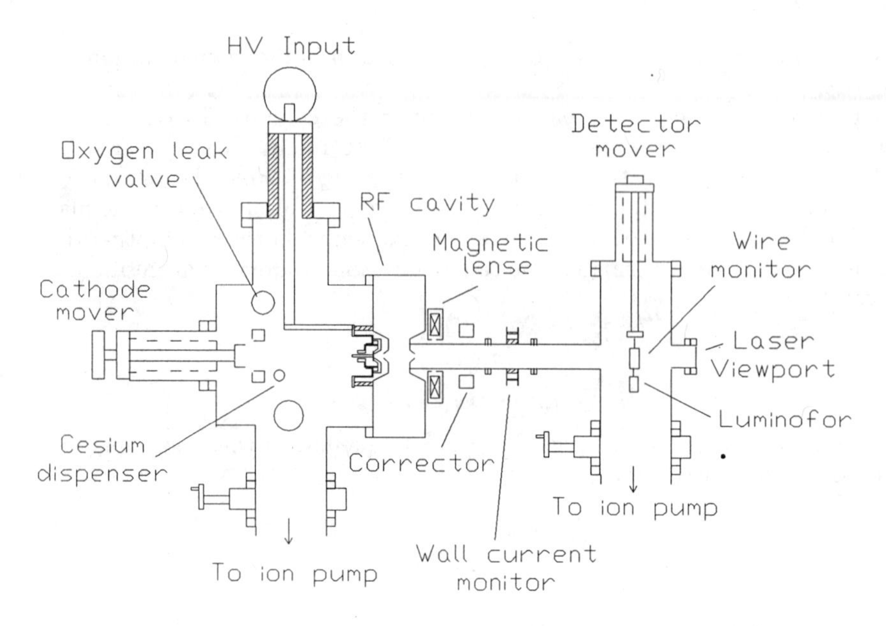

Figure. 1. Experimental setup.

Experimental Setup

The GaAs photocathode (p-doped by Zn, $10^{19}cm^{-3}$) is prepared in NEA condition by depositing Cs and O_2 at its surface with a standard procedure in UHV preparation chamber (2). When activation has been accomplished, the cathode is fastened to the DC gun by manipulator (see Fig.1).

The cathode is negatively biased with a voltage ranging within 0-80 kV. The diameter of the laser spot on the cathode is around 2 mm, and gun perveance is $1.5 \cdot 10^{-3} \frac{A}{kV^{3/2}}$.

Photoemission is excited by a Ar^{+} CW laser of 6 - 75 mW during activation of photocathode, and by pulsed (532 nm, 1.5mJ) Mode-Locked Nd:YLF laser. A pulsed laser provides the minimum laser beam length of 40 ps with a 10 Hz repetition rate. An autocorrelator is used for a control of laser pulse duration (FWHM) with 10 ps accuracy.

The design and final optimization of the cavity shape, taking into account presence of holes for beam pass was made using computer code URMEL. Cavity was optimized to give maximum deflection of the beam with 50KV energy for given input RF power.

The draw of the cavity assembled with DC gun and GaAs crystal holder is shown in Fig.2. The resonant frequency is 2.46GHz, measured unloaded quality factor is 17000. The cavity has two orthogonal loops for RF power input and two piston tuners for adjustment of resonant frequency of each mode in the range +-.5MHz. The scheme of RF part of devise is shown in Fig.3.

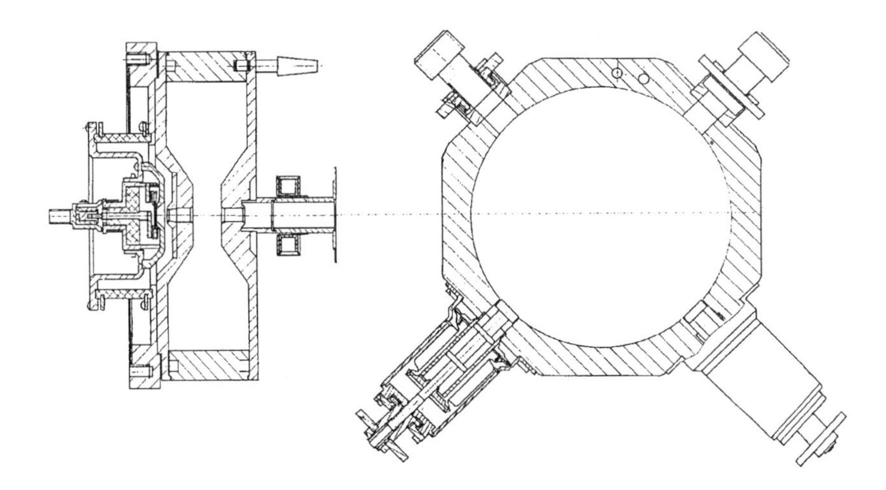

Figure. 2. Deflecting RF cavity.

Pulsed magnetron is used as source of RF power. It's power and frequency can be controlled in some range by amplitude of anode pulse from modulator. The maximum power of 2.0 KW provides deflection angle of .025rad. The stable operation of magnetron is provided by ferrite circulator which decouples magnetron from resonant load. Coaxial 3-dB directional coupler divides RF power between two inputs and provides $\frac{\pi}{2}$ phase shift. Two phase shifters are installed between 3-dB coupler and RF inputs for compensation of phase shift difference in transmission lines.

The adjustment of cavity and phase shifters is performed by using 15ns or continuous electron beam. When polarization is circular, electron beam draws a full circle on the detector surface and makes a uniform charge distribution on the channels of 2π-detector. The direct observation of deflected continuous electron beam is also possible on luminescent screen which can be moved under beam instead of 2π-detector.

For measurement of angular distribution of electrons after drift space we use so called "2π-detector". The 2π-detector is a set of 30 tantalum sectors perpendicular to the beam axis with a hole for laser beam in the center. Each sector acts as a Faraday cup to collect the electrons of the bunch. The maximum resolution of this instrument is 400/30=13.3 ps. Collected charge is measured by 30-chanel integrating ADC with normal electronic noise of 10^6 electrons per each channel.

Experimental Results

Typical distribution of the charge in the bunch is shown in Fig.4. The shape of laser beam measured by autocorrelator is also given at this picture. Shapes

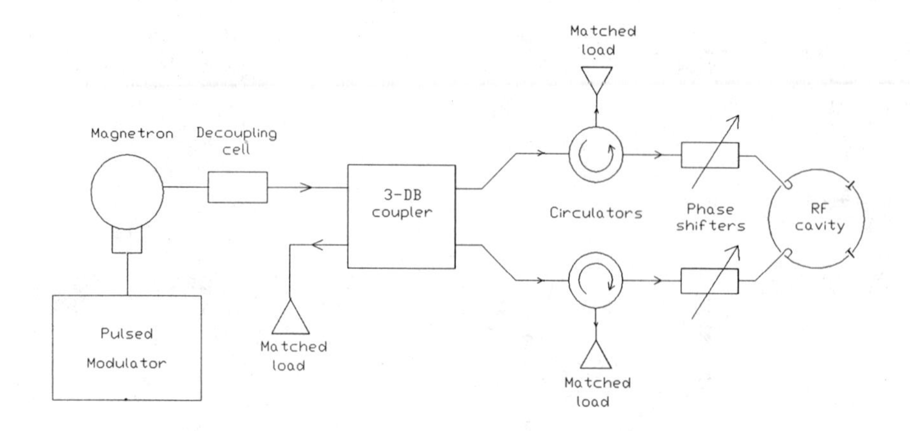

Figure. 3. RF scheme of installation.

of laser beam and electron bunch are comparable if number of electrons is in the range of several units of $\cdot 10^8 e^-$, however with charge increasing bunch length increases. More detailed information on results of measurements may be found in (3).

Bunch Duration Measurement Using Passive Cavity

Another possibility for bunch length measurement is in investigation of wake fields, induced by electron beam in "empty" cavity. Energy W_i stored in i-th mode of the cavity after bunch pass described by expression (4):

$$W_i = q^2 \cdot k_i \cdot F_i$$

where q - total charge in the bunch,
k_i- loss parameter, depending on geometry of cavity and beam velocity,
F_i- bunch form-factor, depending on bunch shape and frequency of mode.

For bunch with Gaussian charge distribution

$$F_i = e^{-(w_i \sigma)^2}$$

where σ is r.m.s. width of distribution. So if energi W_i, charge q and parameters of the i-th mode are known one can calculate σ. By measuring W in two or more different modes bunch charge can be excluded. Output power P_{out} in measuring line connected to cavity with coupling coefficient β_i can be written as:

$$P_{out}(t, \sigma) = q^2 \sum_{i=0}^{\infty} k_i \cdot e^{-(w_i \sigma)^2} \cdot \frac{\beta_i}{\tau_i} \cdot e^{-\frac{t}{\tau_i} \cdot (\beta_i + 1)}$$

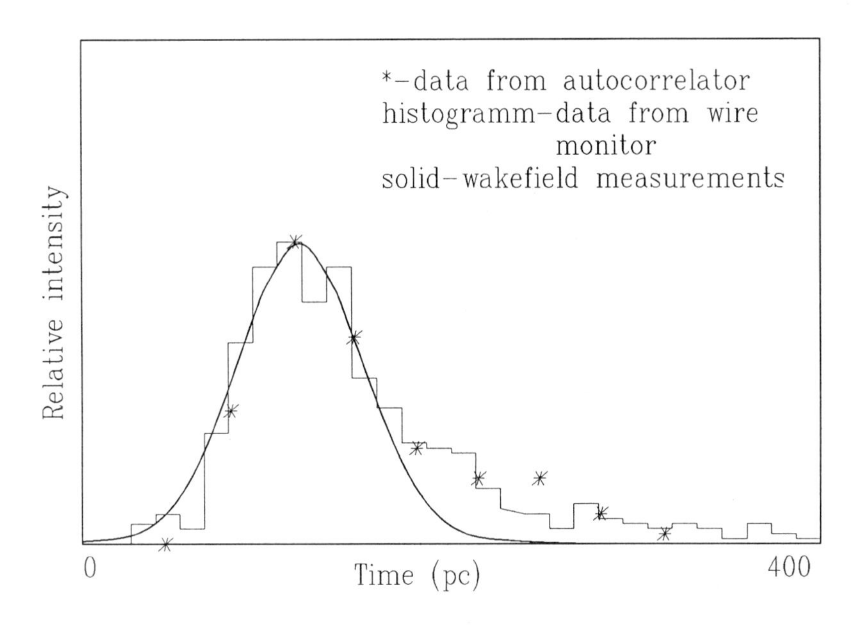

Figure. 4. Laser pulse shape and charge density distribution in the bunch.

where τ_i is unloaded decay time of i-th mode. If the bunch length is in the range when two first modes are excited mainly, then output power can be described by sum of two exponents with different decay times. Analyzing dependence of output power on time it is possible to calculate the amplitudes of excited modes and to determine the bunch length:

$$\sigma = \sqrt{\frac{ln\frac{A_1}{A_0} - ln\frac{U_1}{U_0}}{w_1^2 - w_0^2}}$$

where $\frac{U_1}{U_0}$ - measured ratio of amplitudes of exponents,

$$\frac{A_1}{A_0} = \frac{\tau_0 \cdot \beta_1 \cdot K_1}{\tau_1 \cdot \beta_0 \cdot K_0}$$

Table 1. Frequencies and loss parameters for first eight symmetrical modes

N	1	2	3	4	5	6	7	8
$f[MHz]$	1148	3194	3269	4742	5197	6225	6894	7058
$k[\frac{V}{fC}]$	270	120	0.14	0.67	49	0.17	2.4	0.71

In Table 1. frequencies and loss parameters for first eight symmetrical modes of our cavity are presented. If electron beam shorter than 50pc two modes are mainly excited.

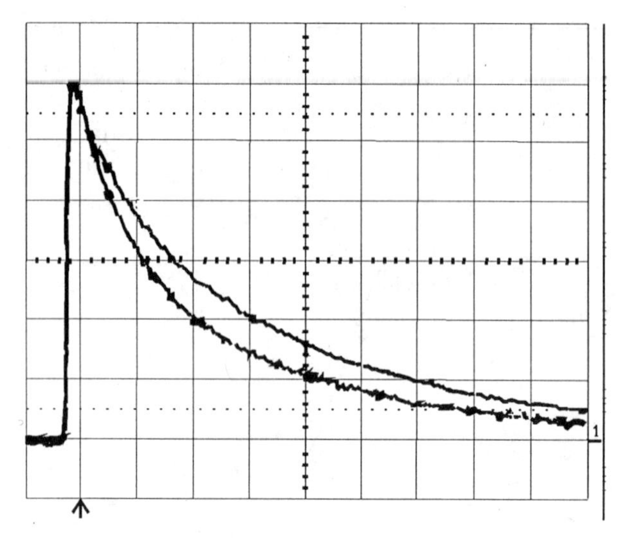

Figure. 5. Signals induced in "empty" cavity by electron bunches with different lengths. (horiz.-200nc/div.,vert.- 5mV/div.)

Oscilloscope pictures of signals induced in cavity by two beams with different length are shown in Fig.5. Measured bunch length in assumption of Gaussian shape is shown in Fig.4.

Conclusion

According to data obtained in this experiment a thick GaAs photocathode has small enough time response. Upper limit for the lengthening of the electron bunch due to GaAs time response is less then 30 ps for electron beam with $2.0 \cdot 10^8 e^-$ particles at energy of 60 kV. More detailed description of obtained results and discussion of physical fenomena behind it will be published elsewhere. The methods used for electron bunch length measurements are proved to be reliable. Both methods can be adjusted for measurement in desirable range of bunch lengths by proper choice of cavity frequency and doesn't need precise synchronization of measuring devise with the beam.

REFERENCES

1. A.V.Novokhatski, A.V.Aleksandrov, A.A.Kulakov, P.V.Logatchov, L.Tecchio. A laser-driven gun for electron-positron factories. NIM A 340(1994) p.237-240, North-Holland.
2. D.T.Pierce, R.J.Celotta, G.-C.Wang, W.N.Unertl, A.Galejs, C.E.Kuyatt and S.R.Mielczarek; Rev. Sci. Instrum. 51 (1980) 478.
3. P.V.Logatchev, M.S.Avilov, etc. Measuring of Time Response of Laser-Triggered GaAs Photocathode, Procs. of EPAC-94, London, (1994).
4. P.B.Wilson. High Energy Electron Linacs; SLAC-PUB-2884,1982.

A Beam Phase Monitor with use of a Micro-Channel Plate for the RIKEN Ring Cyclotron

M. Kase, N. Inabe, I. Yokoyama

The Institute of Physical and Chemical Research (RIKEN)
Wako-shi, Saitama 351-01, Japan

T. Kawama

Sumitomo Heavy Industries Ltd
Niihama, Ehime 792, Japan

Abstract

A time structure of chopped beam from cyclotrons has been continuously monitored with a micro-channel plate (MCP) at the RIKEN Ring Cyclotron (RRC). Its transient structure at the moments of beam-switched-on and beam-switch-off gives us a quantitative information about how single turn extraction in cyclotrons is realized together with a knowledge about how many turns the beam is circulated there. It works also as a phase monitor which tells us how much magnetic fields drift inside cyclotrons. These are so helpful for us to make a high quality beam and to keep it for long duration.

INTRODUCTION

In order to get a high-quality beam from the RIKEN Ring Cyclotron (RRC)[1] which is a post-stage accelerator coupled with an AVF cyclotron (AVF) or a heavy-ion linac (RILAC), it is essential to realize the single turn extraction in each cyclotrons. And, after tuning, in order to maintain a such beam for a long period, it is necessary to adjust precisely a beam phase constant on each accelerator, namely to stabilize the magnetic fields inside cyclotrons.

In the case of a polarized dueteron beam, the single turn extraction is so specially required for during the experiment. A spin orientation, which is determined just after the ion source with a Wien filter, is mixed up in multi-turn extraction.

For this purpose, a beam phase probe with use of a micro-channel plate (MCP) has been developed[2]. It has a good sensitivity for a various kind of beams, as for energies, ion masses and intensities. Since the beam loss due to the measurement is low of less than 1 %, a continuous measurement is possible. It can be a good monitor of a field drift inside cyclotron as low as several ppm, and, by mean of this, we can maintain the isochronous fields. When this monitor is used together with a beam chopper, the spectrum measured has many information about a performance of cyclotron.

DEVICE

The structure of the new phase probe is shown in Fig. 1, and its details are described in Ref. 2. Electrons and/or photons produced by beam irradiation on the target of 0.3 mmϕ tungsten wire are accelerated to the MCP entrance surface. During the measurement, the target can be scanned remotely in perpendicular to the beam axis, keeping a constant distance between the wire and the MCP. Electronics are shown in Fig. 2. As a delay generator up to 100μs, a preset counter is used, since it has very stable performance compared to the analog ones. Two sets of the devices are installed inside a beam-diagnostic chamber located in the extracted beam line of the AVF cyclotron the RRC as shown in Fig. 3.

Three beam choppers are installed in 500kV injection line of RILAC, in the injection line to AVF cyclotron from ECR ion source and in that from the polarized ion source. Their electrodes, of a parallel plate type, is 5 ~ 8 cm in length along the beam line and has a beam space of 2~4 cm. A rise and fall time of voltage applied to these electrode are about 15 ns and repetition cycle is ~20kHz. A peak voltage is 500V at maximum for AVF and 3kV at maximum for RILAC. A pulse duration is determined according to the harmonic number, 200 ns~ 1.2μs. The beam loss due to the chopping is kept less than 1 %.

RESULT

Figure 4 illustrates the double-turn extraction of a general cyclotron. In this case, a beam bunch is divided into two parts at a septum of deflector. When a beam is switched on, the fragments of beam bunch whose number is equal to the harmonic number of the cyclotron, go first, and after that the full-filled beam bunches follow them. When a beam is switched off, after the full-filled beam bunches, the same number of another fragments, which is paired with the former ones, follow them and then beam disappears completely. If the single turn extraction is achieved, the time structure of extracted beam is similar to that of the injection beam. The typical time structure in multi-turn extraction case is shown in Fig. 5. The fraction of eleven small peaks gives a quantitative status of multi-turn extraction. This method can be applied to the cases of AVF itself and the RRC when it is coupled with AVF.

In case when the RRC is coupled with AVF, more convenient method to know the single turn extraction can be applied to the RRC. Figure 6 illustrates this case. The method becomes available due to the combination of two things. One is that the RRC, a post-accelerator, is operated at a frequency which is doubled to the frequency of the AVF, that is an injector. From standpoint of the RRC rf phase, a beam comes every two periods from the injector. The second is that a harmonic number of the RRC is always five, i.e. a odd number. In the double extraction case, a partial beam bunch appears at half way between two main bunch. Figure 7 shows the result of 100MeV/n ^{18}O beam. The small peaks due to the double (at least) turn extraction are observed. One can always know how much degree the single turn extraction is achieved in the RRC without a beam chopper.

The number of n_1 and n_2 which are input to a digital delay circuit in Fig. 2 are very important. The value of delay time, Δt=niTinj when the chopped beam can be observed on MCA screen as a spectrum in Fig. 5, has a information about the turn number inside cyclotrons, if the delay time in cables and drift time of beam outside a cyclotron are considered.

The position of peak in Fig. 5 and Fig. 7 is always relating to a beam phase to an rf phase. This allows the device to work as a phase probe. The shift of the peak is linearly related to a drift of magnetic field in the cyclotron. Its sensitivity is so high that a fields drift of several ppm can be known. As the probe works for a very low intensity beam down to several 10 ppA, it is more useful than a traditional capacitive pick-up probe. An operator can stabilize the magnetic field by adjusting a magnet current so as to keep this peak at the same position.

CONCLUSION

The beam monitor with MCP is very useful not only for the measurement of beam phase but also for judgment of single turn extraction. And as other application, the measurement of turn number inside cyclotron is available. These monitor will help us to maintain beam stable and also to tune a beam. All these measurement can be done with a slight beam loss (almost beam non-destructive). It means that we can use it any time or continuously. Now we are preparing to build an automatic field stabilizing system for cyclotrons using this monitor.

REFERENCES

1) Y. Yano: "Current Status and Future Scope on the RIKEN Accelerator Facility", The 9th Symposium on Accelerator Science and Technology, Tsukuba, Japan, August (1993) pp.23-25.

2) M. Kase, T. Kawama, T. Nakagawa, N. Inabe, I. Yokoyama, A. Goto and Y. Yano: "Development of a Beam Phase Monitor with a Micro-Channel Plate for the RIKEN Ring Cyclotron", The 9th Symposium on Accelerator Science and Technology, Tsukuba, Japan, August (1993) pp. 474-476.

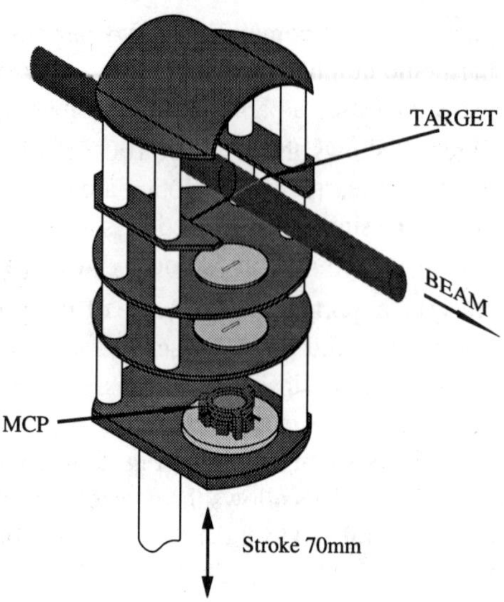

Fig. 1 A structure of beam probe with an MCP. All electrodes are made of stainless steel and suported by ceramic insulator rods. Each electrode is biased so that a uniform field is produced from a wire to the MCP.

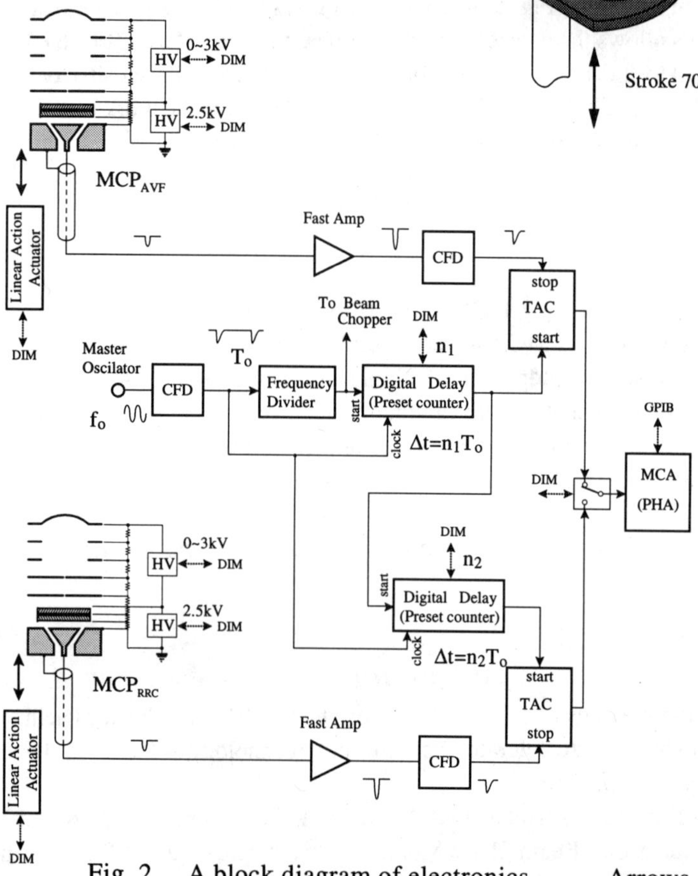

Fig. 2 A block diagram of electronics. Arrows, ↔ show a remote connection with a host computer (Mitsubishi M60/500), and DIM is an original interface linked to CAMAC via an optical fiber.

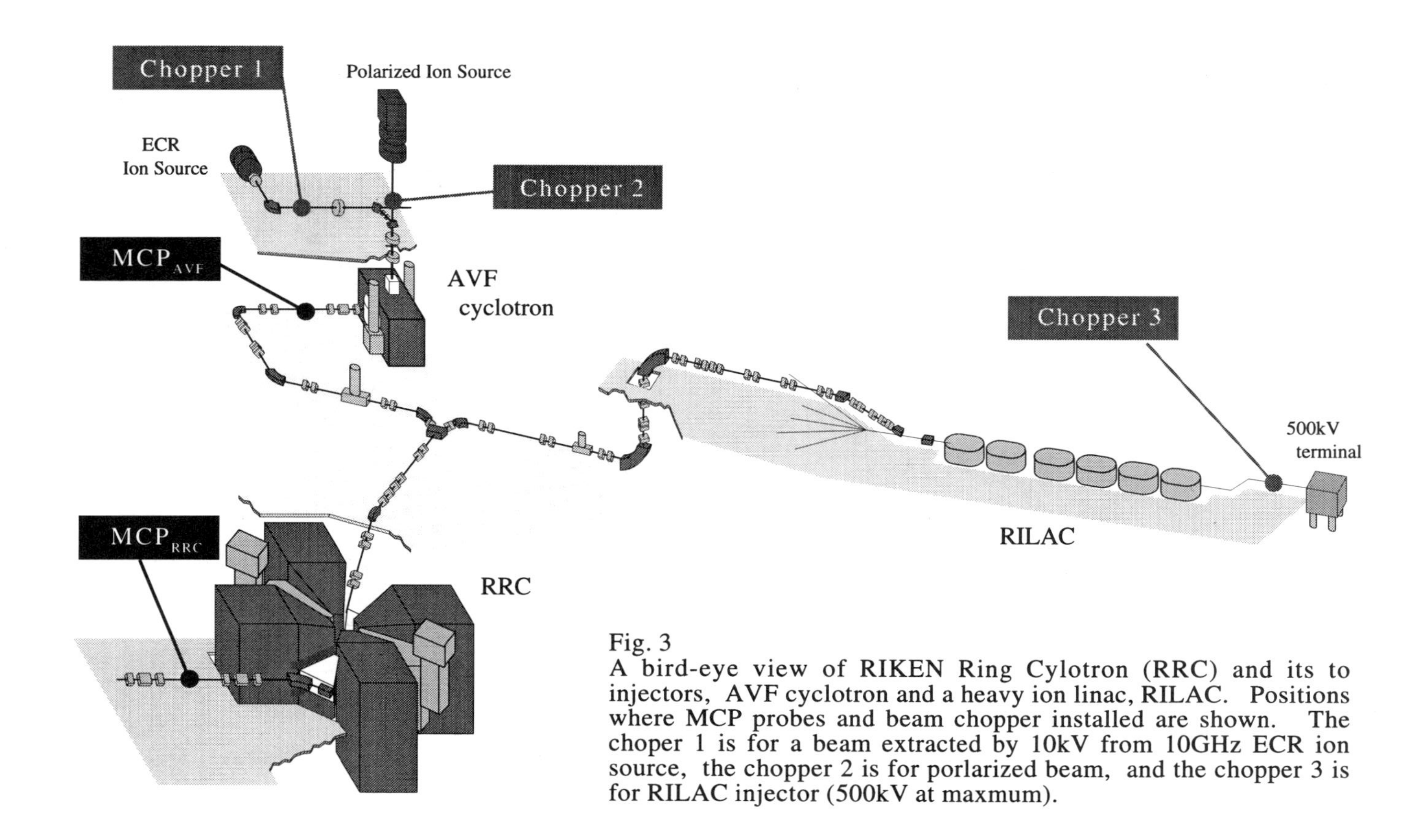

Fig. 3
A bird-eye view of RIKEN Ring Cylotron (RRC) and its to injectors, AVF cyclotron and a heavy ion linac, RILAC. Positions where MCP probes and beam chopper installed are shown. The choper 1 is for a beam extracted by 10kV from 10GHz ECR ion source, the chopper 2 is for porlarized beam, and the chopper 3 is for RILAC injector (500kV at maxmum).

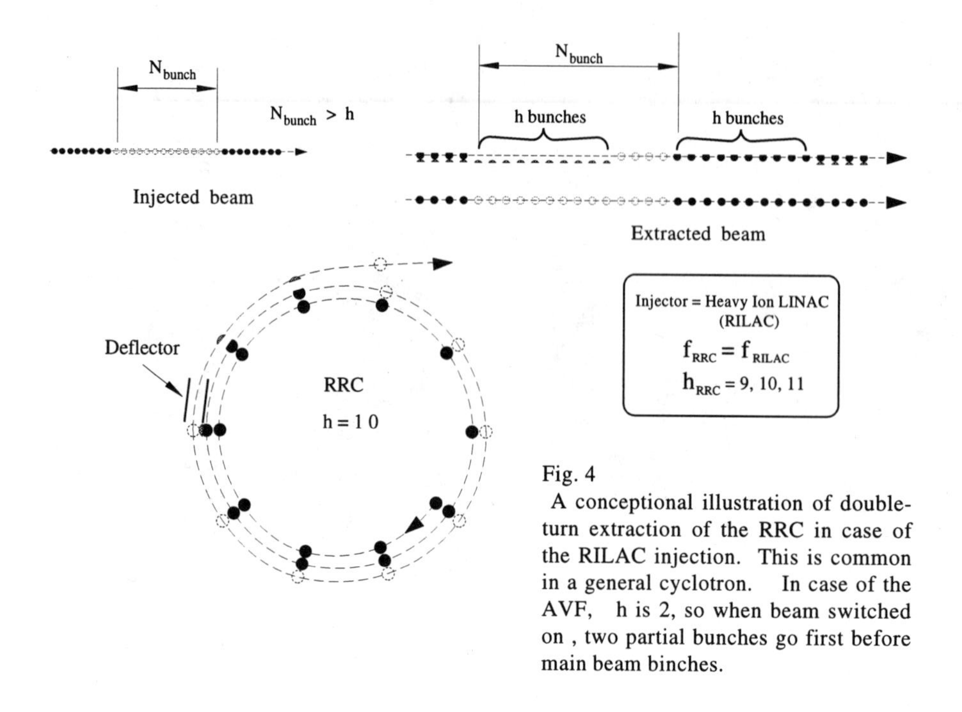

Fig. 4
A conceptional illustration of double-turn extraction of the RRC in case of the RILAC injection. This is common in a general cyclotron. In case of the AVF, h is 2, so when beam switched on , two partial bunches go first before main beam binches.

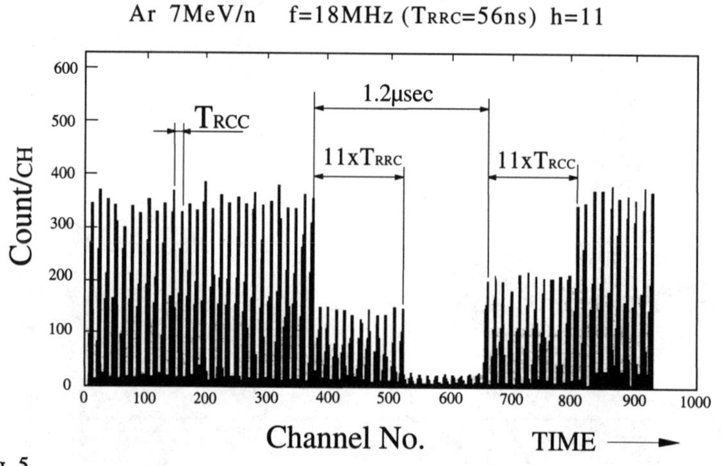

Fig. 5.
A time spectrum of a beam bunches when the beam is cut partially (1.2μs) by a beam chopper installed in the RILAC injection line. The start of the TAC is given by the trigger signal for the chopper which is sychronized to the rf phase. The stop for the TAC is the MCP signal. Here it is evident that more than two turns are extracted.

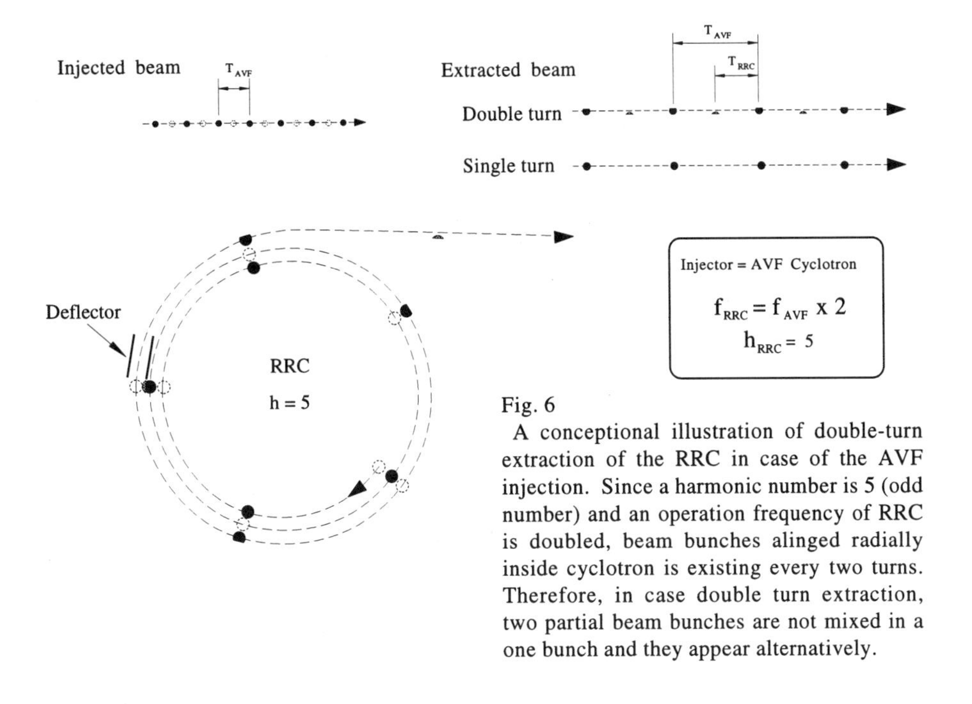

Fig. 6
A conceptional illustration of double-turn extraction of the RRC in case of the AVF injection. Since a harmonic number is 5 (odd number) and an operation frequency of RRC is doubled, beam bunches alinged radially inside cyclotron is existing every two turns. Therefore, in case double turn extraction, two partial beam bunches are not mixed in a one bunch and they appear alternatively.

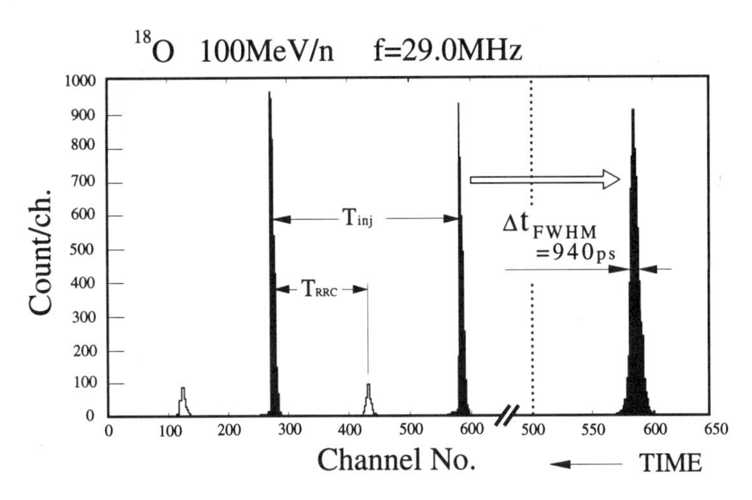

Fig. 7
Example of time spectrum of signals from the MCP for a 100MeV/n 18O beam. These signals are considered to originate from photons produced at the wire by the beam, because no accelration voltage is applied of the electrode during this measurement. In this case, a start signal for the TAC is given by the MCP signal and a stop signal is an rf reference. No beam chopping is done. Large peaks (shadowed) with a time interval of T_{inj} are due to a main beam, and small peaks separated from the main peaks by T_{RRC} are due to the effect of multi-turn extraction. Here at least two turns are extracted obviously.

A Longitudinal Bunch Monitoring System Using LabVIEW and High-speed Oscilloscopes

E.L. Barsotti, Jr.

*Fermi National Accelerator Laboratory, P.O.Box 500, Batavia, IL 60510**

Abstract

A new longitudinal bunch monitor system has been installed at Fermilab for the Tevatron and Main Ring. For each machine, a signal from a broadband wall current monitor is sampled and digitized by a high-speed oscilloscope. A Macintosh computer, running LabVIEW-based software, controls the scopes and CAMAC timing modules and analyzes the acquired data. The resulting bunch parameters are used for a variety of purposes, including Tevatron collider luminosity calculation and injection analysis. This paper examines the system in detail.

INTRODUCTION

Accurate measurements of bunched beam intensities and lengths are necessary for the calculation of luminosity during collider operations. At the relativistic energies of the Fermilab Main Ring and Tevatron accelerators, the signal from a broadband, resistive wall current monitor(1) closely approximates the beam longitudinal profile. These signals can be digitized and processed to obtain the required parameters. An earlier "Sampled Bunch Display" (SBD) system(2,3) used a 1 GHz bandwidth, repetitive-sampling oscilloscope to digitize the signal, which has a maximum 20 dB bandwidth of 250 MHz. Code using a Motorola 68000 microprocessor processed the profile and reported results to the accelerator control system network, or Acnet. This system has operated since 1988.

Technological advances made it possible to upgrade this useful system. In late 1992, the 2 GSample/sec, 500 MHz Tektronix TDS620 oscilloscope(4) was chosen, along with the National Instruments software platform LabVIEW.(5) Longer time slice records of single turns could now be more extensively analyzed at higher speeds. Data refresh rate for the 12 proton and antiproton bunches of a Tevatron store decreased from 90 to 15 seconds. A single bunch mode updates every 1 or 2 seconds, depending on whether profile graphing is on. The longer record captures satellites of the main bunch, or even an entire batch (13 bunches) of uncoalesced beam. Multiple memories and longer records enable acquisitions at many key times during a Main Ring-to-Tevatron injection event, from 13 bunches of 8 GeV beam through the coalesced Tevatron bunch. Single turn acquisitions allow first turn and synchrotron oscillation measurement. During its first eight months of operation, the upgraded system has proven to be quite flexible, while providing results with high accuracy.

---

* Operated by the Universities Research Association under contract with the U. S. Dept. of Energy

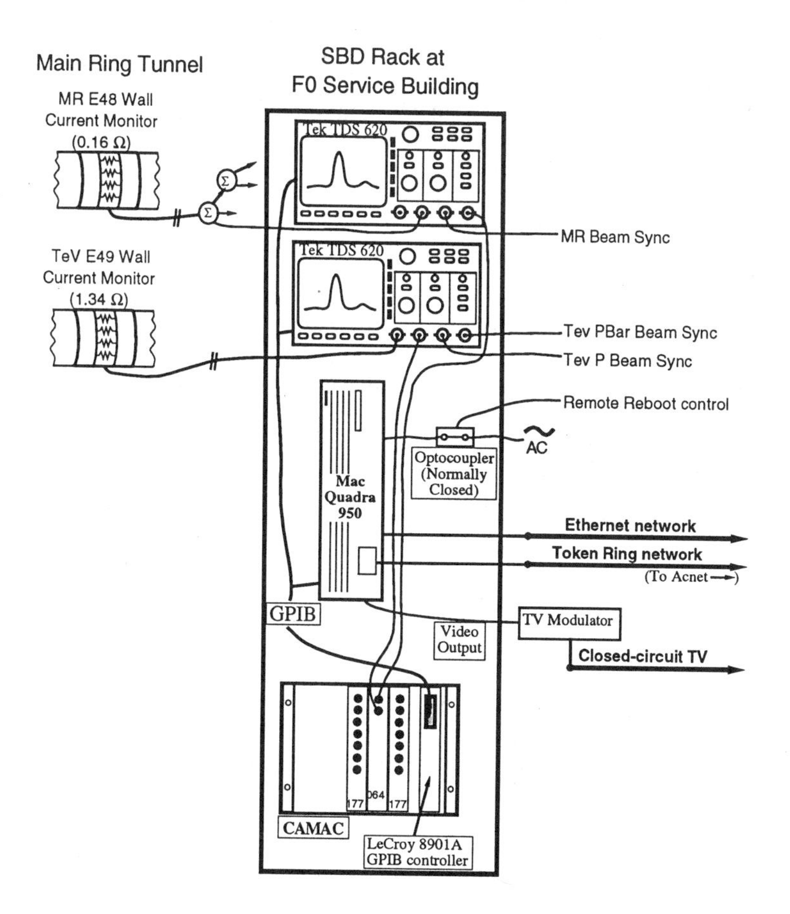

Figure 1. Overview of the new SBD system hardware.

SYSTEM HARDWARE

The heart of the new SBD system (Fig. 1) is a Macintosh Quadra 950 computer running a program written with the LabVIEW graphical language. The software controls and reads two TDS620 oscilloscopes, one for both the Main Ring and the Tevatron. Besides the high sampling rate and analog bandwidth, the scopes have five memory locations, record lengths of at least 500 samples, and a versatile triggering system.(4) The Main Ring signal comes from a 3 pickoff point, 1.5 kHz-2 GHz wall current monitor, which is shared with other functions. The Tevatron signal comes from a dedicated wall current monitor, an improved 4-point, 3 kHz-6 GHz model.(1) Each signal travels over 100' through low-loss cable to a rack in a service building.

Oscilloscope triggering depends on the mode of the SBD. The default mode cycles through the 12 Tevatron bunches during a store. Two Tevatron beam synchronization triggers are used with bunch-specific delays. When beam is injected into the Tevatron during a store setup, the SBD is put into an injection mode. Specific timing delays off an accelerator clock event are necessary for the 5 Main Ring and up to 3 Tevatron acquisitions. Each of 8 channels of a CAMAC Tevatron clock decoder pulses a programmable time after an injection event, and these channels are wire-OR combined into two outputs. The main triggers for the scopes are derived from two such decoder modules and a CAMAC S/R flip-flop module. The beam sync pulses act as the delayed triggers, and acquisitions occur after bunch-specific delays.

The Macintosh acts as controller for GPIB communication with the oscilloscopes and CAMAC decoders. The Quadra 950 has a built-in Ethernet interface, while an Apple TokenTalk card allows communication with Token Ring. An in-house interface between the LabVIEW program and the Acnet control network occurs over Token Ring.(6) The video output of the Macintosh is modulated onto Fermilab's closed-circuit TV system.

A remote computer reboot can be accomplished through Acnet by toggling an optocoupler relay and the computer's AC power. The program restarts itself completely when a reboot has occurred.

SYSTEM SOFTWARE AND MODES

Software development was simplified by using LabVIEW. Besides the graphical programming itself, an extensive analysis library is readily available along with GPIB drivers for the TDS620 scopes. The LabVIEW-Acnet interface, popular among recent Fermilab accelerator instrumentation projects, was very easy to use. Programming the beam profile "waterfall" plots, displayed on the TV system, was trivial with LabVIEW graphical output. The entire program itself is modular, allowing for future additions.

Control system communication to the SBD program takes place through Acnet devices, either as control or reading variables. For example, some variables control the hardware after initialization. Scope gain can be set to automatic or manual, and the number of averages can be selected. CAMAC delays for acquisition timings can be changed. Program calibration types, processing algorithms, number of injection acquisitions, and the program mode can be controlled. Acnet reading variables include all processed data for both cycling Tevatron and injection modes.

During a 6 X 6 store, the SBD continuously cycles through all 12 bunches in 15 seconds. The correct scope gain, trigger, and delay is set for each bunch. The program then waits for completion of a 250 ns, 500 sample acquisition with the specified number of averages. Data is retrieved over GPIB, processed, and made available to Acnet.

The beam can be either uncoalesced (13 bunches) or coalesced (main bunch with satellites). Different processing algorithms exist for each type, and the choice is made through a control variable. In both cases, the boundaries of the 19 nsec RF buckets are carefully defined first. For each bucket, a baseline defined by the edges is subtracted to account for the low frequency cutoff of the wall current monitor. Parameters for coalesced beam include: main bunch intensity, RMS sigma, % area width, longitudinal emittance, and centroid and peak times; leading and trailing satellite intensities and widths; and sum of 5 bunch (main and 4 satellites) intensities. The uncoalesced data are: sum of intensities of 13 bunches;

average width, emittance, and RMS sigma, weighted by intensities; and arrays of 13 elements for the individual intensities, sigmas, widths, and bunch centroid times. The most important parameters above have companion variables with their boxcar averages and standard deviations.

After each cycle, the mode control variable is checked. Mode changes are made either manually or through the sequencer, an Acnet program which orchestrates the entire Tevatron injection process during store setup. Another mode is the aforementioned Tevatron mode on a single, selected bunch. An update rate of 1 second is achieved, or 2 seconds with plotting.

Besides these two Tevatron store modes, other modes acquire both Main Ring and Tevatron data during a proton or antiproton injection. These modes can make up to 5 total acquisitions (coalesced and/or uncoalesced) per machine. The times of the acquisitions can be set for each event, with a minimum spacing of 0.7 seconds. After injection of that specific bunch is complete, an entire Tevatron cycle is done in time for the next bunch injection. The injection acquisitions are processed and the results update the appropriate Acnet reading variables.

CALIBRATION

The following procedure, based on the beam spectrum and system frequency response, has been used to calibrate the intensities and widths of the SBD. For each accelerator, the frequency response of the wall current monitor, cable, splitter (if any), and oscilloscope path is measured. The wall current monitor can be readily measured only before tunnel installation. Afterwards, and on a periodic basis, the rest of the path is measured. A calibrated frequency synthesizer injects a set of discrete frequencies into the cable in the tunnel, and the resulting powers are measured on the scope for all relevant gain settings. The power ratios become the transfer function with two dimensions, frequency and scope gain. An off-line simulation program then transmits an ideal gaussian, representing beam, through this system transfer function. The usual SBD algorithms are performed to find the resulting intensities and widths. The ratio of these values to the input gaussian parameters become the intensity and width calibration factors. Also, the ability to perform an "adaptive" calibration exists. The actual acquired waveform is deconvoluted with the transfer function to obtain the theoretical beam profile. Both waveforms are processed, and the intensity and width ratios become the adaptive calibration factors. The adaptive method was found to be too susceptible to noise, so the simulated calibration factors are now used instead.

Variations of 1.5% sigma can primarily be attributed to the effect of noise and 8-bit ADC truncation on the baseline subtraction algorithm. Averaging decreases this fluctuation. The SBD Tevatron intensity sum has been compared with the DC current transformer, which measures with ±1% absolute accuracy.(7) Agreement has been within ±1% over many months of operation.

OUTPUTS AND USES

The data is presented in a number of formats for monitoring and analysis. All Acnet control and reading variables are listed on an Acnet console parameter page, specific to the SBD. Fermilab cable channel 13 (Fig. 2) displays the running averages of the 12 Tevatron store bunch intensities. Cable channel 22

```
Time in Supercycle= 141.5  Store #=  5013        7-JUL-94  09:12:56
Main   Ring   Beam   Evento   (20, 21, 29, 2A, 2B, 2D, 2E)
9999999999999999999999999999999999999999999999999999999999999


P1= 135.1    E9  A1=  28.5  E9(pbars)   Pbar stack=  19.83 E10
P2= 152.8    E9  A2=  28.4  E9(pbars)   stack rate=   4.16 E10/hr
P3= 143.0    E9  A3=  27.7  E9(pbars)    prod. eff=  14.56 A/P
P4= 139.5    E9  A4=  20.9  E9(pbars)      MR Beam=   2.02 E12
P5= 145.4    E9  A5=  18.2  E9(pbars)     Tev Beam=   1.06 E12
P6= 145.6    E9  A6=  12.0  E9(pbars)   Tev Energy=  899.7 Gev
B0  3.23 E30  STI =    19  WKI =     84    Str dur=   1.53 Hrs
D0  3.81 E30  STI =    22  WKI =     93   Out Temp=   83.3 Deg-F
```

Figure 2. Lab-wide closed-circuit TV channel 13, showing Tevatron stored bunch intensities and other accelerator status information.

actually shows the LabVIEW front panel screen, including two "waterfall" plots (Fig. 3). The most recent waveforms of the 12 bunches, including a selectable number of satellites, are displayed on the left. Many of the Acnet Tevatron parameters are stored in a long-term circular buffer for store lifetime analysis.

The Shot Data Analysis program allows study of Tevatron injections during store setups. At appropriate times after each bunch injection, the Acnet injection parameters are read. Results for the entire store setup are displayed. The right Channel 22 waterfall plot (Fig. 3) shows the time progression of the most recent injection, from Main Ring injection through Tevatron injection.

Besides these main functions, various features allow further accelerator studies. For example, the 1 Hz Fast Tevatron mode has been used as a bunch length monitor up the acceleration ramp. Also, the standard deviation of centroid times has been a rough, undersampled measure of synchrotron oscillations. A comparison of the SBD intensity sum and the DC current transformer indicates the amount of unbunched beam. Injection into the wrong Tevatron bucket can be viewed on the waterfall plot or read with the peak or centroid time parameters. Main Ring coalescing and other injection event studies can take place during a store. A capability to measure the first Main Ring turn has helped with injection problems. A large amount of information is presented for these and other possible future uses.

EXPERIENCES

The new SBD system is presently in operation, after a commissioning process that began in December 1993. During that time, much work was done debugging, adding features, improving output formats, and automating the system within the accelerator controls structure. Much has been learned about the many possible states of the accelerators! The choice of a LabVIEW-based platform has

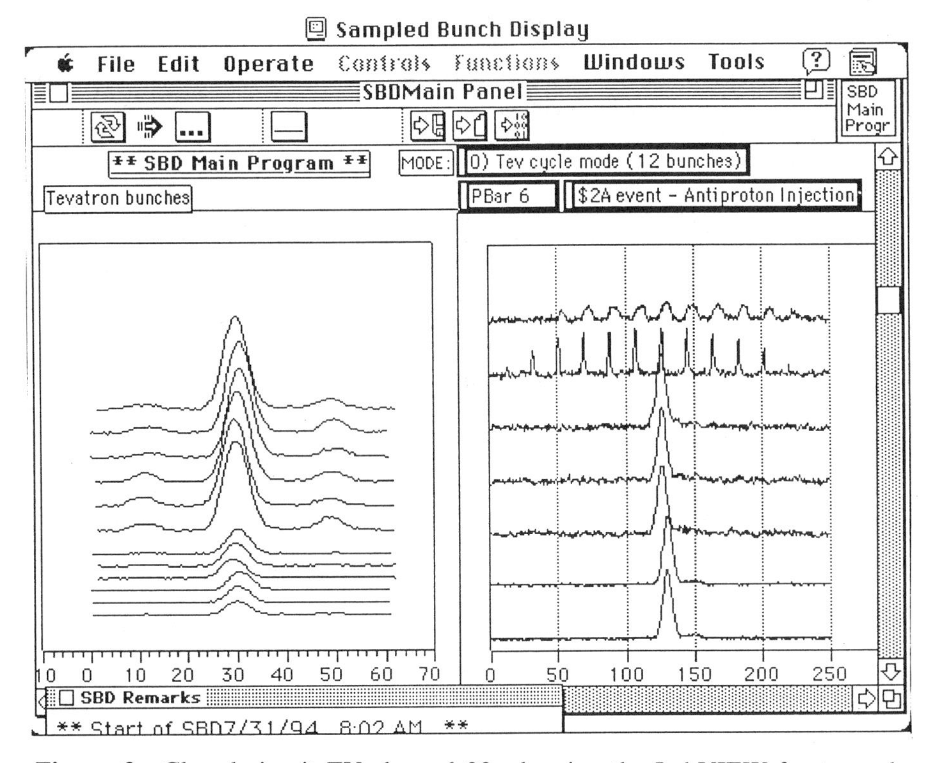

Figure 3. Closed-circuit TV channel 22, showing the LabVIEW front panel. Left: plot of all 12 Tevatron bunches, from Proton 1 (top) to Antiproton 6 (bottom). Right: waterfall plot showing time progression of the most recent bunch injected into the Tevatron, from Main Ring injection (top) to Tevatron.

simplified the work. Since the time that the TDS620 oscilloscopes were purchased, even higher performance digitizers have been developed. Hopefully, the SBD program was written to allow for easy replacement, if desired in the future. The system now runs with good reliability, providing high accuracy longitudinal data for Fermilab colliding beam physics.

ACKNOWLEDGEMENTS

I would like to thank A. Hahn for the project inspiration and much guidance throughout. Also, I would like to thank R. Meadowcroft, W. Blokland, B. Fellenz, D. McConnell, and others who helped with this work.

REFERENCES

1. Webber, R.C., "Longitudinal Emittance: An Introduction to the Concept and Survey of Measurement Techniques Including Design of a Wall Current Monitor," in *Proc. of the Accelerator Instrumentation Wksp.* (AIP, New York, 1990) pp. 85-126.
2. Moore, C.D. et al., "Single Bunch Intensity Monitoring System Using an Improved Wall Current Monitor," in *Proc. of the 1989 IEEE Particle Accelerator Conf.* (IEEE, Piscataway NJ, 1989) pp. 1513-5.
3. Payne, J.J., "Sampling Scope (SBD) Operating Guide," Operations Bulletin #1179, (Fermilab Internal Note, 1989).
4. Tektronix, Inc., *TDS620A, 640A, & 644A Digitizing Oscilloscopes User Manual,* (Tektronix, Inc., 1993, Beaverton OR).
5. National Instruments, Inc., *LabVIEW® 2 User Manual* (National Instruments, Inc., Austin TX).
6. Blokland, W., "An Interface from LabVIEW to the Accelerator Controls NETwork," in *Proc. of the 4th Accelerator Instrumentation Wksp.* (AIP, New York, 1993) pp. 320-9.
7. Vogel, G.L. and Arthur, J.A., "Intensity Monitor Calibrations and Useful Ranges," Operations Bulletin #1248, (Fermilab Internal Note, 1993).

Instrumentation for On-line Mountain Range Displays*

W.K. van Asselt and L.A. Ahrens
AGS Department, Brookhaven National Laboratory
Upton, Long Island, New York 11973

Abstract

A method to obtain and process 'mountain range' displays of beam signals is described. A custom-made trigger generator and a digital oscilloscope are used for the data acquisition and the graphical interface package LabVIEW is used to process the data. High resolution displays of wall monitor signals updating every AGS cycle have proven very powerful as a beam diagnostic.

INTRODUCTION

The development of the longitudinal structure of a bunch can be obtained by observing beam pickup signals, typically from a wall monitor, from a given bunch at regularly spaced intervals. This is because the longitudinal structure evolves relatively slow, since the synchrotron frequency is in general small compared to the revolution frequency. By overlaying a series of beam traces all starting synchronized to the accelerating rf but with slight vertical offsets for successive traces some aspects of the motion become very easy to visualize. This is the mountain-range display. The use of these displays for beam observations in the time domain has been very common in the accelerator community using analog storage oscilloscopes and the appropriate trigger and ramp generators. This paper describes this same basic technique, customized for data acquisition with a digital oscilloscope. There are three main ingredients composing the present method: a trigger generator chassis, sequential digital data acquisition and transfer and display of the data at a host computer. These ingredients will be described, followed by a discussion of the performance of the system.

THE TRIGGER GENERATOR

When acquiring data with an analog oscilloscope, the trigger sequence to the scope completely defines the pattern observed. Although capturing the data digitally opens the possibility for further processing, the triggers give an essential reference to the rf buckets in the machine. Further, selective triggering allows a preprocessing which reduces the time requirements before the data can be displayed. It has been considered critical in the present application to be able to

*Work performed under the auspices of the U.S. Department of Energy.

present the display in real time - within one accelerator cycle. For these reasons a trigger generator chassis similar to that used in analog mountain-range displays is still necessary. This chassis generates a sequence of triggers, which are fundamentally synchronized to the rf accelerating voltage. A representation of this voltage, usually a "vector sum" signal, is converted from a sinusoidal waveform to an rf pulse train. This train is counted down by the harmonic number - the ratio of the rf frequency to the revolution frequency - to generate the basic train for the acquisition triggers. A reset pulse for the harmonic count down circuit allows this train to be locked to a particular bunch. Finally this train is gated to allow a pulse only after a user specified time interval. This last sets the time scale of interest. An additional option in the trigger scheme allows an explicit violation of the synchronization to a particular bunch. Everything applies as before except every trigger slips by one rf cycle relative to the previous one. This allows a quick comparison of the shapes of all the bunches in the machine in a single mountain-range display.

SEQUENTIAL DATA ACQUISITION

Digital oscilloscopes are normally slow with respect to their rearming time. The sequence trigger mode, available on some models, is being used for this application. In this mode the acquisition memory of the oscilloscope is divided in a number of segments and each segment is filled at each subsequent trigger at the sampling rate of the oscilloscope. In this way rearming rates over 10 kHz have been obtained. This does not provide for turn-by-turn observation of beam bunches, but standard single shot data acquisition will allow for these observations within the limitations of the available acquisition memory.

DATA TRANSFER AND DISPLAY

The oscilloscope is controlled from application codes or Virtual Instruments (VIs), written by the graphical programming language LabVIEW, which runs on a SUN-sparc station. This workstation is connected by ethernet to the AGS Distributed Controls System (AGSDCS). Data from the oscilloscope are transferred from the GPIB port through a GPIB-ENET controller. A typical VI will arm the oscilloscope to acquire a waveform, wait for a service request signaling that data acquisition is completed, transfer the sequenced waveform to the workstation and display it. This sequence being built in a while loop, which repeats itself until being stopped by the user.

DISCUSSION

The longitudinal structure of the beam at the AGS is both rich and critical, especially in the effort to accelerate very high proton intensities. The Booster synchrotron batch fills the AGS ring over four Booster cycles in 400 ms. The bunches are "flattened" longitudinally twice during the acceleration cycle. The acceleration of the beam through transition, lately with a "jump" system, is also sensitive in the longitudinal space. Finally the slow extraction of the beam is sensitive to the longitudinal structure of the beam.

Figure 1 shows a typical display obtained with the mountain-range system, in this case an overview of the injection process of beam from the Booster into the AGS during high intensity proton operation. Higher traces come later in time in the graph; the trace separation is 6 ms. The full horizontal scale is 3.4 μs, slightly more then one revolution of the beam in the AGS, which operated at the eighth harmonic. The AGS is seen to be filled by the four consecutive batches of the Booster operating at a 7.5 Hz repetition frequency. With the AGS executing one cycle every 3 second displays like Figure 1, involving the acquisition of 16 kbytes of data, can be obtained on a cycle by cycle basis. The limiting factor for the information flow in our configuration is the rate at which our (first generation) 1 Gs/s oscilloscope transfers the data through the GPIB bus. More recent models digital oscilloscopes have been tested and have shown data transfer rates of over 100 kbytes per second.

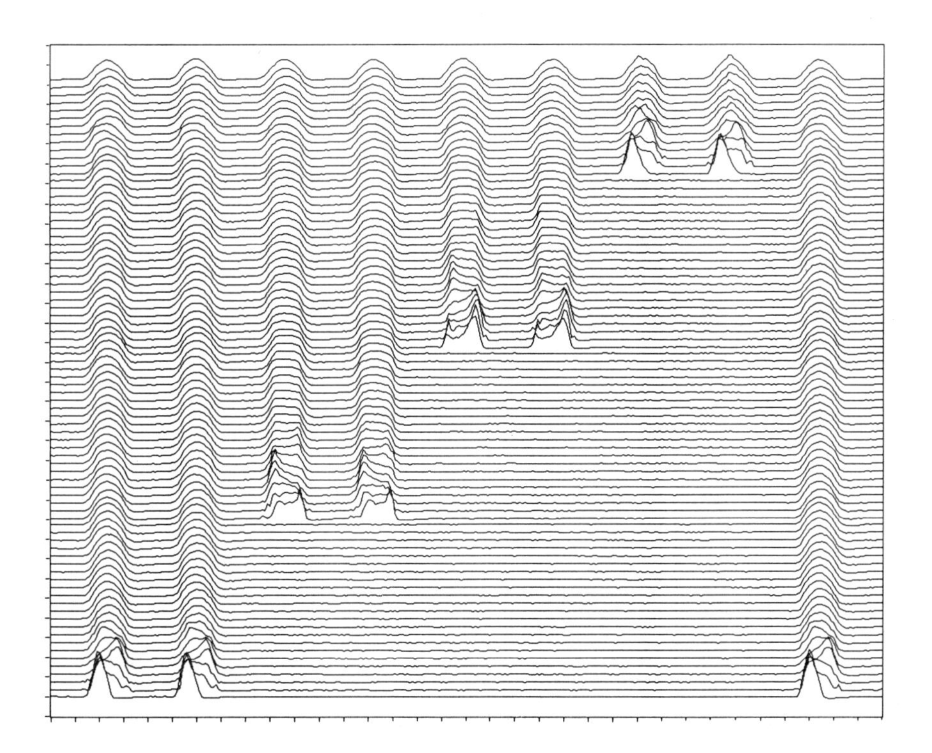

Figure 1. Overview of the injection process from the Booster into the AGS during high intensity proton operation.

Figure 2 shows details of the injection of the first batch of protons from the Booster into the AGS. The display at left has a trace every ms, while at right the trace separation is 100 μs. The full horizontal scale for both displays is 750 ns. The bunches are injected mismatched with respect to the rf buckets and since the phase loop is turned off they execute coherent synchrotron oscillations. These oscillations are flattened by the very high frequency (VHF) cavity in such a way that a controlled longitudinal dilution of the bunches results. This process is repeated for every consecutive transfer.

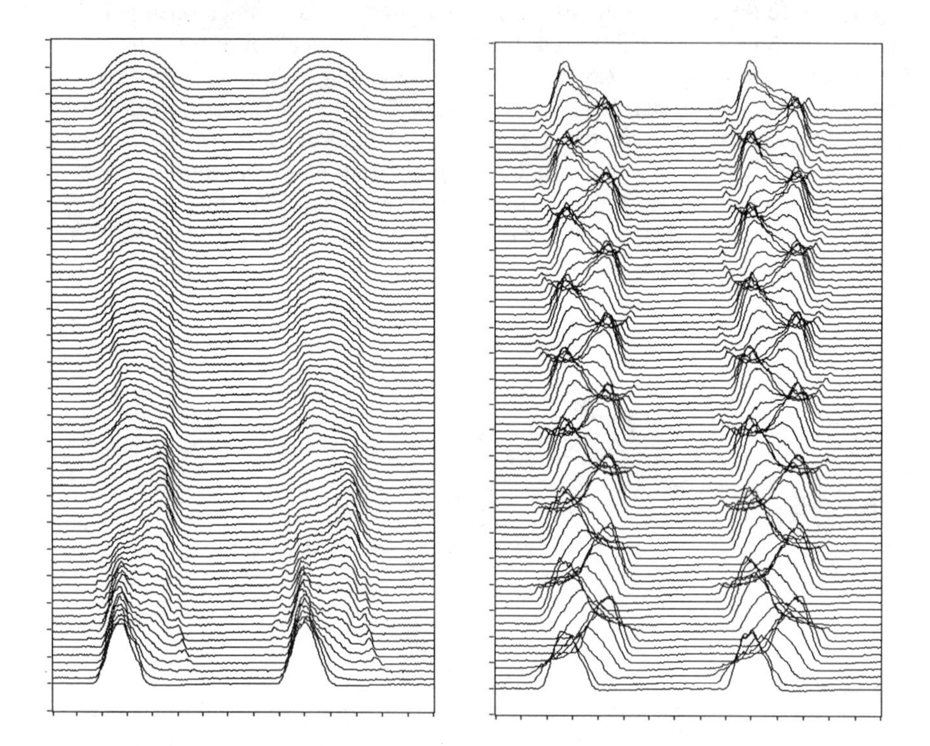

Figure 2. Displays showing the controlled longitudinal dilution of the bunches in the AGS.

An example of a turn-by-turn display is shown in Figure 3. In this case, 50 kbytes of data have been acquired and the maintain-range display has been constructed in LabVIEW. The figure shows ion beam bunches in the Booster accelerated on the 9th harmonic, which are rotating in 3rd harmonic buckets (3 bunches per bucket). This method improves the transfer efficiency from the Booster to the AGS. For this figure, the extraction from the Booster was turned off. The bunches are not lined up through rf synchronization anymore, the ability of lining up the bunches being determined by the sampling frequency, 100 MHz in Figure 3.

Figure 3. Turn-by-turn display of Gold ions in the Booster.

The advantage of the present method with respect to the acquisition with analog storage scopes are manifold. The much larger display of a workstation allows much more information to be presented. This method eliminates the need for "persistence" adjustments, which are so tedious necessary when adjusting the sweep speeds with analog storage scopes. The whole CRT of the workstation can be used for the display of the data, which means that considerably more details of the development of the longitudinal structure of the beam can be displayed usefully. Also, the VIs that run the mountain-range displays can be started from all workstations of the AGSDCS, allowing beam observations at remote locations. This is in contrast with the analog system, where beam observations are tied to the location where the high bandwidth signals, typically from a resistive wall monitor, are available. For the AGS, with rf frequencies in the 1-5 MHz range, the horizontal resolution of 1 ns matches well with the bandwidth of the wall monitor.

This method distinguishes itself from approaches where one digitizes beam signals continuously, writes Megabytes of data to disk and reconstructs beam spectra afterwards. These methods do not allow for the on-line beam observations, which the present method does, by acquiring only those data, which are deemed to be relevant.

VELOCITY MEASUREMENT BY MEANS OF MECHANICAL MOVEMENT OF A DETECTOR

A.V.Feschenko, P.N.Ostroumov
Institute for Nuclear Research, 117312 Moscow, Russia

Abstract

Recently several techniques to measure absolute value of the ion beam velocity have been proposed and tested. All of these methods are based on a mechanical movement of a detector in order to provide time of flight measurements. The detector enables us to deliver an rf signal induced by beam on a fundamental frequency of bunch sequence or higher harmonics. The phase of the induced signal is analyzed before and after the detector translation along the beam pipe with respect to the signal induced by the beam in the other detector. Also the induced signal can be compared relative to the rf reference line. In different measurements the resonator monitor or bunch shape monitor have been used as a detector. To provide more precise phase difference measurement the longitudinal translation of the detector is equal to βλ. The velocity measurements carried out at the INR Linac as well as at the CERN Heavy Ion Linac indicate that the accuracy is 0.1% - 0.3% but the main contribution is due to beam instability during the procedure.

INTRODUCTION

An absolute energy or velocity measurement is one of the most important and complicated problems to be solved during commissioning and operation of accelerators. In addition to magnetic spectrometers the time of flight methods have found application for the resonant accelerators. Generally, two current harmonic monitors located at a known distance are used (1). To measure an average energy of the beam, an absolute value of the phase difference of induced by beam signals is measured. The accuracy of the measurements is aggravated when the energy increases. To provide a reasonable accuracy the drift distance between the monitors has to be increased. That results in appearance of $2\pi n$ uncertainty in the phase measurements and thus in possible systematic errors. A method to avoid this uncertainty reported in ref.(2) requires special equipment to produce single beam bunch. Also, the beam debunching on the drift space contributes to the measurement errors. Typically, to provide an accuracy of a fraction of a percent for β=0.5-0.8 the distance must be more than ten meters, which is not always acceptable. To avoid these limitations we developed and tested a time of flight method utilizing a mechanical movement of a bunched beam detector along the beam pipe. The first results have been reported at the 1989 PAC (3). When moving the detector one can observe changing of a signal phase by a value $\Delta\varphi_0 = \omega t = 2\pi\frac{d}{\beta\lambda}$, where β is the relative velocity of the beam being studied, λ is the wavelength of the rf field induced in the detector and d is the distance of the monitor translation. Measuring $\Delta\varphi_0$ and d one can find beam velocity β. Different types of detectors can be used for this purpose: cavity type current harmonic monitors, pick up rf loops and bunch length detectors. The proposed method has also been proven recently by another team for the works on the cyclotron beam (4).

The first section of the paper describes the method and the results of energy measurements at the injection line of the INR linac (3). One of the two buncher cavities has been used as a beam current harmonic monitor. Energy measurements executed at the CERN Heavy Ion Linac (5) with the help of a Bunch Length Detector (BLD) are presented in the second section. Application of the methods for higher beam velocities is discussed in the third section.

MEASUREMENTS AT THE INR LINAC INJECTION LINE

The INR linac injector comprises a 750 kV pulse transformer. Generally the beam energy is determined by measuring of the HV pulse at the output of the transformer. An accuracy of these measurements is rather poor due to calibration difficulties. The accurate measurements and calibration have been accomplished with the help of two buncher cavities, i.e. the first cavity has been used for its designated purpose as a klystron buncher, the second one - as a current harmonic monitor. The block diagram presented in fig. 1 shows the procedure.

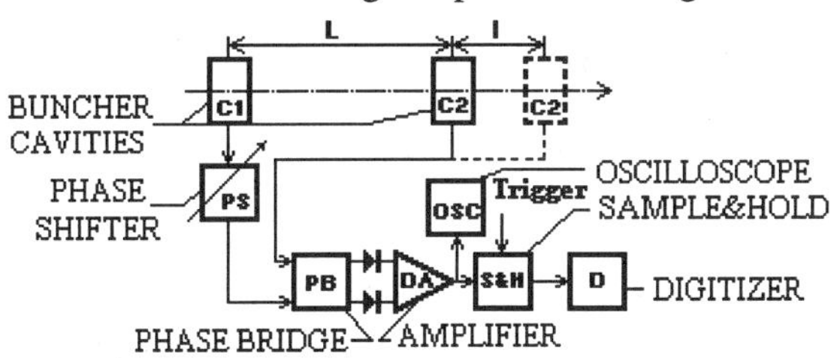

Figure 1. Block diagram of the apparatus.

The variation of the beam relative energy along the beam pulse is proportional to the phase of the induced signal in the second cavity with respect to the reference line. In a linear approximation the phase difference $\Delta\varphi$ is proportional to a beam energy variation $\Delta W / W$: $\Delta\varphi = -\dfrac{\pi L}{\beta\lambda}\dfrac{\Delta W}{W}$.

To measure $\Delta\varphi$ a phase detector (PD) consisting of a phase bridge, two amplitude detectors and a differential amplifier has been set. A phase shifter (PS) is adjusted to obtain a zero signal at the PD output for a nominal beam energy. A PD signal U_{pd} is proportional to the phase difference ($U_{pd} = k\Delta\varphi$) of input signals or to the relative energy variation of the beam: $U_{pd} = -\dfrac{k\pi L}{\beta_0\lambda}\dfrac{\Delta W}{W}$. Measuring U_{pd} one can find a relative beam energy variation:

$\frac{\Delta W}{W} = -\frac{\lambda\beta_0}{k\pi L} U_{pd}$. The PD signal was visually observed by the oscilloscope and was measured by a slow digitizer D with the help of a sample and hold circuit. Figure 2 shows the behavior of the $\Delta W(t)/W$ function and rms instability from pulse to pulse σ_w within the beam pulse. The results of two sets of measurements carried out within one hour for fixed injector parameters are presented. A value of PD sensitivity was equal to 20 deg/V.

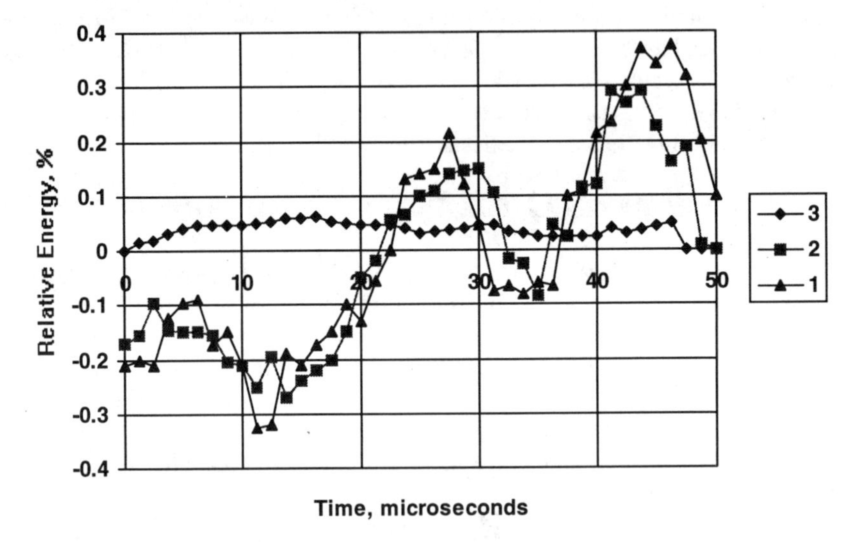

Figure 2. Variation of average energy (1,2) and rms instability (3) within the beam pulse.

To measure the absolute beam energy the second cavity was mechanically translated by a distance l. The accuracy of the energy determination depends on measurement errors of both distance l and of output signal variation. The latter can be avoided if the value of l is taken to provide $\Phi = \pi n$ (n=1,2,...). In this case the signals U_{pd} before and after the translation must be equal. The phase shifter PS was adjusted to have a zero U_{pd} signal before the translation. If n is odd and there are systematic errors of U_{pd} the measurements of the value of Φ differs from πn. To avoid this error n must be even and the value of cavity translation must be equal to or a multiple of $\beta\lambda$. Practically, mainly due to instabilities of the beam energy it is impossible to provide the same signals before and after the cavity translation. Using statistical processing of the obtained data, we have been able to obtain a value for the $\beta\lambda$ translation which is provided when U_{pd} signals before and after cavity translation are exactly equal to each other.

The process of measurements includes the following steps: timing of the S&H circuit is adjusted to make measurements in the most stable point within the beam pulse; the phase shifter is adjusted to get U_{pd} as close to zero as possible; multiple measurements of U_{pd} (N=100-300) are executed and the average value $\langle U_1\rangle=\langle U_{pd}\rangle$ and dispersion σ_1 is calculated; the second cavity is moved by a distance of $l \approx \beta\lambda$ till U_{pd} becomes close to $\langle U_1\rangle$; the exact value of l is measured and the $\langle U_2\rangle$ and σ_2 values are found; and the cavity is displaced again by a distance $\Delta l << l$ and $\langle U_3\rangle$ and σ_3 are found. The values Δl and l are measured with an accuracy about $2.0 \cdot 10^{-5}$. The value of absolute velocity can be found using the following linear expression:

$$\beta = \frac{l}{\lambda} + \frac{\Delta l}{l} \frac{\langle U_1\rangle - \langle U_2\rangle}{\langle U_3\rangle - \langle U_2\rangle} \qquad (1)$$

An accuracy of energy finding depends on the errors of measurements of the Δl , l and $\langle U_i\rangle$ values. The measurements at the INR injection line have been made for the following values L=0.6m, λ=1.512492m, $\delta(\Delta l)=\delta l = 2.10^{-5}$ m, $\sigma_i \approx$ 0.1V, Δl =1-2mm, N=100-300. Random errors are varied from 0.1% to 0.4% depending on experimental conditions.

Besides the random errors there is a systematic one $\delta W_1/W$ due to slow variation of the energy being measured $\delta W_0/W$ during the cavity translation about 2 minutes : $\dfrac{\delta W_1}{W} = \dfrac{L}{\beta\lambda}\dfrac{\delta W_0}{W}$. Special measures enabled us to decrease $\delta W_0/W$ down to 0.02% thus providing $\delta W_1/W$=0.2%. To decrease this error the distance between cavities L must be as small as possible.

It should be mentioned that a signal from another point in the accelerator rf system can be used as a reference signal for phase measurements and the energy measurements can be executed with only one cavity. In this case, however, additional systematic errors are possible.

ENERGY MEASUREMENTS AT THE CERN HEAVY ION LINAC

Energy measurements at the CERN Heavy Ion Linac (5) have been carried out with the help of a new device - a bunch length detector which has the capability of measuring the average velocity of a bunched beam (6,7). The principle of operation, as well as operational experience, of the bunch length detector has been reported elsewhere (8). A new device the Bunch Length and Velocity Detector (BLVD) uses an additional means which enables the device as a whole to be displaced in a longitudinal direction - a fine mechanical glider. This feature enables the combination of bunch shape and time of flight measurements in a single device. When moving the detector one can observe a displacement of the bunch longitudinal profile position with respect to the rf reference line.

The detector was initially used at the entrance of the first IH cavity downstream from the RFQ and buncher and then moved to the exit of the accelerator. The beam velocity was about 2% and 9% accordingly. Two modes of operation have been used for these two locations.

In the first case the value of detector displacement was taken to be $\beta_0\lambda$ (about 38 mm), where β_0 is the nominal velocity of ions. If the velocity differs from the nominal one by a value of $\Delta\beta$ a displacement of bunch centres $\Delta\varphi_0$ can be observed. Measuring $\Delta\varphi_0$ one can find a velocity difference of $\Delta\beta/\beta_0 = \Delta\varphi_0/2\pi$.

In the second case due to larger β the value of detector movement to provide a 2π bunch centre phase displacement becomes too large. The real value of detector displacement d was approximately the same as in the first case and to compensate, for a beam centre displacement an additional precise mechanical phase shifter was used. In this case the difference of the beam average velocity from the nominal value β_0 can be found as $\dfrac{\Delta\beta}{\beta_0} = \Delta\varphi_0 \dfrac{\beta_0\lambda}{2\pi d}$.

Using these compensation methods, in both cases of velocity measurement, an influence of inaccuracy by the phase shifter used for bunch shape measurements on the errors of velocity measurements is avoided. Other factors responsible for the error of energy measurements are the following:

1. The error of the displacement d measurement. A contribution of this error to the finite accuracy is less than 0.1%.

2. The error of phase compensation by the mechanical phase shifter. This error exists in the second mode of measurements (β=0.09) only. The error consists of two fractions. One of them is due to inaccuracy of the mechanical adjustment of the phase shifter, another reason - nonlinearity of its phase response because of reflections in a transmission line. The contribution of the first reason is negligible (less than 0.02%). To decrease nonlinearity, decoupling rf circulators were used at the input and output of the phase shifter. A contribution of this factor produced a reduction of 0.2%.

3. The error due to the changing of the secondary electron trajectories after the detector displacement. Visual observations of the thermal electrons have shown that the location of image of the electrons after the detector movement changes by about 0.4 mm. There are two possible reasons for this effect: a nonuniform external magnetic field and the slight changing of the boundary conditions for the electrostatic field in the target region. Though the error of energy measurement due to this effect is about 0.21%, it can be easily taken into account. We suppose that in this way we reduced it down to 0.05%.

4. The random errors. The value of these errors depends mainly on the stability of accelerator operation. In contrast to the above listed systematic errors an influence of these random statistical errors can be decreased by the factor of $\sqrt{N}$ if multiple measurements are executed. The random error of a single measurement is varied from 0.17% to 0.24% for β=0.09 and was less than 0.1% for β=0.02.

The total error of a single energy measurement was about 0.15% for the measurements at the entrance of the first IH cavity and about 0.3% at the exit of the accelerator.

For various sets of RFQ and buncher voltages, as well as buncher phases, the mean velocity has been measured. The specified input velocity to the first IH cavity was adjusted by the buncher rf phase to be equal to β=0.02308. The measurements at the exit of the accelerator have shown that the beam velocity exceeds the design value by 1.5%. Typical experimental bunch shapes and their relative locations are presented in fig. 3.

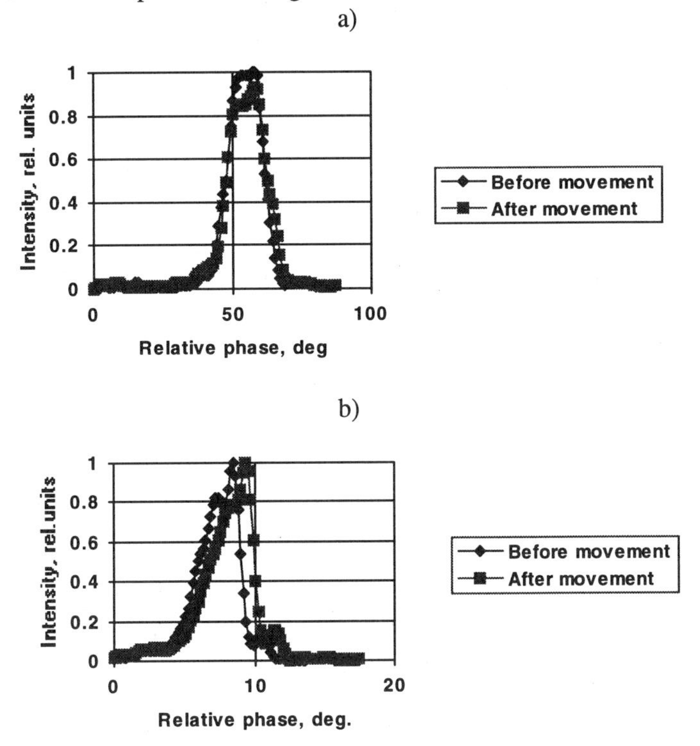

Fig. 3 Experimental bunch shapes and their relative locations.
a) W=250 keV/n, bunch center displacement $\Delta\varphi_0$=0.39^{o}.
b) W=4.3 MeV/n, bunch center displacement $\Delta\varphi_0$=0.61^{o}.

APPLICATION FOR HIGHER BEAM ENERGIES

The method of mechanical movement of the detector and the apparatus described were used for rather low velocities. The method utilizing current harmonic monitors can be extended to $\beta\approx$0.5-0.8 if a higher harmonics n of a fundamental bunch array frequency are used (3). The measurements are possible if $\beta\lambda/2n > l_b$, where l_b is a bunch length. The analysis made in ref. (9) have shown that for the proton beam with an energy from 160 MeV to 600 MeV the 14th harmonic of the 200 MHz frequency can be used and the accuracy of the velocity measurements can be about 0.03%.

CONCLUSION

A new method to measure precisely the mean velocity of the bunched beam by means of the mechanical translation of a detector has been developed and applied in several linacs. A measurement accuracy of about 0.1% has been achieved. As a detector the harmonic monitor as well as the bunch length monitor have been used. The latter allows one to carry out the velocity measurements for the beam intensities in the range of nanoamps to hundreds of milliamps. Work to extend the velocity range of the device up to $\beta = 0.8$ is under progress.

REFERENCES

1. "*Linear Ion Accelerators*", Ed. by Murin, B.P., **v.2**, Atomizdat, 1978 (in Russian)
2. Popovic, M.B., et al., "Time-of-Flight Measurements of Absolute Beam Energy in the Fermilab Linac", *Proc. of the 1993 PAC*, pp. 1689-1690.
3. Feschenko, A.V., et al., "Precise Energy Measurement Of the Continuous Proton Beam", *Proc. of the 1989 IEEE Particle Accel. Conf.*, Chicago, 1989, **v. 2**, pp. 1409-1410. See also *Pribory i Technika Experimenta*, 2,1990, pp.55-57.
4. Kormany, Z. "A New Method and Apparatus for Measuring the Mean Energy of Cyclotron Beams", *Nucl. Instr. and Meth.* A337(1994), pp.258-264.
5. Amedola, G., et al., "A Heavy Ion Linac for the CERN Accelerator Complex", *Proc. of the EPAC-92*, Berlin, 24-28 March, 1992, pp.536-538.
6. Ostroumov, P.N., "Average Velocity Measurement of Accelerated Beam by Means of an Upgraded Bunch Length Detector", *INR Preprint-812/92*, May 1993.
7. Bylinsky, Yu.V., et al, "Bunch Length and Velocity Detector and its Application in the CERN Heavy Ion Linac", *Proc. of the EPAC-94*, London, June 27-July 1,1994.(to be published)
8. Feschenko, A.V. and Ostroumov, P.N. "Bunch Shape Monitor and Its Application for an Ion Linac Tuning", *Proc. of the 1986 Linac Conf.*, Stanford, June 2-6, pp.323-327.
9. Novikov, A.V. and Ostroumov, P.N. "Precise Average Velocity Measurement of Bunched Ion beam", *INR Preprint-844/94*, January 1994.

Noninterceptive Beam Energy Measurements in Line D of the Los Alamos Meson Physics Facility*

J. D. Gilpatrick, H. Carter, M. Plum, J. F. Power, C. R. Rose, R. B. Shurter
Los Alamos National Laboratory, M. S. H808, Los Alamos, NM, 87545

Abstract

Several members of the Accelerator and Operations Technology (AOT) division beam-diagnostics team performed time-of-flight (TOF) beam-energy measurements in line D of the Los Alamos Meson Physics Facility (LAMPF) using developmental beam time. These measurements provided information for a final design of an on-line beam energy measurement. The following paper discusses these measurements and how they apply to the final beam energy measurement design.

INTRODUCTION

The purpose of the beam measurements performed was to provide information for the design and implementation of an on-line beam-energy measurement system with sufficient bandwidth, resolution, and repeatability to monitor the beam energy as beam is injected into the proton storage ring (PSR). Using existing probes, cables, and ground test accelerator (GTA) processing electronics, we were able to monitor the relative beam energy using TOF techniques. This simple technique measures the time that a particular beam bunch travels between two image-current beam probes separated by a known distance. The time measurement is accomplished by measuring the phase difference of the 201.25-MHz probe signals. This technique has the advantage that it does not intercept or perturb the beam and has a wide overall bandwidth.

As the beam measurements were performed, three questions were answered.

1.) With the approximate 25-m drift distance in a straight portion of the LAMPF Line D, is there sufficient energy-measurement resolution for monitoring the PSR injected beam using a TOF measurement system?
2.) What influence does the beam chopping have on the design of the TOF processing electronics?
3.) Does changing the beam position or trajectory angle upstream of this portion of Line D change the flight path of the beam enough to limit the TOF measurement system repeatability?

EXPERIMENT DESCRIPTION

The measurement system used was a combination of microstrip- or stripline-style beam position monitors (BPM) already installed in Line D, 3/8" heliax already installed into the facility, and two altered microstrip-measurement-system electronics used in the GTA (1). The probes were a standard 10-cm aperture, 36°-

* Work supported in part by the US Department of Energy, Office of Defense Programs and in part by the Los Alamos Laboratory Institutional Supporting Research.

subtended angle, 34-cm long lobe probes that have been used for some time at LAMPF/PSR. The distance between probes is 24.77 ± 0.01 m.

Using the existing heliax cables, two sets of GTA-based microstrip processing-electronics were connected to these cables. Each of the probes then had horizontal and vertical beam positions, beam intensities, and a single TOF channel. A block diagram of a typical GTA microstrip processing electronics is shown in Fig. 1.

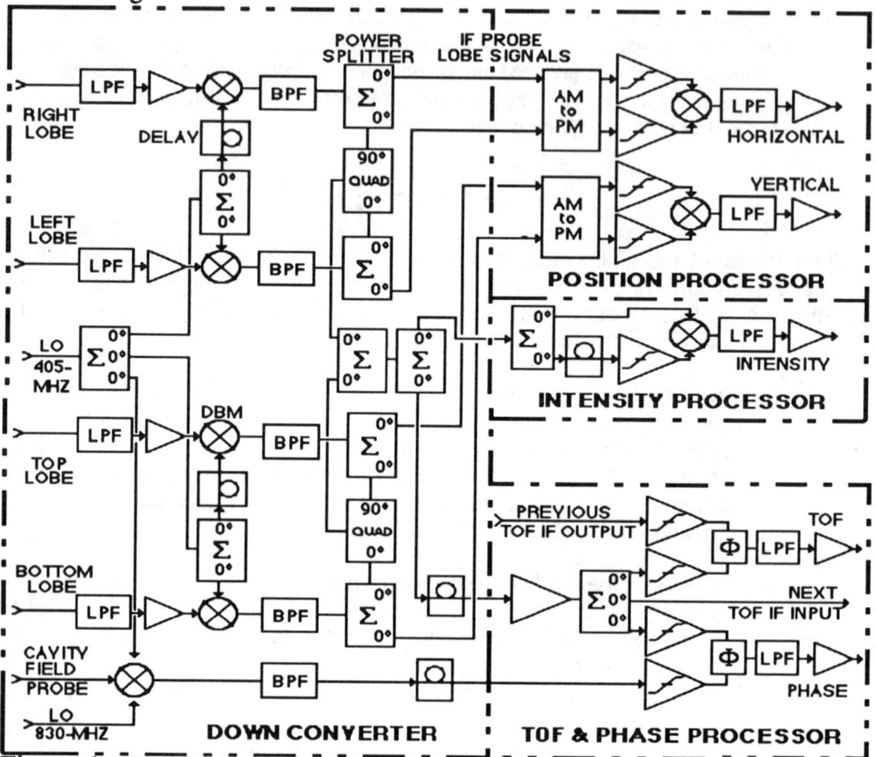

Figure 1. Typical block diagram of the GTA microstrip system showing the beam output phase, energy/TOF, position, and intensity measurements. The component identified as BPF, LPF, LO, and DBM are bandpass filters, lowpass filters, local oscillator, and double balanced mixers, respectively.

The GTA bunching and local oscillator frequencies were 425- and 405-MHz, respectively. These radio frequencies produced an intermediate frequency (IF) of 20 MHz. The overall system bandwidth was 2 MHz. For our tests, we changed the input lowpass filter corner frequency to 240 MHz and the local oscillator to 181.25 MHz so that the rest of the system (i.e., IF bandpass filters, etc.) did not need to be changed. For these measurements, the output-phase portion of this circuitry was not used. The data were acquired locally with the combination of a LeCroy 9400 oscilloscope connected to a Macintosh computer running National Instrument's Labview data acquisition and analysis software. Since the peak beam current was typically 7.7 mA, the peak signal power during the macropulse was 10 dB from the top of the 34-dB dynamic range of the TOF processing electronics. Table 1 lists the beam parameters used during the experiment.

Table 1. Beam parameters used during the three experimental measurement sets.

Parameter	Set #1	Set #2	Set #3
Average Current during Macropulse (mA)	7.7	5.3 to 1.3	7.7
Chopping Pulse Length (μs)	none	0.25	none
Chopping Pulse Repetition Rate (kHz)	none	2795	none
Countdown Number	none	1 to 4	none
Macropulse Length (ms)	0.65	0.65	0.15
Macropulse Repetition Rate (Hz)	1	1	1

EXPERIMENTAL DATA

The experiment consisted of three sets of measurements:
1) TOF measurements versus linac module-48 cavity-phase as the cavity phase is varied;
2) TOF processor output signal acquired and analyzed as the countdown is varied; and
3) TOF measurements versus beam position and trajectory-angle as the beam position and trajectory angle are varied.

During the first set of measurements, the relative beam energy was monitored with the TOF energy measurement and a wire scanner beam profile measurement. The wire scanner horizontal-profile measurement is in the middle of an 89° bend in the beam transport where the dispersion is high (i.e., 50±0.5 mm per % of $\Delta p/p$). Therefore, as the beam energy changes, the centroid of the beam's horizontal distribution changes. The wire scanner measurement is upstream of the first probe of the TOF energy measurement and both measurements are downstream from the last linac accelerating structure, module 48.

In order to get a general sense of the operational characteristics of the TOF measurement, the central beam energy was changed and the measured beam energy and single particle simulation was compared. This technique is commonly known as a "phase scan". Following a typical phase scan procedure, the cavity phase of module 48 was changed by ±10° about the synchronous phase in 2° steps. At each step, wire scanner and TOF energy measurement data were acquired. These data were then normalized to 799.795 MeV and plotted in Fig. 2 and 3. The important information is in the slope and shape of the fitted polynomial curves. The TOF and wire scanner measurements appear to track each other fairly well. The relationship between the two measurements may be further compared if the TOF data are plotted as a function of the wire scanner data (see Fig. 3). Note that the two energy measurements agree within approximately 3%.

Noise data were also taken using an external 201.25-MHz oscillator as a beam-induced input-signal replacement to the down-converter and TOF processing electronics. A power meter whose bandwidth lies between 3-Hz and 300-kHz was used to measure the TOF output-signal noise. These bench data indicated that the precision of the measurement processing electronics is approximately ±0.2 mV or ±0.1 ps over a 15-dB portion of the dynamic range. This is equivalent to ±4 keV of beam energy over a 4- to 22-mA beam current range. However, the expected signal-to-noise ratio must also be estimated and added appropriately to the measurement resolution. If reasonably conservative estimates of the noise floor

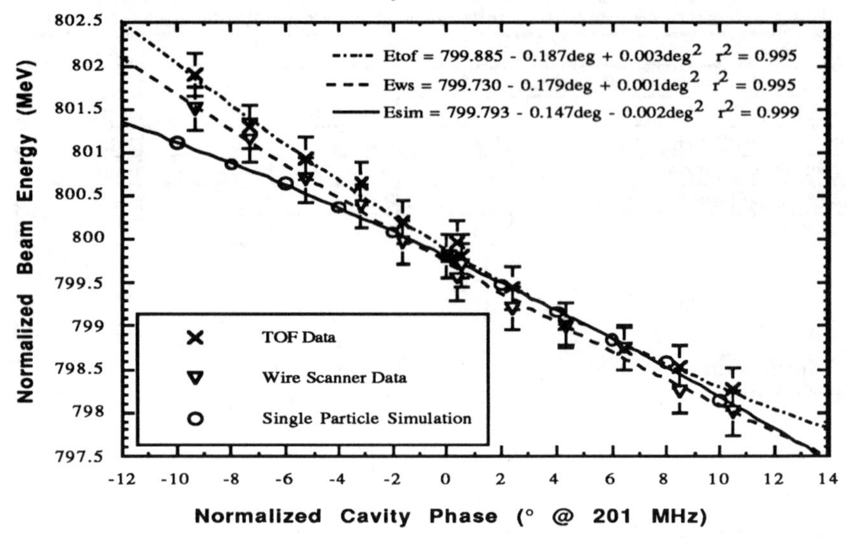

Figure 2. TOF and wire scanner energy measurement data plotted as a function of normalized cavity phase. Both energy measurements were normalized to a single-particle-simulation synchronous energy (also plotted). Error bars indicate the amount of beam energy jitter. TOF energy-measurement resolution is approximately ±7-keV and wire scanner energy-measurement resolution is approximately ±62-keV. The lines are a least-squares second-order fit to the data where each fitted equation and its associated correlation coefficient, r^2, is given.

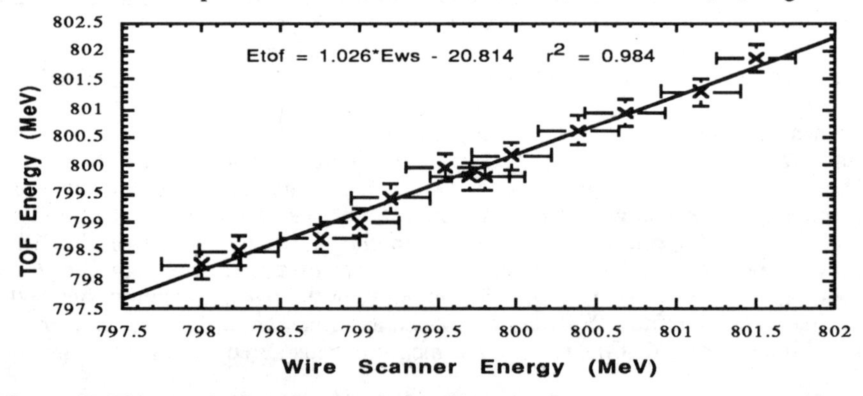

Figure 3. This graph plots the correlation between the two energy measurements.

(-91-dBm) are assumed, the measured -15-dBm probe-signal power produces an overall resolution estimate of ±7 keV. This is better than what will likely be required for the final energy measurement.

For the second set of measurements, the TOF, intensity, and beam-position output signals from both probes were digitized. The important TOF information is contained in a very small bandwidth about 201.25 MHz probe-signal components. As can be seen in Fig. 1, this information is down-converted to 20 MHz and then detected to an overall bandwidth of DC to 2 MHz. The output signals were sampled with 1000 points across a 5-μs section of the macropulse approximately 250 μs into the macropulse. Chopping countdowns from 1 through 4 were applied to the beam. This chopping pattern is generated to provide proper timing for the PSR injected beam. The chopping frequency is the 72nd sub-harmonic of the 201.25-MHz beam-bunching frequency (or approximately 2.8 MHz). The chopped beam structure for a countdown of 1 is 250 ns of beam and a 110-ns gap. As the chopping countdown is increased, 250-ns chopped beam pulses are removed. For example, a countdown of 3 produces 250-ns chopped beam pulses every 1080 ns (i.e., 830 ns gap between chopped beam pulses). The data were then windowed with a Hamming data window and padded with zeros so that the vector length was 1024 points. These time domain data were then transformed to the frequency domain using a fast Fourier transform (FFT). The complex vector was then transformed to magnitude information and the resulting low-frequency portion of the transformed information is displayed in Fig. 4.

One technical design aspect for the final energy measurement is how to deal with the beam chopping characteristic of the PSR-injected beam. The beam chopping produces gaps within the macropulse, during which there is no beam. In the TOF electronics, when there is no beam, the limiters do not have any signal to "lock-on" and will randomly oscillate between their plus and minus power supply voltages. These oscillations would drastically compromise the resolution and repeatability of the measurement. However, it is possible to choose the IF bandpass filter characteristics (i.e., bandwidth, center frequency, etc.) so that the 2.8-MHz chopping frequency does not effect the TOF output signal.

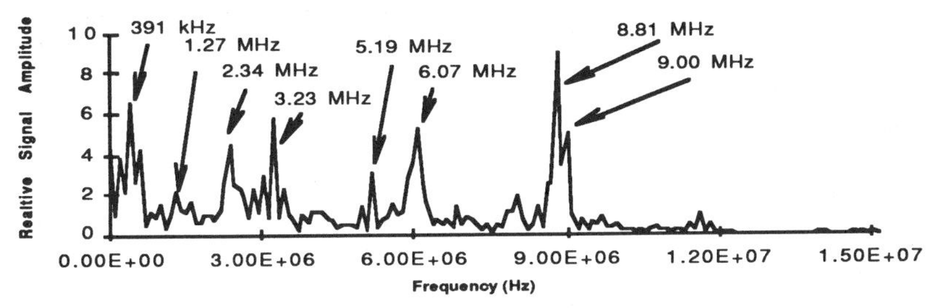

Figure 4. FFT of TOF output signal for countdown-of-1 data.

This figure shows that not only do we need to be concerned with the fundamental 2.8-MHz component, we also need to consider the harmonics of this chopping frequency. For example, a modulation product from the 7th harmonic (i.e., 7x2.8-MHz=19.56-MHz) of the 2.8-MHz chopping frequency and the 20-MHz IF frequency component produces a 434-kHz frequency component at the TOF output-signal port. Since the frequency domain resolution for Fig. 4 is approximately 98 kHz, the 434-kHz component is reflected by noting the presence of a 391-kHz component. Because the bandpass and lowpass filters in the processing electronics have limited bandwidths of below 3 MHz, the higher frequency components between 3- and 12-MHz appearing in Fig. 4 were not initially expected. However, it was discovered that these components are due to the presence of two primary modulation products from the down-converted probe signals, 19.56- and 22.36-MHz. The TOF phase detector demodulates these two products and produces many "daughter" products of sizable magnitude relative to the fundamental beam information. Therefore, there are two separate demodulation processes occurring that produce the spectrally rich signals shown in Fig. 4. The first process produces the fundamental modulation products below 3 MHz and the second process produced by the phase detector in the TOF electronics processor manufactures "daughter" or second generation products. Because the lowpass filters TOF processing electronics were not designed for these chopped beam signals, these spectrally rich signals are seen at the TOF output-signal ports. It is clear that for the final measurement it will be important to pick the IF bandpass filter center frequency, bandwidth, etc. so that all of the modulation products are outside the filter's bandwidth for all operational countdown conditions.

For the final measurement set, the beam was steered in the horizontal and vertical axes with steering magnets upstream of the two line-D probes. A change in these steering magnet settings will result in a change in the beam trajectory in the magnetic transport elements between the two beam probes. There are no steering or dipole magnets between the two probes but there are three sets of quadrupole doublets. The procedure used to change the beam position and angle was that the operators changed a magnet setting until the beam loss measurement between the two microstrip probes reached a maximum acceptable level. This was done for both directions from the nominal trajectory and for both axes. The steering data are shown in Fig. 5. The vertical data shows little vertical position change. The horizontal magnet steered the beam slightly under 15 mm at the first probe and approximately 6 mm at the second probe.

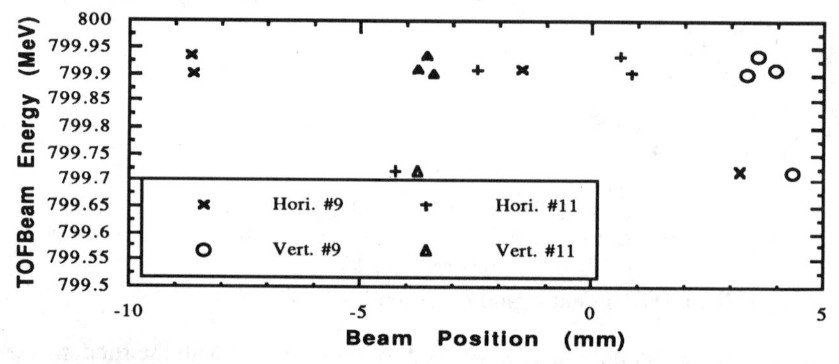

Figure 5. This graph shows the measured beam energy versus the horizontal and vertical steered beam for both probes. Data are averaged over 20 macropulses.

The amount of energy change for the horizontal data shown in Fig. 5 is approximately 200-keV. How does this compare with first order theory? Assume a worst case condition in which at each of the doublets between the two probes, the beam trajectory is offset 30 mm from beamline center (beampipe diameter is 100 mm) in a zig-zag fashion. This offset beam trajectory changes the path length from a centered-beam trajectory by approximately 0.65 mm. Substituting this error into Eq. 1 and solving for the energy error,

$$\Delta E = -\frac{(\beta\gamma)^3 mc^2 \lambda_0}{L}\left[\frac{\Delta\varphi}{2\pi} - \frac{\Delta L}{\beta\lambda_0}\right] \quad (1)$$

where β is the relative beam velocity, γ is the Lorentz contraction factor, L is the drift distance, mc^2 is the rest energy of the accelerated particle, ΔL and $\Delta\varphi$ are the distance and phase errors, and λ_0 is the free space wavelength of the bunching frequency.(2) The resultant error is 110 keV. The measured steering data shown in Fig. 5 are within a factor of 2 of this calculated energy error.

Several indications throughout the experiment corroborate the observation that beam energy jitter larger than the steered-beam energy change was present during the experiment. Both of the energy measurements shown in Fig. 2 have approximately the same amount of error or jitter - approximately ±250 keV. During this initial measurement set, the module 48 cavity power and field amplitude were recorded several times for the nominal set points. The field amplitude varied by 0.3% and the cavity and waveguide power (i.e., forward minus the reflected power) varied by 10-kw out of 830-kw. These variations in module 48 cavity power alone could account for variations in beam energy of 60- to 120-keV. In total, all of these observations seems to indicate that an energy jitter of approximately 300- to 500-keV was present throughout the experiment.

CONCLUSION

Using the existing probes, cables, and GTA TOF processing-electronics, resolutions of 7 keV were estimated based on the 25-m drift distance, bench noise measurements and expected beamline noise. This estimated resolution is much better than will likely be required for the final beam energy measurement. Phase scans were acquired using TOF- and wire-scanner-based energy measurements with good agreement between the two types of measurements. Under chopped beam conditions, good agreement was found between TOF output signal data and theory. These chopping effects will guide the design of the processing electronics and the choice of the bandpass and lowpass filter specifications. TOF flight-path lengths do change as a function of beam steering in this particular beamline. However, the data and simple theory indicate that the change of energy due to the path length change is less than will be required for the final measurement. Based on these initial measurements, an 800-MeV TOF beam-energy measurement in this facility is feasible.

REFERENCES

1. J. D. Gilpatrick, et. al., "Experience with the Ground Test Accelerator Beam-Measurement Instrumentation," presented at the Beam Instrumentation Workshop, Santa Fe, NM, October, 6-12, 1993.
2. J. D. Gilpatrick, et. al., "Beam Energy Measurements in Line D," Los Alamos National Laboratory Technical Note, AOT-3:TN:94-18, September, 1994.

Rotating Scanning Polarization Profile Monitor

J. Soukup, P.W. Green, L. Holm, E. Korkmaz, S. Mullin, G. Roy, T. Stocki
Dept. of Physics, University of Alberta, Edmonton, AB, Canada T6G 2N5

A.R. Berdoz, J. Birchal, J.R. Campbell, A.A. Hamian, S.A. Page, W.D. Ramsay, S.D. Reitzner, A.M. Sekulovich, W.T.H. van Oers
Dept. of Physics, University of Manitoba, Winnipeg, MB, Canada R3T 2N2

J.D. Bowman, R.E. Mischke
Los Alamos National Laboratory, Los Alamos, New Mexico, 87545, USA

C.A. Davis, D.C. Healey, R. Helmer, C.D.P. Levy, P.W. Schmor
TRIUMF, 4004 Wesbrook Mall, Vancouver, BC, Canada V6T 2A3

N.A. Titov, A.N. Zelenskii
Institute for Nuclear Research, Russian Academy of Sciences, Moscow, SU 117312

Abstract

A polarimeter capable of determining transverse component polarization profile of the 222 MeV proton beam both in horizontal (x) and vertical (y) directions with faster than 1 Hz scanning frequency has been developed and built as part of the preparations in the TRIUMF Experiment 497 measuring the flavour conserving hadronic weak interaction (parity violation measurement). The design features and test performance results are presented.

INTRODUCTION

The level of accuracy required in this experiment (A_z to $\pm 2 \times 10^{-8}$) calls for very precise diagnostics and beam control. Besides accurate beam profile, beam intensity and beam energy monitors and their feedback, an accurate periodical measurement of the beam transverse polarization profile both in horizontal and vertical directions at two locations upstream of the LH_2 target is needed. After testing and improving the first prototype of the scanning polarization profile monitor, a second one was built.

The design spin (beam helicity) flip frequency is now about 5 Hz dictated largely by the data acquisition capabilities. It is desirable to obtain (typically) one transverse polarization profile sample in y direction $<y\ Px>_A$ and in x direction $<x\ Py>_A$ for each beam helicity (+ and -) in one spin flip cycle at location A and similarly the same four profiles $<y\ Px>_B^{+}$, $<y\ Px>_B^{-}$, $<x\ Py>_B^{-}$, $<x\ Py>_B^{+}$ in the second polarimeter location B in the next spin flip cycle. (See Fig. 1 for the timing sequence.)

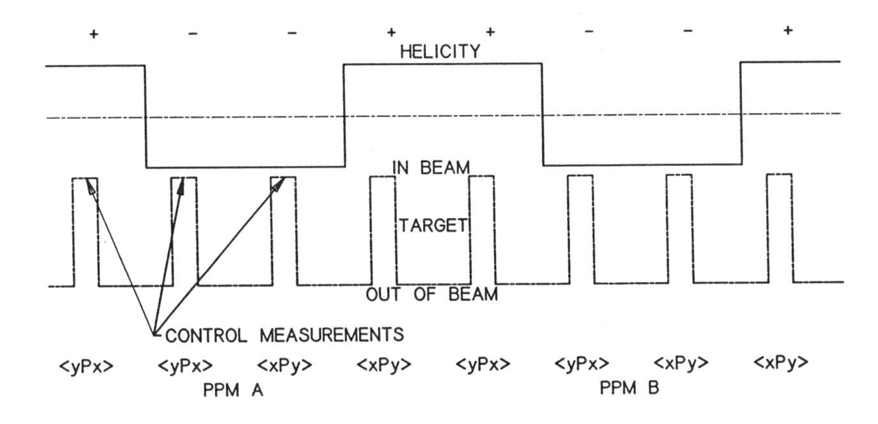

Figure 1. Polarization profile measurement sequence.

DESIGN FEATURES

After careful consideration of all the problems involved, such as the spatial resolution, counting rates, scanning speed control and target position readout, a decision was made to accomplish the scanning of the target through the beam by mounting thin target blades on two synchronized wheels that rotate with constant speed. One wheel is moving its targets horizontally across the beam while the second wheel interleaves its two targets vertically across the beam similarly to a polarimeter designed by Haeberli et al. (1) (see Fig. 2). Such arrangement eliminates reciprocating motion of the target so that the moving mass does not change its momentum. This is a great advantage in controlling the speed and position of the target. Each wheel contains two target blades 1 mm wide x 5 mm thick mounted with their center on a 215.4 mm radius of rotation and 45° apart. The two target wheels, one for horizontal and the second for vertical scans are connected by a flexible toothed belt to assure synchronism. One wheel is driven by a motor positioned outside the vacuum together with the belt and the rest of the driving mechanism. The rotation motion is transferred to the target wheels inside vacuum by the means of two ferrofluidic feedthroughs (2).

Both sets of scattering targets for measuring vertical Py and horizontal Px components of the transverse polarization of the beam have their corresponding scattering scintillator detector telescopes (left/right and up/down respectively). Hydrogen was chosen as the polarization analyzing target, considering the proton beam energy and the fact, that the recoiling proton can be detected as an added constraint in removal of accidental background. Thus, the forward scattering telescope arms are positioned at a 17.5° scattering angle (left, right, up and down from the beam axis) which corresponds to the maximum in analyzing power. Each forward arm is complemented by a corresponding recoil detector arm positioned at

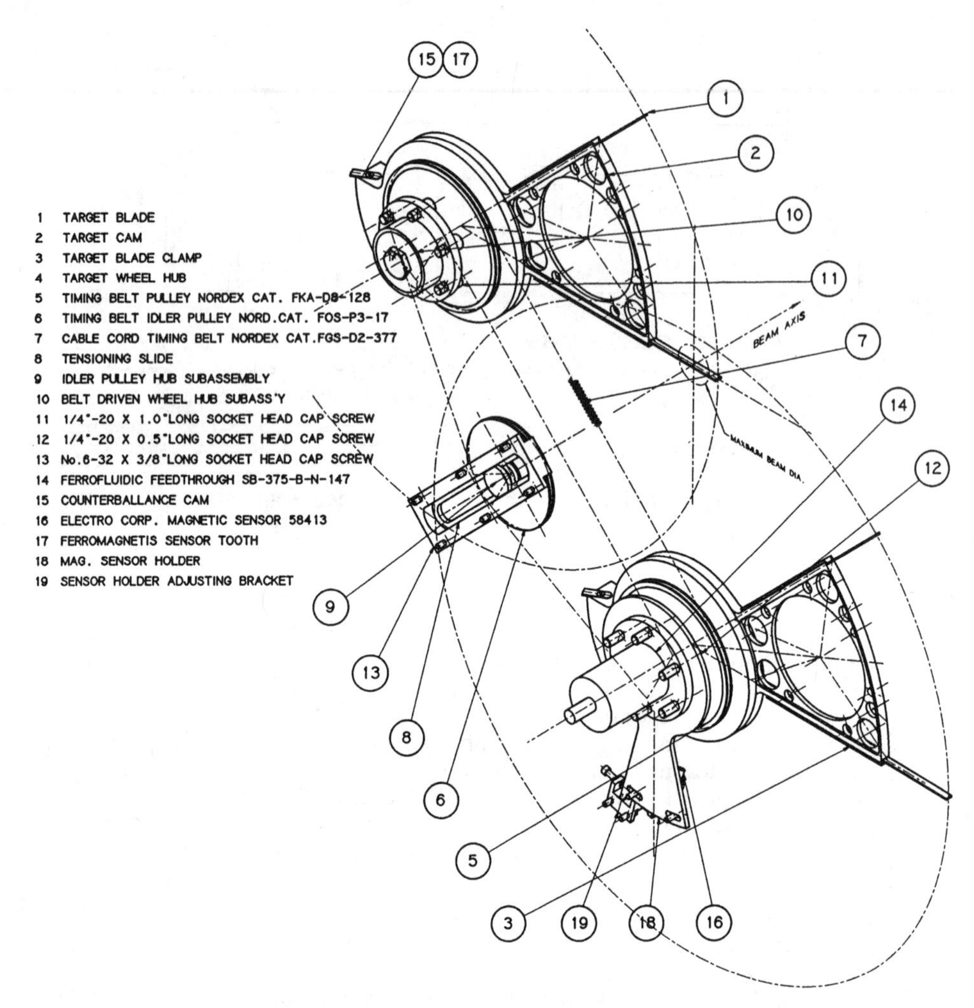

Figure 2. Target wheel mechanism.

70.6° on the opposite side of the beam. Each forward telescope arm consists of a solid angle defining scintillator (<1 msr dictated by counting rates from the chosen target exposed to the realistic beam) and a direction coincidence counter preceding it. The solid angle defining counter is tilted through 49° from being perpendicular to 17.5° scattering centre line to minimize polarization measurement sensitivity to the beam particle displacement from the central target position (see memo by C.A. Davis (3)).

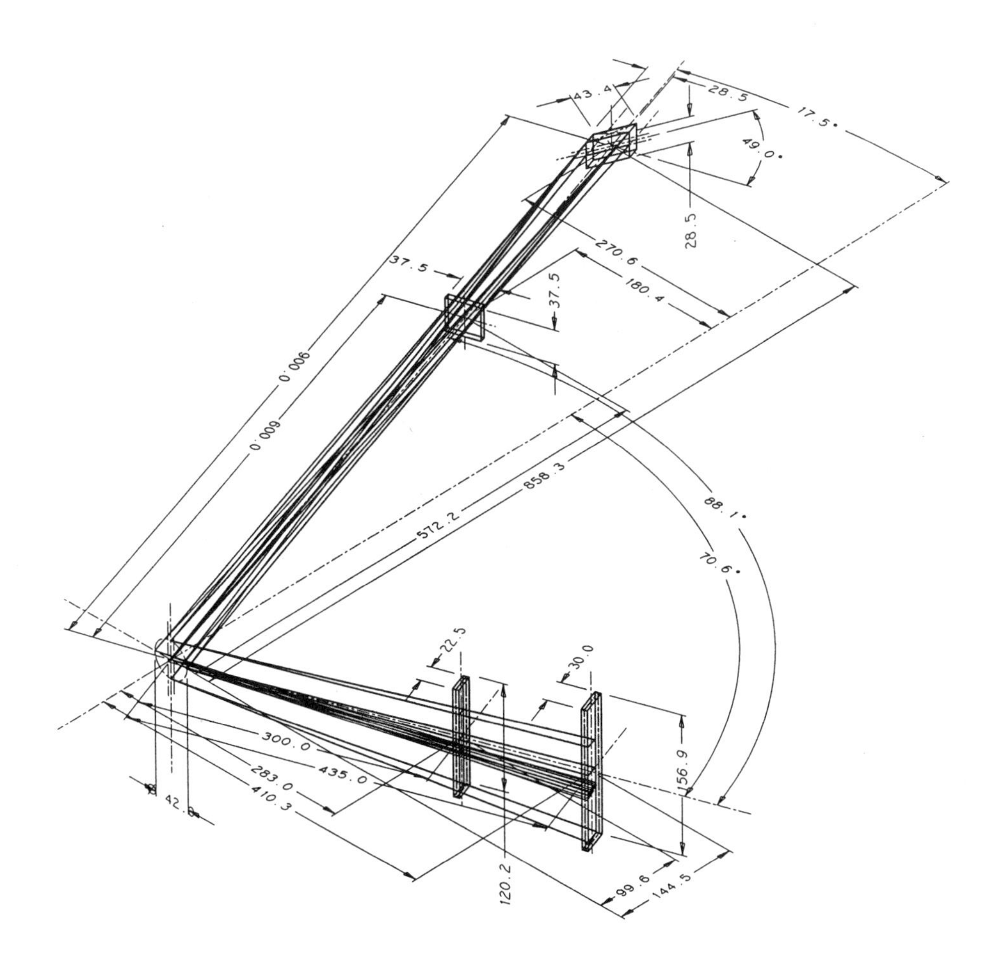

Figure 3. Polarimeter geometry.

The mean recoil proton energy corresponding to the chosen configuration is 22 MeV and thus the recoil arm consisting of a 6.35 mm thick stopping scintillator followed by a veto scintillator has to share common vacuum enclosure with the target. Fig. 3 shows the detector/target configuration and dimensions. The size of the recoil scintillators was obtained graphically by enclosing the envelope of recoil rays corresponding to the forward scattered rays. These originated in the extremes of: the target ($z = \pm 2.5$ mm), the beam spot (42 mm diameter) and they went through the extremes of the solid angle defining forward scintillator. This was accomplished with the use of Computervision's Personal Designer CAD software (4) as an integral part of the overall mechanical design of the polarimeter.

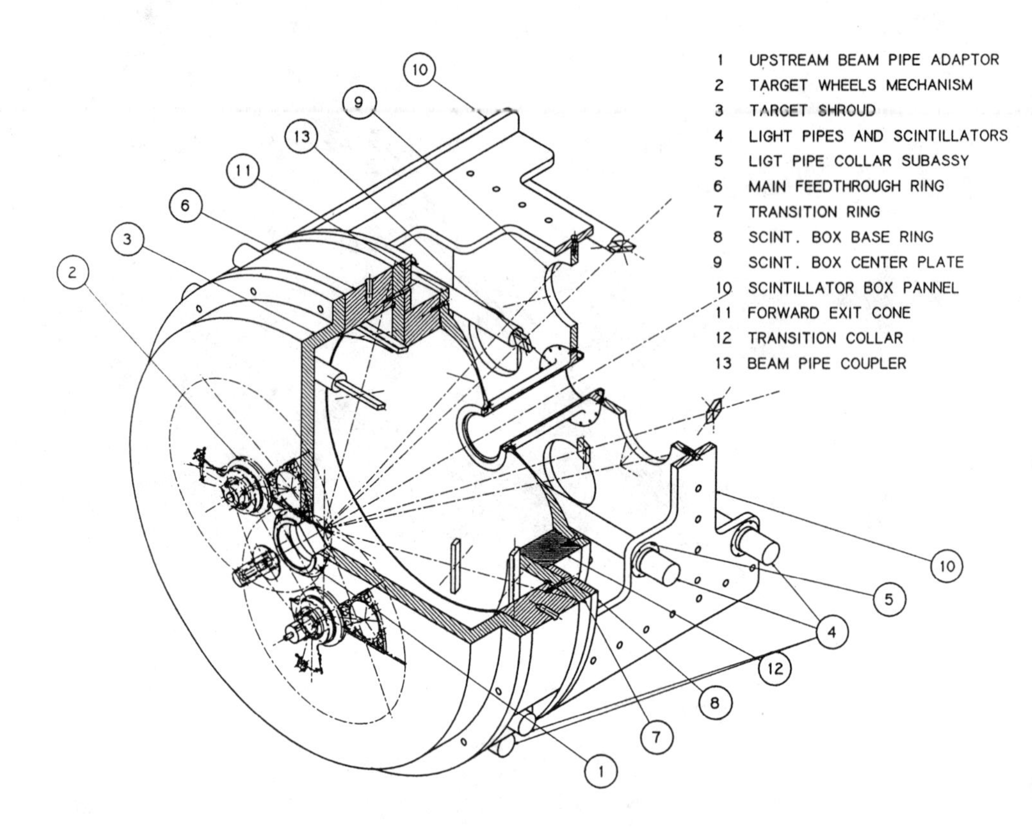

Figure 4. Cut-out view of the polarization profile monitor.

The form of the hydrogen target material has up to now been high density polyethylene. The 1 x 5 mm strips of this material might have been satisfactory for the preliminary tests of the polarimeter, but we hope to find a better alternative in the future as these can be easily bent or twisted. Polyethylene balls suspended in an epoxy matrix has been one suggestion.

Each target strip is complemented by a mild steel chisel edge located exactly opposite on the target wheel (180°). The passage of these steel edges is detected by a magnetic pick-off sensor (5) (one on each wheel). The pick-off signals are used as a target position readout. A DC motor of the kind used on magnetic tape drives with an optimum of its speed control at 15 Hz has been used so far. The actual position signal used by the data acquisition routine and also used to synchronize the two PPMs is obtained from an optical encoder coupled to the axle of each of the belt driven target wheels. An electronic control module built at the University of Alberta allows for variation of the target rotating frequency between 1 and 15 Hz.

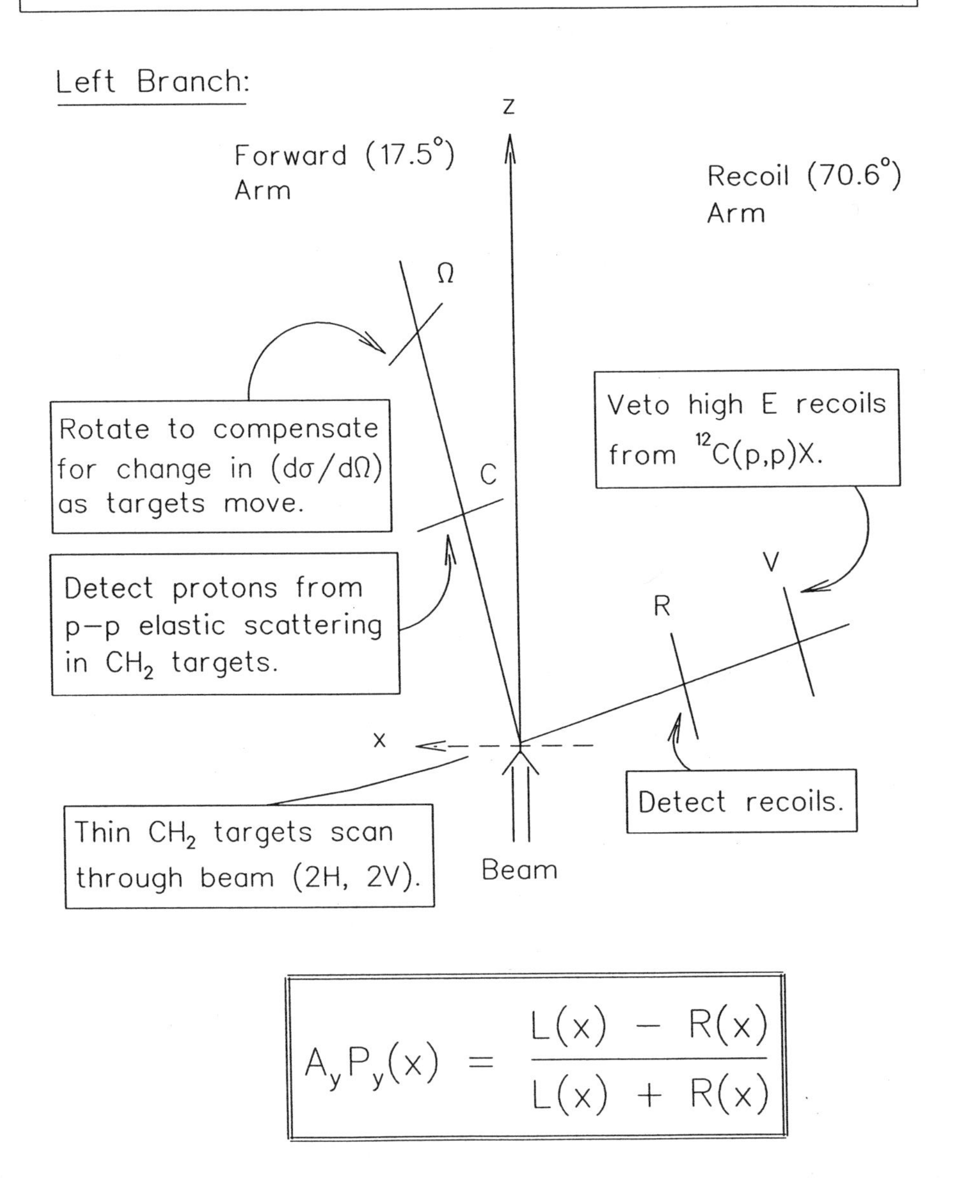

Figure 5. Polarimeter Systematics.

PPM Systematics

Ω counters rotated such that:

$$\Delta\Omega \times \frac{d\sigma}{d\Omega}(\theta) = \text{constant over PPM acceptance.}$$

$\Longrightarrow$ Instrumental asymmetry is small and rate independent.

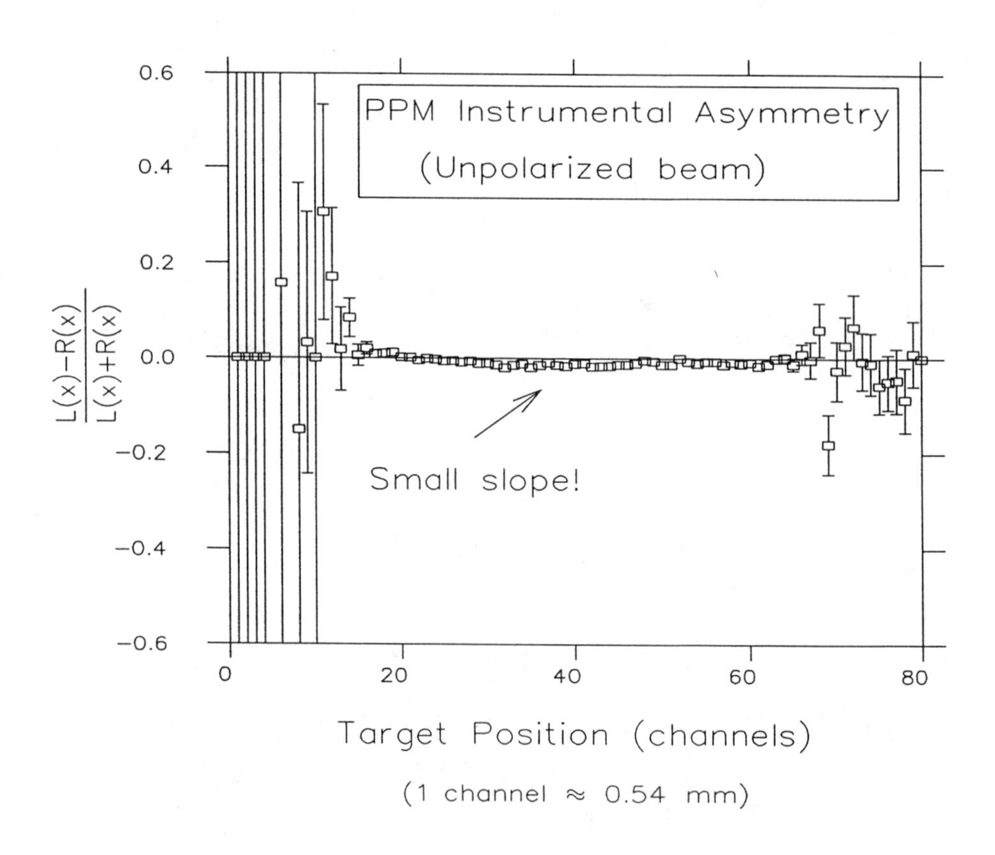

Figure 6. Correcting effect of the rotation of solid angle defining forward scintillators.

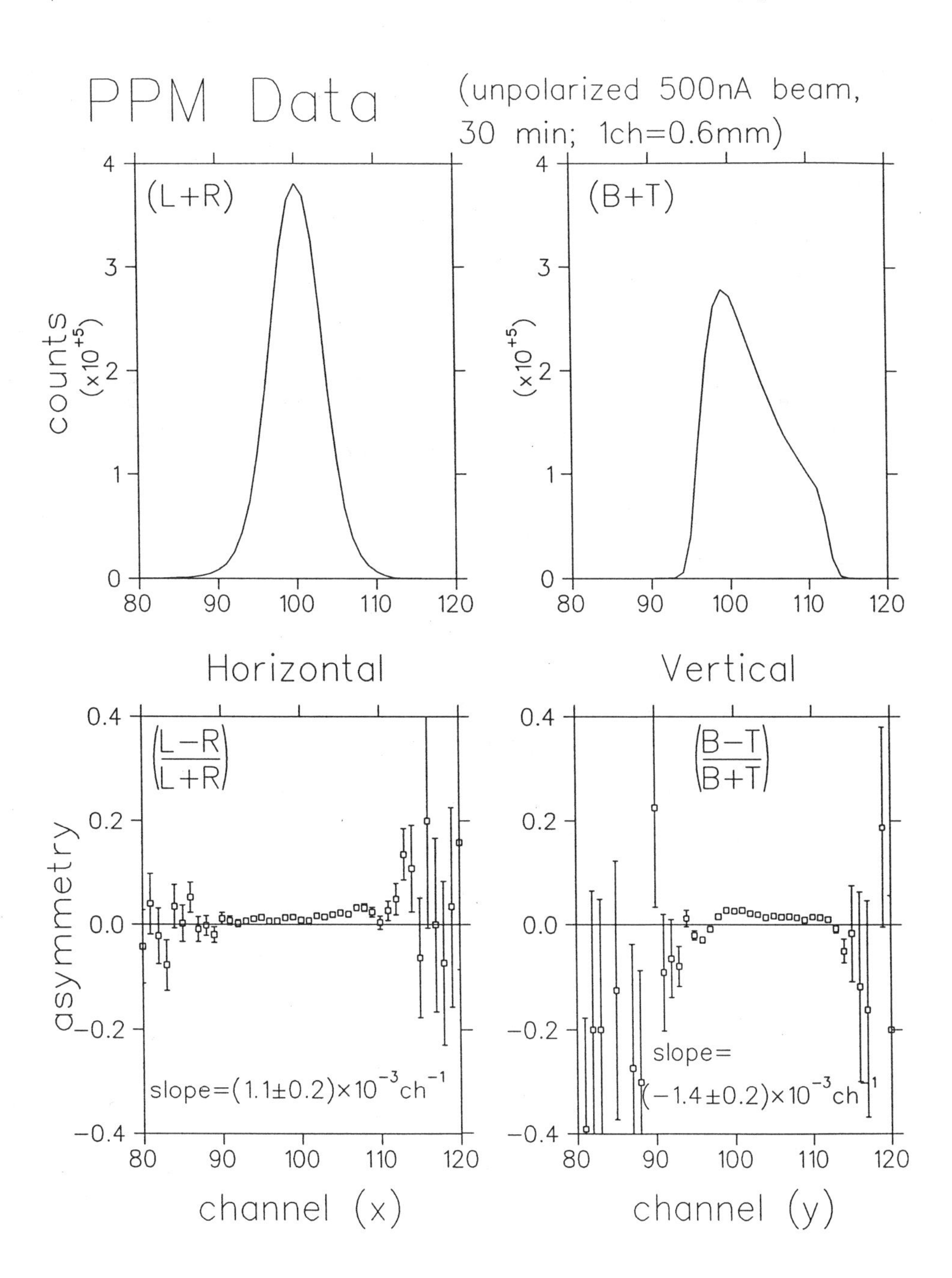

Figure 7. Beam profile and polarization assymetry of the PPM in the horizontal (L,R) and the vertical (B,T) directions.

TEST PERFORMANCE

The systematics of the PPMs is summarized in Figs. 5 and 6. The PPM performance data shown in Figs. 6 and 7 were obtained in preliminary test running using 500 μA of unpolarized 230 MeV proton beam at TRIUMF in March 1994. The data show very small instrumental asymmetry. Figure 7 reflects real characteristics of the used beam tune superimposed onto the small instrumental asymmetry. The tune and stripping foil effect have since been improved.

CONCLUSIONS

Since the time the first PPM prototype was completed, it has undergone an extensive testing and improvement process. A number of additions and changes have been made, such as fitting the vacuum vessel with viewing ports to facilitate visual inspection of the running target, holding the external long light pipes with additional arms to define their position and moving the driving belt mechanism from the inside to the outside of the vacuum vessel. The latest PPM performance data give us confidence that these instruments will support the experiment to the required accuracy.

REFERENCES

1. W. Haeberli et al., Nucl. Instr. and Meth. **163**, 403 (1979).
2. "Considerations for a Scanning Counting Four-branch Profile Polarimeter for Parity", C.A. Davis, TRIUMF, memo, July 24, 1989.
3. Ferrofluidics Corporation, 40 Simon Street, Nashua, NH 03060, U.S.A., (603) 883-9800, Ferrofluidic Feedthrough, Model #SB-375-B-N-147.
4. Prime, Computervision Division, 100 Crosby Drive, Bedford, MA, 01730, U.S.A., (617) 275-1800.
5. Electro Corporation, 1845 - 57th Street, Sarasota, FL 34243, U.S.A., (813) 355-8411, Digital Magnetic Speed Sensor #58413.

The LBL Advanced Light Source (ALS) Transverse Coupled-Bunch Feedback System - Recent Commissioning Results*

W. Barry, J. Byrd, J. Corlett
Center for Beam Physics
Lawrence Berkeley Laboratory
Berkeley, California, 94720

Abstract

The ALS transverse coupled-bunch feedback system is described along with some recent commissioning results. Results presented include transfer function measurements, demonstrations of multi-bunch damping, and demonstrations of simultaneous transverse and longitudinal systems operation.

INTRODUCTION

The LBL Advanced Light Source (ALS) is a third generation 1.5 GeV storage ring for producing synchrotron radiation in the .5 - 10000 eV range [1]. Because of the high average beam current (400 mA) and large number of bunches (328 in buckets separated by 2 nsec), a broad spectrum of transverse coupled-bunch modes can be excited by higher-order cavity resonances and the resistive wall impedance. In order to control growth of this coupled-bunch motion, a 250 MHz bandwidth bunch-by-bunch feedback system has been designed and partially commissioned at ALS.

The feedback system specifications and design are described in references [2] and [3]. In this paper, recent commissioning results are emphasized with only a brief overview of the system presented for orientation.

SYSTEM OVERVIEW

A block diagram of the ALS transverse feedback system is shown in Fig. 1. The system bandwidth is nominally 250 MHz which corresponds to half the maximum bunch rate of 500 MHz. In this case, horizontal and vertical beam moment signals ($I\Delta X$) are detected at the pickups on a bunch-by-bunch basis and used to correct the transverse beam motion. As shown, moment signals from two sets of pickups are summed in proper proportion to obtain a correction signal which is in quadrature with beam position at the kickers. This technique results in optimal damping and is adjustable to allow for arbitrary changes in tune.

The beam moment signals are detected as amplitude modulation of the n=6 (3 GHz) harmonic of the bunch rate in order to exploit the good sensitivity of the button pickups at this frequency. In contrast to the pickups, the stripline kickers operate at baseband (10 kHz - 250 MHz) where their efficiency is greatest. The moment signals are demodulated to baseband with microwave receivers which can be configured

*This work was supported by the Director, Office of Energy Research, Office of Basic Energy Sciences, Materials Sciences Division, of the U.S. Department of Energy under contract No. DE-AC03-76SF00098.

for either heterodyne or homodyne detection. The nominal detection mode is heterodyne. In this case, the moment signals are demodulated with a 3 GHz local oscillator signal which is phase-locked to the storage ring RF. At times, it may be desirable to operate the transverse feedback system in the presence of large synchrotron oscillations, i.e., in the absence of longitudinal damping. In this case, the oscillating arrival time of the bunches with respect to a fixed-phase local oscillator causes reduction and a possible sign change in the average feedback gain [4]. For this operating scenario, homodyne demodulation which employs a local oscillator signal derived from the sum of the four button signals is used.

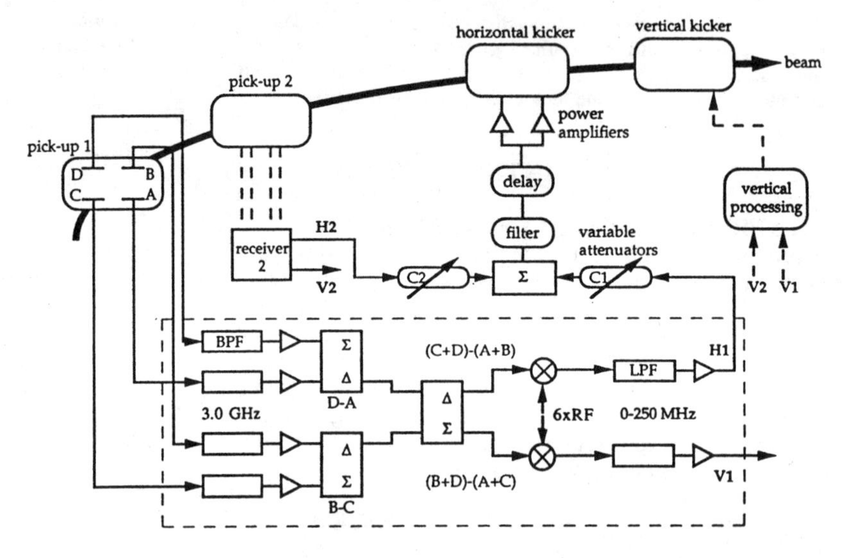

Figure 1. ALS transverse feedback system.

Baseband processing includes proportional mixing of the two pickup signals with attenuators, C1 and C2, and simple two-tap coaxial-cable notch filters for rejecting orbit harmonic signals. In addition, simple coaxial delays are used for timing the kick signals with respect to the bunch arrival times at the kickers. Four 150 W, 10 kHz - 220 MHz, class-A, commercial power amplifiers are used to drive each electrode of each kicker separately (300 W per kicker). The amplifier/kicker combination provides per-turn kicks ranging from 2.3 kV at 100 kHz to 1.6 kV at 220 MHz. At the nominal betatron tunes, these voltages and frequency range are sufficient to control any expected transverse coupled-bunch motion.

INITIAL COMMISSIONING RESULTS

At present, the pickups, kickers, receivers, one of two notch filters, and two of four power amps are installed and operating at ALS. The chassis for mixing the two pickup signals, the remaining notch filter, and the controls interface are under construction and nearing completion. The results reported here were obtained with a partial system consisting of a single pickup and receiver operating in the horizontal or vertical plane. In this case, a single electrode of the horizontal or vertical kicker was driven as part of the feedback system. The other electrodes were used for driving the beam with the tracking generator of a spectrum analyzer to make transfer function measurements.

Setup and timing of the system is typically done in single-bunch mode. In addition, the first transfer function measurements were done with a single bunch in the ring. Figure 2a shows the beam transfer function amplitude for a single 2 mA bunch about a particular horizontal betatron line with and without the feedback system turned on. In this case, the receiver was configured in heterodyne mode. The open-loop gain of the feedback system, given approximately by the ratio of the feedback on/off responses, in this case, is about 20 dB. This is in fact a lower bound on the gain because the line widths are artificially broadened by the resolution bandwidth of the spectrum analyzer and tune jitter. The same measurement was performed with the receiver operating in homodyne mode at a lower gain as shown in Fig. 2b. Note that in both cases, the resonance is shifted upwards in frequency with the feedback on.

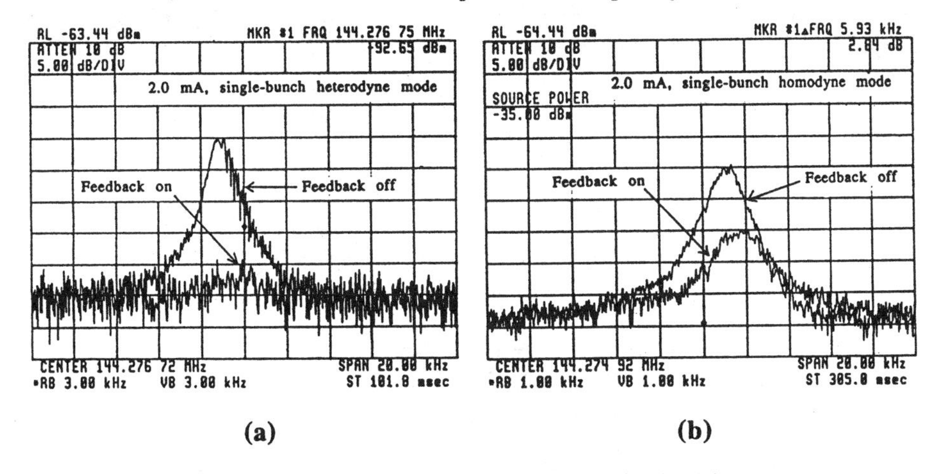

Figure 2. Single-bunch horizontal transfer functions.

This is because the single-pickup system can not meet the kick/position quadrature condition at the kicker. The result is a change in complex resonant frequency which manifests itself as a reduction in Q and a tune shift. The full two-pickup system can be adjusted to utilize the full gain to address only the damping component of the complex resonant frequency for any tune setting. A demonstration of gain control in single-bunch-heterodyne mode appears in Fig. 3a. Finally, a single-bunch demonstration of damping in the vertical plane is shown in Fig. 3b.

An example of multi-bunch damping is shown in Fig. 4. In this case, a 220 mA beam in 82 equally spaced (every fourth) buckets gives rise to significant spontaneous coupled-bunch motion as evidenced by the betatron sideband spectrum with the feedback turned off. With the feedback system on in homodyne mode, the motion is reduced to levels below the noise floor of the spectrum analyzer. The horizontal axes in Fig. 4 are in units of revolution frequency and have a range of 41 which covers one fourth of the 250 MHz baseband and all possible coupled-bunch modes for this fill pattern. Similar results under these conditions were obtained in heterodyne mode. In order to exercise the full bandwidth of the system, the measurement of Fig. 4 was repeated with 320 bunches at 385 mA which is approximately the nominal ALS beam. The results, shown in Fig. 5, indicate complete damping of all but possibly one low-frequency mode. At present, investigations as to whether this line is actually a coupled-bunch mode or a spurious signal are ongoing.

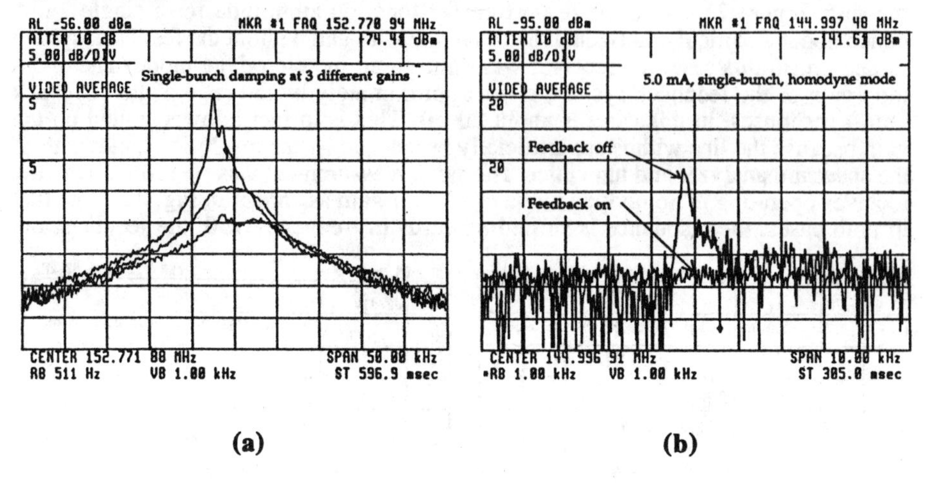

Figure 3. Gain control and vertical transfer function.

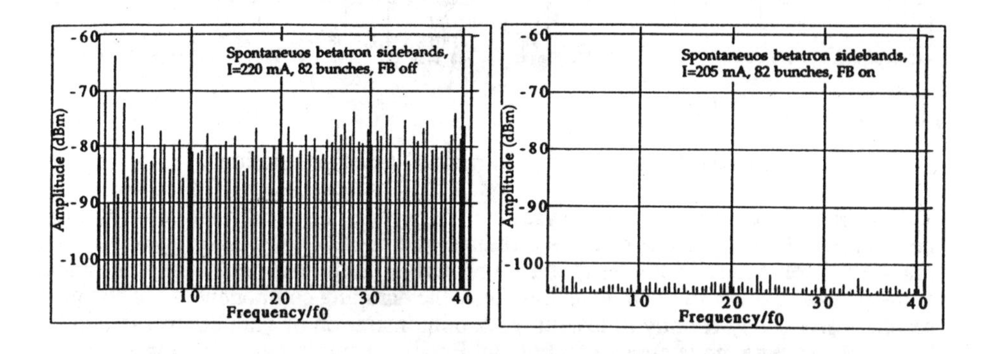

Figure 4. Horizontal multi-bunch damping, 82 bunches, 220 mA.

When transverse coupled-bunch modes are driven by oscillatory impedances such as higher-order modes in the RF cavities, large synchrotron oscillations can have a strong damping effect. Basically, this effect is the same as the gain dilution effect in the heterodyne demodulation technique. That is, the oscillating arrival time of a bunch at the impedance causes the phase of the excitation to be different on every turn resulting in a reduction in the average kick received by the bunch. In addition, with non-zero chromaticity, energy oscillations cause a bunch-by-bunch spread in tune which weakens bunch-to-bunch coupling in the transverse plane. If the synchrotron oscillations are large enough, these effects can cause a significant decrease in the growth rates of transverse modes and possibly prevent growth of modes which would otherwise be unstable. These effects have been clearly seen at ALS during testing of

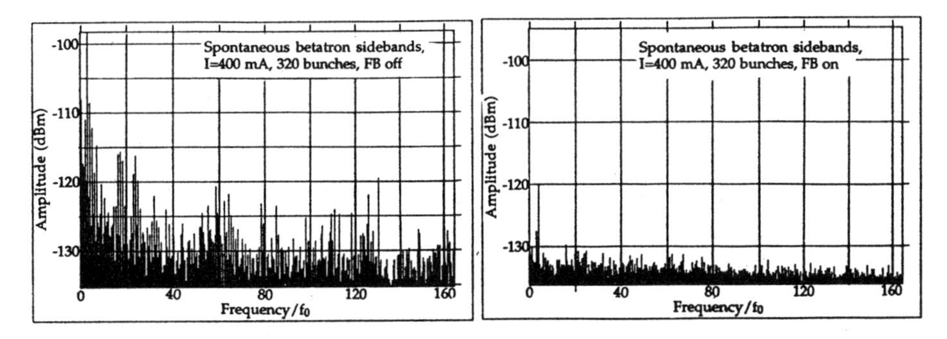

Figure 5. Horizontal multi-bunch damping, 320 bunches, 385 mA.

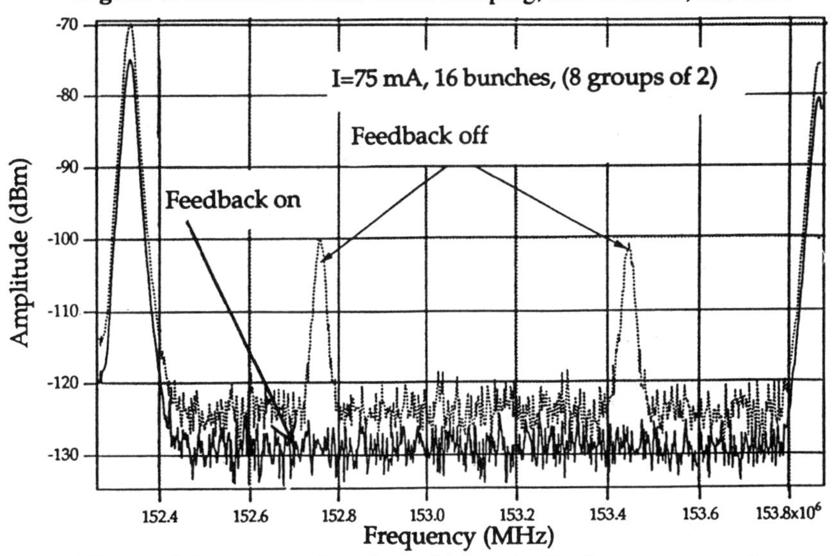

Figure 6. Transverse damping with longitudinal system operating.

the longitudinal feedback system [5]. At moderate currents (> 70 mA), strong betatron sidebands are present when the longitudinal system has damped the longitudinal coupled-bunch motion. In the absence of longitudinal damping, these lines are weak or not present at all. Therefore, an important aspect of the transverse system is it's performance in conjunction with the longitudinal system. Presently, efforts are under way to characterize the simultaneous performance of the longitudinal and transverse systems with the first results shown in Fig. 6. Here, strong spontaneous betatron sidebands which appeared while operating the longitudinal system are completely damped by the transverse system. The fill pattern consisted of 16 bunches in 8 groups of (2 separated by 4 nsec) spaced equally around the ring with an average current of 75 mA.

CONCLUSION

Significant progress in the commissioning of the ALS transverse feedback system has taken place. In particular, damping of transverse oscillations in both planes has been demonstrated with a single pickup system. In addition, testing of the system in conjunction with the longitudinal system has begun with very encouraging initial results. Present efforts are being directed towards the continued testing of the two systems simultaneously and the commissioning of the full two pickup transverse system.

ACKNOWLEDGMENTS

The authors wish to thank the many members of the LBL/SLAC - ALS/PEP-II feedback systems collaboration for many enlightening discussions and useful suggestions.

REFERENCES

1. "1 - 2 GeV Synchrotron Radiation Source", CDR, LBL Pub-5172 Rev., July, 1986.
2. W. Barry, J. Byrd, J. Corlett, G.R. Lambertson, and C.C. Lo, "Transverse Coupled-Bunch Feedback in The Advanced Light Source (ALS)", Proceedings of the 1994 European Particle Accelerator Conference, London, England, June 27 - July 1, 1994.
3. W. Barry, J. Byrd, J. Corlett, J. Hinkson, J. Johnson, and G. Lambertson, "Design of The ALS Transverse Coupled-Bunch Feedback System", Proceedings of the 1993 IEEE Particle Accelerator Conference, Washington, D.C., May 1993.
4 W. Barry, "Stability of a Heterodyne-Based Transverse Feedback System in the Presence of Synchrotron Oscillations", CBP Tech Note-58, LBL, July, 1994.
5. J. Fox, H. Hindi, I. Linscott, J. Olsen, G. Oxoby, L. Sapozhnikov, D. Teytelman, A. Drago, M. Serio, J. Byrd, and J. Corlett, "Operation and Performance of a Longitudinal Damping System Using Digital Signal Processing", Proceedings of the 1994 Beam Instrumentation Workshop, Vancouver, B.C., October 1994 (to be published).

Operation and Performance of a Longitudinal Feedback System Using Digital Signal Processing*

D. Teytelman, J. Fox, H. Hindi, J. Hoeflich, I. Linscott,
J. Olsen, G. Oxoby, L. Sapozhnikov
Stanford Linear Accelerator Center, Stanford University, Stanford, CA 94309
A. Drago, M. Serio
INFN Laboratori Nazionale, Frascati, Italy
W. Barry, J. Byrd, J. Corlett
Lawrence Berkeley Laboratory, 1 Cyclotron Road, Berkeley, CA 94720

Abstract

A programmable longitudinal feedback system using a parallel array of AT&T 1610 digital signal processors has been developed as a component of the PEP-II R&D program. This system has been installed at the Advanced Light Source (LBL) and implements full speed bunch by bunch signal processing for storage rings with bunch spacing of 4ns. Open and closed loop results showing the action of the feedback system are presented, and the system is shown to damp coupled-bunch instabilities in the ALS. A unified PC-based software environment for the feedback system operation is also described.

INTRODUCTION

The PEP-II machine will require feedback to control multibunch instabilities [1]. A longitudinal feedback system prototype has been installed and tested at the Advanced Light Source at the Lawrence Berkeley Laboratory. This system uses a bunch by bunch processing scheme and employs digital signal processing to calculate a correction signal for each bunch. As shown in Fig. 1, signals from four button-type pickups are combined and fed to the stripline comb generator. The generator produces an eight cycle burst at the sixth harmonic of the ring RF frequency (2998 MHz). The resultant signal is phase detected, then digitized at the bunch crossing rate. The detector is designed to have 400MHz bandwidth which allows measurement of the each bunch's synchrotron motion independently for the 4 ns bunch spacing. A correction signal for each bunch is computed by a digital signal processing module and applied to the beam through a fast D/A, an output modulator, a power amplifier and a kicker structure [2, 3].

The signal processing is implemented by four AT&T 1610 processors operating in parallel. These 16 bit processors are equipped with 16K of dual port memory on-chip and allow 25 ns instruction cycle time for cached instructions [4]. The feedback algorithm is downloaded to the prototype through JTAG interface from a IBM PC-compatible computer. This approach allows the user to quickly alter the feedback parameters as well as run a multitude of diagnostics, signal recorders and

*Work supported by the U.S. Department of Energy contract DE-AC03-76SF00515

other programs.

As the synchrotron oscillation frequency in the ALS (10 kHz) is much less than the revolution frequency (1.5 MHz) the processing is implemented as a downsampled system, in which a correction signal for each bunch is computed once every n revolutions, where n is a downsampling factor. The four-processor prototype system allows control of up to 84 bunches when using a six-tap FIR filter algorithm. The maximum number of bunches varies depending on the filter processing time

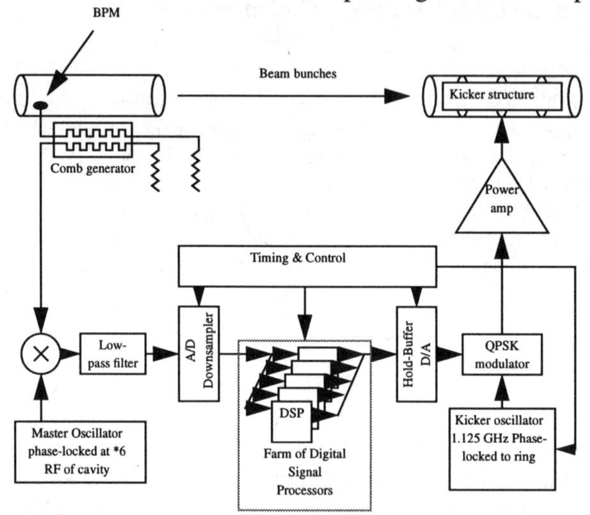

Figure 1. Block diagram of the longitudinal feedback system.

and fill pattern [5]. Design of the signal processing hardware and the front-end electronics has been addressed in the earlier publications [6]. Two important system components; the QPSK (quad phase shift keyed) modulator and the support software have not been described previously and are presented in the following sections.

QPSK MODULATOR OPERATION

The QPSK modulator function is implemented in the back-end signal processing and is used to transfer the baseband computed correction signal into a modulation on a kicker oscillator signal. The need for such a a modulator arises from the design of the kicker structure, which produces a maximum in longitudinal impedance at 1125 MHz, or 2.25 times the ring RF frequency (this choice minimizes the impedance presented at the bunch crossing frequency and higher harmonics) [7]. The QPSK modulator is implemented using a 2 GHz bandwidth gilbert multiplier, 500 MHz ECL counter circuitry, and a passive 90 degree hybrid. The QPSK circuit acts to shift the phase of an 1125 MHz carrier by -90 degrees every 2 ns to align the kick phase for the next bucket. Figure 2 shows the QPSK modulated carrier waveform as well as unmodulated 1125 MHz signal while Fig. 3 illustrates the resultant carrier spectrum. Most of the power is at 1 GHz with a strong component at 1.25

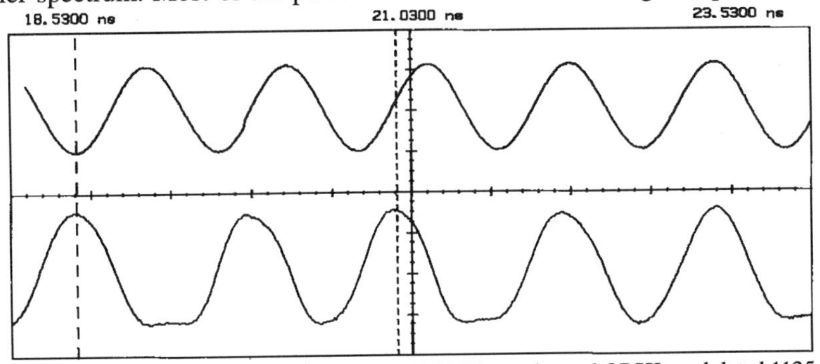

Figure 2. Oscilloscope photograph of the 1125 MHz carrier and QPSK modulated 1125 MHz carrier. The two cursors are 2ns apart to show the spacing of two adjacent bunches at 500 MHz.

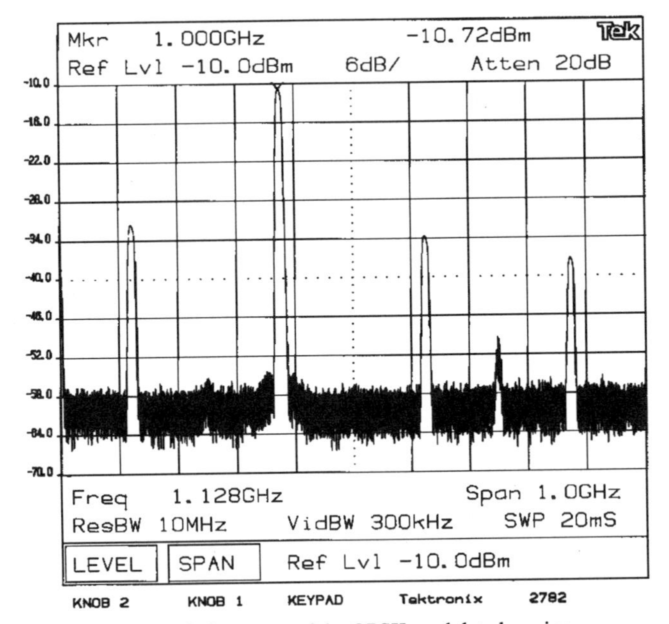

Figure 3. Spectrum of the QPSK modulated carrier.

GHz. Other spectral lines such as 0.75 GHz and 1.5 GHz fall outside the kicker bandwidth and do not affect the beam. This QPSK modulated 1125MHz signal, if applied to the beam, would produce a DC correction signal - the final function in the QPSK modulator is an amplitude modulator which multiplies the QPSK'ed signal by the baseband correction signal from the output D/A. This modulation adjusts the magnitude of the kicker drive signal every 4 ns to provide bunch by bunch correction signals (negative kicks require phase inversion of the kicker signal). The resulting output spectrum for multi-bunch operation fills in the 250 MHz bandwidth between 1000 and 1250 MHz and covers all coupled-bunch modes in the storage ring. The circuitry as implemented has a 48 dB dynamic range and can be operated at any RF/4 ring harmonic up to 2 GHz with the full 500 MHz QPSK modulation rate.

UNIFIED SOFTWARE ENVIRONMENT

During the quick prototype development as the number and the sophistication of the DSP programs grew, management of the many configurations and feedback filter programs became a serious concern.

To coordinate the development of various operational programs and accelerator diagnostics a unified software environment has been created. This environment uses a text-based parameter file to specify the operational modes of the quick prototype system. All of the variables for a given experimental configuration, such as the machine revolution time, synchrotron frequency, filter gain, filter phase, etc. are contained in the parameter file. The file is read in turn by a number of relatively simple C programs which generate binary tables for downloading into the DSP memory and include files for the assembly language DSP code. The DSP code and tables are downloaded through the JTAG interface using the AT&T DSP1610 development system. All of these activities are coordinated by the UNIX make program. Using file timestamps and dependencies defined in a makefile make program ensures that tables and code downloaded to the DSP correspond to the variables in the parameter file.

SYSTEM TESTS AT ALS

The prototype system including a high-gain longitudinal kicker has been installed at the ALS and is being used to gain operational experience and to verify the system design for the PEP-II system. Figure 4 shows the longitudinal transfer function of a single bunch measured with no feedback, positive feedback, and negative feedback with two different loop gains. The action of the feedback system is seen in the higher or lower Q of the synchrotron resonance for positive or negative feedback respectively [8]. The graph shows that for a gain change of 8 (18dB) we get a change in damping of about 15dB.

Presently the ALS kicker is driven by a 10W power amplifier. This power limits the total current which can be controlled. It is interesting to note that relatively high ring currents (up to 125mA) can be controlled with relatively low voltage correction kick as long as the feedback system is turned on during injection, and the injection process injects only a single bunch at a time. This injection method allows the feedback system to damp the excitations caused by the injected bunch in the existing stored beam, and damp the resulting motion before the next injection cycle. If the feedback system is turned off for any substantial current (above 10mA) the bunch motion becomes very large (greater than 10 degrees at the 500 MHz RF frequency) and turning the feedback system back on does not control the synchrotron motion. This happens because the feedback system saturates and cannot generate enough voltage to control the large amplitude motion once it grows from the quiescent state. Figure 5 shows the bunch spectrum obtained from a BPM for 8 groups of 2 bunches equally spaced around the ring at 100mA. Data shows that the longitudi-

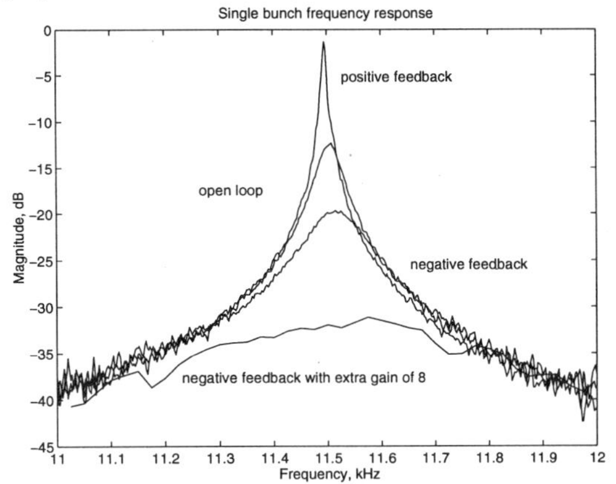

Figure 4. Single bunch transfer functions measured at the ALS. The open loop synchrotron resonance at 11.5 kHz can be damped or excited via negative or positive feedback respectively.

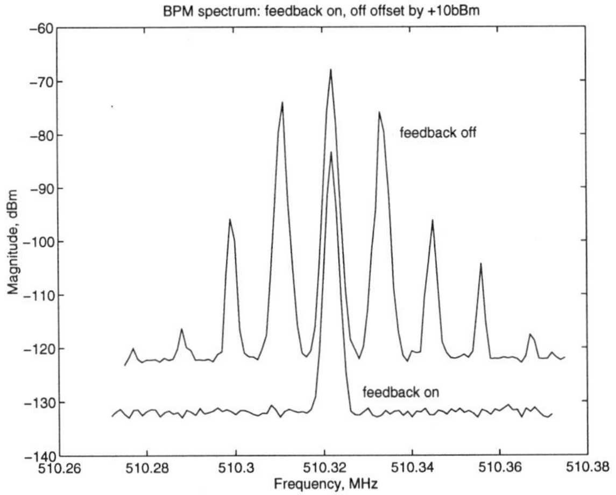

Figure 5. Spectrum of a BPM signal.

nal feedback suppresses the synchrotron oscillations (manifested as 10 kHz sidebands) from -73dBm to the noise floor of spectrum analyzer, i.e. suppression of 50dB.

Since ALS is a light source machine is it possible to utilize optical diagnostics to investigate the performance of the longitudinal feedback system. Experiments have been conducted at the ALS to measure the optical spectrum with and without feedback. Figure 6 presents an undulator spectrum taken at a 108mA ring current with 84 bunch fill pattern. The feedback system increases the optical intensity by a factor of 2.5 and narrows the peak width to almost 1/4 of that of the undamped system. For synchrotron light users who are conducting narrowband spectroscopic measurements such an improvement in machine performance is very desirable. It is interesting to speculate on the bimodal structure visible in the "feedback off" spectrum which appears to be due to the coherent dipole mode longitudinal oscillations.

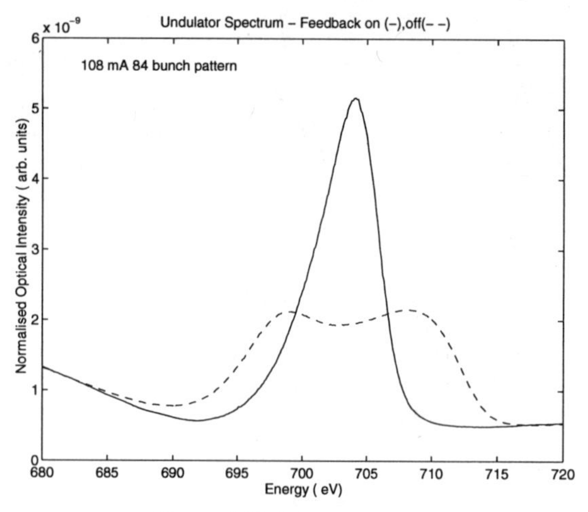

Figure 6. Undulator spectrum.

SUMMARY

The longitudinal bunch-by-bunch feedback system quick prototype is operated at the ALS at Lawrence Berkeley Laboratory. It includes all of the subsystems required for the PEP-II machine. The quick prototype system is used for algorithm development and various accelerator measurements. Closed-loop feedback operation has been demonstrated and longitudinal instabilities have been controlled for an 84 bunch fill pattern with 125mA ring current. We expect to be able to damp longitudinal motion at the 400mA design current when the high-power output amplifier is installed. The information gained from the quick prototype system has been incorporated in the PEP-II system design [9]. A complete PEP-II prototype for ALS operations is in construction and should be installed and commissioned at the ALS in early 1995.

ACKNOWLEDGMENTS

The authors thank the ALS staff of LBL for their hospitality and interest in this hardware development program and the SLAC PEP-II Group and Technical division for their support. The authors particularly appreciate the help of Tony Warwick for the optical spectrum measurement.

REFERENCES

1. "PEP-II, An Asymmetric B Factory - Design Update," Conceptual Design Report Update, SLAC, 1992.
2. Pedersen, "Multi-bunch Feedback - Transverse, Longitudinal and RF Cavity Feedback," Proceedings of the 1992 Factories with e+/e- Rings Workshop, Benalmadena, Spain, November 1992.
3. Fox, et al., "Operation and Performance of a Longitudinal Damping System Using Parallel Digital Signal Processing," Proceedings of the 1994 European Particle Accelerator Conference, London, England.
4. "WE DSP1610 Digital Signal Processor Information Manual," AT&T Microelectronics Corporation, Allentown PA.
5. Hindi et al., "Down-Sampled Bunch by Bunch Feedback for PEP-II," B Factories: The State of Art in Accelerators, Detectors, and Physics, SLAC Report 400, p. 216.
6. Sapozhnikov, et al., "A Longitudinal Multi-Bunch Feedback System Using Parallel Digital Signal Processing," Proceeding of the 1993 Beam Instrumentation Workshop, Santa Fe, NM, AIP Conference Proceedings 319.
7. Corlett, et al., "Longitudinal and Transverse Feedback Kickers for the ALS," Proceedings of the 1994 European Particle Accelerator Conference, London, England.
8. Hindi, et al., "Measurement of Multi-Bunch Transfer Functions Using Time-Domain Data and Fourier Analysis," Proceedings of the 1993 Beam Instrumentation Workshop, Santa Fe, NM, AIP Conference Proceedings 319.
9. Oxoby, et al., "Bunch by Bunch Longitudinal Feedback System for PEP-II," Proceedings of the 1994 European Particle Accelerator Conference, London, England.

Design of the APS Transverse and Longitudinal Damping System*

W. Sellyey, D. Barr, E. Kahana, and A. Votaw
Advanced Photon Source, Argonne National Laboratory, Argonne, IL 60439

INTRODUCTION

The main sources of instabilities in the Advanced Photon Source (APS) storage ring are expected to be higher-order modes (HOMs) of the accelerating cavities and the resistive wall impedance of the small insertion devices beam tubes (12 mm vertically). Extensive efforts are being made to reduce the Qs of HOMs.(1) The maximum operating current of the ring will be 300 mA. At this current, analysis of measurements on cavity prototypes (2,3) shows that the transverse growth rates will be less than 500/sec above radiation damping. The longitudinal growth rate due to HOMs is predicted to never exceed the radiation damping of 213/sec. The largest transverse resistive wall growth rate is calculated to be 2720/sec when 54 evenly spaced rigid bunches are used to produce 300 mA.(4) There will be 26 additional unstable modes. The sum of these growth rates is 17,163/sec. Thus, it is clear that an effective transverse damping system will be needed and that the strength of this damper will be dominated by the resistive wall modes. A longitudinal damper system will also be built. This will provide damping about 2/3 times that due to synchrotron radiation.

The most serious disturbances which can initiate instabilities will take place at injection. Typically, each bunch in the ring will be accumulated by injecting 1/5 of the final charge five times. A standard mode of operation is used in this paper in which there will be 54 evenly spaced bunches around the ring. During the ring filling process, the highest growth rates will occur when the last fifth of a bunch is injected into the last bunch. The largest expected vertical excursion of 1/5 of a bunch is about 5 mm. Anything larger will cause the bunch to scrape in the insertion device sections. Thus, the average excursion of the last bunch will be 1 mm. For longitudinal motion, it is expected that the maximum fractional energy excursion will be 1% for the injected bunchlet. Thus, the largest average excursion of the last bunch will be 0.2% of 7 GeV. The harmonic number of the APS storage ring is 1296. Only a small fraction of the 1296 buckets will ever be filled. To help alleviate bandwidth-related problems in the damper system, bunch spacing will never be more frequent than every second bucket. Since the bucket frequency is 352 MHz, bunch repetition rates are limited to 176 MHz, and amplifier bandwidths are kept to 88 MHz.

---

*Work supported by the U.S. Department of Energy, Office of Basic Energy Sciences, under contract no. W-31-109-ENG-38.

SYSTEM OVERVIEW

Figure 1 shows an overview of the damper systems. Three stripline units will be used: one as a pickup, one as a transverse damper, and one as a longitudinal damper. They will each contain four strips with two strips located symmetrically about the vertical center plane at the top of an elliptical beam tube and two located at the bottom.

Each pickup stripline signal will be split into two parts by a diplexer. Power below 800 MHz will pass to the transverse hybrid/pulse shaper section while power above 800 MHz will pass to the phase detector system. A set of low pass Gaussian-like filters and wide-band hybrids will be used to produce Δx, Δy, and Σ intensity signals for each bunch. These will pass through a gain control and limiting section. The primary function of the Σ signal will be to generate a timing signal. All three signals will be digitized up to a 176-MHz rate by 8-bit ADCs and all will be occasionally sampled by a relatively slow VME system. The Δx and Δy signals will be stored in a fast, dual-ported memory that will be able to store up to 375 turns of data in five groups. One port of the memory will be used to write data into memory from the ADCs and to pass data to the next group. The other port will be used to pass data to a transversal filter (DSP). At any one time, each group will pass the data from the same bunch from earlier turns to the filter. The filter will be a combined digital-analog system to produce the sum of five products of weights times position signals. The resulting analog signals will be used to generate rf drive signals for the 150-W power amplifiers. These will drive the transverse damper stripline.

The high frequency output of the diplexer will be combined by a power splitter, and then split to supply signals to a 3rd and 7th harmonic bunch phase measurement systems. The reason for two different systems is that the range of phase variation of the injected bunch will exceed 360° in the 7th harmonic system; however, the 7th harmonic system will be useful for keeping the phase noise low during normal operation. Thus, the 3rd harmonic will be used at injection, and the 7th harmonic will take over after the injection transients have been damped.

Each system will contain a Gaussian-like filter centered at the harmonic. The bandwidth of the 3rd harmonic filter will be 524 MHz, and that of the 7th harmonic will be 352 MHz. Harmonic multiplier outputs of the accelerating cavity 352 MHz will be mixed with the filter outputs. After low-pass filtering, gain and limiting sections, the signal is treated in the same way as in the case of the transverse processing. It passes through memory, filter, rf generator, and to the four power amplifiers and strips of the longitudinal damper.

A very important part of the system will be a relatively slow feedback system which will keep the average Δx, Δy, and $\Delta\Phi$ signals near zero. This is necessary to prevent the damping system from introducing unwanted motion and thus increasing apparent beam size and emittance. To accomplish this, a processor in a VME crate will read position and phase data and generate corrections continuously. In the transverse case, electronic attenuators in the signals coming from the four strips will be adjusted. For the phase detectors, the reference phase will be adjusted by electronic delays.

STRIPLINES

Figure 2 shows one quadrant of a stripline unit cross section. This is what will be used for the pickup and longitudinal kickers. Each strip will be 50 ohms when all four strips carry identical in-phase signals. The transverse kicker cross section is similar except it is optimized for vertical kicking. The impedance of each strip is 50 ohms when the four strips carry equal amplitude signals, but the top set is 180° out of phase with the bottom set.

The pickup stripline will be placed in the diagnostics straight section. The strips will be 8.4 in long, 1/4 wavelength at 352 MHz. The kickers will be put in a 10.81-in space located just before the second dipole in each sector originally intended for expansion bellows. Now two of these locations will accommodate this function and l = 6.5-in long strips.

The longitudinal kicker constant (5) as applied to our case is

$$K_{\parallel} = \delta E/eV_k = 2v_0 \sin(\omega l/c), \tag{1}$$

where δE is the change in beam energy, V_k is the amplitude of the kicker voltage, and ω refers to the driving frequency. v_0 is the voltage at the center of the stripline unit when 1 volt is put on each stripline electrode, and the beam tube body of the unit is grounded. Using 352 MHz, $K_{\parallel} = 1.30$.

For the transverse case, the magnitude of the kicker constant as applied to our case is

$$K_{\perp} = \Delta pc/eV_k = 2E_y l \sin(\omega l/c)/\omega l/c). \tag{2}$$

Here Δp is the change in transverse momentum when 180° out-of-phase signals of amplitude V_k are applied to opposing pairs of strips. $E_y = 39.5$ volts/m is the electric field at the center of the stripline in the transverse kicker geometry when +1 volt is put on each of the top electrodes, -1 volt on the bottom, and the body of the unit is grounded. Most of the damping power will be used to control resistive wall instabilities. The bandwidth needed for this is heavily dependent on the bunch pattern. In the case of evenly spaced bunches, the driving frequency will be restricted to about 27 times the revolution frequency. This is 7.3 MHz, resulting in $K_{\perp} = 13$. At 88 MHz, $K_{\perp} = 12.8$. In the x-direction, $K_{\perp} = 12$ at 7.3 MHz.

POWER AMPLIFIER REQUIREMENTS

To calculate the required amplifier power, three things need to be known. For the transverse case, these are the maximum transverse velocity at the kicker location, the growth rates of the unstable modes, and the transverse kicker constant. For the longitudinal case, these are the maximum fractional energy deviation, the growth rates, and longitudinal kicker constant.

For the transverse case, the value of the vertical ß function in the insertion devices is about 10, and at the kicker location it is 13 m. This gives a velocity amplitude of v = 26310 m/sec at the kicker for a 1-mm vertical displacement amplitude inside the insertion device. We will make an estimate of how much kick, Δv, will be needed to cancel the resistive wall growth effect. The injected particle will participate in all modes equally. Of these, 27 modes will have an average growth rate $1/T_G$ = 635.7/sec. Only half of the transverse velocity will participate in growth. The other half participate in naturally damped modes. Using $\Delta v/v = T_0/T_G$, where T_0 = 3.68 μs is the revolution frequency, one gets Δv = 30.8 m/sec.

In order to put the above estimate on a solid quantitative footing, the damping process was simulated. The coordinates of each particle were expressed in terms of the 54 normal modes with resistive wall growth and decay constants. The amplitudes of the normal modes were calculated in terms of the initial coordinates and velocities of each particle. One fifty-fourth of a turn later, the coordinates and velocities of all particles were calculated. The velocity of the particle at the damper was reduced by some maximum velocity Δv_m or by a velocity less than Δv_m such that its transverse velocity went to zero. Using the resulting new set of velocities and coordinates, the normal mode amplitudes are recalculated and a new set of velocities and coordinates is calculated, 1/54 of a turn later. The next particle is then kicked, etc. The process is typically continued for 1500 to 3000 turns until it is obvious whether the injection disturbance will be damped or not. In this way it was found that Δv_m = 33 m/sec is just sufficient to damp the beam with a 5-mm injection amplitude for the injected bunchlet. Using $\Delta p = \gamma m_0 \Delta v_m$ and γ = 13697 for APS in Eq. (2), one gets V_K = 59 volts. This gives a power of 35 watts into a 50-Ω stripline. Four 150-watt amplifiers are being procured to have a factor of two safety factor. This safety factor goes to four if the charge of each bunchlet is reduced by a factor of two near the end of the APS fill cycle.

For the longitudinal case, there is a zero predicted growth rate. To be on the safe side, four 250-W amplifiers will be used to drive one strip each of a stripline unit to provide damping. We will calculate how much damping these can provide. Using the kicker constant $K_{\parallel}$ = 1.3 in Eq. 1, one gets δE = 205.5 e-volts. Using the 0.2% energy excursion amplitude at injection, one gets ΔE_0 = 14 MeV for the energy oscillation amplitude. Assuming that only one growth mode is present, only one 1/54 of this needs to be damped, or ΔE = 259 KeV. Using $(\delta E/\Delta E)$ = (T_0/T_g) one gets $1/T_g$ = 215.4/sec. A simulation gives $1/T_g$ = 134/sec. The synchrotron radiation damping rate $1/T_s$ = 213/sec, and this is just enough to cancel the predicted growth rate due to cavity HOMs. Therefore, the 250-W amplifiers will provide damping 2/3 above $1/T_s$.

The bandwidth of the amplifiers is determined by the 176-MHz bunch rate. In the worst case, adjacent bunches are 180° out of phase with respect to each other as they pass a kicker. Thus, the drive signal will need to change phase also. This defines a half period of 1/176 MHz. The frequency is thus 176/2 = 88 MHz. It is better to use a larger bandwidth. Simulations indicate that a 132-MHz, 3-dB bandwidth with a Gaussian-type response can result in only a few percent variation in kick amplitudes for a constant input amplitude. For the longitudinal case, a 35-kHz to 150-MHz constant delay and gain amplifier will be purchased. A filter on the input will be used to achieve an overall Gaussian response. The 35 kHz is picked to accommodate the horizontal fractional tune of 0.22 (the revolution frequency is 271.6 kHz). For the longitudinal, the operating frequency will be 352 MHz, and the amplifier will be of constant delay and gain from 200 to 500 MHz. Again, a Gaussian filter will be used. If it turns out to be practical, delay and gain errors as a function of frequency in the amplifiers will also be compensated for by filters.

TRANSVERSAL FILTER

The main goal of the digital signal processing (DSP) is to develop a transversal filter to process the incoming data. The filter should be adaptive in order to deal with changes in the beam. Specifically tune shifts could warrant an update to the filter. The main goal of the filter is to provide the proper phase and amplitude shift to the incoming signal that will produce the desired output for the kicker. Any DC offset must also be minimized. The input signal is of the form

$$x_i[n] = D_i \cdot \cos(2\pi n\nu + \phi_i) + E_i, \tag{3}$$

where i is the bunch number, n is the turn number, ν is the fractional tune, ϕ is the reference phase, and D and E are constants. It is desired for the filter to produce some output $x'_i[n]$ such that

$$x'_i[n] = D_i \cdot F \cdot \sin(2\pi n\nu + \phi_i + \Delta\phi). \tag{4}$$

F is known *a priori* and so is $\Delta\phi$. The value of $\Delta\phi$ is related to the change in the betatron phase from the pickup to the kicker. It is possible to synthesize $x'_i[n]$ independently from $x_i[n]$ if ϕ_i is known accurately enough. This would allow for perfect DC offset cancellation (E_i in Eq. 3). The problem is that D_i in Eq. 3 is not known and must be derived from multiple measurements of $x_i[n]$. At least two measurements would be required and probably more would be used in practice. Unfortunately, it will take too long to solve for D_i and synthesize $x'_i[n]$. This leaves some sort of real-time filtration of $x_i[n]$ to produce $x'_i[n]$.

The transversal filter will take the form

$$x'_i[n] = \sum_{j=0}^{N-1} b_j x_i[n-j], \tag{5}$$

where N is the number of filter weights. The filter operates on data from past turns as well as the present turn (assuming N>1). Each bunch in the ring will have to be dealt with separately, but will use the same filter. The value for N has not yet been chosen, but it will be no greater than five.

The goal now is to design the filter weights (the b_js in Eq. 5) in order to model Eq. 2. The filter must effectively implement the gain, the phase shift (this also includes transforming the cosine to a sine), and the DC offset rejection. Some sort of bandpass filter with a given phase shift will be required. A simple form of this is just a sine wave in the filter coefficients. Take the weights to be

$$b_j = K \cdot \sin(2\pi j\nu + \theta). \tag{6}$$

Using Eqs. 3 and 6, Eq. 5 becomes

$$x'_i[n] = D_i \cdot K \sum_{j=0}^{N-1} \sin(2\pi j\nu + \theta)\ [\cos(2\pi(n-j)\nu + \phi_i) + E_i]. \tag{7}$$

If Eq. 7 is simulated using a digital computer, it is seen that the chosen filter effectively accomplishes the desired goal. We are currently seeking effective techniques to determine K and θ in Eq. 6. These values will be calculated offline and updated periodically to keep up with changing accelerator parameters.

ELECTRONICS DETAILS

A simulation of the signal from one stripline passing through 30 feet of cable and a 264-MHz low-pass Gaussian filter produces a peak voltage of 65.8 volts. In the transverse pulse shaper hybrid system, diplexers will protect the subsequent electronics from excessive power. Electronic attenuators will be used to keep the average Δx and Δy signals at zero. A set of wide-band hybrids will be used to produce the Δx, Δy, and sum signals. These will then be shaped into a positive and negative lobed signal by a 264-MHz Gaussian low-pass filter. The peak sum signal output will be about 84 volts. The Δx and Δy signals will be 4.23 volts for 1 mm displacement.

It is assumed that the analog-to-digital converter (A/D) is 256 bit with a range of ± 1 volt. A limiter/amplifier with output range of about ± 1 volt will be used. With a gain of 3.7, the digital granularity would be 2 bits per micron. A programmable attenuator would allow this to vary to 1 bit per 16 microns. Assuming an overall noise figure of 7, the FWHM noise will be 1/3 bit when the granularity is 2 bits per micron. Thus, beam motion could be kept to about 0.5 microns.

The sum signal will be split in two. One leg will be attenuated by about 50 dB, limited, and digitized in the same way as the Δ signals. The other leg will be used to generate precise timing signals for the A/D converters.

For the phase detector system, the four stripline signals are combined, attenuated, and split. One part is passed through a 2.464-GHz band pass filter of 352 MHz bandwidth. This will be mixed with the seventh multiple of accelerating frequency. An electronically controlled phase shifter will keep the average of the mixer output at zero. After filtering and limiting, the phase information is digitized. With a gain of 30, the sensitivity will be 0.3 degrees per bit at 2.464 GHz. This is a resolution of 0.34 ps. Noise FWHM will be one bit. Since the bunch lengths are of the order of 60 ps, this will be more than adequate.

When a 1/5 bunch gets injected with no charge as yet present in the bucket, the phase at 2.464 GHz will vary over a range of ± 225°. The phase detector output will then yield incorrect polarity information. This may be remedied by using a second, 3rd harmonic system. Its output would range over ± 96°, and thus no incorrect polarity information would result.

Figure 3 represents a functional description of the memory. It is shown in a configuration which would typically be used for damping synchrotron oscillations. There are five 49152-byte, dual-port memories (ECL). Each memory can store up to 75 turns of information containing 648 buckets. As shown, 28 turns are stored. As each new bucket of information is digitized by the A/D, the address to which it is to be written in the first memory block is read using one port on all blocks of memory. These are stored in registers for all but the last memory block. The A/D information is also written into a register. Then all five registers are written into the next memory group at the memory address which was just read. For the next bucket, the next memory location is treated the same way, etc. After 28 turns (in this case) the process is restarted at location 1 in each of the five memory groups. The second port is used to read out information from the same location within a memory section, and therefore for the same bunch for five earlier turns. The number of turns of delay will be determined by the number of turns stored per memory section. The shortest possible delay will be one turn, and this is usually what will be used for transverse motion.

Most of the 648 buckets will be empty. For the longitudinal case, these will be filled with numbers obtained from memory. The numbers will be chosen so that full rf drive is supplied to the class AB amplifiers. This is necessary because the amplifiers derive their bias from the drive signal. It takes about one microsecond for this drive to build up or decay. Thus, during the relatively brief times when buckets are filled, full rf drive will be available, even if some of the bunches call for near zero drive.

The numbers read from memory will be latched into a buffer whenever a bucket is filled, or when a number has been written into memory for AB operation. The latch output goes to the digital multiplier. This will be a look-up table. Each table will consist of a 256 x 12-bit memory. Each location will contain the product of a weight times the 8-bit bunch information number equal to its address. The multiplier outputs will go to 12-bit digital-to-analog converters. The resulting five outputs will be added by an analog adder and used to generate drive signals for the power amplifiers. To change a filter weight, the RAM is reloaded with new numbers. The memory multipliers and adder constitute the transversal filter.

The analog adder outputs will be limited such that the power amplifiers never saturate, i.e., they will never exceed the power for which they are rated. In the longitudinal case, this signal will be the input to an as yet undesigned device. This device will put out a 352-MHz signal proportional to the input. The signal phase will reverse when the polarity reverses.

For x and y, the transversal filter outputs will be combined to generate the amplifier drives. Both x and y signals will enter 0/180° hybrids. The outputs of the hybrids will be split and added by op-amp adders to produce the final amplifier drive signals.

Extensive damping simulations were done which were essentially the same as what was described earlier but took noise, signal quantization, signal limiting, and the transversal filter into account. In one simulation of vertical damping, the maximum drive power was limited to 50 watts. The quantization was 2 bits per micron, and noise which varied randomly over 1 bit was added. The beam damped down into a range of ± 0.25 microns. Then the simulation was repeated without noise. The beam damped down as rapidly, but the final beam noise was twice as great as when noise was put into the feedback circuits. This is because, without noise, the ADC resolution is limited to 0.5 μm, but with noise, the effective resolution improves, on average.

REFERENCES

1. L. Emery, "Coupled Bunch Instabilities in the APS Storage Ring," Proceedings of the 1991 Particle Accelerator Conference (IEEE, San Francisco, CA 1991), p. 1713-1715.
2. J. Song, Y.W. Kang, R. Kustom, "HOM Test of the Storage Ring Single-Cell Cavity with a 20-MeV e- Beam for the Advanced Photon Source," Proceedings of the 1993 Particle Accelerator Conference (IEEE, Washington, DC, 1993), p. 1057-1059.
3. L. Emery, "Required Cavity HOM DeQing Calculated from Probability Estimates of Coupled Bunch Instabilities in the APS Ring," Proceedings of the 1993 Particle Accelerator Conference (IEEE, Washington, DC, 1993), p. 3360-3362.
4. Yong-Chul Chae, APS, Argonne National Laboratory, private communication.
5. G. Lambertson, "Dynamic Devices, Pickups, and Kickers," AIP Conference Proceedings 153, Physics of Particle Accelerators, SLAC, 1985, FNAL 1984, V2, p. 1413-1442.

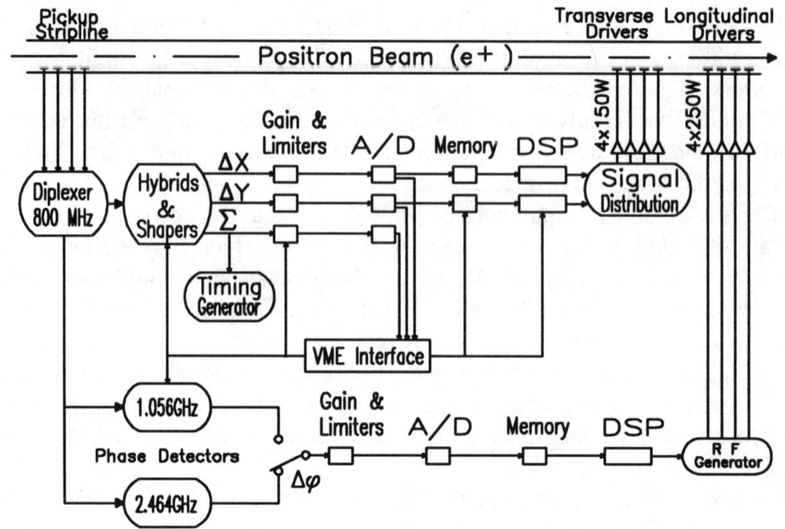

Figure 1. System Overview

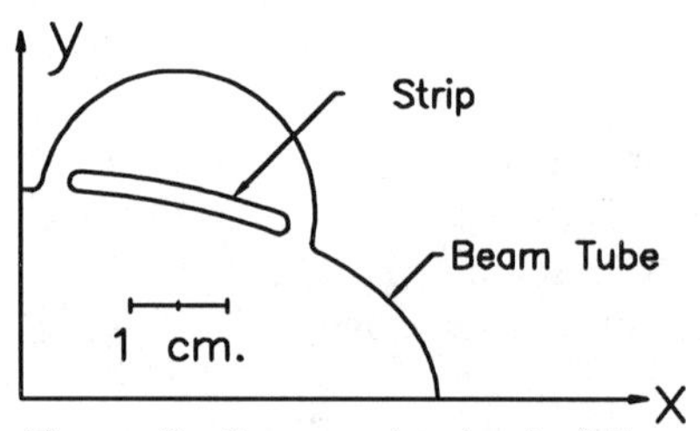

Figure 2. One quadrant of pickup & kicker stripline transverse geometry.

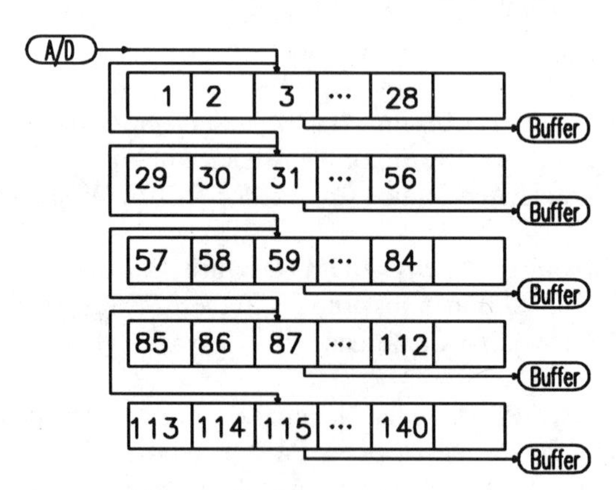

Figure 3. Memory

On the Use of Iterative Techniques for Feedforward Control of Transverse Angle and Position Jitter in Linear Particle Beam Accelerators*

Dean S. Barr
Advanced Photon Source, Argonne National Laboratory
9700 South Cass Ave., Argonne, IL 60439

Abstract

It is possible to use feedforward predictive control for transverse position and trajectory-angle jitter correction. The control procedure is straightforward, but creation of the predictive filter is not as obvious. The two processes tested were the least mean squares (LMS) and Kalman filter methods. The controller parameters calculated offline are downloaded to a real-time analog correction system between macropulses. These techniques worked well for both interpulse (pulse-to-pulse) correction and intrapulse (within a pulse) correction with the Kalman filter method being the clear winner. A simulation based on interpulse data taken at the Stanford Linear Collider showed an improvement factor of almost three in the average rms jitter over standard feedback techniques for the Kalman filter. An improvement factor of over three was found for the Kalman filter on intrapulse data taken at the Los Alamos Meson Physics Facility. The feedforward systems also improved the correction bandwidth.

INTRODUCTION

In the design of many pulsed linear particle beam accelerators, it is necessary to control transverse angle and position. Simple feedback systems have been used for years, but it is now desired to design a new type of feedforward controller. This system will increase the accuracy and bandwidth of correction over that of currently used feedback systems. The feedforward controller is trained offline using an iterative learning process. The feedforward control concept is similar to that in (1), but iterative techniques are used here as opposed to the autocorrelation and covariance methods used in (1).

Figure 1 shows the standard setup for a feedforward transverse jitter control system. Note that one pickup #1 and two kickers are needed to correct beam position jitter, while two pickup #1's and one kicker are needed to correct beam trajectory-angle jitter.

It is assumed that the beam is fast enough to beat the correction signal (through the fast loop) to the kicker. The fast loop will be completely analog for maximum speed. H will take the form of an analog transversal filter with digitally

* Work supported by The University of California under contract no. W-7405-ENG-36.

adjustable gains. Such devices are commercially available. Figure 2 shows a typical transversal filter.

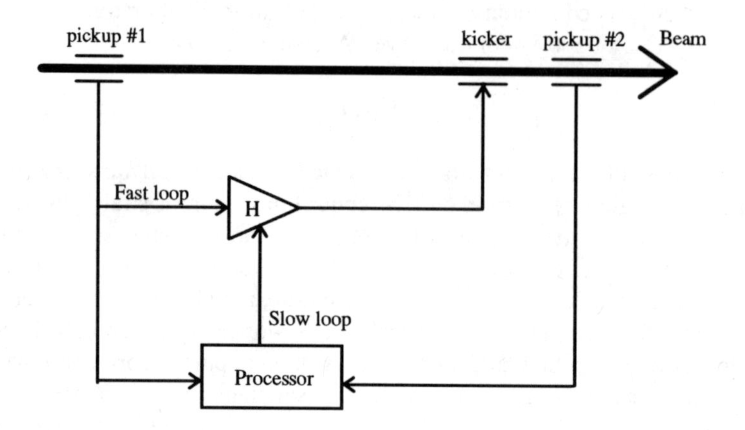

Figure 1. Feedforward Transverse Jitter Control System

The slow loop will use digitized inputs and process the values for H in an offline computer. The updated values for H will be downloaded to the filter between macropulses. These values will be chosen so that the output of the transversal filter, φ, will best predict the beam position at the kicker at some point in the future. The feasibility of this scheme depends on the degree of determinism (as opposed to stochasticity) of the beam jitter data.

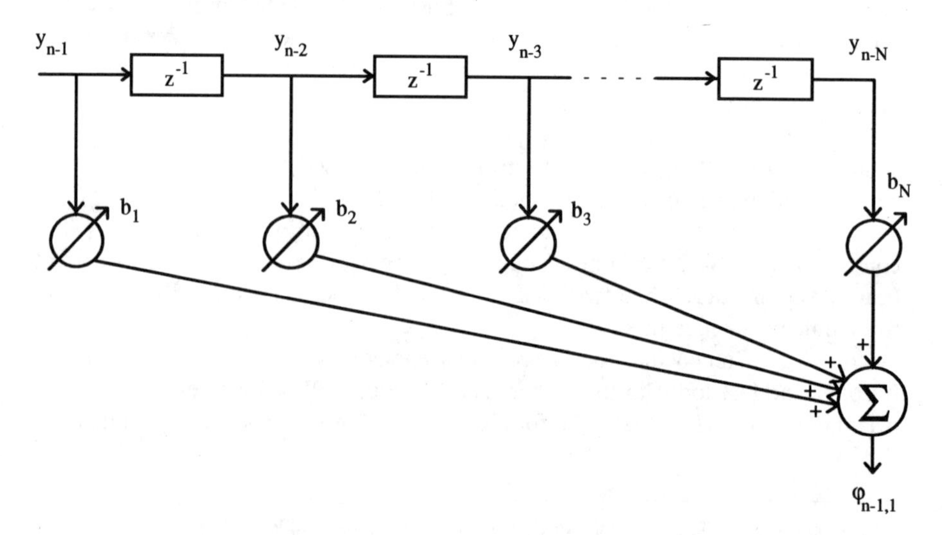

Figure 2. Transversal Prediction Filter (ν=1)

It will be shown that for at least two sets of data (from actual accelerators) there exist enough deterministic frequency components to both increase the accuracy of correction and extend the bandwidth of correction over that of a standard negative feedback correction system. The first is based on interpulse data (pulse-to-pulse jitter) obtained from the Stanford Linear Collider (SLC) at the Stanford Linear Accelerator Center (SLAC) in Palo Alto, California. The second is based on intrapulse data (jitter within a macropulse) taken at the Los Alamos Meson Physics Facility (LAMPF) at Los Alamos National Laboratory (LANL).

PREDICTION THEORY AND METHODS

The most general form of a linear filter is found in Eq. 1

$$y_n = \sum_{k=0}^{M} a_k x_{n-k} + \sum_{j=1}^{N} b_j y_{n-j} \tag{1}$$

where x_n is the filter input and y_n is the filter output. The general prediction equation can be found by using M=0 in Eq. 1

$$\varphi_{n-v,v} = a_0 x_n + \sum_{j=v}^{N} b_j y_{n-j} \tag{2}$$

where $\varphi_{n-v,v}$ is the prediction for y_n based on values of y up to and including y_{n-v}. The variable v is the number of points into the future into which the prediction is being made. Figure 2 shows a prediction filter with a value of v=1 (prediction one point into the future). The $a_0 x_n$ term is the error between the prediction and the actual value and is not used as part of the prediction.

Referring to Fig. 1, the values of y are taken from pickup #1. It is assumed that any jitter incurred between pickup #1 and the kicker is completely deterministic and can be removed without error by information gained at pickup #2. Thus, since the beam lattice is known in this region and any extra jitter is known, any predicted value of y at pickup #1 can be translated to a value for correction at the kicker. Thus it is only necessary to develop a method for determining the b_j's in Eq. 2 in order to perform predictive correction.

Referring to Fig. 2 and Eq. 2, the output φ, can be expressed as

$$\varphi_{n-v,v} = \sum_{j=v}^{N} b_{jn}\, y_{n-j}. \tag{3}$$

The second subscript on the b's denotes updated versions of the weights.

In vector notation,

$$\mathbf{B}_n = \begin{bmatrix} b_{vn} & b_{(v+1)n} & b_{(v+2)n} & \cdots & b_{Nn} \end{bmatrix}^T \tag{4}$$

and

$$\mathbf{Y}_n = \begin{bmatrix} y_{n-v} & y_{n-v-1} & y_{n-v-2} & \cdots & y_{n-N} \end{bmatrix}^T. \tag{5}$$

Equation 3 becomes

$$\varphi_{n-v,v} = \mathbf{Y}_n^T \mathbf{B}_n = \mathbf{B}_n^T \mathbf{Y}_n. \qquad (6)$$

The desired output corresponding to $\varphi_{n-v,v}$ is d_n. Define an error $\varepsilon_n = d_n - \varphi_{n-v,v}$. Using Eq. 6,

$$\varepsilon_n = d_n - \mathbf{Y}_n^T \mathbf{B}_n = d_n - \mathbf{B}_n^T \mathbf{Y}_n. \qquad (7)$$

Square Eq. 7 to get the instantaneous squared error,

$$\varepsilon_n^2 = d_n^2 + \mathbf{B}_n^T \mathbf{Y}_n \mathbf{Y}_n^T \mathbf{B}_n - 2 d_n \mathbf{Y}_n^T \mathbf{B}_n. \qquad (8)$$

Assume that ε_n, d_n, and $\mathbf{Y}_n$ are stationary to order two. Also assume for now that $\mathbf{B}_n$ is held fixed. Take the expected value of ε_n^2,

$$E[\varepsilon_n^2] = E[d_n^2] + \mathbf{B}_n^T E[\mathbf{Y}_n \mathbf{Y}_n^T] \mathbf{B}_n - 2E[d_n \mathbf{Y}_n^T] \mathbf{B}_n. \qquad (9)$$

In general y_n and d_n are not independent. It is convenient to define an input correlation matrix $\mathbf{R}_n = E[\mathbf{Y}_n \mathbf{Y}_n^T]$, and a vector $\mathbf{P}_n = E[d_n \mathbf{Y}_n^T]$. E[] is the expectation operator. Use these definitions in Eq. 9 to define the mean-square error (MSE) as ξ,

$$\text{MSE} \equiv \xi \equiv E[\varepsilon_n^2] = E[d_n^2] + \mathbf{B}_n^T \mathbf{R}_n \mathbf{B}_n - 2\mathbf{P}_n \mathbf{B}_n. \qquad (10)$$

The MSE defines a function in the filter weights, $\mathbf{B}_n$, for a given $\mathbf{Y}_n$ and d_n. If $\mathbf{Y}_n$ and d_n are stationary to order two or higher, the MSE is a quadratic function in $\mathbf{B}_n$. Thus ξ will be a multidimensional paraboloid in $\mathbf{B}_n$. The minimum point of this paraboloid or "bowl" is the optimal or Wiener solution for $\mathbf{B}_n$. This is denoted by $\mathbf{B}_n^*$.

LMS Algorithm and Kalman Filter

It is now desired to determine $\mathbf{B}_n$ as closely as possible to $\mathbf{B}_n^*$. The LMS algorithm (least mean square) will be used to accomplish this task. It is a form of a steepest descent method. The filter weights are adjusted in the direction of the gradient of $\xi = E[\varepsilon_k^2]$. This corresponds to

$$\mathbf{B}_{n+1} = \mathbf{B}_n + \mu(-\nabla_n) \qquad (11)$$

where ∇_n is the gradient of y, and μ is a parameter that controls the rate and stability of adaptation. The speed at which the optimal solution is approached depends on μ. The larger the value used, the quicker the routine converges to a solution. Unfortunately, if this value is too large, a correct solution may never be reached. Obviously this must be chosen with care.

There are many ways of determining ∇_n. The LMS algorithm uses ε_n^2 itself as an estimate for ξ_n rather than attempting some sort of short-time average. Now

$$\nabla_n = \begin{bmatrix} \dfrac{\partial \xi}{\partial b_{vn}} \\ \vdots \\ \dfrac{\partial \xi}{\partial b_{Nn}} \end{bmatrix} = \begin{bmatrix} \dfrac{\partial \varepsilon_n^2}{\partial b_{vn}} \\ \vdots \\ \dfrac{\partial \varepsilon_n^2}{\partial b_{Nn}} \end{bmatrix} = 2\varepsilon_n \begin{bmatrix} \dfrac{\partial \varepsilon_n}{\partial b_{vn}} \\ \vdots \\ \dfrac{\partial \varepsilon_n}{\partial b_{Nn}} \end{bmatrix}. \qquad (12)$$

Using Eq. 7,

$$\frac{\partial \varepsilon_n}{\partial \mathbf{B}_n} = -\mathbf{Y}_n. \tag{13}$$

Now Eq. 12 becomes

$$\nabla_n = -2\varepsilon_n \mathbf{Y}_n. \tag{14}$$

Plugging this into Eq. 11,

$$\mathbf{B}_{n+1} = \mathbf{B}_n + 2\mu\varepsilon_n \mathbf{Y}_n. \tag{15}$$

This is the LMS algorithm in vector form. Since the algorithm uses imperfect gradient estimates it does not follow the true line of steepest descent. Its advantage comes in its ease of implementation. For more details on the LMS algorithm and selection of the proper value of μ, see (2) and (3).

The Kalman filter improves on the LMS algorithm by making full use of all information available to it, whereas the LMS algorithm does not. The Kalman filter uses a set of difference equations to obtain a solution recursively. It uses a state-space approach to formulate a solution. It is trained and used in the same manner as the LMS algorithm in the previous section. Due to the complexity of the algorithm, and the space required to cover its operation even minimally, it will not be covered in this paper. The reader is referred to (2) and (4) for details.

RESULTS

Two sets of data were used to show the superiority of the iterative (specifically the Kalman filter) over previously used negative feedback systems. In both cases, a standard closed single-loop negative feedback system was used. The Nyquist stability criterion was applied to ensure stability. The details of the feedback system will not be given. (2,5)

The first data set is comprised of interpulse data and was collected at the SLC. The data consists of various sequences of 220 consecutive macropulses with a 60-Hz repetition rate. Each macropulse was passed through an analog low-pass filter before digitization in order to get its average value. The first 100 of the data points were used as a training set for the iterative filters (LMS and Kalman). The LMS algorithm was allowed one iterative step between pulses. This was possible due to the length of time between the pulses and the simplicity of the LMS algorithm. Due to its complexity, the Kalman filter was not allowed to retrain. The data was taken at three beam position monitors (BPMs) at different locations in the beamline. The locations correspond to different particle energies. The energies are 1.3 GeV, 17 GeV, and 42 GeV. For each energy, x and y position data was taken at six different currents. These were (in units of 10^9 e^- particles per macropulse) 13, 18, 30, 36, 40, and 45. The analyzed data is given in Table 1. At higher charge (36, 40, and 45), problems occurred in the standard feedback correction system for medium and high energies (17 and 42 GeV) in the x-plane. Slight problems occurred in the y-plane at medium energy and higher charges.

Table 1 - Jitter Reduction Values for SLC Interpulse Data

Data Set Name	Energy (GeV)	Charge (10^9 e^- /pulse)	Standard Feedback R_{orig} (dB)	LMS Method R_{feed} (dB)	Kalman Method R_{feed} (dB)
x1	1.2	13	-18.94	-10.79428	-11.45157
x2	1.2	45	-17.36	-9.34704	-10.62972
x3	42	13	-14.88	-5.50787	-4.47929
x4	42	45	4.39	-4.81796	-5.98826
y1	1.2	13	-20.43	-16.95955	-18.57240
y2	1.2	45	-20.47	-19.07728	-19.61238
y3	42	13	-16.01	3.35405	-8.67695
y4	42	45	-16.73	-7.65893	-7.69607
x5	17	40	2.98	-5.11643	-6.95194
x6	17	30	-4.09	-6.99720	-7.29789
x7	17	45	3.40	-3.72507	-5.36792
x8	42	30	-9.60	-5.57590	-5.50660
x9	42	40	5.04	-4.42966	-6.18646
y5	17	45	-1.82	-4.68803	-6.15240
y6	17	40	-3.95	-5.95581	-8.60337

The values for R_{orig} (standard feedback system only) are the rms jitter reduction amounts over those with no correction applied. The values for R_{feed} (iterative methods) are the rms jitter reduction amounts over those for the standard feedback system in column four. Therefore, R_{feed} is improvement beyond that given by the standard feedback system. Also note that the larger negative the value, the better the correction. The average jitter reduction value for the LMS method over the standard feedback is -7.4 dB and for the Kalman method is -9.4 dB. This corresponds to an improvement factor of 2.95 over the standard feedback system.

Table 2 - Jitter Reduction Values for LAMPF Intrapulse Data

Data Set Name	Standard Feedback R_{orig} (dB)	LMS Method R_{feed} (dB)	Kalman Method R_{feed} (dB)
x2	-12.2689	-0.81506	-10.4571
x3	-12.6459	-1.33137	-8.49451
x4	-13.5622	-1.65347	-9.17455
x5	-12.0185	-0.72877	-10.2226
y2	-16.2005	-3.21156	-11.1358
y3	-16.1898	-2.86748	-10.7281
y4	-17.5858	-2.15140	-8.97009
y5	-17.4924	-1.84464	-8.91657

The second data set is comprised of intrapulse data and was obtained from LAMPF. The beam consisted of P^- particles (polarized H^- particles) at 733 MeV. The sampling time was 0.4 μsec. Each macropulse was 1000 points long. Before sampling, the data was pre-filtered using a 600 kHz, 9th-order Chebychev low-pass filter. The beam was running at a 36-Hz repetition rate, but because of equipment limitations, macropulses were gathered at 90-second intervals. Table 2 shows the analyzed data. Macropulse x1 was used as a training pulse for predicting jitter within macropulses x2, x3, x4, and x5 for both the LMS and Kalman methods. The y-plane worked similarly. As in Table 1, the definitions of R_{orig} and R_{feed} are the same. The average jitter reduction value for the LMS method over the standard feedback is -1.8 dB and for the Kalman method is -9.7 dB. This corresponds to an improvement factor of 3.05 over the standard feedback system.

CONCLUSION

For both the interpulse and intrapulse data the iterative methods far exceeded the standard feedback system. In some cases, the predictive methods worked considerably better indicating a high level of deterministic frequency components in the data. An analysis of the correction bandwidth for the feedforward systems showed a vast improvement over that of the standard feedback system. For the LAMPF intrapulse data, the average correction bandwidth for the feedback system was around 10 kHz compared to 600 kHz for the Kalman filter method. In the SLC interpulse case, the bandwidths of the standard feedback and iterative systems were comparable, but the iterative systems had increased stability at higher frequencies.

REFERENCES

1. Barr, D.S., "On the Use of the Autocorrelation and Covariance Methods For Feedforward Control of Transverse Angle and Position Jitter in Linear Particle Beam Accelerators," presented at The Accelerator Instrumentation Fifth Annual Workshop in Santa Fe, NM 1993.
2. Barr, D.S., *Beam Position and Angle Jitter Correction in Linear Particle Beam Accelerators*, Ph.D. Dissertation, Texas Tech University, 1992, UMI order #9226291, pp. 125-132, 207-218, 255-256, 259-261.
3. Widrow, B. and Stearns, S., *Adaptive Signal Processing*, (Prentice-Hall, Inc., N.J., 1985).
4. Haykin, S., *Adaptive Filter Theory*, (Prentice-Hall, Inc., N.J., 1986).
5. Barr, D.S., "An Adaptive Feedback Controller For Transverse Angle and Position Jitter Correction in Linear Particle Beam Accelerators," in *Proceedings of the Accelerator Instrumentation Fourth Annual Workshop in Berkeley, CA 1992* (AIP Press, NY, 1993) pp. 204-214.

Dynamic Beam Based Alignment

I. Barnett, A. Beuret, B. Dehning, P. Galbraith, K. Henrichsen, M. Jonker, M. Placidi, R. Schmidt, L. Vos, J. Wenninger
CERN, 1211 Geneva 23, Switzerland

I. Reichel, F. Tecker
Phys. Inst. III A, RWTH Aachen, Germany

Abstract

A new method is presented to measure the relative offset of beam position monitors with respect to the magnetic center of quadrupole magnets.

Slow unavoidable orbit drifts lead to changing beam positions. The beam position is detected by modulating the strength of the magnetic field of a quadrupole and measuring the amplitude of the induced closed orbit oscillation. The amplitude of the orbit oscillation depends linearly on the modulation strength and on the beam offset in the quadrupole magnet.

INTRODUCTION

The method of beam based alignment is already used in some accelerators and storage rings (1–3). Using a modulation of the quadrupole strength was first used in LEP in 1992 and presented on the Third Workshop on LEP Performance (4).

The measurements are carried out during normal luminosity runs. The modulation of the magnetic field is done by modulating the current of the magnets at fixed frequencies between 0.8 and 17 Hz. The relative change of the magnetic field is of the order of 10^{-4}. The resulting oscillation amplitude of the beam is measured by calculating the Fourier spectrum of a directional coupler signal (Fig. 1). This allows simultaneous excitation and amplitude measurements of several magnets with different frequencies (Fig. 3). These measurements are recorded for several hours, and the offset is extracted by evaluating the minimum response and comparing it to the nearby position monitor.

INSTRUMENTATION

The magnetic field of the quadrupoles is changed in two different ways. The quadrupoles near to the IPs have their own power supply thus they can be modulated by changing the current in the power supply. All other quadrupoles are powered with one power supply for at least two quadrupoles. These magnets were equipped with additional windings (back legs) which can be powered by a separate generator.

In LEP 16 magnets are modulated by their own power supply and 88 are modulated with back leg windings. The equipped magnets are located in the

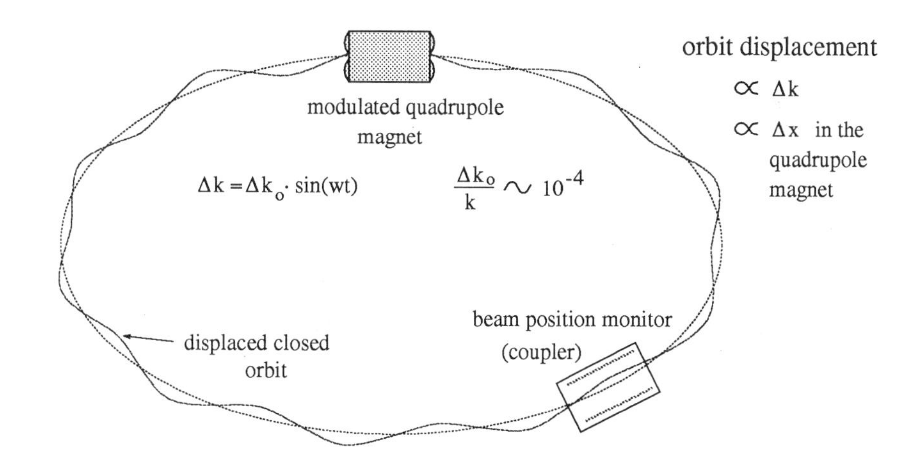

Figure. 1. Principle of dynamic beam based alignment

straight sections to both sides of the four even IPs of LEP where the experiments are located (Fig. 2). For the quadrupole magnets with back leg windings the magnetic field can be modulated at a frequency of up to 1.6 Hz. This limit is given by the closed loop break frequency of the quadrupole power supply. The current change in the back leg winding induces a voltage in the main winding of the magnet. This voltage has to be counterbalanced by the regulation of the main power supply to avoid current changes in this circuit. The coupling between two magnets powered by the same power supply was found to be smaller than 0.1 by modulating one magnet and measuring the field changes in both of them.

The oscillation of the beam is detected by two directional couplers installed at locations with different betatron phases. This arrangement allows measurements of beam oscillations induced by any quadrupole around the ring.

A data acquisition system collects the signals from the couplers: the pulse from one bunch passage is integrated in a gate of about 500 ns and digitized using a 12 bit ADC. The ADC is read out with a Digital Signal Processor which adds the values of 200 or 400 bunch passages and writes up to 17000 sum values into memory. LEP operates with a revolution frequency $f_{rev} = 11.246$ kHz and eight bunches per beam. Due to the electronics only every second bunch can be measured. The sampling frequency is given by : $f_{rev} \times N_{bunches}/200(400)$, this results in a sampling frequency of 224.92 Hz or 112.46 Hz. Each coupler gives values for both planes and in each plane signals from electrons and positrons which are all digitized at the same time.

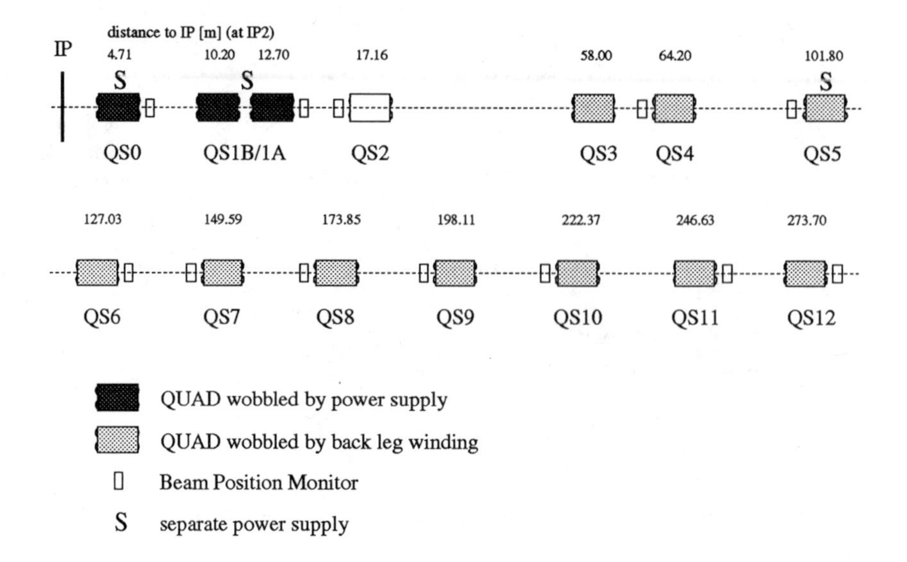

Figure. 2. Location of quadrupole magnets and beam position monitors on the right side of IP2. This arrangement is symmetrical around all even LEP IPs.

CALIBRATION OF THE BEAM POSITION MONITORS

For each of the eight superconducting quadrupole magnets (QS0s) the modulation frequency can be selected in a range between 12 Hz and 15 Hz. A spacing of 1 Hz between two frequencies was used. Four quadrupoles were modulated at the same time. Figure 3 shows a FFT spectrum with the modulation of four quadrupoles switched on.

The amplitude of the peaks shown in Fig. 3 are stored together with the orbit measured with the orbit monitors at approximately the same time. If the beam is centered in the quadrupole, the peak height is at a minimum. If the beam is off center the peak height rises linearly with the displacement. We now plot the measured peak height versus the orbit measured with the orbit monitors. Results for some quadrupoles around an interaction point are shown in Fig. 4.

From these plots it is possible to get the offsets for the different pickups. The offsets can be different for different particles and also for different gains in the electronics of the orbit monitors. Preliminary results for the offsets using the recently installed orbit monitor electronics are given in Table 1. If the offset is calculated by doing a fit, the error on the offset is between 30 and 50 μm. The reason for the – quite large – errors at the moment is, that the new electronics have just been installed and a more accurate analysis is not finished yet.

Table 1. Preliminary results for determination of the absolute offset of pick-ups located close to the superconducting quadrupoles (QS0s). All values are estimated from the plots and given in mm. Large errors occur, if the beam was always on the same side of the quadrupole and never went through the center.

QS0		electrons		positrons	
		gain 0dB	gain 10dB	gain 0dB	gain 10dB
IP 2	left		.1 ± .2		.2 ± .2
	right		-.8 ± .2		-.8 ± .1
IP 4	left	-1.4 ± .3	-1.75 ± .1	-1.55 ± .1	-1.75 ± .1
	right	-1.3 ± .3	-1.4 ± .1	-1.3 ± .1	-1.3 ± .1
IP 6	left		-1.7 ± .1		-1.75 ± .1
	right		-1.2 ± .1		-1.5 ± .1
IP 8	left		-.2 ± .2		-.7 ± .1
	right		-.25 ± .1		-.5 ± .1

CONCLUSION

It is possible to measure the offsets of beam orbit monitors using the modulation of the quadrupole strength. The measurement can be done during luminosity runs and has no influence on the quality of the beam. The accuracy of about 0.1 mm is achieved if the beam approached the magnetic center from both sides. If it was just on one side, it is more difficult to determine the minimum (Fig. 4). Fitting a V-shaped function to the data gives errors between 30 and 50 μm.

ACKNOWLEDGMENT

We thank C. Bovet and M. Tonutti for their support of this project. We are deeply indebted to G. Morpurgo and the other collegues working on the beam orbit measurement system for their help. Thanks to P. Castro for helping us with the DSP - software. It was a pleasure to collaborate with the LEP operation team to perform the measurements.

REFERENCES

1. Rice, D. et al., "Beam Diagnostic Instrumentation at CESR", IEEE Transitions on Nuclear Science **Vol. NS-30, No. 4**, August 1983, pp.2190.
2. Bulos F. et al. "Beam-Based Alignment and Tuning Procedures for e^+e^-Collider Final Focus Systems", Contributed to the IEEE Particle Accelerator Conference, San Francisco 1991
3. Röjsel, P., "A beam position measurement system using quadrupole magnets magnetic centra as the position reference", Nuclear Instruments and Methods in Physics Research **A 343 (1994)**, pp. 374.

4. Schmidt R., "Misalignments from k-modulation", Proceedings of the Third Workshop on LEP Performance, CERN **SL/93-19 (DI)**, 1993, pp.139.

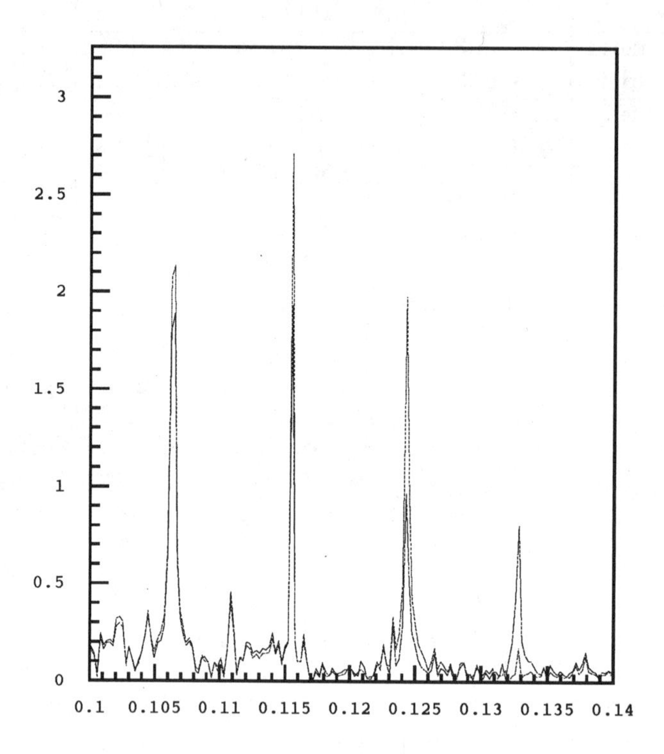

Figure. 3. Spectrum of the beam oscillations. The sampling frequency is 112.46 Hz and the peaks correspond to a quadrupole magnet excitation frequencies of 12, 13, 14 and 15 Hz.

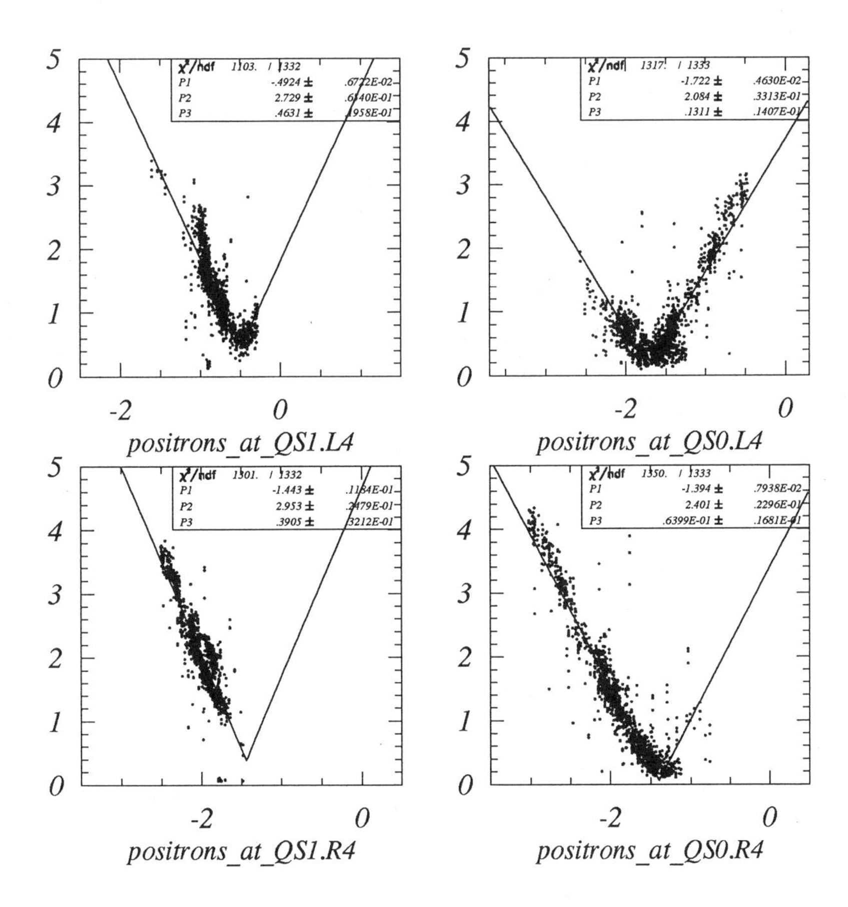

Figure. 4. Modulation amplitudes of four magnets symmetrically located around LEP interaction point 4 as a function of the orbit measured with the orbit monitors. The minimum indicates the center of the magnet. The offsets are -.5 mm, -1.7 mm, -1.4 mm and -1.4 mm. P1 gives the offset with errors from fitting.

Simulation and steering in the Intertank Matching Section of the Ground Test Accelerator*

V. W. Yuan, G. O. Bolme, J. L. Erickson†, K. F. Johnson,
C. T. Mottershead, O. R. Sander, M. T. Smith
Los Alamos National Laboratory, Los Alamos NM 87545

1 Introduction

The Intertank Matching Section (IMS) of the Ground Test Accelerator (GTA) is a short (36 cm) beamline designed to match the Radio Frequency Quadrupole (RFQ) exit beam into the first Drift Tube LINAC (DTL) tank. The IMS contains two steering quadrupoles (SMQs) and four variable-field focussing quads (VFQs). The SMQs are fixed strength permanent magnet quadrupoles on mechanical actuators capable of transverse movement for the purpose of steering the beam. The upsteam and downstream steering quadrupoles are labelled SMQ1 and SMQ4 respectively. Also contained in the IMS are two RF cavities for longitudinal matching. A schematic of the IMS beam line is shown in figure 1.

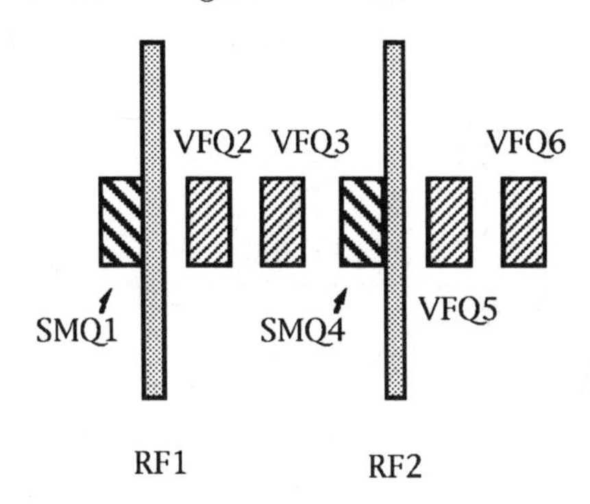

Figure.1. Schematic diagram of IMS

The settings of the Variable-field quadrupoles, named VFQ2, VFQ3, VFQ5, and VFQ6, are called the "tune" of the IMS, expressed as the GL product (in Tesla) of each VFQ. The tunes of the IMS were chosen for DTL matching, but have a significant impact on the steering sensitivities. A diagnostics plate containing a slit and collector diagnostic (ES5) was located just downstream of the IMS and was used for measuring changes in positional and angular distributions of the beam when SMQ positions were varied. These measured values are compared with the changes predicted by TRACE3D(1) simulations.

*Work supported and funded by the US Department of Defense, Army Strategic Defense Command, under the auspices of the Department of Energy.
†Industrial partner, Grumman Energy Systems Division, Bethpage, NY

2 Steering Coefficients for a Single Quad

The transport through a quadrupole with transfer matrix **Q**, displaced by a steering offset $+s$ may be written as

$$\begin{bmatrix} x_e - s \\ p_e \end{bmatrix} = \begin{bmatrix} Q_{11} & Q_{12} \\ Q_{21} & Q_{22} \end{bmatrix} \begin{bmatrix} x_0 - s \\ p_0 \end{bmatrix} \quad \text{or} \quad \begin{array}{l} x_e = Q_{11}\,x_0 + Q_{12}\,p_0 + (1 - Q_{11})\,s \\ p_e = Q_{21}\,x_0 + Q_{22}\,p_0 - Q_{21}\,s \end{array}$$

where (x_0, p_0) are the coordinates of the centroid ray entering the quad and (x_e, p_e) are its coordinates exiting the quad. The terms that remain when $s = 0$ are the normal application of the transfer matrix of the quadrupole.

We are normally interested in the beam centroid steering movements at a particular downstream monitor location. The beam coordinates at the monitor (x_m, p_m) are obtained by propagating the steering-quad exit coordinates (x_e, p_e) through the R-matrix between the two points:

$$\begin{bmatrix} x_m \\ p_m \end{bmatrix} = \begin{bmatrix} R_{11} & R_{12} \\ R_{21} & R_{22} \end{bmatrix} \begin{bmatrix} x_e \\ p_e \end{bmatrix} ,$$

or

$$\begin{array}{l} x_m = (RQ)_{11}\,x_0 + (RQ)_{12}\,p_0 + \{R_{11}(1 - Q_{11}) - R_{12}Q_{21}\}\,s \\ p_m = (RQ)_{21}\,x_0 + (RQ)_{22}\,p_0 + \{R_{21}(1 - Q_{11}) - R_{22}Q_{21}\}\,s \end{array} \tag{1}$$

From this we read off the steering coefficients for the monitor in question:

$$C \equiv \frac{\partial x_m}{\partial s} = R_{11}(1 - Q_{11}) - R_{12}Q_{21} \,, \tag{2}$$

$$A \equiv \frac{\partial p_m}{\partial s} = R_{21}(1 - Q_{11}) - R_{22}Q_{21} \,. \tag{3}$$

Again, for $s = 0$, we are left with just the usual first order propagation of the incoming beam, beginning at the input surface of the steering quad.

3 Case of Two Steerers

Now what happens when a second steering quad (SMQ4) is added downstream of the first one (SMQ1)? The result becomes more complicated, but still remains linear in the steering offsets. To see this, consider Eq.1 applied to the second steerer, but with the input coordinates replaced by the (x_m, p_m) resulting from the first steerer. The result is of the same matrix form as Eq. 1, but with additional terms linear in the offset s_4 of the second quad. The R-matrix can be redefined to carry the beam from wherever (x_0, p_0) is specified to the point of interest (x_f, p_f). The general expression for the final beam centroid coordinates may then be written

$$\begin{bmatrix} x_f \\ p_f \end{bmatrix} = \begin{bmatrix} R_{11} & R_{12} \\ R_{21} & R_{22} \end{bmatrix} \begin{bmatrix} x_0 \\ p_0 \end{bmatrix} + \begin{bmatrix} C_1 & C_4 \\ A_1 & A_4 \end{bmatrix} \begin{bmatrix} s_1 \\ s_4 \end{bmatrix} \tag{4}$$

where the displacement of the first quad is now labelled s_1. The final beam coordinates are thus seen to be a sum of two contributions. The first depends only on the incoming beam (x_0, p_0) as propagated by the usual R-matrix. The second depends only on the steering offsets (s_1, s_4) as propagated by a new steering, or S-matrix, built as shown from the steering sensitivity coefficients

$$C_k \equiv \frac{\partial x_f}{\partial s_k} \text{ and } A_k \equiv \frac{\partial p_f}{\partial s_k} \tag{5}$$

There are, of course, corresponding expressions in the y-plane.

4 Effective pivot points

The optics beyond VFQ6 is simply a drift space in which particle trajectories are straight lines at a constant angle. The angular sensitivity coefficients A_k are therefore independent of longitudinal coordinate z in the drift space, and thus the same for all monitoring points beyond VFQ6. The position steering coefficients C_k have a linear z dependence in a drift, which may be seen by transporting the output coordinates from Eq. 4 through an additional drift: from the reference point z_f where C_k and A_k were evaluated to the final point of interest at z

$$\begin{bmatrix} x(z) \\ p(z) \end{bmatrix} = \begin{bmatrix} 1 & z - z_f \\ 0 & 1 \end{bmatrix} \left\{ \begin{bmatrix} R_{11} & R_{12} \\ R_{21} & R_{22} \end{bmatrix} \begin{bmatrix} x_0 \\ p_0 \end{bmatrix} + \begin{bmatrix} C_1 & C_4 \\ A_1 & A_4 \end{bmatrix} \begin{bmatrix} s_1 \\ s_4 \end{bmatrix} \right\} \tag{6}$$

When this is multiplied out one sees that $C_k(z) = C_k + (z - z_f)A_k$ is linear in z, while $A_k(z) = A_k$ is constant. The position coefficient $C_k(z)$ passes through zero at $z_k \equiv z_f - C_k/A_k$, which we call the effective pivot point for converting centroid angle swings, due to steering-quad movements, into position changes. This means that the steering matrix anywhere in the drift is completely specified by an angular sensitivity A_k and a pivot point coordinate z_k for each steering quad:

$$\mathbf{S}(z) = \begin{bmatrix} (z - z_1)\, A_1 & (z - z_4)\, A_4 \\ A_1 & A_4 \end{bmatrix} \tag{7}$$

Obviously, a steering quad will not affect the transverse position of the beam at its own pivot point coordinate.

5 Measured steering coefficients

Steering studies were performed for 3 different tunes of the IMS variable-field quadrupoles. These tunes are designated tune 1, tune 2, and tune 3. Table 1 summarizes the nominal quadrupole field settings for the three different tunes used in the steering studies.

Tune	VFQ2	VFQ3	VFQ5	VFQ6
1	-5.583	4.884	6.288	-6.136
2	-4.954	4.674	5.644	-5.891
3	-4.380	5.635	5.667	-6.174

Table 1. Quadrupole (VFQ) field settings of the three IMS tunes used for our experimental measurements. Fields are given in Tesla.

The measured steering coefficients were determined from sets of runs incorporating all three beam tunes and in which the steering quadrupoles were singly displaced along either the x or y axis. The resultant changes in beam position and angle were monitored at ES5. For the displacements of each quadrupole, the data were linearly fitted using MINUIT(2), a minimum chi-squared routine, to determine the steering coefficients With separate coefficients for each of the x and y directions, and for position and angle sensitivities, this results in 8 steering coefficients for each tune and 24 coefficients overall for the entire set of 3 beam tunes.

The measured steering coefficients appear along with MINUIT-determined uncertainties in Table 2. These uncertainties were determined by adjusting the error on each individual data point until the fit using MINUIT resulted in a chi-squared of one per degree of freedom (DOF). There is a wide spread in the uncertainty values due in part to the limited number of points that comprised each data set and in part to the mechanical limitation on the range of travel of the ES5 slit and collector. When the steering quadrupoles were displaced and the beam was not centered at ES5, the slit and collector could not always capture the full phase space of the beam, and centroid determination in those cases had to be made based on high level contours of the beam distribution.

6 Comparison of measured steering coefficients with simulation

For comparison with measured steering coefficients, a set of coefficients was calculated using the TRACE3D simulation. These calculated coefficients were determined assuming nominal quadrupole field strengths, and they are compared to measured values in Table 2. In calculating the simulation values, we treated the steering quadrupoles as permanent-magnet quadrupoles (PMQs) with fringe fields extending into the neigboring drift spaces. The VFQs were treated as quadrupole magnets with hard-edged fields that cut off abruptly at the magnets' physical boundaries.(3) The average difference between calculated and measured values is 18%, and accepting the uncertainties of Table 2, the chi-squared for the comparison of calculated values to measured values is 20/DOF. Such a large chi-squared value is indicative of poor agreement

	Tune 1			Tune 2		
Coeff	measured	calc	fit	measured	calc	fit
CX1	-20.50 ± 0.65	-23.08	-19.90	-11.31 ± 0.51	-14.88	-10.87
CY1	4.59 ± 0.22	3.95	4.51	4.68 ± 0.15	3.81	4.62
CX4	-8.98 ± 0.58	-7.71	-7.94	-10.98 ± 0.39	-10.22	-10.18
CY4	-17.05 ± 0.77	-19.84	-17.70	-15.44 ± 0.39	-16.98	-15.12
AX1	-52.90 ± 1.81	-61.11	-54.09	-30.50 ± 1.20	-40.75	-31.63
AY1	11.51 ± 0.87	9.70	11.73	11.24 ± 0.41	8.54	11.43
AX4	-19.78 ± 1.57	-17.55	-18.22	-23.86 ± 0.98	-23.71	-23.71
AY4	-51.73 ± 2.77	-60.01	-53.84	-48.37 ± 1.31	-52.05	-46.64

Tune 3

Coeff	measured	calc	fit
CX1	-11.00 ± 2.84	-15.74	-11.22
CY1	10.74 ± 0.75	10.05	10.80
CX4	-10.24 ± 0.59	-11.94	-11.81
CY4	-16.46 ± 0.76	-19.60	-17.51
AX1	-28.05 ± 6.84	-40.54	-30.32
AY1	29.18 ± 2.45	26.73	29.50
AX4	-23.00 ± 1.59	-28.01	-27.78
AY4	-50.65 ± 2.76	-58.77	-52.77

Table 2. Measured, calculated and fit coefficients for Tunes 1,2, and 3. Calculated coefficients are based on the nominal quadrupole fields given in Table 1.

magnet	GL (old)	Δ GL	% change
SMQ1	4.742	-0.184	-3.88
VFQ2	-5.583	0.076	-1.37
VFQ3	4.884	-0.594	-12.16
SMQ4	-5.499	0.389	-7.08
VFQ5	6.288	-0.202	-3.22
VFQ6	-6.136	0.114	-1.86

Table 3. Changes to quadrupole field values (in Tesla) determined by FIELDFIT program when quadrupole strengths are varied to achieve the best match of calculated to measured steering coefficients.

between measured and calculated values for the uncertainties assumed. Three possible reasons for the disagreement could be: 1) the uncertainties are larger than assumed, 2) the quadrupole field values are not what is assumed, or 3) the simulation model is incorrect.

7 Fitting program results

A fitting program, FIELDFIT, was written that allows quadrupole field strengths to differ from their nominal values in order to optimize the agreement of measured to calculated steering coefficients. It therefore searches for a set of deviations in quadrupole field strengths which results in a significant reduction in chi-squared between measured and calculated steering coefficients.

The fitting program FIELDFIT incorporates TRACE3D beam-tracking and matrix propagation routines to calculate steering coefficients. This allows the correct calculation of steering coefficients when more than one of the quad fields are varied. If FIELDFIT is used to fit simultaneously the sets of data for the 3 different tunes, a fit with a much-improved chi-squared of 1.9/DOF can be obtained with the restriction that the same fractional quadrupole deviations are assumed for each tune. The fit values for such a fit are listed in Table 2 under the column heading "fit". The fit-determined steering coefficients have an average of 4.9% deviation from the measured values. This is approximately 3.5 times smaller than the average 17.7% deviation of the nominal-field-determined coefficients from the measured values.

The field deviations determined by the fit are listed in Table 3. The largest two deviations are at the 12.1% and 7.1% levels, and the other deviations are at the 3.9%, 3.2%, 1.9%, and 1.4% levels. The average deviation is 4.9% which is 2.5 times larger than the 2% uncertainty in GL estimated by our magnet mapping group. The largest deviation, 12.1%, corresponds to a field deviation of .6 T which seems unreasonably large. It is interesting that all the field deviations determined by the fit are in the direction to decrease the magnitude of each field strength.

With a set of programs written using the ADLIB(4) modular format, we estimated the error in each steering coefficient which results from a 2% standard deviation in each quadrupole field value. For each of the three IMS tunes, a Gaussian random number generator was used to produce a file of 2000 possible samples of the tune, such that the ensemble average strength of each quad was its nominal value, with a standard deviation of 2% of the nominal value. The result of these Gaussian field-strength errors is an approximate 9% average uncertainty in the steering coefficients. We see from the FIELDFIT analysis that field deviations which average 4.9% and which contain individual deviations of as large as 12% will change the steering coefficients by an average of approximately 13%.

8 Conclusions

A comparison of measured to calculated steering coefficients has been made for data taken in 3 different tunes of the IMS transport line. Using a TRACE3D-based calculation, calculated values differ from measured values by an average 17.7%. Estimated uncertainties in the measured values are too small to explain the observed discrepancy. Changes to the effective field strengths of the quadrupoles were found that result in a significant improvement between measured and calculated values. The field strength deviations required, however, are on average 2.5 times larger than what is estimated by our magnet mapping group as the actual uncertainty of each magnet. The model calculation is not perfect. For instance, we know that the VFQ fringe fields are not correctly modelled by TRACE3D. However, tests with approximations to the field shape effect the results by only 20%. The fact that changing the field strengths of the quadrupoles can significantly improve the agreement between measured and calculated coefficients may well indicate the presence of some other as yet unknown systematic problem with the model calculation.

REFERENCES

[1] K. R. Crandall and D. P. Rusthoi, "TRACE 3-D Documentation", Los Alamos report LA-UR-90-4146, second edition (1990).

[2] General minimization program contained in CERNLIB codes, CERN, Geneva, Switzerland.

[3] In actuality the VFQ fields have fringe fields that lies somewhere between the field of an electromagnet and that of a permanent-magnet quadrupole (PMQ). This type of hybrid fringe field is not expressly treated in TRACE3D. However an approximation to this fringe field shape can be made by treating the VFQ as a PMQ with an artificially small inner-bore radius. Using such an approximation results in a 20% improvement in the agreement between the measured and calculated steering coefficients for the nominal quadrupole fields.

[4] C. T. Mottershead, "ADLIB- A Simple Database Framework for Beamline Codes", Computational Accelerator Physics, AIP Conference Proc. 297 (1993).

Optics-Driven Design Criteria for BPMs

Yu-Chiu Chao
Continuous Electron Beam Accelerator Facility
12000 Jefferson Avenue, Newport News, Virginia 23606

Abstract

We study the impact of BPM resolution on optics measurements at various levels of complexity: (1) Formula linking a given distribution of generalized BPM resolutions to the degree of precision to which a beam trajectory can be determined. (2) Formula for the precision achievable in a generalized experimental scheme measuring transfer matrices in the presence of potentially coupled orbit errors. (3) The results from (1) and (2) are combined to give the formula relating the precision of the transfer matrix measurement to the signal-to-noise ratio of the BPM used. (4) A criterion is developed summarizing how well the overall optical behavior of a large modular beam transport system can be quantified. The results from (1), (2) and (3) are used to derive the final analytical expression for a generic criterion on BPM resolution for such systems. Examples are briefly discussed.

INTRODUCTION

It is time-tested wisdom that in the process of building an accelerator, diagnostics should be designed in at the lowest level. Retrofitted diagnostics rarely deliver optimal results. With respect to optics-related measurements, usually the critical parameters and required degree of accuracy are known at the design stage. A formulation translating the latter into the former is therefore potentially valuable to the designer. On the other hand, a qualitative argument leading to a design criterion no more accurate than an order of magnitude is practically useless. The purpose of this report is therefore to present a highly accurate formulation of the criterion for BPM resolution under various optics measurement schemes.(1)

Besides monitoring beam orbit, BPM's are collectively used in trajectory determination for feedback systems or correction programs in the control system. The trajectories in turn can be collectively used to determine the transfer matrices across a section of the beam line. This is illustrated in Fig. 1. The symbol $\mathbf{m}^{ab}{}_{ij}$ stands for the ij-th transfer matrix element from point **a** to **b**, while x^p stands for the orbit vector (x,x') at the point **p**.

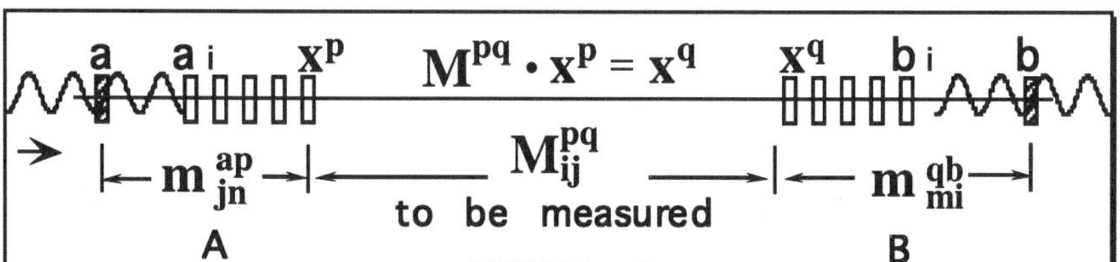

$$M^{pq} \cdot x^p = x^q$$

FIGURE 1

1. PULSE-TO-PULSE TRAJECTORY MEASUREMENT

Using the notation of Fig. 1, we study the achievable precision in determining the pulse-to-pulse trajectory at point **p** using the BPM's in beam line section **A** upstream of the unknown section. The difference between two orbits can be determined by fitting the difference in all the BPM's, represented by a vector $\mathbf{X}^A$, to the known optical model of **A**:

$$\mathbf{X}^{A}=\begin{pmatrix} x_1^1 \\ x_1^2 \\ \vdots \\ x_1^{N_B} \end{pmatrix}=\begin{pmatrix} m_{11}^{p1} & m_{12}^{p1} \\ m_{11}^{p2} & m_{12}^{p2} \\ \vdots & \vdots \\ m_{11}^{pN_B} & m_{12}^{pN_B} \end{pmatrix}\bullet\begin{pmatrix} x_1^p \\ x_2^p \end{pmatrix}=m\bullet\mathbf{x}^p,\quad \mathbf{x}^p=m^{-1}\bullet\mathbf{X}^A, \qquad \textbf{(1.1)}$$

where N_B is the number of BPM's used in section **A**. The matrix inverse represents the least square fit. The covariance error matrix for the fitted orbit vector at **p**, $\langle\delta_x^{pi}\delta_x^{pj}\rangle$, can be derived, using symplectic conditions, as a function of the optics and the resolution σ_B^q for the BPM's, with q indexing the BPM:

$$\sigma_X^{p1^2}=\langle\delta_X^{p1}\cdot\delta_X^{p1}\rangle=\frac{2}{N_B}\cdot\frac{\langle m_{12}^{pa}\rangle_s^2}{\langle m_{12}^{aa}\rangle_s^2},\quad \sigma_X^{p2^2}=\langle\delta_X^{p2}\cdot\delta_X^{p2}\rangle=\frac{2}{N_B}\cdot\frac{\langle m_{11}^{pa}\rangle_s^2}{\langle m_{12}^{aa}\rangle_s^2},$$

$$S^{p12}=\langle\delta_X^{p1}\cdot\delta_X^{p2}\rangle=\frac{-2}{N_B}\cdot\frac{\langle m_{11}^{pa}\cdot m_{12}^{pa}\rangle_s}{\langle m_{12}^{aa}\rangle_s^2},$$

$$\begin{aligned}\langle m_{12}^{aa}\rangle_s^2&=\sum_{a^i=1}^{N_B}\sum_{a^j=1}^{N_B}\left(\frac{m_{12}^{a^ia^j}}{\sigma_B^{a^i}\sigma_B^{a^j}}\right)^2, & \langle m_{12}^{pa}\rangle_s^2&=\sum_{a^j=1}^{N_B}\left(\frac{m_{12}^{pa^j}}{\sigma_B^{a^j}}\right)^2,\\ \langle m_{11}^{pa}\cdot m_{12}^{pa}\rangle_s&=\sum_{a^j=1}^{N_B}\left(\frac{m_{11}^{pa^j}m_{12}^{pa^j}}{\sigma_B^{a^j}\sigma_B^{a^j}}\right), & \langle m_{11}^{pa}\rangle_s^2&=\sum_{a^j=1}^{N_B}\left(\frac{m_{11}^{pa^j}}{\sigma_B^{a^j}}\right)^2.\end{aligned} \qquad \textbf{(1.2)}$$

This result can be used in feedback systems or other control program designs. We have assumed that all BPM's have different resolutions in Eqn. 1.2. If all BPM's have the same resolution, we have

$$\sigma_X^{p1^2}=\langle\delta_X^{p1}\cdot\delta_X^{p1}\rangle=\frac{2\sigma_B^2}{N_B}\cdot\frac{\langle m_{12}^{pa}\rangle^2}{\langle m_{12}^{ab}\rangle^2},\quad \sigma_X^{p2^2}=\langle\delta_X^{p2}\cdot\delta_X^{p2}\rangle=\frac{2\sigma_B^2}{N_B}\cdot\frac{\langle m_{11}^{pa}\rangle^2}{\langle m_{12}^{ab}\rangle^2},$$

$$S^{p12}=\langle\delta_X^{p1}\cdot\delta_X^{p2}\rangle=\frac{-2\sigma_B^2}{N_B}\cdot\frac{\langle m_{11}^{pa}\cdot m_{12}^{pa}\rangle}{\langle m_{12}^{ab}\rangle^2},$$

$$\begin{aligned}\langle m_{12}^{aa}\rangle^2&=\sum_{a^i=1}^{N_B}\sum_{a^j=1}^{N_B}\left(m_{12}^{a^ia^j}\right)^2, & \langle m_{12}^{pa}\rangle^2&=\sum_{a^j=1}^{N_B}\left(m_{12}^{pa^j}\right)^2,\\ \langle m_{11}^{pa}\cdot m_{12}^{pa}\rangle&=\sum_{a^j=1}^{N_B}\left(m_{11}^{pa^j}m_{12}^{pa^j}\right), & \langle m_{11}^{pa}\rangle^2&=\sum_{a^j=1}^{N_B}\left(m_{11}^{pa^j}\right)^2.\end{aligned} \qquad \textbf{(1.3)}$$

Partitioning the Double Sum

In many cases discussed below we can partition the BPM's into subgroups and simplify the double sum in Eqn. 1.3. These subgroups can be identical cells or all the BPM's identically located in each cell. The double sum then is reduced to

$$\frac{1}{G^2}\cdot\left\langle m_{12}^{aa}\right\rangle^2 = \sum_{k=1}^{G}\left\langle m_{12}^{aa}\right\rangle_k^2 + 2\cdot\sum_{\substack{m>n\\ m=1,n=1}}^{G}\sum_{a=1}^{N_m}\sum_{b=1}^{N_n}\left(m_{12}^{ab}\right)^2 . \qquad \textbf{(1.4)}$$

G above is the total number of subgroups, indexed by m and n. The subscript k indicates double sum only within a subgroup.

Simple Rule of Thumb

The sums in Eqn. 1.3 are actually very easy to calculate.(1) If one wants even more immediate estimates, the following rules of thumb can be a substitute. Notice that the last three equalities break down for small number of BPM's.

$$\left\langle m_{12}^{aa}\right\rangle^2 = \left\langle\beta\beta\right\rangle_{\sin}^2 = \frac{1}{2N^2}\sum_{i=1}^{N}\sum_{j=1}^{N}\beta^i\cdot\beta^j\cdot\sin^2\left(\varphi^i-\varphi^j\right),\ \left\langle m_{11}^{pa}\right\rangle^2 \xrightarrow{N>>1} \frac{\gamma_p\left\langle\beta\right\rangle}{2} = \frac{\gamma_p}{2N}\sum_{j=1}^{N}\beta^j ,$$

$$\left\langle m_{12}^{pa}\right\rangle^2 \xrightarrow{N>>1} \frac{\beta_p\left\langle\beta\right\rangle}{2} = \frac{\beta_p}{2N}\sum_{j=1}^{N}\beta^j,\ \left\langle m_{11}^{pa}\cdot m_{12}^{pa}\right\rangle \xrightarrow{N>>1} \frac{-\alpha_p\left\langle\beta\right\rangle}{2} = \frac{-\alpha_p}{2N}\sum_{j=1}^{N}\beta^j . \qquad \textbf{(1.5)}$$

The subscript p labels the observation point **p**.

2. TRANSFER MATRIX MEASUREMENT

A scheme for measuring the unknown transfer matrix $M^{pq}{}_{ij}$ is devised in Fig. 1. A total of N_O trajectories are sent through beam line sections **A** and **B**, where the orbit vectors are determined in the fashion discussed above at observation points **p** and **q**. These two sets of orbit vectors are sufficient for unfolding the unknown $M^{pq}{}_{ij}$. This scheme is better than the commonly adopted method relying only on knowledge of upstream kickers, in that it is immune to kicker errors and incoming orbit/energy jitters, that the beam line structure affords more exact error analysis, and that the flexibility in expanding the upstream section frees us from the limit on overall precision occurring otherwise.(1) The fitting problem now takes on the form

$$\begin{pmatrix}M_{11}^{pq} & M_{12}^{pq}\end{pmatrix}\bullet\begin{pmatrix}x_1^{p1} & x_1^{p2} & \cdots & \cdots & x_1^{pNo}\\ x_2^{p1} & x_2^{p2} & \cdots & \cdots & x_2^{pNo}\end{pmatrix} = \begin{pmatrix}x_1^{q1} & x_1^{q2} & \cdots & \cdots & x_1^{qNo}\end{pmatrix},$$

$$\begin{pmatrix}M_{21}^{pq} & M_{22}^{pq}\end{pmatrix}\bullet\begin{pmatrix}x_1^{p1} & x_1^{p2} & \cdots & \cdots & x_1^{pNo}\\ x_2^{p1} & x_2^{p2} & \cdots & \cdots & x_2^{pNo}\end{pmatrix} = \begin{pmatrix}x_2^{q1} & x_2^{q2} & \cdots & \cdots & x_2^{qNo}\end{pmatrix}. \qquad \textbf{(2.1)}$$

Fitting for $M^{pq}{}_{ij}$ is more involved now that the orbit vectors on both sides of Eqn. 2.1 have random errors, most likely coupled in the manner of Eqn. 1.3.

Similar problem involving <u>uncorrelated</u> orbit errors in normalized coordinates has been addressed.(2) They correspond to eigenvectors of the covariance matrix constructed as follows:

$$\mathbf{C}_{ij} = \sum_{k=1}^{N_o} z_i^k \cdot z_j^k , \quad \mathbf{C}_{ij} \cdot \mathbf{N}_j = \lambda \mathbf{N}_i , \quad \mathbf{u}_1 = \frac{\mathbf{N}}{|\mathbf{N}|} , \quad |\mathbf{u}_r| = 1, \; r = 2, \ldots .n. \tag{2.2}$$

The $z^k{}_i$'s are the orbit vectors normalized by the uncoupled errors. The eigenvectors $\mathbf{N}_i$ contain the fitted transfer matrix elements, which are then normalized to $\mathbf{u}_i$'s. The error covariance for the $\mathbf{u}_i$'s is given by

$$\left\langle (\delta \mathbf{u}_1)_i (\delta \mathbf{u}_1)_j \right\rangle = \sum_{r=2}^{n} \frac{\lambda_r + \lambda}{(\lambda_r - \lambda)^2} (\mathbf{u}_r)_i (\mathbf{u}_r)_j . \tag{2.3}$$

To make Eqn. 2.3 applicable, we need to take the following steps.

- <u>Diagonalization</u>: We find the transformations diagonalizing the orbit error covariance matrices in <u>both</u> upstream and downstream sections. This is accomplished with symplectic matrices of the form

$$O_{p,q} = \frac{1}{\sqrt{2}} \begin{pmatrix} 1 & \mp\sqrt{\frac{A_{p,q}}{D_{p,q}}} \\ \pm\sqrt{\frac{D_{p,q}}{A_{p,q}}} & 1 \end{pmatrix} , \qquad \begin{aligned} A_{p,q} &= \left\langle \delta x_1^{p,q} . \delta x_1^{p,q} \right\rangle , \\ B_{p,q} &= \left\langle \delta x_1^{p,q} . \delta x_2^{p,q} \right\rangle , \\ D_{p,q} &= \left\langle \delta x_2^{p,q} . \delta x_2^{p,q} \right\rangle . \end{aligned} \tag{2.4}$$

  In doing this we introduce extra couplings among the orbit vectors at **p** and **q**.
- <u>Application of Eqn. 2.3</u>: This gives the error covariance between the fitted matrix elements in the diagonalized coordinates.
- <u>Cross coupling between rows:</u> The two equations of Eqn. 2.1 appear uncorrelated. They nonetheless are coupled through sharing the same set of incoming orbits. This coupling has nontrivial effects when we restore to the undiagonalized coordinates. This effect, not addressed by Eqn. 2.3, has to be calculated.
- <u>Un-normalizing the unit vectors $\mathbf{u}_i$</u>: This gives the covariance in $\mathbf{N}_i$'s.
- <u>Un-diagonalization</u>: This gives the final error covariance in the physical coordinates, summarized as follows:

$$\left\langle \delta M_{ij}^{pq} \cdot \delta M_{kl}^{pq} \right\rangle = \sum_m \sum_n \sum_r \sum_s (O_q)_{im}^{-1} \cdot (O_q)_{kr}^{-1} \cdot \left\langle \delta M'^{pq}_{mn} \cdot \delta M'^{pq}_{rs} \right\rangle \cdot (O_p)_{nj} \cdot (O_p)_{sl} . \tag{2.5}$$

One quantity $\delta^{qi}{}_{Em}$, defined by

$$\begin{aligned} \delta_{Em}^{q1} &= M_{11}^{pq} \cdot \delta_{Xm}^{p1} + M_{12}^{pq} \cdot \delta_{Xm}^{p2} - \delta_{Xm}^{q1} , \\ \delta_{Em}^{q2} &= M_{21}^{pq} \cdot \delta_{Xm}^{p1} + M_{22}^{pq} \cdot \delta_{Xm}^{p2} - \delta_{Xm}^{q2} , \qquad m = 1, 2, \cdots N_O , \end{aligned} \tag{2.6}$$

stands out in the final expression. Notice that $\delta^{qi}{}_{Em}$ has an index m for the trajectory number. It represents the error at the exit point **q** when the difference between the exit orbit and the properly propagated entrance orbit from **p** is calculated. The error covariance in the fitted matrix elements then takes on an intuitive form:

$$\left\langle \delta M_{ij}^{pq} . \delta M_{km}^{pq} \right\rangle = \frac{1}{N_O} \cdot \frac{\left\langle \delta_E^{qi} \cdot \delta_E^{qk} \right\rangle}{\left\langle x_j^p \cdot x_m^p \right\rangle_{(d)}}, \qquad i, j, k, m = 1, 2.$$

$$\left\langle x_j^p . x_m^p \right\rangle_{(d)} = \begin{cases} \left\langle x_j^p . x_m^p \right\rangle \cdot \left(1 - R_p^2\right), & j = m \\ \left\langle x_j^p . x_m^p \right\rangle \cdot \left(1 - R_p^{-2}\right), & j \neq m \end{cases}, \qquad R_p^2 = \frac{\left\langle x_1^p . x_2^p \right\rangle^2}{\left\langle x_1^p . x_1^p \right\rangle \cdot \left\langle x_2^p . x_2^p \right\rangle}. \qquad \textbf{(2.7)}$$

3. OVERALL FORM FACTOR

Combining Eqns. 1.3 and 2.7, we can calculate the overall error covariance in the fitted matrix elements in terms of the signal-to-noise ratio of the BPM's. We need to use the generalized symplectic condition:

$$m_{12}^{pai} = -\frac{P_p}{P_{ai}} \left(m^{pai}\right)_{12}^{-1} = -\frac{P_p}{P_{ai}} m_{12}^{aip}, \quad m_{11}^{pai} = -\frac{P_p}{P_{ai}} \left(m^{pai}\right)_{22}^{-1} = -\frac{P_p}{P_{ai}} m_{22}^{aip}, \qquad \textbf{(3.1)}$$

in case the momenta are different at **p** and **q**. This allows us to propagate all the orbits from **p** to **q**. The overall error covariance is then given by

$$\left\langle \delta M_{ij}^{pq} . \delta M_{km}^{pq} \right\rangle = 2 \cdot \frac{1}{N_O} \cdot S_B^{jm} \cdot \frac{1}{T_{(d)}^{jm}} \cdot \left[\frac{1}{N_{Bq}} \cdot \mathcal{M}_b^i \cdot \mathcal{M}_b^k + \frac{1}{N_{Bp}} \cdot \mathcal{M}_a^i \cdot \mathcal{M}_a^k \cdot \left(\frac{P_p}{P_q} \right) \right], \qquad i, j, k, m = 1, 2.$$

$$S_B^{jm} = \frac{\sigma_B^2}{\sigma_O^{pj} \cdot \sigma_O^{pm}}, \qquad T_{(d)}^{jm} = \begin{cases} \left(1 - R_p^2\right), & j = m \\ \left(1 - R_p^{-2}\right) \cdot R_p, & j \neq m \end{cases}, \qquad R_p^2 = \frac{\left\langle x_1^p . x_2^p \right\rangle^2}{\left\langle x_1^p . x_1^p \right\rangle \cdot \left\langle x_2^p . x_2^p \right\rangle},$$

$$\mathcal{M}_a^1 = \frac{\left\langle m_{12}^{qa} \right\rangle}{\left\langle m_{12}^{aa} \right\rangle}, \quad \mathcal{M}_a^2 = \frac{\left\langle m_{11}^{qa} \right\rangle}{\left\langle m_{12}^{aa} \right\rangle}, \quad \mathcal{M}_b^1 = \frac{\left\langle m_{12}^{qb} \right\rangle}{\left\langle m_{12}^{bb} \right\rangle}, \quad \mathcal{M}_b^2 = \frac{-\left\langle m_{11}^{qb} \right\rangle}{\left\langle m_{12}^{bb} \right\rangle}. \qquad \textbf{(3.2)}$$

The physical significance of these quantities deserves some elaboration:

1. Factors determined by experimental parameters:

- $S_B{}^{jm}$ is the generalized signal to noise ratio. It may take on a dimension of meter or meter^2 in some cases.
- $T_{(d)}{}^{jm}$ characterizes the position-angle coupling at the observation point **p**, nearly inevitable in real experiments. When R_p=0, this term makes some of the correlation terms disappear.
- N_O is the sample size, i.e., the number of orbits used.

2. Factors determined by machine parameters:

- $N_{p,q}$ are the number of BPM's used to determine each trajectory at the observation points **p** and **q** respectively. It's evident by Eqn. 3.1 that the overall precision can not be improved indefinitely by increasing the number of BPM's only on one side of the measured section.

- $P_{p,q}$ are the momentum values at the observation points **p** and **q**.
- $\mathcal{M}_{a,b}^{1,2}$ are the RMS ratios defined in Eqn. 1.3. Their evaluation is easier than appears.(1) Notice the minus sign in the last equation.

For a BPM system designer, these quantities translate into other machine specifications and have to be taken into account in optimizing the performance. For example, N_O is limited by the speed of the BPM electronics and operation/control interface, S_B^{jm} is limited by the beam pipe radius and transfer properties all around the machine, $\mathcal{M}_{a,b}^{1,2}$ are bound by optical or experimental conditions, while everything else has to conform to cost restrictions. But Eqn. 3.2 does take the guesswork out of the design so far as optical requirements are concerned. All analytic formulas presented above have been numerically verified.

4. OVERALL CRITERION FOR PERIODIC SYSTEMS

We obtained the precision criterion for measuring a specific section of any beam line. In designing large modular machines, the emphasis however can be more on a concise figure of merit for the overall achievable precision, while overlooking minor features from a particular module. Figure 2 conceptualizes such a system. One can conceive an optics verification scheme where the transfer matrix across each module is measured in turn, using any combination of the remaining modules for trajectory determination. The latter thus fulfills an extended notion of the trajectory measurement section discussed earlier. Figure 2 also shows the lattice parameters at each module boundary and the phase advance per module, as well as module numbers **I**, **J** and BPM numbers per module N.

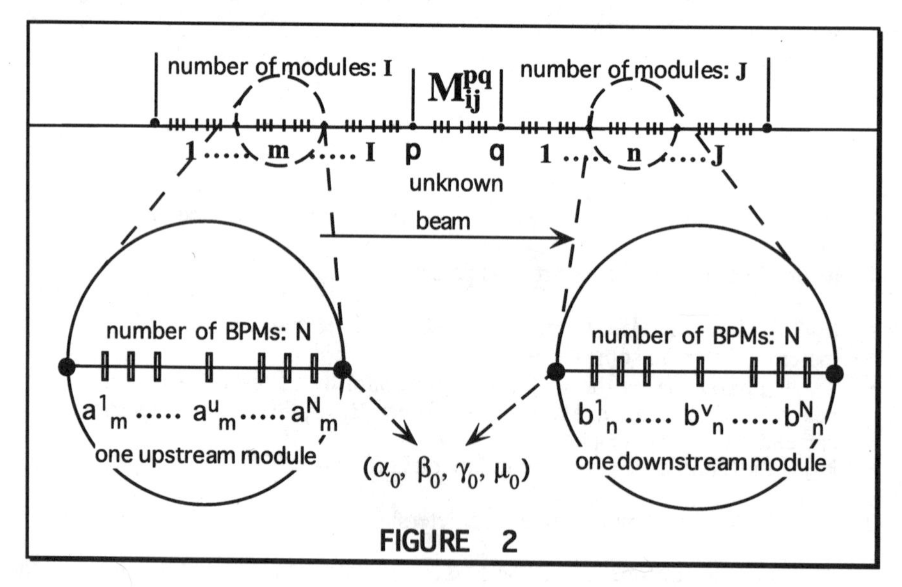

FIGURE 2

Replacing all the RMS matrix elements in Eqn. 3.2 with lattice parameters within each identical module and using the partition formula Eqn. 1.4, we arrive at the following identities (All summations are over BPM's within a single module.):

$$\langle m_{12}^{aa}\rangle^2 = \frac{1}{2}\cdot\left\{\left\{\langle\beta\beta\rangle - \frac{1}{\mathbf{I}}\cdot\langle\beta\beta\rangle_{\sin} - \frac{1}{\mathbf{I}^2}\cdot\left(\frac{\sin(\mathbf{I}\cdot\mu_0)}{\sin(\mu_0)}\right)^2\cdot\left(\langle\beta\beta\rangle_{\cos} - \langle\beta\beta\rangle_{\sin}\right)\right\}\right\},$$

$$\langle m_{12}^{bb}\rangle^2 = \frac{1}{2}\cdot\left\{\left\{\langle\beta\beta\rangle - \frac{1}{\mathbf{J}}\cdot\langle\beta\beta\rangle_{\sin} - \frac{1}{\mathbf{J}^2}\cdot\left(\frac{\sin(\mathbf{J}\cdot\mu_0)}{\sin(\mu_0)}\right)^2\cdot\left(\langle\beta\beta\rangle_{\cos} - \langle\beta\beta\rangle_{\sin}\right)\right\}\right\},$$

$$\langle m_{11}^{qa}\rangle^2 = \frac{1}{2}\cdot\gamma_0\cdot\left\{\langle\beta\rangle + \frac{1}{\mathbf{I}}\left(\frac{\sin(\mathbf{I}\cdot\mu_0)}{\sin(\mu_0)}\right)\cdot\langle\beta\rangle_{c1}\right\},\quad \langle m_{12}^{qa}\rangle^2 = \frac{1}{2}\cdot\beta_0\cdot\left\{\langle\beta\rangle - \frac{1}{\mathbf{I}}\left(\frac{\sin(\mathbf{I}\cdot\mu_0)}{\sin(\mu_0)}\right)\cdot\langle\beta\rangle_{c2}\right\},$$

$$\langle m_{11}^{qb}\rangle^2 = \frac{1}{2}\cdot\gamma_0\cdot\left\{\langle\beta\rangle + \frac{1}{\mathbf{J}}\left(\frac{\sin(\mathbf{J}\cdot\mu_0)}{\sin(\mu_0)}\right)\cdot\langle\beta\rangle_{c3}\right\},\quad \langle m_{12}^{qb}\rangle^2 = \frac{1}{2}\cdot\beta_0\cdot\left\{\langle\beta\rangle - \frac{1}{\mathbf{J}}\left(\frac{\sin(\mathbf{J}\cdot\mu_0)}{\sin(\mu_0)}\right)\cdot\langle\beta\rangle_{c4}\right\},$$

$$\langle\beta\beta\rangle = \frac{1}{N^2}\sum_{i=1}^{N}\sum_{j=1}^{N}\beta^i\cdot\beta^j,\quad \langle\beta\rangle = \frac{1}{N}\sum_{j=1}^{N}\beta^j,\quad \langle\beta\beta\rangle_{\sin} = \frac{1}{N^2}\sum_{i=1}^{N}\sum_{j=1}^{N}\beta^i\cdot\beta^j\cdot\sin^2(\varphi^i-\varphi^j),$$

$$\langle\beta\beta\rangle_{\cos} = \frac{1}{N^2}\sum_{i=1}^{N}\sum_{j=1}^{N}\beta^i\cdot\beta^j\cdot\cos^2(\varphi^i-\varphi^j) = \langle\beta\beta\rangle - \langle\beta\beta\rangle_{\sin},$$

$$\langle\beta\rangle_{c1} = \frac{1}{N}\sum_{j=1}^{N}\beta^j\cdot\cos((3+\mathbf{I})\cdot\mu_0 - 2\cdot\varphi^j + \vartheta_\alpha),\quad \langle\beta\rangle_{c2} = \frac{1}{N}\sum_{j=1}^{N}\beta^j\cdot\cos((3+\mathbf{I})\cdot\mu_0 - 2\cdot\varphi^j),$$

$$\langle\beta\rangle_{c3} = \frac{1}{N}\sum_{j=1}^{N}\beta^j\cdot\cos((1-\mathbf{J})\cdot\mu_0 - 2\cdot\varphi^j + \vartheta_\alpha),\quad \langle\beta\rangle_{c4} = \frac{1}{N}\sum_{j=1}^{N}\beta^j\cdot\cos((1-\mathbf{J})\cdot\mu_0 - 2\cdot\varphi^j),$$

$$\vartheta_\alpha = \tan^{-1}\left(\frac{2\cdot\alpha_0}{1-\alpha_0^2}\right). \qquad \mathbf{(4.1)}$$

The form factor, Eqn. 3.2, can now be computed for any module in the entire machine using only summations within one module. A good strategy is therefore to choose as simple a module as possible without compromising periodicity. Once the precision requirement for a typical optics measurement is defined, the designer can translate it, using Eqns. 3.2 and 4.1, into requirements for the BPM signal-to-noise ratio with relative ease. For example, in a large storage ring with **K** identical modules, the optical performance can be quantified by the transfer matrix measurement of each module using half of the remaining modules each for upstream and downstream trajectory determination. In this case substitution of **K**/2 for both **I** and **J** in Eqn. 4.1 will give the relation between these two requirements. When **K** is large, this relation becomes even more simple.

REFERENCES

1. Complete detail, more specific formulas and numerical examples can be found in Chao, Y., "Optics-Driven Design Criteria for BPM's," CEBAF TN-93-073 (1993).
2. Lohse, T. and Emma, P., "Linear Fitting of Beam Orbits and Lattice Parameters," SLAC-CN-371 (1989).

Beam Distributions Beyond RMS

F.-J. Decker
*Stanford Linear Accelerator Center**,
Stanford University, Stanford, CA 94309, USA

Abstract

The beam is often represented only by its position (mean) and the width (rms = root mean squared) of its distribution. To achieve these beam parameters in a noisy condition with high backgrounds, a Gaussian distribution with offset (4 parameters) is fitted to the measured beam distribution. This gives a very robust answer and is not very sensitive to background subtraction techniques. To get higher moments of the distribution, like skew or kurtosis, a fitting function with one or two more parameters is desired which would model the higher moments. In this paper we will concentrate on an Asymmetric Gaussian and a Super Gaussian function that will give something like the skew and the kurtosis of the distribution. – This information is used to quantify special beam distribution. Some are unwanted like beam tails (skew) from transverse wakefields, higher order dispersive aberrations or potential well distortion in a damping ring. A negative kurtosis of a beam distribution describes a more rectangular, compact shape like with an over-compressed beam in z or a closed to double-horned energy distribution, while a positive kurtosis looks more like a "Christmas tree" and can quantify a beam mismatch after filamentation. Besides the advantages of the quantification, there are some distributions which need a further investigation like long flat tails which create background particles in a detector. In particle simulations on the other hand a simple rms number might grossly overestimate the effective size (e.g. for producing luminosity) due to a few particles which are far away from the core. This can reduce the practical gain of a big theoretical improvement in the beam size.

1. INTRODUCTION

Beam distributions are measured by different techniques. Transverse distributions are generated simply by screens or projections directly by wire scanners. In the longitudinal phase space, the z-distribution is measured by Streak cameras and the energy distribution is measured by the distribution at a dispersive region. These one-dimensional distributions (or the one-dimensional projections) have a Gaussian shape, if the mechanism for generating the shape is purely statistical. An example is the transverse distribution after the radiative damping in a damping ring. Different effects can disturb this shape and can therefore be an indication for the origin of the disturbance. Transverse wakefields kick the tail of the bunch and create an asymmetric distribution. Quantifying this effect with an Asymmetric Gaussian fit function has help to improve the SLC performance. The next chapter discusses this function, the relation to the skew and the causes of different other asymmetric distributions.

* Work supported by the Department of Energy contract DE-AC03-76SF00515.

Then the next higher moment is studied, which can be fitted by a Super Gaussian fit function and gives a value for the kurtosis. These higher moments give hints of how much a size can be reduced by which effect. A simple increase in the Gaussian beam size (or emittance) from one point to another is more difficult to attack.

At the end there are some examples given how an rms number generated by a simulation can lead to wrong estimates and answers. The right effect can be more easily implemented, but might need more CPU time.

2. ASYMMETRIC GAUSSIAN

An asymmetric Gaussian fit function was developed to get a quantitative answer for the asymmetry of a beam spot especially induced by transverse wakefield. Different approaches are discussed elsewhere [1,2], which include a more detailed understanding of the wakefields. A simple additional parameter to a Gaussian fit function can give most of the desired information.

2.1 Fit Function

A Gaussian function is represented by 4 parameters which cover an offset, the peak height, the centering, and the size:

$$A + B \cdot g(x - x_0)\ , \quad \text{with} \qquad g(x) = \frac{1}{\sqrt{2\pi}\sigma} \exp\left(\frac{-x^2}{2\sigma^2}\right).$$

With an additional parameter E the skewness of the distribution can be estimated:

$$g(x) = \frac{1}{\sqrt{2\pi}\sigma} \exp\left(\frac{-x^2}{2(\sigma \cdot (1 + \mathrm{sign}(x) \cdot E))^2}\right).$$

This is like fitting a left and right half of a Gaussian to the distribution with

$$\sigma = \frac{\sigma_r + \sigma_l}{2}, \text{ and } E = \frac{\sigma_r - \sigma_l}{\sigma_r + \sigma_l}.$$

The precise values for the rms and the skew of this distribution can be calculated and are:

$$rms = \sigma \cdot \sqrt{1 + \left(3 - \frac{8}{\pi}\right) \cdot E^2}, \text{ and}$$

$$skew = E \cdot \sqrt{\frac{8}{\pi}} \cdot \left(1 + \left[\frac{16}{\pi} - 5\right] E^2\right)$$

giving a small correction to just σ and E.

E gives roughly the amount of improvement possible, the exact value for the small σ is:

$$\sigma_{\min} = \sigma \cdot (1-|E|).$$

Figure 1 shows a distribution of a beam profile with a Gaussian and an asymmetric Gaussian distribution.

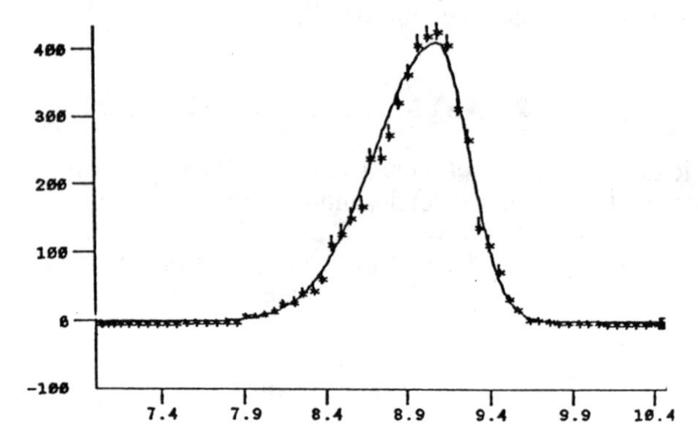

Fig. 1: Asymmetric Gaussian Fit Function

A Gaussian fit to an asymmetric distribution would only indicate the asymmetry, while the asymmetric fit gives an estimate of the beam blow up due to the skewness, ($E = -0.35$ in this case).

2.2 Reasons for Beam Asymmetry

Besides wakefields, higher order dispersion T_{166} and potential well distortion can lead to asymmetries in the bunch shape.

2.2.1 Wakefields

A beam offset in a cavity will generate transverse wakefields, which will kick the tail of the bunch. A betatron oscillation will drive the tail further and further out. Fig. 2 shows a simulation for two betatron oscillations and the resulting distribution for $3 \cdot 10^{10}$ particles with an asymmetric fit. The asymmetry parameter of $T = 18\%$ indicates most of the possible improvement of 24%.

2.2.2 Higher Order Dispersion

Besides the normal linear dispersion η there can be higher order terms like the quadratic T_{166} term which will generate beam tail.

2.2.3 Potential Well Distortion

The longitudinal beam shape in the SLC damping ring is strongly influenced by longitudinal wakefields which distort the focusing potential well, giving an asymmetric distribution which was calculated [3] and measured with an E-parameter of 0.5.

3.2 Reasons for Big Beam Kurtosis

The beam can get a rectangular-like shape by folding it on top of itself in the other dimension of the phase space. By smearing it out like filamentation, the beam distribution gets wide symmetric tails.

3.2.1 Rectangular Distribution

For N bigger than 2 the value $N/2$ gives a factor of how much smaller the beam spot would be if it were a simple Gaussian with the rise and fall slopes of the more rectangular distribution. Fig. 3 shows an example of an over-compressed beam [4] indicating the 2.5 times smaller possible bunch length. (The bunch is purposely formed in that way to compensate longitudinal wakefields giving a small energy spread at the end of the SLC linac.) A double-horned energy distribution and a filamented beam offset are other examples giving an S-shape or respectively donut in phase space.

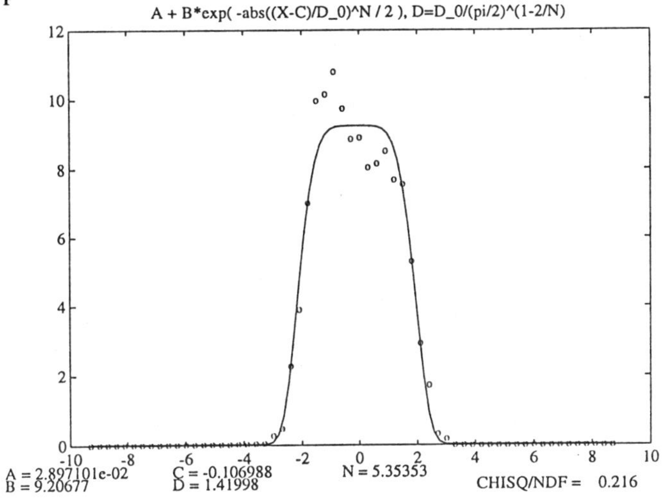

Fig. 3: Simulated Longitudinal Bunch Shape and Fit

The non-linearity of the rf in a compression scheme can be used to form the bunch in such a way that it will give a small energy spread at the end. The fit can quantify this distribution in an analytical way to use it for other studies and comparison with experiments.

3.1.2 Christmas Tree Distribution

If the form parameter N is less than 2 it will describe distributions with a small peak and wide tails (like a Christmas tree). Such distributions were observed earlier with no direct tool to quantify and fix. In simulations such a distribution is easily achieved by a betatron mismatch and filamentation (smearing out the elliptic concentrically in phase space). The distributions and fits for different mismatches are shown in Fig. 4. A measured distribution can now be quantified and compared with simulations to give a prediction for the size of the mismatch.

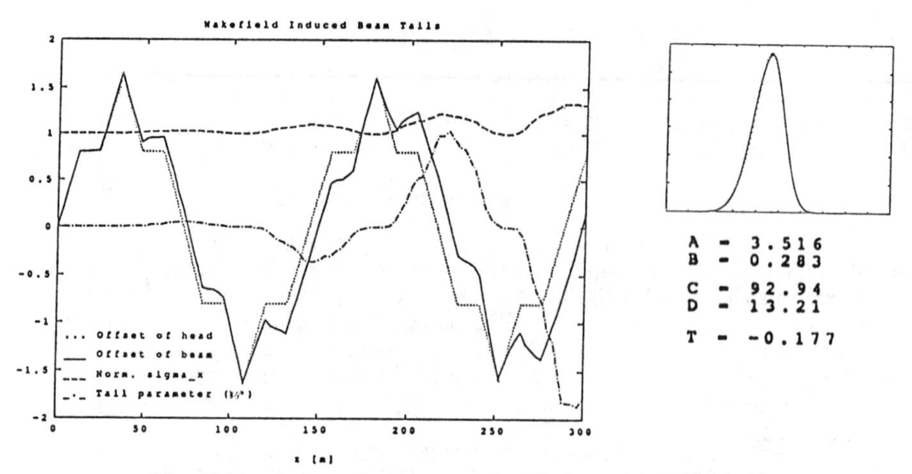

Fig. 2: Simulation of a Betatron Oscillation with Wakefields

A betatron oscillation (left) creates a wakefield tail which blows up the beam size sigma_x. The asymmetry (tail) is well parametrized by an asymmetric Gaussian (right) and gives quantitative values for the possible improvement.

3. SUPER GAUSSIAN DISTRIBUTION

Super Gaussian distributions are used in laser physics to describe higher order beam modes and therefore more rectangular distributions. The steepness of the rectangular shape gives hints for a possible reduction in size and can be quantified by an additional parameter in the exponent. First the mathematical function, then some beam distributions like a longitudinal bunch distribution with over-compression, energy distributions, and special transverse distribution from mismatched beam after filementation are discussed.

3.1 Super Gaussian Function

A distribution which has a symmetric higher moment can be approximated with a Super Gaussian function where the exponent of the Gaussian is a variable N and will give a Gaussian for $N = 2$. For big N the function will describe a more rectangular distribution, while for small N it fits to a distribution with long tails on both sides

$$g(x) = \frac{1}{\sqrt{2\pi}\sigma_0} \exp\left(\frac{-(\mathrm{abs}(x))^N}{2\sigma_0^N} \right) \quad \text{with} \quad \sigma = \sigma_0 \cdot \left(\frac{\pi}{2} \right)^{2/N-1} .$$

The difference between σ and σ_0 helps to keep σ close to the right number of a normal Gaussian fit.

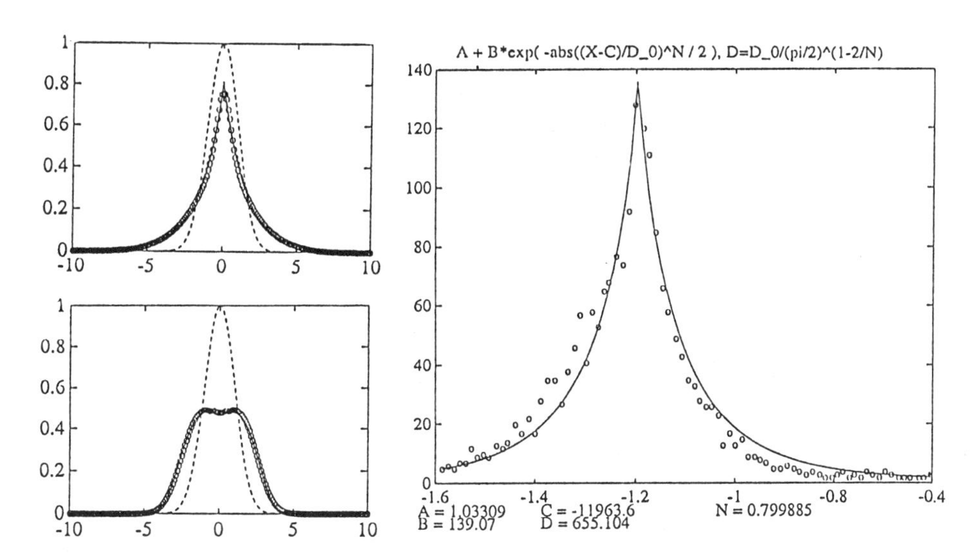

Fig. 4: Filamented Beam Shape after a Betatron Mismatch

A peaked distribution with wide tails give a hint for a mismatch which is already filamented. By fitting a Super Gaussian function to that shape the amount of the mismatch can be measured (Bmag more than 5, left). An offset will filament to a donut shape in phase space which will give a more rectangular shape.

4. RMS IN SIMULATIONS

To get a quick result of a beam size in simulations the rms (root mean square) is used. This number might have not much in common with the effective size of the distribution which will be show in an example and some simulation results.

4.1 RMS Example

Let's assume a Gaussian beam distribution in y. The luminosity is proportional to 1/size. Now we take 2% of that distribution and put it to a big halo around the beam, so that the rms number goes up by a factor of two indicating half the luminosity. On the other hand the real luminosity is only reduced by 4% since 2% of each bunch are more or less not contributing to the luminosity.

4.2 Simulation of Effective Size

The effective size of a distribution depends on the subject you are studying. If the concern is background in the detector more interest is spend on the behavior of tail and halo particles, while the core is relevant for luminosity. The right effective size for luminosity can be calculated by convoluting the distribution of the two colliding beams, which is essentially a simulation of the collision.

Figure 5 shows the effect of a large higher order chromatic term (U_{3466}) [5] in the final focus optics for different angular divergences. The simple rms value would predict a large degradation in spot size, while the effective size enlargement is much more moderate and closer to the linear optics results.

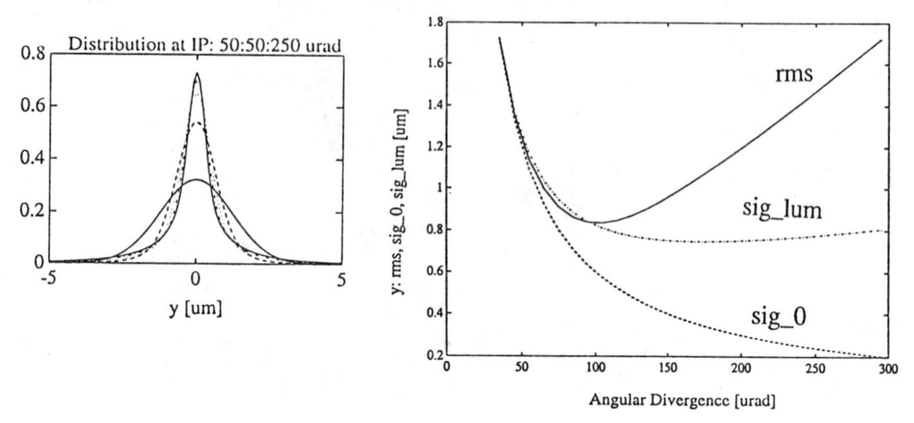

Fig. 5: RMS and Effective Sizes of a Beam Distribution

At the interaction point the beam distribution gets influenced by higher order chromatic terms giving wide tails and therefore a bigger rms value for higher angular divergences. The effective beam size is still going down since it depends on the core where the peak height is still rising.

CONCLUSION

Higher moments in a distribution can be fitted with special functions, Asymmetric Gaussian for the 3rd and Super Gaussian for the 4th moment. They give the advantage of a special form that is robust against varying pedestrial offsets below the measured distribution. Different beam conditions are discussed, which are quite remarkably fitted by these functions which quantify the measured effect. A simple rms value in simulations can lead to wrong conclusions if wide tails are present.

REFERENCES

[1] J.T. Seeman, F.-J Decker, I. Hsu, C. Young, *Characterization and Monitoring of transverse beam tails,* PAC, Washington, May 1993, p1734.

[2] W.L. Spence, F.-J. Decker, M.D. Woodley, *Transverse tails and higher order moments,* PAC, Washington, May 1993, p3576.

[3] K. Bane, private communication

[4] F.-J Decker, R. Holtzapple, T. Raubenheimer, *Over-compression a method to shape the longitudinal bunch distribution for a reduced energy spread,* LINAC94, Tsukuba, Aug. 1994.

[5] N.J. Walker, R. Helm, J. Irwin, M. Woodley, *Third-order corrections to the SLC final focus,* PAC, Washington, May 1993, p92.

Wavelet Analysis and Accelerator Signals

Arnold Stillman*

Brookhaven National Laboratory, Upton, NY, 11973-5000

Abstract

Wavelets have suddenly become very popular in signal analysis applications. They have become a darling topic in several fields: astronomy, image processing, filtering, theoretical physics, and harmonic analysis, to name only a few. This paper focuses on applications for wavelet analysis in accelerator engineering, specifically, signal processing. It is split into three parts; the first contrasts wavelet analysis with Fourier analysis, the second constructs a simple wavelet basis as a tutorial, the third demonstrates the analytic uses of wavelets with the help of a software analyzing tool.

1 Wavelet and Fourier Analyses

To understand why wavelets are not just an extension of Fourier sines and cosines, it is useful first to explore their similarities.

A common definition of the Fourier transform is:

$$\hat{f}(\omega) = \frac{1}{\sqrt{2\pi}} \int_{-\infty}^{\infty} e^{-i\omega x} f(x)\, dx \tag{1}$$

The Fourier coefficients;

$$c_n = \frac{1}{2\pi} \int_0^{2\pi} e^{-in\omega_0 x} f(x)\, dx. \tag{2}$$

represent the distribution of frequencies in the input function. Summing the contribution from each frequency gives a reconstruction of the original signal;

$$f(x) = \sum_{-\infty}^{\infty} c_n e^{in\omega_0 x}. \tag{3}$$

These definitions while only too familiar, are useful to keep in mind, since the analogous wavelet transforms have similar definitions. First, however, some preliminary concepts are important.

Fourier analysis involves an integral transform which uses only one complex variable, $z = e^{i\omega x}$. The fact that the analysis decomposes an input function into both sines and cosines is misleading, since the two functions are the orthogonal imaginary and real parts of a single complex number. There is no choice of basis

*Work performed under the auspices of the U.S. Department of Energy, Contract No. DE-AC02-76CH00016.

functions, and the integral transform encompasses all frequencies. Wavelets are fundamentally different. There is a multitude of basis functions, not necessarily orthogonal, and generally with only local support. The art of wavelet analysis is in choosing the appropriate basis set for the specific task, and the choice of what basis is best depends on the type of analysis desired.

Central to wavelet analysis, but alien to Fourier analysis, is the existence of a scaling function. The wavelets themselves derive from this scaling function, similar to the generating function that is customary in series expansions, for instance. The wavelet initially generated by the scaling function generates in turn all succeeding wavelets in the basis set. Let us start with a general scaling function, $\phi(x)$, so called because its dilations change the analyzing scale of the wavelet transform (just as high frequency sines and cosines select the sharper transients in a signal). Its dilations and translations, which are intimately linked, are

$$\phi_{a,b} = |a|^{-1/2}\phi\left(\frac{x-b}{a}\right); \tag{4}$$

where the indices a and b label the scales and translations, respectively. Note also that a and b are continuous parameters; this definition of the scaling function is the kernel for the continuous wavelet transform. To generate the discrete version of the scaling function, let a and b assume only discrete values using the assignments $a \to a_0^{-m}$ and $b \to nb_0a_0^{-m}$. The discrete version of the scaling function then becomes

$$\phi_{m,n} = a_0^{-m/2}\phi(a_0^{-m}x - nb_0); \tag{5}$$

Now, define $\phi_{0,0} \equiv \phi(x)$. The translation index of zero implies that ϕ exists in the vicinity of the origin, and the scale index of zero defines a central frequency or scale. With this location and scale for ϕ, we have the start of a *multiresolution analysis.* A multiresolution anaysis labels each "window" with a scale, and each succeeding level has a scale factor half the preceeding one. Each set of translations for a given scale form an orthogonal basis set for that scale(1), thus, it is possible to write ϕ as

$$\phi = \sum_n h_n\phi_{-1,n}. \tag{6}$$

The inner product of the orthogonal scale functions extracts the coefficients, h_n,

$$h_n = \langle \phi, \phi_{-1,n} \rangle; \tag{7}$$

which are an orthonormal set;

$$\sum_n |h_n|^2 = 1. \tag{8}$$

To define the initial wavelet, ψ, expand in the scale function in the exact same manner as eq. 6;

$$\psi = \sum_n g_n \phi_{-1,n}. \tag{9}$$

Since this is an expansion of ψ in terms of the functions ϕ, there is some freedom in the definition of the coefficients g_n. A convenient choice(1) for the g_n is,

$$g_n = (-1)^n \overline{h_{-n+1}}. \tag{10}$$

The selection of these coefficients is central to the analyzing properties of the wavelet, and many methods of selecting them exist(1)(2)(3). The function ψ in eq. 9 is the wavelet which generates the full wavelet basis, often called the mother wavelet, and the scaling function is, less often, the father of the wavelets. In the original French literature on wavelets, the wavelets themselves are *ondelettes* and the scaling functions, *pre-ondelettes.*

Choosing the set of wavelet coefficients in eq. 10 defines the wavelet basis; the variously named wavelets in the rest of this paper have different rules for defining the g_n. Immediately, it is apparent that a fundamental difference between wavelet and Fourier analysis is the ability to choose a basis set. There are now many more options than only the sines and cosines. The benefit of this greater freedom in choice of basis sets is the ability of wavelets to analyze on different scales, a sort of mathematical "zoom" function, so to speak. This multiresolution property is a consequence of the local support for each wavelet. By nature of their definition as a sum over translations of a compactly supported scale function, wavelets have finite values only locally, and tend to zero at $\pm\infty$. Sharply localized wavelets can resolve faster transients in a signal, while highly oscillatory wavelets can filter modulated signals quite well. The figures in the third section demonstrate some of these analytic abilities.

2 Constructing the Basis

The simplest choice of scaling function is the Haar scaling function, which is simply, for engineers, the familiar θ function;

$$\theta(x) = \begin{cases} 1, & 0 \leq x \leq 1 \\ 0, & \text{otherwise} \end{cases} \tag{11}$$

To generate the mother wavelet for the Haar system, it is necessary to calculate the g_n and h_n. For the Haar basis, all the h_n are real and $\overline{h_n} = h_n$. For $a_0 = 2$ and $b_0 = 1$, the coefficients are

$$g_0 = h_1 =< 2^{1/2}\theta(x), 2^{1/2}\theta(2x-1) >= (2) \cdot (1/2) = 1, \tag{12}$$

and

$$g_1 = -h_0 = - < 2^{1/2}\theta(x), 2^{1/2}\theta(2x) >= (2) \cdot (-1/2) = -1. \tag{13}$$

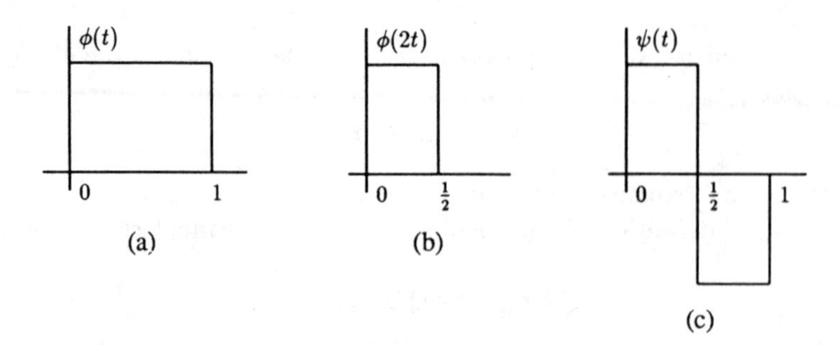

Figure. 1. (a) and (b) are the Haar scaling function and $\phi_{-1,0}$, it's next finer scale version, untranslated. (c) is the Haar mother wavelet, ψ_0.

All other coefficients are zero, since they are inner products of orthogonal functions. They are orthogonal since all the functions $\theta(2x-n)$ for $n \neq 0, 1$ have compact support that exists outside the support of $\theta(x)$. That is, $\theta(x)$ is zero everywhere they are not.

Now, according to the prescription in eq. 9 and the ceofficients g_0 and g_1, the mother wavelet consists of the scaling function of the next finer scale, minus itself translated by one unit;

$$\psi(x) = \phi(2x) - \phi(2x-1) \begin{cases} 1, & 0 \leq x \leq 1/2 \\ -1, & 1/2 \leq x \leq 1 \\ 0, & \text{otherwise} \end{cases} \tag{14}$$

Figure 1 shows the functions $\phi(x), \phi(2x)$, and $\psi(x)$. The Haar wavelet system is sharply localized in time, but spread out in frequency, since the Fourier transform of the Haar scaling function is the broad function $\sin(\omega)/\omega$. An example of the ability of the Haar system to localize properties in time will demonstrate the capabilities of even this simple function.

3 Practical Applications

There are now available adequate sofware tools for wavelet analysis. The familiar routines in(3) now include a Fast Wavelet Transform, and there is also available an on-line analysis and graphing program for use on several platforms.(4) Figure 2 is an example of the output of the program, Xwpl. It shows the wavelet analysis of a sine wave embedded in white noise. The sine wave starts abruptly halfway through the time series in the full signal window. Although there are many zero crossings in the raw signal, the wavelet analysis reconstructs the true zero crossings with no error.

The second example is a filtering experiment. A sine wave abruptly changes to white noise halfway through the time scale. The change is continuous at the midway point. The analyzing wavelet is a relatively smooth one. The smoothness allows the wavelet reconstruction to pick out only the sine wave and ignore the noise, though the white noise *surely has a component at this frequency.* This is the beauty of the scaling and translation properties of the wavelet transform. Though there is energy in the noise at this particular frequency, it is uncorrelated over the second half of the time interval, and the analysis ignores it. The final example is of a real accelerator signal, a PUE signal from the AGS Damper system. The analyzing wavelet now looks like the ripples in the last two peaks of every group of four bunches. It is often worthwhile to start a wavelet analysis with a wavelet that has some property in common with the input signal. The Haar wavelet case took advantage of the abrupt edges in the wavelet to select zero crossings. In the present case, the need is to resolve the bunches in the PUE signal. The phase space window indicates two delta functions in frequency extending throughout the window area. Some experimentation with the best level and best basis expansions is necessary to achieve this resolution, though.

4 Conclusion

Wavelet analysis is a new mathematical procedure that is quickly becoming useful in engineering applications. It does not replace Fourier analysis; its analytic power is in simultaneous time and frequency resolution. It thus requires a thoughtful application of the method and careful consideration of the result. The set of analyzing wavelets, for instance, is quite broad. The engineering insight is that of knowing, via experimentation or derivation, which wavelets work best in a particular application. I have presented examples of the types of signals that are difficult to resolve, and have shown the ability of wavelet analysis to transcend the difficulties in the Fourier analysis of them.

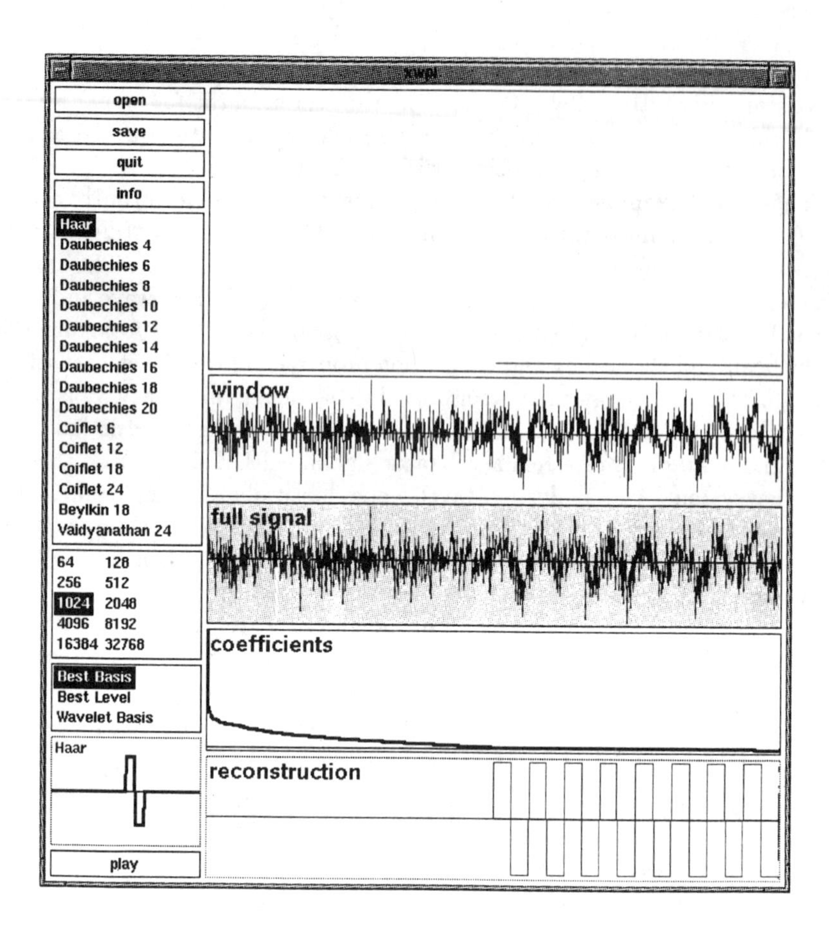

Figure. 2. A wavelet analysis of a sine wave embedded in noise at the midway point in a time series. The input series is in the **window** area. The program orders the coefficients of the reconstruction expansion in decreasing magnitude; there are relatively few large coefficients. The upper window is a phase space portrait. Time increases to the left, scale (or frequency) increases upward. The sudden appearance of the sine wave manifests itself as a Dirac delta function in frequency, extended along the time axis for the duration of the sine signal.

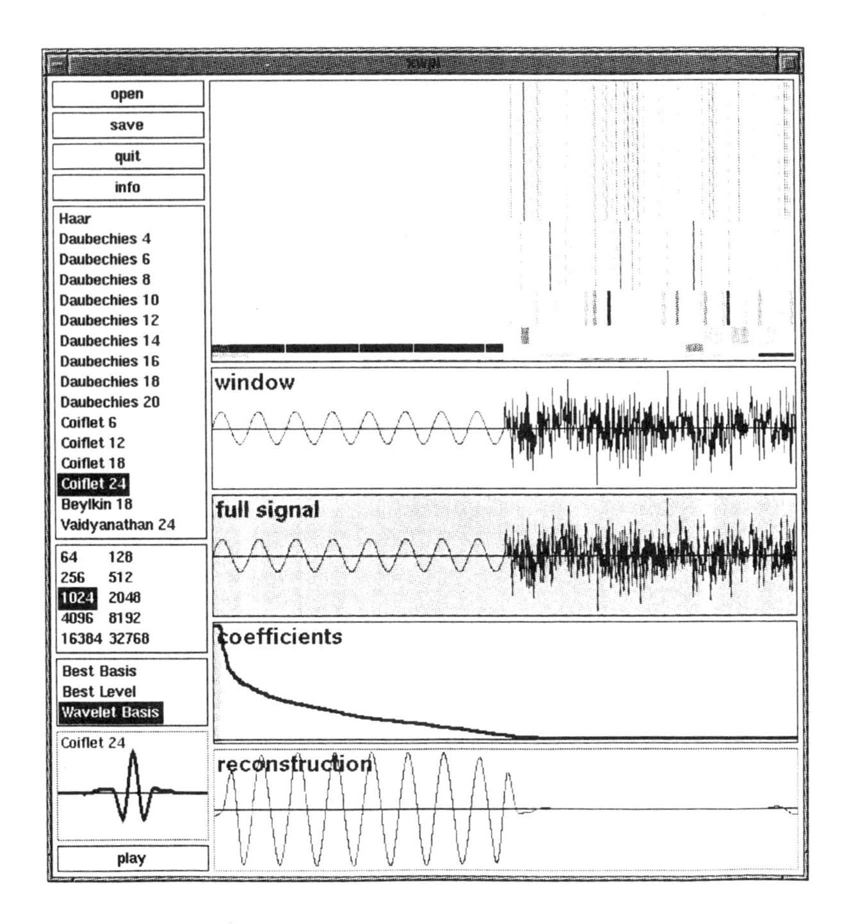

Figure. 3. An analysis of a sine wave embedded in random noise. This is an example of wavelet analysis performing a matched filter function. The presence of the sine signal is apparent, but even more noteworthy is the ability of the filtering to ignore noise which contains uncorrelated power at the same frequency as the original signal.

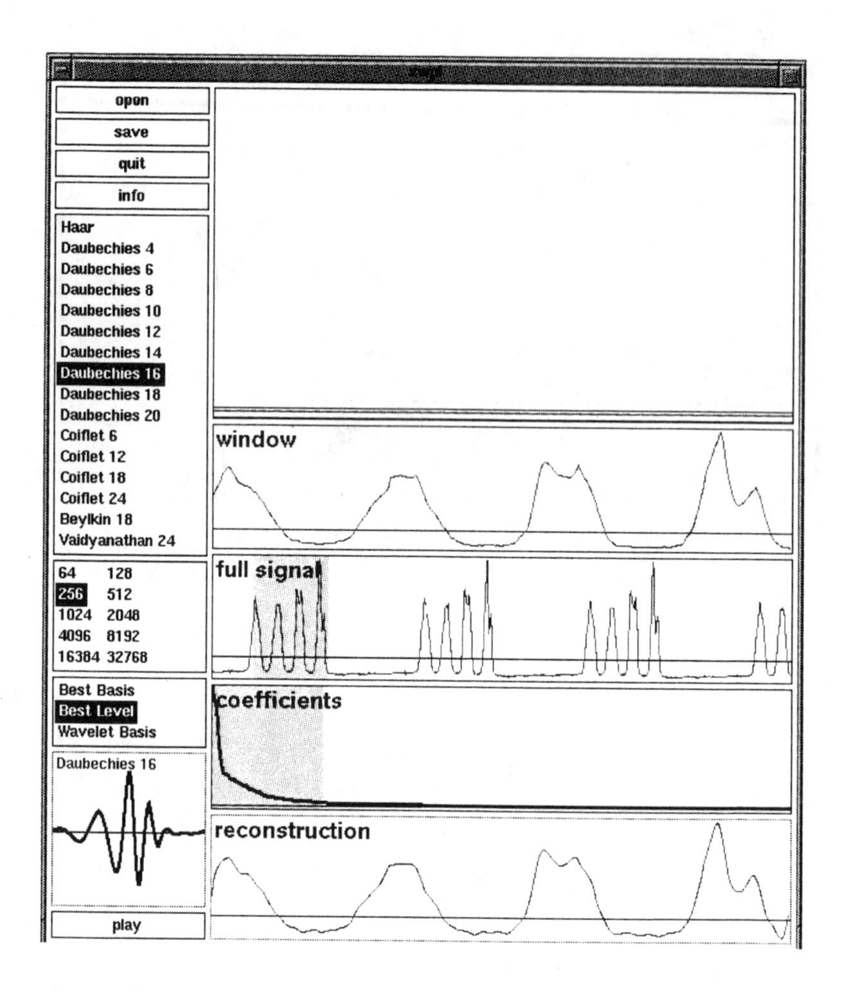

Figure. 4. The resolution of two bunches in a PUE signal in the AGS Damper system. The dirac delta function representations of the two different bunch frequencies appear in the phase space portrait. The **Best Level** analysis forces all the boxes in phase space to have the same aspect ratio, in this case, short and wide. Thus, although the bunch frequencies are visible, the time structure in the phase portrait is very poor, since it is unchanged over the length of the window.

REFERENCES

1. Ingrid Daubechies, *Ten Lectures on Wavelets*, (CBMS-NSF Regional Conf. Series in Appl. Math., Vol. 61. Society for Industrial and Applied Mathematics, Philadelphia, 1992)
2. Wavelets, *Mathematics and Applications*, John J. Benedetto and Michael W. Frazier, eds., (CRC Press, Boca Raton, 1994), Ch. 1, pp. 38-43.
3. William Press, Saul Teukolsky, William Vetterling, Brian Flannery, *Numerical Recipes in C*, (Cambridge University Press, Cambridge, 1992). Ch. 13, pp. 592-594.
4. Fazal Majid, *et al. The Xwpl system*, (version 1.3, available by anonymous ftp from *pascal.math.yale.edu* (128.36.23.1) in the directory /pub/software/xwpl).

APPENDICES

Discussion sessions were held on Monday and Tuesday; participants could choose from three topics at each session. The Chairman of a group introduced the topic, participants were then free to comment, state the practice and problems at their laboratories, and solicit advice. Five of the discussion topics were chosen on the basis of "Interest Questionnaires" returned with the workshop registration forms. The sixth group, on feedback and damping, was organized by Bill Sellyey at the workshop.

The Chairmen gave a brief summary of the discussion in their groups during the plenary "wrap-up" session on Thursday. These summaries, with some editing, are reproduced below. We hope that the editing has not introduced errors.

Instrumentation for Low Current, Radioactive and Medical Beams (Ewart Blackmore, TRIUMF)

These topics were combined since they all involve detection of sub-nano-ampere beams and for the case of radioactive beams there are requirements to cover a range of intensities down to as low as hundreds of particles per second. The design of instrumentation for radioactive beams has additional problems associated with the high radiation environment near the target/ion source, the fact that the beam can consist of several different types of ions some of which are radioactive (β emitters), and that the low energy and range of the ions makes non-destructive monitoring difficult.

CERN (ISOLDE) has a lot of experience in instrumentation for radioactive beams and makes use of scanned slits in front of a Faraday cup to measure intensity profiles. Wire or blade scanners are also used. TRIUMF (TISOL) uses similar techniques and would be interested in some new ideas for their ISAC-1 proposal. A common method of dealing with very low intensity beams is to use an ion beam of a nearby mass that is produced with a larger intensity as a pilot beam for tuning; the magnet settings are then adjusted for the desired beam. There was discussion about the possible use of a residual gas monitor based on micro-channel plate technology as developed at AGOR and elsewhere for protons. The radioactive beams would be heavily ionizing and therefore give a good signal but a potential problem is that this monitor requires a significant electric field which could affect the low velocity beams.

The use of multi-wire proportional chambers based on a fast gas, eg.CF_4-Isobutane, was described as a solution to detecting individual particles in the secondary beams at TRIUMF. The wire spacing is about 1 mm, the timing resolution is better than 10 ns and the count rate is more than 1 MHz per wire.

Medical beams of protons or light ions for cancer therapy are monitored by several types of detector: ion chambers and secondary emission monitors for intensity, and integrating profile monitors or segmented ion chambers for lateral distributions. Common diodes or optical diodes provide a linear dose response and a large signal and are used to measure dose distributions since they have a small active area. They have to be irradiated to about 10^4 to 10^5 rads before the response is stabilized. Miniature ion chambers have a lower signal and are small (<1mm) in only one dimension (thickness) but are useful for absolute dose determination and calibration.

EMI, Grounding and Shielding (Gregory Stover, LBL)

The electro-magnetic interference (EMI) and noise discussion group began with general queries to the audience. Did any participant have a particular noise problem? Would anyone like to discuss their or their laboratory's EMI reduction philosophy? Though there were no volunteers, Jim Zagel (FNAL) posed another general question to the group: "Why don't accelerator facilities identify and attempt to correct their most dominant noise sources rather than continue to over shield sensitive low level systems?" To further promote this idea and generate group discussion, the Chairman presented a basic summary of the EMI philosophy followed during the construction of the Advanced Light Source (ALS) at LBL. The more important methods cited were "reduced area" ground returns, single-point grounds, floating loads, LC filters on all magnet power supplies, full shields on fast pulse magnets, optical isolation of subsystem safety chains, and the complete fibre optic interconnection of the timing and computer control systems. Of particular note, was the ALS philosophy of transmitting a majority of accelerator beam signals (DCCTs, BPMs, etc.) by the way of the computer control system's fibre optic link. This method avoided cable way jam ups and the introduction of excessive conductive and ground loop noise into the control room electronics. At the end of the presentation, the Chairman recommended a well-known book entitled "Noise Reduction Techniques in Electronic Systems" by Henry W. Ott.

This presentation was followed by John Perry's (CEBAF) comments on the successes and failures of CEBAF's program to allocate and enforce cable installation according to signal integrity principles. Dave Peterson (FNAL) discussed the more notable problems and solutions to various EMI problems at Fermilab. These included rf induced trips of the P bar liquid nitrogen source and an oscillating BPM (approx. 1.2 GHz) that was coupling into the stochastic cooling system. Additionally, he noted an on-going program at Fermilab to improve the robustness of the power distribution system to lighting strikes. Bob Webber (FNAL) described how a radio dispatched vehicle was occasionally causing a ground fault trip of the main ring dipole power supplies. Finally, a member of the SLAC technical group described a noise problem that he was having with a seemingly fully shielded resonant beam current pickup system. Several individuals suggested using a solid shield cable or alternatively to employ rf down conversion techniques at the detector location.

High Resolution (Sub-micron/Sub-picosecond) Measurements
(Robert Hettel, SSRL/SLAC)

The discussion group for sub-micron and sub-picosecond measurements was attended by more than half of the Workshop participants. The Chairman proposed several topics for discussion. These included:

1) Types of beams and applications requiring these measurements,
2) Measurement methods
3) "Hot topics", state-of-the-art, and design challenges in measurement technology
4) Methods that should be pursued and further developed for accelerator facilities.

The Chairman presented an overview of the first two topics; several group members offered suggestions for and elucidations of various measurement technologies. A paraphrased summary of the discussion follows.

Sub-micron Measurements

Two types of accelerator beams were identified as potentially requiring sub-micron resolution measurements: colliding lepton beams, particularly those in linear colliders, and third-generation synchrotron radiation (SR) sources. SLAC has beam sizes in the micron range with sub-micron steering requirements; requirements for the Next Linear Collider (NLC) will be one or two orders of magnitude more stringent (a 5 nm steering requirement has been cited). The need to steer beams with an accuracy of about 10% of the beam size is also needed for other colliding beam facilities and SR sources. SR sources are now approaching the 1 μm stability requirement, while the stability of future fourth-generation sources will need to be less than a micron. The SLAC B-Factory will require micron level beam steering at the interaction region.

Instruments to measure collider beam size and position at SLAC include:

1) Wire scanners (~0.1 μm resolution; M. Ross et al.)
2) Beam-strahlung, luminosity, and beam-beam deflection monitors (~0.3 μm)
3) Bremsstrahlung monitor (<0.1 μm)
4) Laser interference pattern + Compton backscattering (measured 60 nm beam with <10 nm resolution; Shintake)
5) Ion probe (~1 μm resolution; Pasour et al).

It was noted that beam emittance can be measured with sub-micron resolution by observing the opening angle of Compton back-scattered photons originating from a laser; this gives a measure of the transverse velocity spread of a particle beam.

Beam position at SR sources is measured routinely with sub-micron sensitivity using X-ray photon beam position monitors. Recently processors for the rf pick-up electrodes used to sense beam position for these sources have been built with sub-micron sensitivity.

The design challenge for all these monitors is not their sensitivity (i.e. their ability to sense small changes on a short time scale), but the degree of mechanical stability needed to have true sub-micron resolution. Motion damping systems and motion and position sensors having sub-micron resolution (accelerometers, proximity gauges, interferometers, etc.) may need to be built into the monitors, much like what has been done with telescope systems and which will be done for the NLC interaction region monitors.

Sub-picosecond Measurements

Accelerator beams used for short-wavelength free electron lasers (FEL), which can generate coherent light ranging from IR to potentially X-ray wavelength, require measurement of timing or bunch length with sub-picosecond resolution. A future fourth-generation low momentum compaction SR source might also have sub-picosecond bunches. It was noted that there are many femtosecond pulsed lasers and that measurement devices have been developed for them.

Methods used to measure sub-picosecond bunch lengths include:

1) Streak cameras (~200 fsec resolution; "synchroscan" models may offer some advantage in obtaining high resolution for rf-bunched beams).
2) Measurement of linac beam momentum spread induced in an accelerating section with near-zero phasing so that the momentum of the centre of the beam bunches remains unchanged (~200 fsec resolution).
3) Autocorrelation using coherent IR transition or Cerenkov radiation in a Michelson interferometer together with a bolometer (10 fsec resolution).

Several methods were mentioned that presently have resolutions of a few picoseconds, but that might be pushed to the sub-picosecond regime:

1) Stroboscopic sampling of the beam by rf-synchronized deflection of either the primary beam or a low energy scattered secondary beam.
2) Transverse rf deflection of the primary beam that spreads the beam out to a width that is a function of bunch length.
3) Measurement of a subset of spectral components of a bunch current; frequencies in the few hundred GHz regime must be measured for sub-picosecond lengths.
4) An optical "picoscope" that has been developed commercially.
5) Ion probe (mentioned in the sub-micron measurement section).
6) A fiberoptic ring that stores a single light pulse produced by the primary beam to permit repetitive sampling (@200 MHz for a ring developed at CERN).

It was speculated that electro-optical materials and devices now used for laser measurements might be used for accelerator beams. Crystals having non-linear or rotatable polarization properties, birefringence, or other special properties are used for autocorrelation measurements. Photoconducting Auston switches and dipoles have sub-picosecond response times and can be used for sampling short pulses.

Beam Shape and Halo Measurements (Richard Witkover, BNL)

While profile measurements are routine at all accelerators, the need to measure "halo" has been appreciated mostly by those with high average current or requiring very clean, bright beams. Loss of only 0.1% of a 100 μA high energy beam can cause considerable radiation. This has led to the need to measure profiles much more precisely than the 1% level typically found, resolutions of 10^{-3} and 10^{-4} are required. The paper by G. Haouat on the halo of the ELSA beam describes several methods of doing this.

R. Fiorito also suggested that optical transition radiation (OTR) or synchrotron radiation could be used with neutral density filters over the bright core to give such a range. He also suggested that OTR might allow the divergence of the halo to be measured.

E. Rossa of CERN described a cadmium telluride photo detector with 100μm spacing which could provide sub-micron resolution of the vertical profiles in LEP. These detectors are fast enough to allow turn-by-turn profiles. The electronics can store 1600 turns which can be analyzed later to display centroid motion.

K. Mess of DESY described work using the halo as a diagnostic to study the sources of beam movement. He used a loss monitor to measure the scraping of the halo on a collimator over a long time period. The data was frequency analyzed and components identified due to the 50 Hz power lines, mechanical shake from vacuum pumps, the 12.5 Hz booster rep-rate, the 16.33 Hz frequency used by the German railroads and a 2 Hz ground ripple. He claimed 3μm spatial resolution, and cited this as an example of how "today's noise is tomorrow's signal".

F.-J. Decker of SLAC described his work with fitting asymmetric gaussians and "super gaussian" distributions to profiles rather than simple gaussians. He stated that some of these distributions could be associated with specific beam problems such as mis-match or magnet mis-alignment.

R. Shafer of LANL described some work by others at LANL, which suggests that there might be a self-limiting effect by which halo's reach a limiting radius.

A. Tron of MEPI described a profile monitor using multi-channel secondary emission grids. He presented data showing measured two-dimensional contour maps.

R. Witkover of BNL showed some results of a phosphor screen - CCD camera profile monitor installed after the Booster. Data was taken at a 30 Hz frame rate which showed that both the phosphor screen and frame grabber were able to acquire data at that rate.

Closed Orbit Measurement and Correction (Shane Koscielniak,TRIUMF)

At present, the study of closed orbit correction schemes is driven by the quest for low emittance and high brilliance in synchrotron light sources. The "state of the art" as of November 1993 is presented in the AIP Conference Proceedings 315 "Orbit Correction and Analysis in Circular Accelerators". Since that time, a variety of schemes, Singular Value Decomposition (SVD), Eigenvector Decomposition, Tikhonov Regularization, etc., have been successfully tried at the NSLS (Brookhaven) and at SPEAR (Stanford). Most recently, the SVD method is being applied at the APS Booster (Argonne). Typically, insertion devices and interaction points have different tolerances from the arcs, so both local and global corrections are applied. The problem of fighting between local and global corrections has been solved by devising a unified inverse response matrix with no coupling from global BPMs to local steering elements. The method relies upon sharing of all beam position data by all correctors and this is facilitated by the new technology of "reflective memory". Beta function measurement was quickly discussed, but nobody present in the discussion group was aware of any new development.

Damping and Feedback (William Sellyey, APS-ANL)

Most of the discussion dealt with the use of class AB amplifiers to damp bunch by bunch longitudinal instabilities. The cost of a class AB amplier is half to one-third that of a class A amplifier with the same power output. Unfortunately, the rise time of a commercial class AB amplifier is usually inadequate. These amplifiers derive their bias from the drive signal and the rise time of this bias is usually advertised as about 100 ns. Active damping, however, usually needs amplifier bandwidths of 150 to 300 MHz, or in other words, rise times of 1 to 2 ns.

The following approaches were suggested during the discussion:

a) Use phase modulation with a class AB supply.
b) Reduce the product of charge and maximum energy error, or charge and maximum phase error, at injection. This will reduce the forces driving the instability and enable a lower power (hence lower cost) class A amplifier to be used.
c) Feed back a signal to one of the accelerating cavity amplifiers if they have enough bandwidth. If practical, this would eliminate the need for a separate damper.
d) Many accelerating cavities are fitted with (passive) damping ports. These may reduce but not eliminate beam instabilities. Once the offending higher order modes have been identified, an active damping signal could be fed into an accelerating cavity via the appropriate port.

LIST OF PARTICIPANTS

ALEKSANDROV, A.V.	Budker Inst., Academy of Sciences of Russia, Siberian Division, Lavrentieva - 11, Novosibirsk, 630090, RUSSIA ALEKSANDROV@INP.NSK.SU
ALEXANDER, John R.	DRAL-Daresbury, Keckwick Lane, Daresbury, Warrington, Cheshire, WA4 4AD, UK ALEXANDER@DL.AC.UK
AVERILL, Robert	MIT Bates Linear Accelerator Center, P.O. Box 846, Middleton, MA 01949, USA
BARRY, Walter C.	LBL, 1 Cyclotron Road, Berkeley, CA 94720 USA WALTER_BARRY@MACMAIL.LBL.GOV
BARSOTTI, Ed L.	Fermilab, P.O. Box 500, MS 308, Batavia, IL, 60510 USA EBARSOTTI@ADCALC.FNAL.GOV
BERG, William Jeffrey	ANL, APS/ASD-411, 9700 S. Cass Ave., Argonne, IL, 60439 USA BERG@ANLAPS.APS.ANL.GOV
BERGOZ, Julien	Bergoz, Inc., 01170 Crozet FRANCE CIP@APPLELINK.APPLE.COM
BLACKMORE, Ewart	TRIUMF, 4004 Wesbrook Mall, Vancouver, BC, V6T 2A3, CANADA EWB@TRIUMF.CA
BLOEMHARD, Rick	TRIUMF, 4004 Wesbrook Mall, Vancouver, BC, V6T 2A3, CANADA BLOEMHARD@TRIUMF.CA
BORDEN, Michael J.	LANL, P.O. Box 1663, AOT-2, MS H838, Los Alamos, NM 87545, USA JBORDEN@.LANL.GOV
BOVET, Claude	CERN, SL Division, CH-1211 Geneva 23 SWITZERLAND CLAUDE_BOVET@MACMAIL.CERN.CH
CANTONI, Paolo	INFN-Modena, Physics Department, University of Modena, Via Campi 213/A, Modena 41100 ITALY CANTONI@IMOAX1.UNIMO.IT
CAPEK, Karel Bruce	CEBAF, 12000 Jefferson Av., Newport News, VA 23606, USA

LIST OF PARTICIPANTS

CARTER, Hamilton	LANL, MS-H838, Los Alamos, NM 87545, USA HCARTER@LAMPF.LANL.GOV
CHAO, Yu-Chiu	CEBAF, MS-12A, 12000 Jefferson Avenue, Newport News, VA 23606, USA CHAO@CEBAF.GOV
CHASE, Brian E.	Fermilab, P.O. Box 500 MS308, Batavia, IL 60510, USA CHASE@ADCALC.FNAL.GOV
CHU, William T.	LBL, 64-227, Berkeley, CA, 94720, USA WTCHU@LBL.GOV
CHUNG, Youngjoo	ANL, Advanced Photon Source, 362-C165, 9700 S. Cass Ave., Argonne, IL 60439 USA YCHUNG@APS.ANL.GOV
CLIFFT, Benny E.	ANL, Physics Division, 9700 S. Cass Avenue, Argonne, IL 60439, USA
CUTTONE, G.	INFN-LNS, V.S. SOFIA 44, Catania 95123, ITALY CUTTONE@LNS.INFN.IT.
DAINELLI, Antonio	INFN-LNL, Laboratori Nazionali di Legnaro, Via Romea, 4 I-35020 Legnaro (Padova), ITALY DAINELLI@LNL.INFN.IT
DECKER, Franz-Josef	SLAC, P.O. Box 4349, M.S. 66, Stanford, CA 94309 USA DECKER@SLAC.STANFORD.EDU
DEHNING, Bernd	CERN, SL Division, CH-1211 Geneva 23 SWITZERLAND DEHNING@VXCERN.CERN.CH
DENARD, Jean-Claude	CEBAF, 12000 Jefferson Ave., Newport News, VA 23606, USA DENARD@CEBAF.GOV
DIETRICH, Jürgen	KFK Jülich, Institut für Kernphysik, Briefpost: 52425 Jülich, Fracht/Paketpost: 52428 Jülich, GERMANY JUDI@SNOOPY.CC.KFA-JUELICH.DE
DRENNAN, Craig C.	Fermilab, P.O. Box 500, Pine St. at Kirk Rd., Batavia, IL 60510 USA CDRENNAN@FNAL.GOV

LIST OF PARTICIPANTS

FIORITO, Ralph B.	NSWC-CD, 10901 New Hampshire Avenue, Silver Spring, MD 20903 USA FIORITO@DASYS.DT.NAVY.MIL
FOCKER, Gerrit Jan	CERN, Division ECP, 1211 Geneva 23, SWITZERLAND GJF@CERNVM.CERN.CH
FRITSCHE, Craig	AlliedSignal Aerospace, 2000 E. 95th Street, P.O. Box 419159, Kansas City, MO 64141, USA
FUJA, Ray	ANL, Advanced Photon Source, Bld. 411 T1, 9700 S. Cass Ave., Argonne, IL 60439 USA
GASSNER, David	BNL, Bldg 911A, Upton NY, NY 11973, USA
GILPATRICK, John D.	LANL, Mail Stop H808, Los Alamos NM 87545 USA GILPATRICK@LANL.GOV
GONZALEZ, José Luis	CERN, PS Division, 1211 Geneva 23, SWITZERLAND GONZALEJ@CERNVM.CERN.CH
GREENWALD, Shlomo	Cornell Univ., Wilson Laboratory, Ithaca, NY 14853, USA
HAHN, Alan	Fermilab, MS 308, P.O. Box 500, Batavia, IL, 60510 USA AHAHN@ADCALC.FNAL.GOV
HAMILTON, Brett J.	IUCF, 2401 Milo B. Sampson Lane, Bloomington, IN, 47408, USA BRETT@VENUS.IUCF.INDIANA.EDU
HAOUAT, Gerard A.	CEA - Service PTN, BP #12, Bruyères-le Châtel, Essonne 91680, FRANCE HAOUAT@BRUYERES.CEA.FR
HARDEK, Thomas W.	LANL, MP Division, MS-H827, P. O. Box 1663, Los Alamos, NM, 87545 USA HARDEK@ATDIR.LANL.GOV
HETTEL, Robert	SSRL/SLAC, SSRL, Bin 69, P. O. Box 4349, Stanford, CA 94309 USA HETTEL@SSRL01.SLAC.STANDFORD.EDU

LIST OF PARTICIPANTS

HINKSON, Jim	LBL, 1 Cyclotron Road, Bldg 46/125, Berkeley, CA 94720 USA JAHINKSON@LBL.GOV
HOFFMAN, Clarence R.	CRNL, AECL Research, Chalk River, Ontario K0J 1J0 CANADA
HUANG, Jung Yun	PLS, POSTECH, Pohang 790-390, KOREA
HURST, Andy	TRIUMF, 4004 Wesbrook Mall, Vancouver, BC V6T 2A3, CANADA HURST@TRIUMF.CA
IWASHITA, Yoshihisa	Kyoto Univ., Accelerator Laboratory, Institute for Chemical Research, Gokanosho, Uji-city, Kyoto-Fu 611 JAPAN IWASHITA@KYTICR.KUICR.KYOTO-U.AC.JP
JORDAN, Kevin	CEBAF, 12000 Jefferson Ave., Newport News, VA 23606, USA JORDAN@CEBAF.GOV
KASE, Masayuki	RIKEN – Wako-shi, 2-1 Hirasawa, Wako-shi, Saitama 351-01, JAPAN KASE@RIKVAX.RIKEN.GO.JP
KEANE, John	BNL, Bldg. 725B, Upton, NY 11973, USA KEANE@BNL.GOV
KELLER, Roderich	LBL, 1 Cyclotron Road, MS 80-101, Berkeley, CA 94720, USA RKELLER@LBL.GOV
KIKUTANI, Eiji	KEK, Oho 1-1, Tsukuba-shi Ibaraki, 305, JAPAN KIKUTANI@KEKVAX.KEK.JP
KOGAN, Michael	MIT - Bates, 21 Manning Rd, Middleton, MA 01949, USA KOGAN@BATES.MIT.EDU
KOSCIELNIAK, Shane	TRIUMF, 4004 Wesbrook Mall, Vancouver, B.C. V6T 2A3 CANADA SHANE@TRIUMF.CA
KUGLER, Erich	CERN, PPE Division, Geneva 23, 1211, SWITZERLAND EKUG@CERNVM.CERN.CH

LIST OF PARTICIPANTS

KUO, Thomas	TRIUMF, 4004 Wesbrook Mall, Vancouver, BC V6T 2A3, CANADA KUO@TRIUMF.CA
LANCASTER, Henry	LBL, MS 46-125, 1 Cyclotron Road, Berkeley, CA 94720 USA
LE GRAS, Marc	CERN, CH-1211 Geneve 23, SWITZERLAND LEGRAS@CERNVM.CERN.CH
LIHN, Hung-chi	Stanford University, SLAC/SSRL Bin 99, P.O. Box 4349, Stanford, CA 94309 USA LIHN@SSRL01.SLAC.STANFORD.EDU
LINSCOTT, Ivan R.	Stanford/SLAC, Center for Radar Astronomy, 235 Durand, P. O. Box 4349, Stanford, CA 94305 USA LINSCOTT@NOVA.STANFORD.EDU
LUMPKIN, Alex H.	ANL, Advanced Photon Source, Building 362, 9700 So. Cass Avenue, Argonne, IL 60439 USA LUMPKIN@ANLAPS.APS.ANL.GOV
MA, Hengjie	Fermilab, P.O. Box 500, Mail Stop 340, Batavia, IL 60510 USA MAHENGJIE@FNALV.FNAL.GOV
MACKENZIE, George	TRIUMF, 4004 Wesbrook Mall, Vancouver, BC V6T 2A3 CANADA GHM@TRIUMF.CA
MASSOLETTI, Dexter J.	LBL, Advance Light Source, One Cyclotron Road, 80-101, Berkeley, CA 94720 USA
MCCORMICK, Douglas	SLAC, 2575 Sandhill Rd., Menlo Park, CA 94025, USA DJM@SLAC.STANFORD.EDU
MCGINNIS, David	Fermilab, P.O. Box 500, MS 341, Batavia, IL, 60510 USA MCGINNIS@ADCALC.FNAL.GOV
MEISNER, Keith	Fermilab, P.O. Box 500, Batavia, IL 60510, USA FNAL::MEISNER
MERLETTI, Robert	EBCO Tech., 4004 Wesbrook Mall, Vancouver, BC V6T 2A3, CANADA

LIST OF PARTICIPANTS

MESS, Karl Hubert	DESY, Notkestr.85, D22603 Hamburg, GERMANY MESS@PKTR.DESY.DE
MORITZ, Peter G.	GSI-Darmstadt, Planckstr. 1, Darmstadt, Hessen, D-64291, GERMANY PMORITZ@V6000A.GSI.DE
MOUAT, Michael	TRIUMF, 4004 Wesbrook Mall, Vancouver, BC V6T 2A3, CANADA MOUAT@TRIUMF.CA
NAOUMENKO, Gennady	NPI-Tomsk, P.O. Box 25, pr. Lenina 2-A, Tomsk 634050, RUSSIA PAP@TSINPH.TOMSK.SU
NAWROCKY, Roman J.	BNL, NSLS Dept. Bldg. 725B, Upton, NY, 11973, USA NAWROCKY@BNL.GOV
NGO, Mai T.	Mission Research Corporation, 8560 Cinderbed Road, Suite 700, Newington, VA, 22122, USA
ORTIZ, Ramon	SSRL, 2575 Sand Hill Rd M/S 69, Menlo Park, CA 94025, USA ORTIZ@SSRL750.BITNET
OSTROUMOV, Petr	INR, 60th October Anniversary Prospect 7A, Moscow 117312, RUSSIA OSTROUMOV@INR.MSK.SU
PASOUR, John A.	Mission Research Corporation, Advanced Beam Technologies, 8560 Cinderbed Road, Suite 700, Newington, VA 22122 USA
PASQUINELLI, Ralph	Fermilab, Mail Stop 341, P. O. Box 500, Batavia, IL 60510 USA PASQUINELLI@FNALV.FNAL.GOV
PATTERSON, Donald	ANL, Advanced Photon Source, Building 362, 9700 South Cass Avenue, Argonne, IL 60439 USA DRP@ANLAPS.APS.ANL.GOV
PEARCE, D. R.	TRIUMF, 4004 Wesbrook Mall, Vancouver, B.C. V6T 2A3 CANADA PEARCED@TRIUMF.CA
PERRY, John	CEBAF, 12000 Jefferson Avenue, MS 16A, Newport News, VA 23606 USA JPERRY@CEBAF.GOV

LIST OF PARTICIPANTS

PETERSON, Dave	Fermilab, P.O. Box 500, MS 341, Batavia IL, 60510 USA PETERSON@ADCALC.FNAL.GOV
PLUM, Michael A.	LANL, AOT-2, MS H838, Los Alamos, NM 87545 USA PLUM@LANL.GOV
POWER, John F.	LANL, P.O. Box 1663, MS H808, Los Alamos, NM 87545 USA JPOWER@LANL.GOV
POWERS, Thomas J.	CEBAF, 12000 Jefferson Ave., Newport News, VA 23606, USA POWERS@CEBAF.GOV
PRUSS, Stan	Fermilab, P.O. Box 500, MS 345, Batavia, IL 60510 USA PRUSS@ADCALC.FNAL.GOV OR FNAL::PRUSS
RAWNSLEY, W. R.	TRIUMF, 4004 Wesbrook Mall, Vancouver, B. C. V6T 2A3 CANADA RAWNSLEY@ERICH.TRIUMF.CA
REECE, R. Kenneth	BNL, AGS Dept. - Bldg. 911C, Upton, NY 11973, USA REECE@BNLDAG.AGS.BNL.GOV
REZZONICO, Luigi	PSI, Würenlingen Villigen, Villigen PSI, CH-5232 SWITZERLAND REZZONICO@CVAX.PSI.CH
RIES, Thomas	TRIUMF, 4004 Wesbrook Mall, Vancouver, B. C. V6T-2A3 CANADA TRRT@ERICH.TRIUMF.CA
ROSE, Chris R.	LANL, P.O. Box 1663, M/S H808, Los Alamos, NM 87545 USA CROSE@LANL.GOV
ROSSA, Edouard R.	CERN, Division SL/BI, Geneva 23, CH-1211, SWITZERLAND ROSSA@CERNVM.CERN.CH
ROVELLI, A.	INFN-LNS, V.S. Sofia 44, Catania 95123, ITALY ROVELLI@LNS.INFN.IT.
RULE, Don W.	NSWC-CD, Mail Stop R36, 10901 New Hampshire Ave., Silver Spring, MD 20903-5640 USA

LIST OF PARTICIPANTS

SANDOVAL, D.	LANL, M.S. H808, Los Alamos, NM 87545 USA DSANDOVAL@LANL.GOV
SANNIBALE, Fernando	INFN-LNF, C.P. 13, 00044 Frascati (Roma), ITALY SANNIBALE@IRMLNF or SANNIBALE@LNF.INFN.IT
SCHAFFNER, Sally K.	CEBAF, 12000 Jefferson Ave., MS 85A, Newport News, VA 23606, USA SCHAFFNER@CEBAF.GOV
SCHIPPERS, J.M.	KVI, Zernikelaan 25, 9747 AA Groningen, THE NETHERLANDS SCHIPPERS@KVI.NL
SCHMICKLER, Hermann	CERN, Division SL-OP, CH-1211 Geneva 23, SWITZERLAND SCHMICKLER@VXCERN.CERN.CH
SEBEK, Jim	SLAC, Stanford Synchrotron Rad. Lab., P.O. Box 4349, Stanford, CA 94309 USA SEBEK@SSRL01.SLAC.STANFORD.EDU
SELLYEY, William	ANL, Advanced Photon Source, Bldg. 362, 9700 S. Cass Avenue, Argonne, IL 60439 USA SELLYEY@ANLAPS
SERIO, Mario	INFN-LNF, C.P. 13, Via E. Fermi 160, Frascati (Roma), 00044 ITALY SERIO@IRMLNF
SHAFER, Robert E.	LANL, MS-H808, Los Alamos, NM 87545 USA RSHAFER@LANL.GOV
SHOJI, Masazumi	RIKEN – SPRING8, Harima Science Garden City, Kamigori, Hyogo, 678-12, JAPAN SHOJI@SP8SUN.SPRING8.OR.JP
SILZER, Richard	SAL, Univeristy of Saskatchewan, Saskatoon, Sask. S7N 0W0, CANADA MARK@SKATTER.USASK.CA
SINGH, Om V.	BNL, Bldg. 725B, Upton NY 11973 USA SINGH@BNL.GOV
SMITH, Gary A.	BNL, Bldg. 911C, Upton, NY 11973 USA SMITH1@BNLDAG.BNL.GOV

LIST OF PARTICIPANTS

SMITH, Robert James	DRAL-Daresbury, Keckwick Lane, Daresbury, Warrington, Cheshire, WA4 4AD, UK RSM@DL.AC.UK
SMITH, Stephen Robert	SLAC, MS 50, P.O. Box 4349, Stanford, CA, 94019 USA SSMITH@SLAC.STANFORD.EDU
SONG, Joshua J.	ANL 371T, 9700 S. Cass Avenue, Argonne IL 60439 USA JSONG@ANLAPS.APS.ANL.GOV
SOUKUP, Jan	Univ. of Alberta, Center for Subatomic Research, Department of Physics, CANADA JSOU@PHYS.UALBERTA.CA
STILLMAN, Arnold	BNL, Bldg. 911B, P.O. Box 5000, Upton, NY 11973-5000 USA STILLMAN@BNLDAG
STOVER, Greg D.	LBL, 1 Cyclotron Road, Bldg 46/125, Berkeley, CA 94720 USA GDSTOVER@LBL.GOV
STREHL, Peter	GSI-Darmstadt, Planckstr. 1, D-64291 Darmstadt, GERMANY STREHL@ALICE.GSI.DE
TAKANO, Shiro S.	RIKEN - SPRING8, Spring-8, Harima Science Garden City, Kamigori, Hyogo, 678-12, JAPAN TAKANO@SP8SUN.SPRING8.OR.JP
TALLERICO, Thomas	BNL, Building 911B, Upton, NY 11973 USA TALLERICO@AGS.BNL.GOV
TASSOTTO, Gianni	Fermilab, MS-222, P. O. Box 500, Batavia, IL 60510 USA TASSOTTO@FNALV.FNAL.GOV
TEYTELMAN, Dmitry	SLAC, MS 33, 2575 Sand Hill Road, Menlo Park, CA 94309, USA DIM@SLAC.STANFORD.EDU
TOBIYAMA, Makoto	KEK, Accelerator Dept., 1-1 Oho, Tsukuba, Ibaraki 305, JAPAN TOBIYAMA@KEKVAX.KEK.JP
TORIKOSHI, Masami	NIRS-Chiba, 9-1 Anagawa-4-chome, Inage-ku, Chiba-shi, 263, JAPAN

LIST OF PARTICIPANTS

TRON, A.M.	MEPhI, Kashirskoe SH. 31, Moscow 115409, RUSSIA PLASMA@ETP.MEPI.MSK.SU
UNSER, Klaus B.	CERN, SL Division, CH-1211 Geneva 23 SWITZERLAND
URŠIČ, Rok	CEBAF, 12000 Jefferson Av., Newport News, VA 23606, USA ROK@CEBAF.GOV
VAN ASSELT, Willem	BNL, AGS, Building 911B, P.O. Box 5000, Upton, NY, 11973 USA VANASSELT@BNLDAG.AGS.BNL.GOV
WACHTER, John	SSRL, 2575 Sandhill Road, Bin 69, Menlo Park, CA 94025, USA WACHTER@SSRL750.BITNET
WALKER, Ian J.	GMW Associates, P. O. Box 2578, Redwood City, CA 94064 USA
WANG, Xucheng	ANL, 9700 S. Cass Ave., Bldg 362/ASD, Argonne, IL 60439 USA XOW@ANLAPS.APS.ANL.GOV
WATSON, Scott A.	LANL, MS-P-940, Los Alamos, NM 87545 USA 102430@M4NEXT.LANL.GOV
WEBBER, Robert C.	Fermilab, MS 341, P.O. Box 500, Batavia, IL, 60510 USA WEBBER@FNAL.GOV
WENDT, Manfred	DESY, Dep. MKI, Nothestr. 85, Hamburg 22607, GERMANY WENDT@PKTR.DESY.DE
WEST, Charlene B.	CEBAF, 12000 Jefferson Ave. MS 85A, Newport News, VA 23606, USA WEST@CEBAF.GOV
WILKE, Mark	LANL, P.O. Box 1663, Group P-15, MS D-406, Los Alamos, NM 87545 USA WILKE@LANL.GOV
WITKOVER, Richard	BNL, Bldg. 911B, Upton, NY 11973 USA WITKOVER@BNLDAG.AGS.BNL.GOV

LIST OF PARTICIPANTS

YANG, Bingxin	ANL, 9700 South Cass Ave., Argonne, IL 60565, USA BXYANG@ANLAPS.APS.ANL.GOV
YIN, Yan	TRIUMF, 4004 Wesbrook Mall, Vancouver, BC V6T 2A3, CANADA YANYIN@TRIUMF.CA
YOKOYAMA, Ichiro	RIKEN – Wako-shi, 2-1 Hirasawa, Wako-shi, Saitama, 351-01, JAPAN KASE@RIKVAX.RIKEN.GO.JP
YUAN, Vincent	LANL, M.S. H818, Los Alamos, NM 87545, USA YUAN@LAMPF.LANL.GOV
ZACH, Milos	TRIUMF, 4004 Wesbrook Mall, Vancouver, BC V6T 2A3, CANADA ZACH@TRIUMF.CA
ZAGEL, James R.	Fermilab, Mail Stop 308, P. O. Box 500, Batavia, IL 60510 USA ZAGEL@ALMOND.FNAL.GOV
ZALTSMAN, Alex	BNL, Bldg 911-B, Upton, NY 11973, USA
ZHAO, Yi	BNL, Bldg. 510A, Upton, NY 11973, USA YI@BNLKU4.PHY.BNL.GOV

AUTHOR INDEX

H

I

J

K

L

M

T

U

V

W

Y

Z